Helicenes

Helicenes

Synthesis, Properties, and Applications

Edited by Jeanne Crassous, Irena G. Stará, and Ivo Starý

WILEY-VCH

Editors

Dr. Jeanne Crassous
Institut des Sciences Chimiques de Rennes (UMR CNRS 6226)
Université de Rennes 1
Campus de Beaulieu
35042 Rennes Cedex
France

Dr. Irena G. Stará
Institute of Organic Chemistry and Biochemistry (IOCB Prague)
Czech Academy of Sciences
Flemingovo náměstí 542/2
166 10 Prague 6
Czech Republic

Dr. Ivo Starý
Institute of Organic Chemistry and Biochemistry (IOCB Prague)
Czech Academy of Sciences
Flemingovo náměstí 542/2
166 10 Prague 6
Czech Republic

Cover Image: Tomáš Bellon

Library of Congress Card No.:
applied for

British Library Cataloguing-in-Publication Data
A catalogue record for this book is available from the British Library.

Bibliographic information published by the Deutsche Nationalbibliothek
The Deutsche Nationalbibliothek lists this publication in the Deutsche Nationalbibliografie; detailed bibliographic data are available on the Internet at <http://dnb.d-nb.de>.

Print ISBN: 978-3-527-34810-7
ePDF ISBN: 978-3-527-82939-2
ePub ISBN: 978-3-527-82940-8
oBook ISBN: 978-3-527-82941-5

Cover Design: ADAM DESIGN, Weinheim, Germany

Typesetting Straive, Chennai, India
Printing and Binding: CPI Group (UK) Ltd, Croydon, CR0 4YY

C9783527348107_080824

Contents

Preface *xv*

1 **The Photochemical Approach to Helicenes** *1*
Jan Storch, Jaroslav Žádný, Vladimír Církva, Martin Jakubec, Jan Hrbáč, and Jan Vacek
1.1 Introduction *1*
1.2 General Features *1*
1.3 Photochemical Preparation of Carbohelicenes *6*
1.3.1 Oxidative Photocyclizations of Unsubstituted Carbo[*n*]helicenes *6*
1.3.2 Oxidative Photocyclizations of Substituted Carbo[*n*]helicenes *9*
1.3.3 Oxidative Photocyclizations of Benzocarbo[*n*]helicenes *11*
1.3.4 Oxidative Photocyclizations of Multiple Carbo[*n*]helicenes *13*
1.3.5 Eliminative Photocyclizations of Carbo[*n*]helicenes *14*
1.4 Photochemical Preparation of Heterohelicenes *15*
1.4.1 Azahelicenes *15*
1.4.2 Thiahelicenes *23*
1.4.3 Phosphahelicenes and Other Heterohelicenes *27*
1.5 Photochemical Preparation of Helicene-Like Molecules *28*
1.6 Photochemical Transformations of Helicenes *31*
1.7 Conclusions *33*
References *36*

2 **Synthesis of Helicenes by [2 + 2 + 2] Cycloisomerization of Alkynes and Related Systems** *53*
Irena G. Stará and Ivo Starý
2.1 Introduction *53*
2.2 Helicenes via [2 + 2 + 2] Cycloisomerization of π-Electron Systems *54*
2.2.1 Intramolecular [2 + 2 + 2] Cycloisomerization *54*
2.2.2 Intermolecular [2 + 2 + 2] Cycloisomerization *59*
2.3 Carbo- and Heterohelicenes *62*
2.3.1 Carbohelicenes *62*
2.3.2 Azahelicenes *63*
2.3.3 Azoniahelicenes *68*

2.3.4 Oxahelicenes *71*
2.3.5 Sila-, Phospha-, and Thiahelicenes *73*
2.4 Functionalized Helicenes *76*
2.5 Asymmetric Synthesis of Nonracemic Helicenes *76*
2.5.1 Stoichiometric Diastereoselective Synthesis *76*
2.5.2 Catalytic Enantioselective Synthesis *82*
2.6 Advanced Helicene Architectures *85*
2.7 Summary and Outlook *89*
2.8 Abbreviations *89*
References *91*

3 Enantioselective Synthesis of Helicenes *103*
Ken Tanaka
3.1 Introduction *103*
3.2 Metal-Catalyzed [2 + 2 + 2] Cycloaddition *103*
3.2.1 Ni- and Co-catalyzed [2 + 2 + 2] Cycloaddition *104*
3.2.2 Rh- and Ir-Catalyzed [2 + 2 + 2] Cycloaddition *105*
3.2.3 Pd-catalyzed [2 + 2 + 2] Cycloaddition *114*
3.3 Metal-catalyzed Hydroarylation *115*
3.4 Metal-catalyzed Oxidative Annulation *120*
3.5 Organocatalyzed Annulation *121*
3.6 Conclusion *124*
References *124*

4 Cationic Triarylcarbenium Helicenes: Synthesis, Resolution, and Applications *127*
Johann Bosson, Niels Bisballe, Bo W. Laursen, and Jérôme Lacour
4.1 Introduction *127*
4.2 Synthesis and Resolution of Cationic [4]-, [5]-, and [6]Helicenes *128*
4.2.1 [4]Helicenes *128*
4.2.2 [6]Helicenes *131*
4.2.3 [5]Helicenes *133*
4.3 Regioselective Functionalization *135*
4.3.1 Generalities: Nomenclature *135*
4.3.2 Late-stage Functionalization *135*
4.3.2.1 North Functionalization of Cationic [4]- and [6]Helicenes *135*
4.3.2.2 South Functionalization of Cationic [4]- and [6]helicenes *137*
4.3.2.3 Tropical Functionalization Cationic [6]Helicenes *139*
4.3.2.4 Central Nucleophilic Additions on [4]-, [5]-, and [6]Helicenes *140*
4.3.3 Early-stage Equatorial Functionalization of [4]Helicene *141*
4.4 Structural Properties *142*
4.4.1 Helical Pitch and Configurational Stability *142*
4.4.2 pK_{R+} and Chemical Stability *143*
4.5 Optical and Chiroptical Properties and Applications of Cationic Helicenes *144*
4.5.1 Absorption and Fluorescence Properties *144*

4.5.2 Chiroptical Properties *148*
4.5.3 Biological Applications *150*
4.5.3.1 Interaction with DNA *151*
4.5.3.2 Cellular Imaging *151*
4.5.4 Applications Based on Redox Chemistry *153*
4.5.4.1 Generalities *153*
4.5.4.2 Electrochemiluminescence *154*
4.5.4.3 Thin Film Voltammetry *156*
4.5.4.4 Stable Radicals *157*
4.5.5 Organometallic Chemistry and Photocatalysis *159*
4.5.6 Spintronics *161*
4.6 Conclusion *161*
References *162*

5 Organometallic and Coordination Chemistry of Helicenes *167*
Rafael Rodriguez, Natalia Del Rio, and Jeanne Crassous
5.1 Introduction *167*
5.2 Coordination Chemistry *168*
5.2.1 Coordination of Monodentate Helicenes *168*
5.2.2 Coordination of Bidentate Helicenes *171*
5.2.3 Coordination of Polydentate Helicenes *177*
5.3 Organometallic Chemistry *179*
5.3.1 Metal-vinyl- and Metal-ethynyl-helicenes *179*
5.3.1.1 Metal-vinyl-helicene Complexes (Ru, Os) *179*
5.3.1.2 Metal-ethynyl-helicene Complexes (Fe, Ru, Au) *179*
5.3.2 Cyclometalated Helicenic Complexes *182*
5.3.2.1 Cyclometalated Helicenic Complexes Bearing C^N Chelate Ligands (Pt, Ir, Os) *182*
5.3.2.2 Cyclometalated Helicenic Complexes Bearing NHCs (Ir, Re) and Related Complexes *184*
5.3.3 Sandwich-type Helicenic Complexes *188*
5.4 Conclusion *192*
Abbreviations *192*
References *194*

6 Tetrathiahelicenes: An Infinite Source of Inspiration *199*
Silvia Cauteruccio, Andreas Dreuw, Emanuela Licandro, and Patrizia Romana Mussini
6.1 **Introduction** *199*
6.2 Synthesis of 7-TH Scaffold *200*
6.2.1 Oxidative Cyclization of Bis(benzodithienyl)ethenes *200*
6.2.1.1 Photochemical Oxidative Cyclization *201*
6.2.2 Annulation of 3,3′-Bis(benzo[1,2-*b*:4,3-*b*′]dithiophene) Derivatives *204*
6.3 Functionalization *206*

6.4 Resolution of Enantiomers *209*
6.4.1 Chiral HPLC *209*
6.4.2 Enzymatic Resolution *210*
6.4.3 Formation of Diastereoisomers *210*
6.4.4 Kinetic Resolution *210*
6.5 Electrochemical Properties and Applications of Tetrathiahelicenes *211*
6.5.1 Investigation of Tetrathiahelicene Electroactivity on a Systematic Compound Series *211*
6.5.2 Investigation of Electronic Properties of Electroactive 7-TH Adducts With Advanced Functional Properties *213*
6.5.3 Application of 7-TH-based Inherently Chiral Films for Enantioselective Electroanalysis *214*
6.6 Computational Studies on Tetrathiahelicenes *216*
6.6.1 General Remarks *216*
6.6.2 Electron-Detached and Attached States of 7-TH Phosphorus Derivatives *216*
6.6.3 Optical Properties of Chiral Tetrathiahelicene-Based Alkyl Phosphine–Borane Complexes *219*
6.6.4 Computational Study on Vibrational Raman Optical Activity of Helicenes *220*
6.7 Materials *220*
6.7.1 NLO *220*
6.7.2 Self-Assembly *222*
6.7.3 Molecular Spring *223*
6.7.4 Photoswitches *223*
6.8 Catalysis *224*
6.8.1 Organometallic Catalysis *224*
6.8.2 Organocatalysis *225*
6.9 Interaction with Biomolecules *226*
6.9.1 Chiral Recognition of Biomolecules *226*
6.9.2 Delivery Systems for 7-TH *227*
6.10 Conclusion *229*
References *229*

7 Synthesis and Properties of Helicenes with Small Numbers of Benzene Rings *237*
Masahiko Yamaguchi
7.1 Helicenes with Small Numbers of Benzene Rings *237*
7.1.1 Helicenes *237*
7.2 Synthesis of Helicene Derivatives *238*
7.2.1 Tetracyclic Helicenes *238*
7.2.2 Bicyclic Helicenes *241*
7.3 Noncovalent Bond Interactions and Chiral Recognition *242*
7.3.1 Complexation and Chiral Recognition by Oligosaccharides in Water *242*
7.3.2 Formation of Langmuir–Blodgett (LB) Films *244*

7.3.3 Charge Transfer (CT) Complexation *244*
7.3.4 Formation of Self-assembly Gel *246*
7.4 Asymmetric Catalysis *246*
7.5 Electronic Properties *247*
7.5.1 Reversible Switching of Charge Injection Barriers Using Helicenedithiol *247*
7.5.2 Ferroelectricity *248*
7.6 Interactions with Biological Macromolecules *249*
7.6.1 Nucleic Acid Binding *249*
7.6.2 Enzyme Inhibitor *250*
7.7 Separation Involving Aggregation of Helicene-Grafted Nanoparticles *251*
7.7.1 Separation of Molecules by Aggregation of Nanoparticles *251*
7.7.2 Chiral Recognition in Aggregation of Helicene-Grafted Gold Nanoparticles *251*
7.7.3 Optical Resolution Using Helicene-Grafted Silica Nanoparticles *251*
7.7.4 Molecular Recognition and Separation of Small Molecules from Metal-Catalyzed Chemical Equilibrium *252*
7.7.5 Molecular Recognition and Separation of Homo- and Hetero-double-helix Complexes by Helicene-Grafted Silica Nanoparticles *252*
7.7.6 Time Event by Helicene-Grafted Silica Nanoparticles *256*
7.8 Conclusion *257*
Acknowledgment *257*
References *258*

8 Multihelicenic Platforms from Halogenated Helicenes and Related Precursors *263*
Myriam Roy and Marc Gingras
8.1 Introduction *263*
8.2 General Synthetic Strategies for Constructing Multihelicenic Platforms from Aromatic Halides *265*
8.3 Palladium-Mediated Cross-Coupling Reactions Involving Aromatic Halides *266*
8.4 Electrophilic Aromatic Substitutions Involving Aromatic Halides *273*
8.5 Metal-Catalyzed Homocouplings with Aromatic Halides *274*
8.6 Conclusion *279*
References *279*

9 Helical Nanographenes: Synthetic and Chiroptical Achievements *283*
Sandra Míguez-Lago, Juan P. Mora-Fuentes, Carlos M. Cruz, and Araceli G. Campaña
9.1 Introduction *283*
9.2 All Carbon Helical Nanographenes *290*

9.2.1 Achiral or Racemic Helical Nanographenes *292*
9.2.2 Chiral and Enantioenriched Helical Nanographenes *297*
9.3 Heteroatom-Containing Helical Nanographenes *305*
9.3.1 Helical Nanographenes Containing Heteroatoms of Group 15 (N, P) *306*
9.3.2 Helical Nanographenes Containing Heteroatoms of Group 16 (O, S) *313*
9.4 Summary Table *322*
Acknowledgments *322*
References *322*

10 Helicene-based Electron Acceptors *329*
Xiao Xiao and Colin Nuckolls
10.1 Introduction *329*
10.2 Helicenes with Remote Electron-withdrawing Functionalities *330*
10.3 Electron-accepting Heterohelicenes *335*
10.3.1 Azahelicenes *335*
10.3.2 Incorporation of Oxygen Atoms in Helicenes *341*
10.3.3 Borahelicenes *343*
10.3.4 Cationic Heterohelicenes *343*
10.4 Electron-accepting π-extended Frameworks Embedded with Helicenes *345*
10.4.1 Extended Helicenes with Naphthalimide Units *346*
10.4.2 Perylene Diimide (PDI)-based Helicenes *348*
10.4.3 Terrylene-based Helicenes *363*
10.5 Conclusion and Outlook *365*
References *365*

11 Chiroptical Properties of Helicenes *373*
Sergio Abbate, Giovanna Longhi, and Tadashi Mori
11.1 Introduction *373*
11.2 Chiroptical Spectroscopies of Model Hexahelicenes: A Brief History *373*
11.3 Helical Length Dependence of Chiroptical Properties of Helicenes *376*
11.4 Impact of Symmetry in Multiple Helicenes on Their Chiroptical Properties *378*
11.5 Relevance of Substituents, of Vibronic Couplings, and of Charge Transfer Effects in the ECD Spectra of Helicene Systems *385*
11.6 Concluding Remarks *389*
References *389*

12 Photophysical and Chiroptical Properties of Metal–Organic Helicenic Systems: Experiment vs. Theory *395*
Monika Srebro-Hooper, Jeanne Crassous, and Jochen Autschbach
12.1 General Aspects *395*
12.1.1 Introduction *395*
12.1.2 Theory Developments in the Context of Helicene Optical Activity *396*

12.2 Case Studies *400*
12.2.1 Azahelicene-Phosphole Pd(II) and Cu(I) Complexes: OR and ECD Analyses *400*
12.2.1.1 Complexes with Extended π-Conjugated Helicenic P,N Bidentate Ligands *400*
12.2.1.2 Assemblies with Multitopic Helicenic Ligands *402*
12.2.2 Ru- and Fe-[6]helicene Redox-Triggered Chiroptical Switches: UV-vis, OR, ECD, VCD, and ROA Analyses *404*
12.2.2.1 Ruthenium-vinyl-helicenes *404*
12.2.2.2 Iron-Ethynyl-Helicenes *406*
12.2.3 Platinahelicenes: UV-vis, OR, ECD, Non-polarized Emission and CPL Analyses *407*
12.2.3.1 Mono- and Bis-cycloplatinated Helicenes *407*
12.2.3.2 Bipyridine-Helicene-Based Multifunctional pH-Switchable Platform *408*
12.2.4 Re(I) and Ir(III) Helicenic Phosphors: OR, ECD and Non-polarized Emission Analyses *410*
12.2.4.1 Bipyridine-Helicene-Rhenium Complexes *410*
12.2.4.2 Cycloiridiated Complexes with Helicenic N-Heterocyclic Carbene *411*
12.3 Concluding Remarks *413*
Abbreviations *414*
References *414*

13 Helicene Derivatives with Circularly Polarized Luminescence *423*
Wen-Long Zhao, Meng Li, Hai-Yan Lu, and Chuan-Feng Chen
13.1 Introduction *423*
13.2 Carbohelicenes with CPL Properties *424*
13.3 Heterohelicenes with CPL Properties *434*
13.3.1 Silahelicenes with CPL Properties *435*
13.3.2 Borohelicenes with CPL Properties *435*
13.3.3 Azahelicenes with CPL Properties *438*
13.3.4 Oxalohelicenes with CPL Properties *440*
13.3.5 Thiohelicenes with CPL Properties *441*
13.4 Metal Helicene Derivatives with CPL Properties *442*
13.5 Regulation of CPL for Helicene Derivatives *446*
13.6 Summary and Outlook *448*
Acknowledgments *448*
References *448*

14 Helicenes and Nonlinear Optical Properties: A Good Match? *453*
Benoît Champagne
14.1 Introduction *453*
14.2 Electric Second-Order NLO Responses *456*

14.2.1 Effect of the Nature and Position of the Substituents *456*
14.2.2 Effect of the Nature of the Helicene Backbone, Including Oxidation Effects *461*
14.2.3 Effect of the Size of the Helicene *464*
14.2.4 Three-Wave Mixing *465*
14.2.5 Other Helicenes and Related Helical Systems *465*
14.3 Third-Order NLO Responses *466*
14.4 Beyond the Electric Dipole Approximation *466*
14.5 Helicene Aggregates and Self-Assembled Monolayers *467*
14.6 Conclusions and Outlook *468*
Acknowledgments *469*
References *470*

15 Helicenes for Optoelectronic Applications and Devices *473*
Seán T. J. Ryan and Matthew J. Fuchter
15.1 Introduction *473*
15.2 OLED Applications *474*
15.2.1 Helicene Racemates as Emissive OLED Materials *474*
15.2.2 Helicene Racemates as OLED Device Components *477*
15.2.3 Helicenes as Enantiopure Materials for OLEDs *480*
15.2.3.1 Helicenes as Small-Molecule Dopants for CP-OLEDs *482*
15.2.3.2 Helicene Complexes for Circularly Polarized Phosphorescent Light- Emitting Diodes (CP-PHOLEDs) *484*
15.3 Enantiomers vs. Racemates: Bulk Charge Transport of Helicenes and Application to Optoelectronic Devices *487*
15.3.1 Helicene Bulk Charge Transport and OFETs *487*
15.3.2 Helicenes for CP Photodetectors *489*
15.3.3 Helicenes for Organic Photovoltaics *490*
15.4 Helicenes and the Chiral-Induced Spin Selectivity Effect *494*
15.5 Conclusion *496*
Abbreviations *496*
References *498*

16 Miscellaneous Applications of Helicenes: Surface Science and Asymmetric Catalysis *505*
Jeanne Crassous
16.1 Introduction *505*
16.2 Self-Assembly of Helicenes onto Surfaces *505*
16.2.1 Self-Assembly of Heptahelicene *506*
16.2.2 Self-Assembly of 5-Amino-Hexahelicene *507*
16.2.3 Self-Assembly of 6,13-Di-Cyano-[7]Helicene *509*
16.3 Applications of Helicenes in Enantioselective Catalysis *510*
16.3.1 Asymmetric Transition Metal Catalysis *510*
16.3.1.1 P-containing Helicenes in Enantioselective Rh(I)-promoted Hydrogenation *510*

16.3.1.2 P-containing Helicenes in Enantioselective Pd(II)-Catalyzed Allylic Substitution *510*
16.3.1.3 Helicenic Phosphite in Enantioselective Ir(I)-Catalyzed Allylic Amination *513*
16.3.1.4 Pentahelical Phosphines in Enantioselective Suzuki–Miyaura Coupling *513*
16.3.1.5 Helical NHCs in Enantioselective Ni(0)-Catalyzed [2+2+2] Cycloisomerization *514*
16.3.1.6 Helical NHCs in Enantioselective Ru-Catalyzed Metathesis *514*
16.3.1.7 Phosphahelicenes in Enantioselective Au(I) Cycloisomerization of Enynes *516*
16.3.1.8 Helicenes as Chiral Inducers in the Autocatalytic Soai Reaction *517*
16.3.2 Asymmetric Organocatalysis *519*
16.3.2.1 Helicene N-oxide Catalyzed Enantioselective Desymmetrization of *Meso* Epoxides *519*
16.3.2.2 Helical Amino-Pyridium Catalyzed Asymmetric Addition to Nitroalkenes *520*
16.3.2.3 Helical Bipyridine *N*-monoxide-Catalyzed Enantioselective Propargylation of Aldehydes *522*
16.3.2.4 Azahelicene-Catalyzed Enantioselective Acyl Transfer Reaction *522*
16.3.2.5 Phosphahelicene-Catalyzed Enantioselective [3+2] Annulation *522*
16.4 Conclusion *524*
List of Abbreviations *524*
References *525*

Index *527*

Preface

Helicenes are the archetypal helical molecules simply composed of *ortho*-condensed aromatic rings whose helical shape derives from the steric repulsion of terminal cycles or substituents. It leads to configurationally stable helices with minus (*M*) or plus (*P*) handedness. These chiral polycyclic hydrocarbons have attracted the attention of chemists for decades, not only because of their beautiful 3D helical topology artificially mimicking natural helical structures such as double-stranded DNA or snail shells but also because of their many appealing physicochemical properties and consequent applications in various fields of research. Over time, the structural diversity of helicenes has grown considerably and many synthetic approaches have been developed to access a variety of helical scaffolds decorated with a decent range of heteroatoms, functional groups, or metal ions. With advances in chiral high-performance liquid chromatography (HPLC) methodologies, non-racemic helicenes have become accessible via racemate resolution. Alternatively, new stereoselective synthetic procedures have been developed to receive helicenes in a nonracemic form.

These helically chiral extended π-conjugated molecules strongly interact with circularly polarized (CP) light and as a result exhibit remarkable chiroptical properties such as unusually high values of specific optical rotation, very intense electronic and vibrational circular dichroism spectra, CP emission, and nonlinear optical responses. From a technology perspective, helicenes can behave as semiconductor materials and as such can constitute attractive optoelectronic devices, in particular chiral photovoltaic cells, transistors, and CP organic light emitting diodes (OLEDs). Moreover, their helical topology combined with extended π-conjugation makes them unique models for studying the basis of optical activity, the role of chirality in 2D self-assembly of molecules on surfaces, and principles of chirality transfer in enantioselective catalysis.

Undoubtedly, helicene research is undergoing a dynamic development, and the number of groups focusing on this topic is steadily increasing. This monograph collects chapters written by internationally recognized experts in their field and covers all of the aforementioned aspects.

Chapters 1–9 describe manifold facets of the synthesis of helicene architectures and discuss selected properties derived from their molecular structure. In Chapter 1, one of the most widely used and historically important synthetic methods for accessing helicene derivatives, "the photochemical approach to helicenes" is reviewed by Jan Storch, Jaroslav Žádný, Vladimír Církva, and Martin Jakubec from the Institute of Chemical Process Fundamentals of the CAS, Prague; Jan Hrbáč from Masaryk University, Brno; and Jan Vacek from Palacký University Olomouc, Czech Republic. In Chapter 2, another popular and efficient method for obtaining racemic or enantioenriched helicenes, "synthesis of helicenes by [2 + 2 + 2] cycloisomerization," is presented by Irena G. Stará and Ivo Starý, coeditors, from the Institute of Organic Chemistry and Biochemistry of the CAS, Prague, Czech Republic. In Chapter 3, Ken Tanaka from the Tokyo Institute of Technology, Japan, describes the most efficient methods for "enantioselective synthesis of helicenes." Chapter 4 reports on a class of charged helicene scaffolds, "cationic triarylcarbenium helicenes," written by Johann Bosson and Jérôme Lacour from University of Geneva, Switzerland, and Niels Bisballe and Bo W. Laursen from University of Copenhagen, Denmark. "Organometallic and coordination chemistry of helicenes" is the subject of Chapter 5, authored by Rafael Rodriguez, Natalia Del Rio, and Jeanne Crassous, coeditor, from University of Rennes, France, while Chapter 6 focuses on "tetrathiahelicenes" by Silvia Cauteruccio, Emanuela Licandro, and Patrizia Romana Mussini from University of Milan, Italy, and Andreas Dreuw from Heidelberg University, Germany. In Chapter 7, Masahiko Yamaguchi from Tohoku University, Japan, describes the "synthesis and properties of helicenes with small numbers of benzene rings" to introduce the smallest configurationally stable [4]helicene derivatives. The next two chapters, 8 and 9, focus on multihelicene architectures, specifically on "preparation of multihelicenic platforms from halogenated helicenes" by Myriam Roy from Sorbonne Université, Paris, and Marc Gingras from Aix Marseille University, France, and on "helical nanographenes and their synthetic and chiroptical achievements" by Sandra Míguez-Lago, Juan P. Mora-Fuentes, Carlos M. Cruz, and Araceli G. Campaña from Granada University, Spain.

After the first part of the monograph discusses mainly the chemical aspects of helicene research (Chapters 1–9), the second part presents primarily the properties and applications of helicenes (Chapters 10–16). Accordingly, in Chapter 10, Xiao Xiao and Colin Nuckolls from Columbia University, New York, United States, describe the properties of "helicene-based electron acceptors." Chapters 11 and 12 focus on the "chiroptical properties of helicenes" through a "historical perspective and structure-property relationships" by Sergio Abbate and Giovanna Longhi from Università degli Studi di Brescia, Italy, and Tadashi Mori from Osaka University, Japan, and on "photophysical and chiroptical properties of helicene-based systems: experiment vs. theory" by Monika Srebro-Hooper, Jagiellonian University, Krakow, Poland; Jochen Autschbach, University of Buffalo, United States; and Jeanne Crassous, coeditor. Chapter 13 discusses "helicene derivatives with circularly polarized luminescence" written by Meng Li from Beijing National Laboratory for Molecular Science and Wen-Long Zhao, Hai-Yan Lu, and Chuan-Feng Chen from University of Chinese Academy of Sciences, Beijing, China. In Chapter 14, Benoît Champagne from University of Namur, Belgium, describes the "nonlinear optical properties" of helicenes, and in Chapter 15, the "helicenes for optoelectronic devices" research area is reported by Seán T.J. Ryan and Matthew J. Fuchter from Imperial College London, United Kingdom. Finally, this book could not be completed without addressing the important application fields of helicenes, which are surface science and asymmetric catalysis, as very briefly reviewed in the last Chapter 16 by coeditor Jeanne Crassous.

Jeanne Crassous would like to warmly thank the Centre National de la Recherche Scientifique (CNRS), which employs her, and the University of Rennes and the Institut des Sciences Chimiques de Rennes for giving her a pleasant and enriching environment to work in, with so many friendly and professional colleagues from both the scientific and administrative parts. She also thanks all the students and coworkers with whom she has shared her passion for chirality and helicenes during the past 25 years. She expresses special thanks to her former and current close collaborators, Régis Réau and Ludovic Favereau. She dedicates this book to the memory of her mentors André Collet and François Diederich, who passed away too early.

Irena G. Stará and Ivo Starý would like to express their sincere thanks to the Institute of Organic Chemistry and Biochemistry of the Czech Academy of Sciences for creating an inspiring environment and for all financial and technical support. They still owe enough thanks to their great teachers Pavel Kočovský, Jiří Závada, Wolfgang Oppolzer, Peter E. Kündig, and Josef Michl, who fundamentally influenced their scientific career. Last but not least, they are grateful to all their former students whom they have had the privilege of teaching, so that they can now learn from them themselves.

This book has been written under the spiritual influence of many chirality-related communities, especially those from international symposia (Chirality and Chiroptical Spectroscopy conference series) and national and international consortia, including the HEL4CHIROLED ITN project Grant Agreement (H2020-MSCA-ITN-2019 n° 859752).

1

The Photochemical Approach to Helicenes

Jan Storch[1], Jaroslav Žádný[1], Vladimír Církva[1], Martin Jakubec[1], Jan Hrbáč[2], and Jan Vacek[3]

[1] *Department of Advanced Materials and Organic Synthesis, Czech Academy of Sciences, Institute of Chemical Process Fundamentals, Prague, Czech Republic*
[2] *Department of Chemistry, Faculty of Science, Masaryk University, Brno, Czech Republic*
[3] *Department of Medical Chemistry and Biochemistry, Faculty of Medicine and Dentistry, Palacky University, Olomouc, Czech Republic*

1.1 Introduction

In this chapter, the authors have decided to follow the nomenclature recommendations of IUPAC [1] for class names of organic compounds, which classifies *helicenes* as "*ortho*-fused polycyclic aromatic or heteroaromatic compounds in which all rings (minimum five) are angularly arranged so as to give helically shaped molecules, which are thus chiral." Therefore, the following text includes [n]helicenes where $n \geq 5$ and the photoreaction is the very last step of their preparation, unless stated otherwise.

Two basic photo-approaches (oxidative photocyclodehydrogenation and photoinduced elimination; for details, see Section 1.2) were used for the preparation of all helicenes listed in this chapter. The photochemically created bonds are highlighted in red in all figures. Non-oxidative photochemical approaches are discussed in specific cases.

At the beginning of the chapter, some general features are mentioned, including the mechanism of the photocyclization, reaction conditions, attempts at asymmetric photosynthesis, and the synthetic approach to starting materials. The appropriate helical structures and their preparations are described in sections on carbo-, aza-, thia-, and phosphahelicenes and other helicenes. Helicene-like molecules, including dihydrohelicenes, are discussed separately, as well as photochemical transformations of helicenes.

1.2 General Features

Historically, the photocyclization of stilbenes [2, 3] was discovered during the investigation of their *cis/trans* photoisomerization [4], but the reaction was not used synthetically until Mallory found that iodine catalyzed this reaction in 1964 [5, 6].

Helicenes: Synthesis, Properties, and Applications, First Edition.
Edited by Jeanne Crassous, Irena G. Stará, and Ivo Starý.

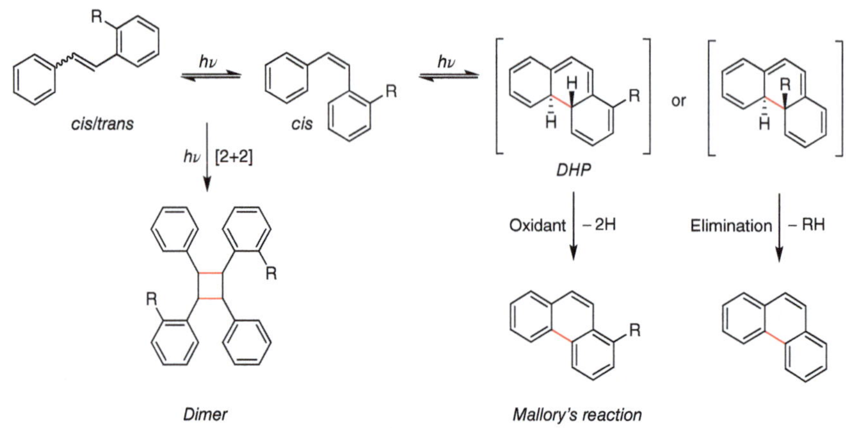

Figure 1.1 Photochemical reaction pathways for stilbene derivatives.

From the mechanistic point of view, the *cis*/*trans* photoisomerization of stilbenes is very fast with a high quantum yield, allowing stilbenes to be used as an isomeric mixture (Figure 1.1), although only the *cis*-isomer is capable of cyclization. The symmetry-allowed photoreaction typically takes place from the singlet S_1 state by a conrotatory process according to the Woodward–Hoffmann rules. Thus triplet sensitizers do not sensitize the photocyclization, and triplet quenchers (such as oxygen) do not quench it. The unstable 4a,4b-dihydrophenanthrene (DHP) intermediate possesses *trans*-configuration [7] and can, unless trapped, relax back to the stilbene. In the presence of an oxidant, DHP forms a phenanthrene derivative. This type of photocyclization is called the *Mallory reaction* [8]. If the stilbene contains suitable leaving groups (R = OMe, Cl, Br, etc.) in *ortho*-position, the elimination reaction producing cyclization product can take place (in the absence of an oxidant) [9–11]. Photocyclizations are typically carried out at concentrations of 10^{-3} M and lower to avoid the competing photodimerization [12]. The proposed photocyclization mechanism is also applicable to aza-, thia-, and other stilbene derivatives.

Originally, air was used as an oxidant until Mallory discovered that oxidative trapping occurs much faster when iodine (5 mol%) is used together with air [5]. It was proposed that iodine is photochemically cleaved into radicals reacting with hydrogen to form hydrogen iodide, which is then reoxidized to iodine by oxygen [13]. Other oxidants (e.g. selenium radicals, TCNE, TCNQ, chloranil, etc.) were investigated by Laarhoven without any practical significance for photocyclizandation of helicenes [14].

Higher amounts of hydrogen iodide may contribute to side reactions including double bond saturation of stilbenes [13]. In 1986, Katz developed new photocyclization conditions using propylene oxide as a hydrogen iodide scavenger under inert conditions [15, 16]. As a consequence, the iodide could not be reoxidized by air, and its stoichiometric amount is needed. Accordingly, it allows for a higher concentration of starting material in the reaction mixture without the undesired formation of dimers. When speaking about photocyclodehydrogenation leading to [*n*]helicenes, these conditions are sometimes familiarly referred to as *Katz's conditions*. Other cyclic ethers (such as THF) are often used as HI scavengers too.

The previously mentioned conditions provide [*n*]helicenes in strictly racemic mixtures (1 : 1 ratio of (*P*)- and (*M*)-enantiomer). Although attempts to lead photocyclizations asymmetrically using circularly polarized light sources [17–23], chiral solvents [24–26], or cholesteric liquid crystals [27, 28] were made, the obtained results (% ee) were more or less at the level of experimental error and did not have any practical importance. Thus, the photochemical approach has to be followed by an optical resolution to obtain helicenes in their optically pure forms. The nonracemic helicenes are often photochemically accessible as corresponding diastereomers with (photo-stable) chiral auxiliaries (providing up to >99% de), which can be synthetically cleaved or transformed after cyclization [29–31]. Eventually, such diastereomers might be separated using standard chromatographic methods. When an enantiomerically pure [6]helicene moiety was a part of the precursor, nonracemic [*n*]helicenes (n = 8–11, 13) were obtained [32]. Other asymmetric photosyntheses to enantioenriched metallocene helicenes were developed by Katz [33–35]. Some studies suggested that only one chiral auxiliary is not sufficient, and a better result might be obtained with the chiral substitution at the most sterically hindered position [36]. The same phenomenon was observed by Carbery and Pearson [37]. For an overview see Ref. [26] and the references therein.

Conventional (Figure 1.2a) and transition metal-catalyzed (Figure 1.2b) methods of preparation of stilbene-like molecules as a common starting material for the photochemical synthesis of helicenes are described in the literature [38, 39], including several other synthetic procedures. In practice, the Wittig reaction (and its variations) and Pd-catalyzed cross-coupling reactions belong to commonly used methods.

The most widely used sources of UV–vis light for continuous irradiation in laboratory experiments are commercially available mercury lamps (low, medium, and high pressure) [40]. Their spectral irradiance is strongly dependent on the mercury vapor pressure. The lamp also produces a considerable amount of infrared radiation and heat. Therefore, cooling-water circulation must be utilized to protect the reaction solution from heating. Recently, new energy-efficient light sources like light-emitting diodes (LEDs) [41–43] (Figure 1.3d) became available, thus avoiding the use of optical filters and reducing consumption costs.

Photoreactors with an *immersion well* (Figure 1.3a) and *external chamber* (*merry-go-round*, Figure 1.3b) are the most common types of photochemical equipment on a preparative laboratory scale. Both reactor types are well established and in widespread utilization. The use of quartz allows light of all wavelengths above about 200 nm to enter the reaction mixture. For some photoreactions, higher yields can be obtained by employing Pyrex glass. This excludes from the reaction mixture light of wavelengths below about 300 nm and thereby protects the forming products from further photochemical degradation.

Following the experiments by Mallory [6], Scholz [44], and Martin [45], the batch setup of photocyclization of stilbene derivatives under UV–vis irradiation has become one of the most popular methods for the synthesis of helicenes [13]. This is

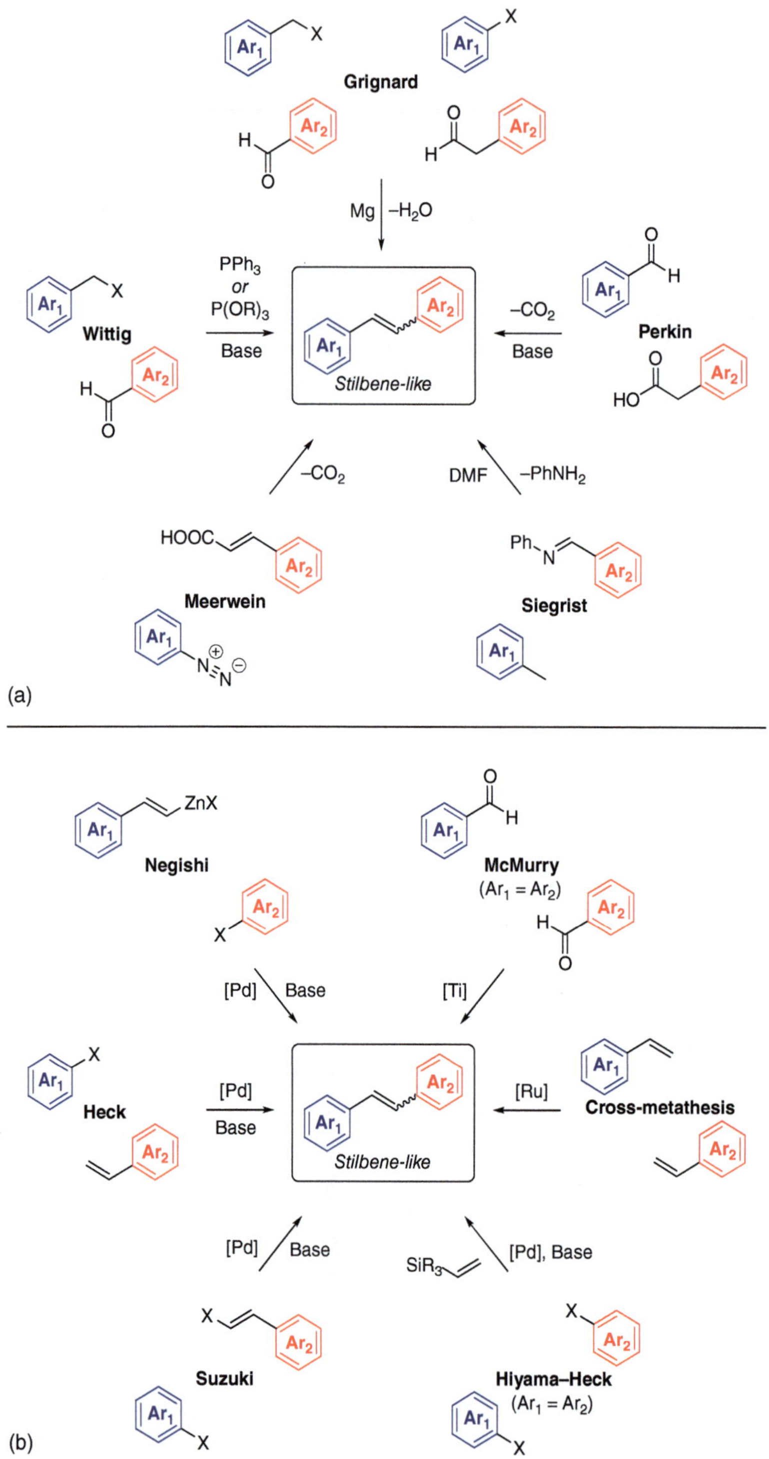

Figure 1.2 Commonly used preparations of stilbene-like molecules by conventional (a) and metal-catalyzed (b) procedures.

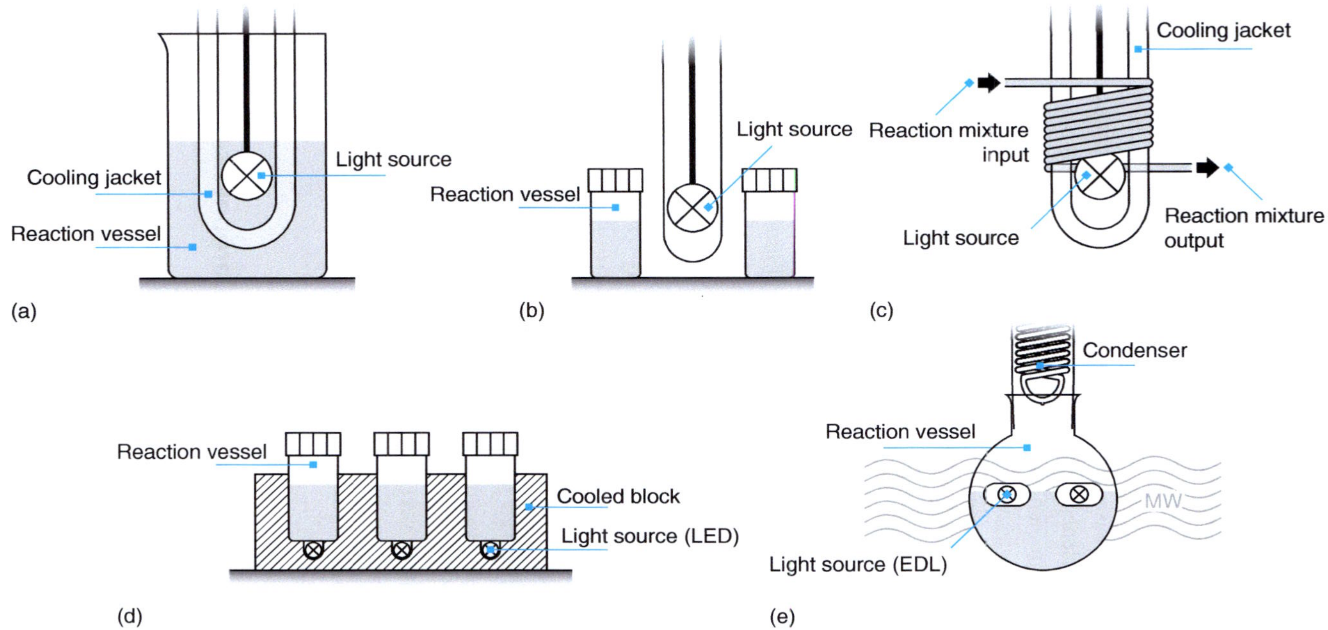

Figure 1.3 Common photoreactor types: (a) immersion well, (b) external chamber, (c) continuous-flow, (d) LED-type, and (e) electrodeless discharge lamps.

mainly due to the synthetic accessibility of the stilbene precursors and functional group tolerance under the reaction conditions (pH, temperature, etc.). However, the development of this setup is limited by the necessary dilution of the reaction mixture (~10^{-3} mol·l^{-1}) to prevent the undesired [2+2] cycloaddition, requiring large volumes of pure solvents and prolonged reaction times (>15 hours).

Therefore, finding an efficient protocol for the photocyclization of stilbene derivatives in a flow mode (Figure 1.3c) would be useful for the synthesis of helicenes on a gram scale. The photochemical transformations conducted under continuous-flow conditions are commonly characterized by shorter reaction times, higher yields, increased selectivities, easier purification, improved productivities, and high photonic and energy efficiencies compared with the conventional batch methods [46]. The equipment requirements are a special tubing (fluorinated ethylene propylene [FEP], which is highly transmittable, flexible, and anticorrosive) wound around the light source and a diaphragm pump.

Collins et al. reported the first example of the [5]helicene preparation by the photochemical continuous-flow strategy under visible light in the presence of Cu complex formed in situ [47]. In comparison to the traditional setup in a batch reactor, the flow method under photocatalytic conditions prevented the product from overannulation and enabled its gram-scale preparation. The same group used it also for the synthesis of pyrene–helicene dyads [48]. Rueping et al. [49] used a photo-flow methodology for the synthesis of [5]- and [6]helicenes with different substitution patterns. Murase et al. [50] accomplished the synthesis of amide-type aza[6]helicene in high yield by photocyclization of boron hydroxamate complex using a continuous-flow reactor in order to minimize its decomposition. Sýkora et al. [51] prepared 2-bromo[6]helicene as a key intermediate for helicene functionalization utilizing a continuous-flow photoreactor. Finally, Církva et al. [52] synthesized [6]helicenes fluorinated at terminal rings in the same setup.

A special photochemical reactor utilizing electrodeless discharge lamps (EDLs; Figure 1.3e) was also designed [53, 54]. It generates UV irradiation when placed in a microwave (MW) field. This methodology was used for the preparation of [6]helicene derivatives. Next to low costs, the arrangement in which the EDLs are placed inside the reaction vessel has several other advantages, such as simultaneous UV and MW exposure of the sample, enabling the performance of photochemistry at high temperatures.

1.3 Photochemical Preparation of Carbohelicenes

This part is further divided into oxidative and eliminative photocyclization according to the mechanism of the reactions. More details on photocyclizations toward carbohelicenes can be found in reviews [8, 13, 14, 26, 55–60] and a book chapter [61].

1.3.1 Oxidative Photocyclizations of Unsubstituted Carbo[*n*]helicenes

The first preparation of helicene molecule by photocyclization reaction was reported by Martin in 1967 on the preparation of [7]helicene (**3**) [45]. Since then,

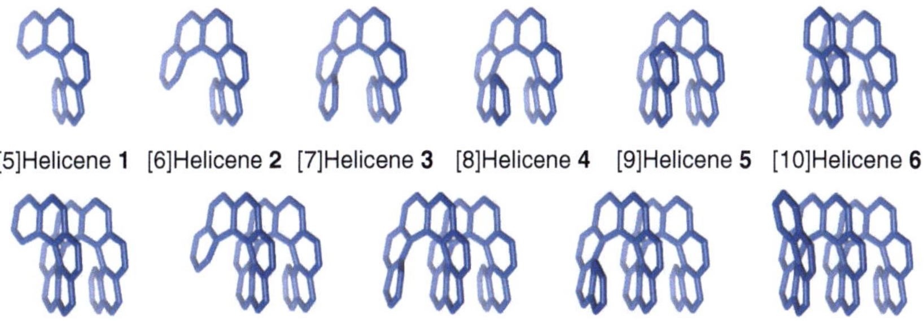

[5]Helicene **1** [6]Helicene **2** [7]Helicene **3** [8]Helicene **4** [9]Helicene **5** [10]Helicene **6**

[11]Helicene **7** [12]Helicene **8** [13]Helicene **9** [14]Helicene **10** [16]Helicene **11**

Figure 1.4 Unsubstituted carbo[*n*]helicenes.

photocyclization has become one of the most important methods for the synthesis of helicene homologs, from [5]- to [16]helicenes (Figure 1.4) and, of course, their derivatives.

[*n*]Helicenes can be photochemically formed from olefinic precursors of [*a*]+[*b*]+[*c*] structure, where "[*a*–*c*]" denotes the number of *ortho*-fused benzene subunits and "+" stands for the vinylene linker(s).

In the case of [5]helicene (**1**), great attention must be paid to the design of precursors due to possible benzo[*ghi*]perylene formation (see Section 1.6), which was observed by Dietz [44, 62], who attempted the synthesis of **1** using the [2] + [2] or [3]+[1] strategy. Using the [1]+[1]+[1] photocyclization strategy fails as well, as Liu and Katz prepared [5]helicene only in small amounts (6% yield) [63]. A possible solution to this problem was recently shown by Collins [47], who used copper(I) complexes as photocatalyst and [2]+[2] under visible light irradiation to prepare [5]helicene in 57% yield.

In the case of [6]helicene (**2**), several strategies have been investigated. Martin reported on [1]+[2]+[1], [2]+[1]+[1], and [4]+[1] cyclization leading to **2** with yields up to 80% [64, 65]. On the other hand, photocyclizations of [3]+[2] precursors by Laarhoven [66] and Kagan [22] provided low yields due to poor regioselectivity of the reaction.

[7]Helicene (**3**) reported by Martin [64, 65] and Laarhoven [66] employed [2]+[1]+[2] and [4]+[2] precursors, respectively; however, due to poor regioselectivity of the reaction, only low yields of the desired products were obtained. Consequently, Laarhoven irradiated [3]+[3] stilbene under an inert atmosphere, yielding up to 50% of **3** [66]. However, the best result was obtained in the case of a [1]+[3]+[1] cyclization published by Ben Hassine and Marinetti, who prepared [7]helicene in 90% yield [67].

The preparation of [8]helicene (**4**) included several strategies, such as [4]+[1]+[1], [4]+[3], and [6]+[1], all regioselectively yielding helicene in up to 85%, as published by Kagan and Martin [17, 23, 32, 65, 68]. The same authors furnished also [9]helicene (**5**) using [4]+[1]+[2], [4]+[4], or [6]+[2] starting materials with yields up to 74% [17, 23, 32, 65]. More recently, a triple photocyclization was proposed for a precursor containing two phenylene and two naphthylene units linked together with vinylene spacers ([2] + [1] + [1] + [2]) [69]. The resulting [9]helicene was isolated in 67% yield. [10]Helicene (**6**) was prepared by [4] + [1] + [3], [4] + [3] + [1], and [6] + [3] photodehydrogenation in 79% yield [17, 32].

Martin [32, 70] reported a regioselective photocyclization of [3] + [3] + [3] and [6] + [4] stilbene-like molecules that yielded [11]helicene (**7**) in up to 54%, which was later increased to up to 80% using the [4] + [1] + [4] or [4] + [3] + [2] strategy [17, 70]. In the same works, [2] + [6] + [2], [3] + [4] + [3], and [4] + [3] + [3] starting materials served as precursors for the synthesis of [12]helicene (**8**), which was obtained in 42% yield. A comparable yield was obtained in the case of higher [13]helicene (**9**) using regioselective [4] + [3] + [4] and [8] + [4] precursors, respectively [17, 32, 71]. Moreover, it was found that [6] + [6] stilbene [72] is not synthetically practical, probably due to attractive π–π dispersive interactions. [14]Helicene (**10**) was obtained in moderate yields using regioselective irradiation of [3] + [6] + [3] or [4] + [4] + [4] precursors [70].

The recently published [2] + [1] + [1] + [2] + [1] + [1] + [1] + [2] method proposed by Murase and Fujita to synthesize the longest carbohelicene to date, the [16]helicene (**11**), was based on sextuple photocyclization of a single-strand oligo(arylene-vinylene) precursor [69]. The key to the successful synthesis of **11** was the rational design of the precursor, in which the phenylene and naphthylene units are arranged in a way that disfavors side reactions (benzoanthracene formation or overannulation). Additionally, only readily available aromatic precursors were used, further improving the practicality of this approach. However, the bulky TIPSO (triisopropylsilyl ether) substitutions at the peripherals of the olefin were necessary to increase the solubility of the precursor prior to photocyclization. **11** was then prepared by their subsequent removal in 7% yield.

The Mallory reaction was studied using Hückel molecular orbital (HMO) calculations, which provided a remarkably accurate description of its regioselectivity [44]. Usually, only one product is formed, even when the formation of several others looks plausible. The reaction favors such ring structures that are curved toward helicenes. Laarhoven evaluated reactivity parameters such as free valence numbers (F^*_{rs}) [14, 73] and localization energies (L^*_{rs}) [44] for a large number of examples. He found a good correlation between these two parameters, although the use of just free valence numbers was more convenient as only one calculation was needed to evaluate all cyclization modes of one particular compound. Upon the analysis of the excited state calculations, Laarhoven derived three simple rules [14] to determine the product and regioselectivity of the proposed reaction, in which the sum of the free valence numbers of atoms *r* and *s* is involved ($\sum F^*_{rs}$):

(i) Photocyclizations do not occur when $\sum F^*_{rs} < 1.0$.
(ii) When two or more ways of cyclization are possible in a single compound, only one product arises if $\Delta(\sum F^*_{rs}) > 0.1$; otherwise more products are formed.
(iii) The second rule stands when either only planar or only nonplanar products (penta- or higher helicenes) can arise. When both planar and nonplanar products can be formed, the planar aromatics is generally the main product ($\sum F^*_{rs} > 1.0$).

When helicenes become larger than five benzene rings, they become nonplanar. The formation of planar S-shaped molecules becomes the major side reaction or even the main reaction. This led Katz's group to the development of various strategies to facilitate the synthesis of helicenes by photocyclization. A bromine substituent was used as a protecting group preventing a photocyclization in this position [15].

1.3.2 Oxidative Photocyclizations of Substituted Carbo[*n*]helicenes

Substituted carbo[*n*]helicenes have been studied from the very beginning of helicene chemistry to the present day, and it was found that the substituents can in some cases have a significant influence on the course of the photocyclization reaction. For example, it was found that nitro, acetyl, or dimethylamino substituents are not suitable for the photocyclization because they enhance intersystem crossing to the triplet excited state [74].

Substituted [*n*]helicenes can be synthesized in the same way as unsubstituted ones. With a *meta*-substituent at the styryl group (Figure 1.5), two isomers can be formed (1- and 3-substituted helicenes) [75], while *meta*-substituted distyryl-naphthalene derivatives lead to a mixture of three products [76]. To increase the formation of 1-substituted helicenes, the blocking group, e.g. an additional *ortho*-bromo-substituent, has to be introduced to the styryl group [30]. Finally, 2-substituted [*n*]helicenes [19, 51, 75] can be synthesized from *para*-substituted derivatives and 4-substituted [*n*]helicenes [19, 30, 77, 78] from *ortho*-substituted derivatives with no obstacles for regioselectivity of the reaction.

Like [5]helicene, its derivatives may be poorly photochemically available due to rapid overannulation into benzo[*ghi*]perylenes. This phenomenon can be hindered by the introduction of functional groups at C_1 [79–82] or C_2 [37, 63, 67, 83–85] positions of the [5]helicene backbone (see Figure 1.6). In some cases, however, rearrangement or elimination may occur [79]. To prevent the overannulation, Matsuda found that the introduction of cyano groups to ethylene moieties using the [1] + [1] + [1] cyclization strategy could eliminate the degeneracy of unoccupied molecular orbitals; therefore 5,10-dicyano[5]helicene derivatives **13a–c** have been prepared in high yields [86]. This phenomenon was also observed by Durola [88]

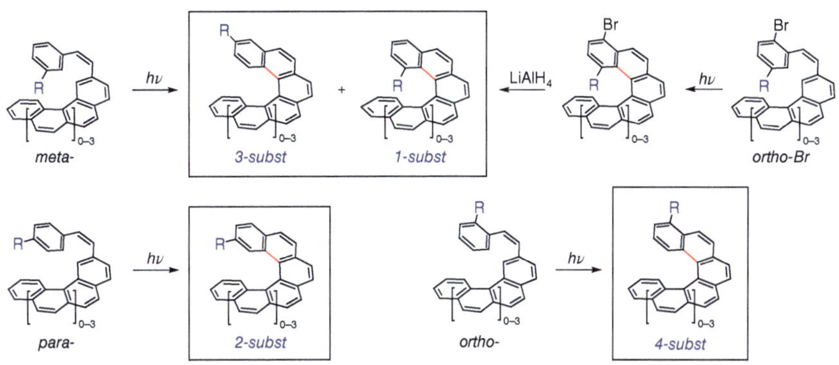

Figure 1.5 Photocyclization of substituted carbo[*n*]helicenes.

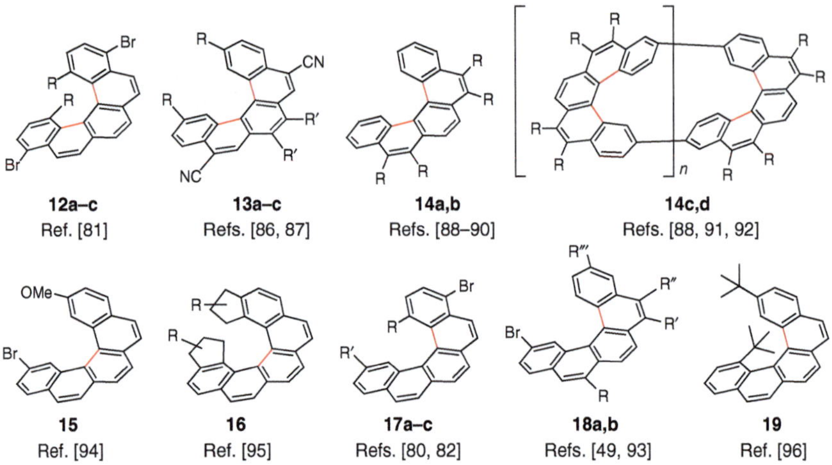

Figure 1.6 Representative examples of substituted carbo[5]helicenes. **12a**: R = F; **b**: R = Me; **c**: R = OMe; **13a**: Ref. [86] R, R′ = H; **b**: Ref. [86] R = H, R′ = OMe; **c**: Ref. [87] R = OMe, R′ = H; **14a**: Refs. [88, 89] R = CO_2Et; **b**: Ref. [90] R = anhyd; **c**: Refs. [88, 91] R = CO_2Et, *n* = 1; **d**: Ref. [92] R = CO_2Et, *n* = 2; **16**: R = $OSiMe_2$*t*-Bu; **17a**: Ref. [80] R = F, R′ = Br; **b**: Ref. [80] R = F, R′= Me; **c**: Ref. [82] R = CO_2Me, R′ = H; **18a**: Ref. [49] R = Me, R′ = OMe, R″ = OMe, R‴ = H; **b**: Ref. [93] R′ = CN, R, R″ = H, R‴ = OMe.

and Frimer [97] for the preparation of [5]helicene tetraesters **14a** and dianhydride **14b**. This finding then led to the preparation of cyclobis[5]helicene macrocycles **14c,d** with stable lemniscular, or figure eight, shapes [37, 88, 91, 92]. Helicenes **15** and **16** were used for the preparation of the first helically chiral phosphane ligands [85], helical donor–acceptors [94], and helical metallocenes [95]. Recently, during photocyclization, Mallory discovered an unusual skeletal rearrangement of 1-styrylphenanthrene with *tert*-butyl groups to form [5]helicene derivative **19** in addition to the expected *tert*-butyl-substituted [5]phenacene [96].

The same photocyclization strategies as for carbo[6]helicene (**2**) were chosen for the preparation of its derivatives (Figure 1.7) [49, 85, 108, 109]. The choice of strategy depends on the availability of starting compounds and the method of preparation (see Section 1.2). Based on the literature, the most common photocyclization strategy is [4] + [1] [19, 30, 51, 78, 106, 110–117]. Meier used oxidative photocyclization [1] + [2] + [1] to prepare [6]helicenes **20b–e** and **23a,b** bearing four to six alkoxy chains with good solubility and processability for applications in materials science [99, 102]. The same reaction led to [6]helicenophanes **20f,g** with methylenedioxy chains [100]. Církva synthesized racemization-stable [6]helicenes **21** and **26** fluorinated at terminal rings to study their physicochemical properties [52]. It is worth noting that a [3] + [2] method for the preparation of helicene **22** with alkoxy groups [101] in 3,3′ positions has been successful, although this method is not very common. Other substances, such as optically active 1,8-naphthalimide with [6]helicene scaffold [103] **24**, diastereopure [6]helicene binaphthol hybrid [104] **25**, 2,2′-bis-[6]helicenyl [106] **27**, and paracyclophane [6]helicene derivative [107] **28** have been also successfully synthesized.

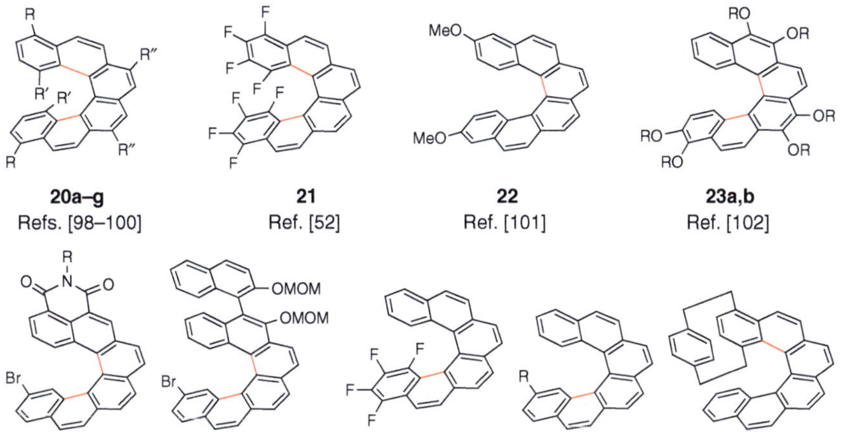

20a–g
Refs. [98–100]

21
Ref. [52]

22
Ref. [101]

23a,b
Ref. [102]

24
Ref. [103]

25
Ref. [104]

26
Refs. [52, 105]

27
Ref. [106]

28
Ref. [107]

Figure 1.7 Representative examples of substituted carbo[6]helicenes. **20a**: Ref. [98] R = Br, R″ = OMe, R‴ = H; **b**: Ref. [99] R = R″ = OMe, R′ = H; **c**: Ref. [99] R = R″ = *O*-dodecyl, R′ = H; **d**: Ref. [99] R = R″ = *O*-(2-ethylhexyl), R′ = H; **e**: Ref. [99] R = R″ = *O*-(2-butylhexyl), R′ = H; **f**: Ref. [100] R = $O(CH_2)_8O$, R′ = R″ = H; **g**: Ref. [100] R = $O(CH_2)_{10}O$, R′ = R″ = H; **23a**: R = C_6H_{13}; **b**: R = $C_{12}H_{25}$; **24**: R = CH_2*t*-Bu; **27**: R = [6]helicen-2-yl.

[7]Helicene derivatives are usually prepared by the same photocyclization strategies as unsubstituted ones (Figure 1.8) [33, 98, 105, 115, 125]. Notable examples include [7]helicene phosphines **29b** [118], methoxy- and cyano[7]helicenes **30a–c** [119, 120], [7]helicene fused with hexafluorocyclopentene **31** [123], the important synthetic synthon bromo[7]helicene **32a** [15], and diastereopure [7]helicene binaphthol hybrid **33a,b** [104]. Also, a simple method for producing dimethyl bifunctionalized [7]helicene precursor **34** [124] was discovered by Howarth. It is important to mention that the [3] + [1] + [1] to **35a,b** [122] and [5] + [1] to **36a–j** [78, 110] strategies were newly used, while the [4] + [2] method for synthesis of [7]helicene derivatives has so far not been utilized. In the case of [8]helicene derivatives, the tetrafluoro[8]helicene **37** [105] and dimethoxy-dibromo[8]helicene **38** [16] were prepared using [2] + [3] + [1] and [1] + [4] + [1] strategies, respectively.

1.3.3 Oxidative Photocyclizations of Benzocarbo[*n*]helicenes

To investigate the effect of fused benzene rings on helicene helicity, several benzocarbo[*n*]helicenes (*n* = 5–7) were synthesized [14]. At the same time, the influence of benzene rings on the course of the photocyclization reaction was also studied (Figure 1.9).

Several benzocarbo[*n*]helicenes differing in the location of the benzo fusion on helicene have been prepared. Martin and Laarhoven prepared a series of benzo[5]helicenes, such as **39a** [126, 131, 132]. Recently, the substituted derivative of maleic anhydride **39b** was synthesized by Wang [90]. Many other aromatic

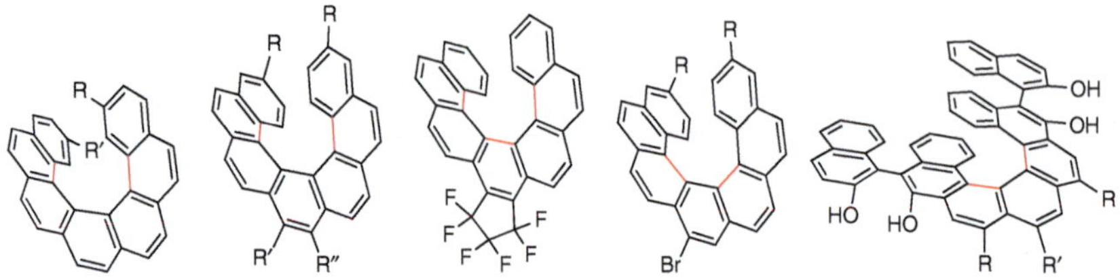

29a–d Refs. [67, 118, 119] | **30a–c** Refs. [119, 120] | **31** Ref. [123] | **32a,b** Refs. [15, 121] | **33a,b** Ref. [104]

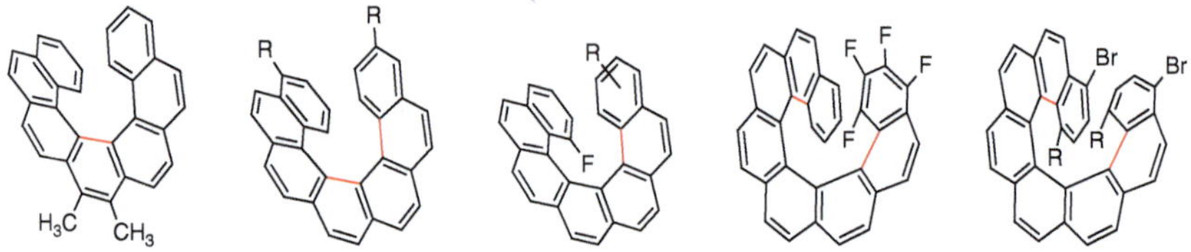

34 Ref. [124] | **35a,b** Ref. [122] | **36a–j** Ref. [78] | **37** Ref. [105] | **38** Ref. [16]

Figure 1.8 Representative examples of substituted carbo[7]- and [8]helicenes. **29a**: Ref. [67] R = OMe, R′ = H; **b**: Ref. [118] R = $POPh_2$, R′ = H; **c**: Ref. [67] R = R′ = Me; **d**: Ref. [119] R = R′ = OMe; **30a**: Ref. [119] R = OMe, R′ = R″ = OEt, **b**: Ref. [120] R = H, R′ = R″ = CN; **c**: Ref. [120] R = R′ = H, R″ = CN; **32a**: Ref. [15] R = H; **b**: Ref. [121] R = CH_2OTHP; **33a**: Ref. [104] R = CN, R′ = H; **b**: Ref. [104] R = H, R′ = Br; **35a**: Ref. [122] R = Me; **b**: Ref. [122] R = CO_2Et; **36a**: Ref. [78] R = H; **b–d**: Ref. [78] R = 2′ to 4′-OMe; **e–g**: Ref. [78] R = 2′ to 4′-CO_2Me; **h–j**: Ref. [78] R = 2′ to 4′-CN; **38**: Ref. [16] R = OMe.

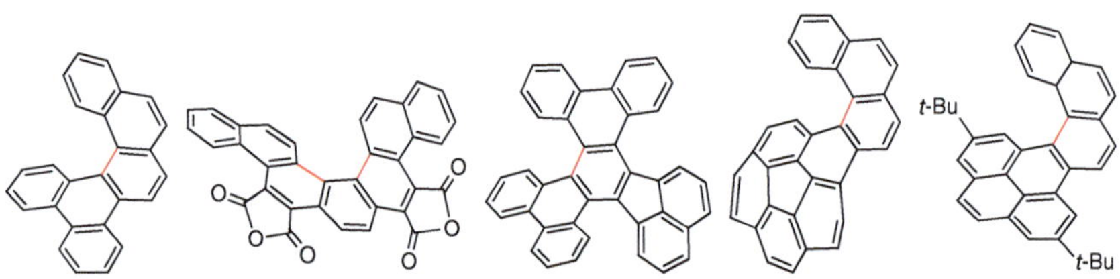

39a Ref. [126] | **39b** Ref. [90] | **40a** Ref. [127] | **40b** Ref. [128] | **40c** Refs. [48, 129]

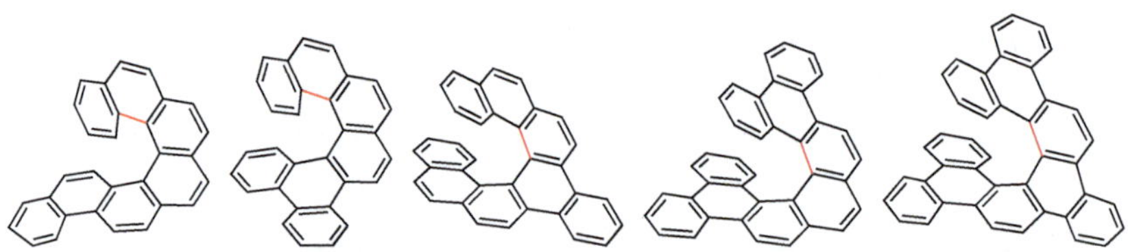

41a Ref. [66] | **41b** Ref. [66] | **42a** Ref. [130] | **42b** Ref. [130] | **42c** Ref. [130]

Figure 1.9 Representative examples of benzocarbo[*n*]helicenes.

building blocks can be introduced into the benzo[5]helicene structures. For example, acenaphthylene derivative **40a** [127], corannulene hybrid **40b** [128], and pyrene-cored helicenes **40c** were investigated [48, 129]. Also, Laarhoven prepared several benzo[6]helicenes **41a,b** [66] and benzo[7]helicenes like **42a–c** [130] with varying numbers (one to three) of benzo groups on the helicene skeleton.

1.3.4 Oxidative Photocyclizations of Multiple Carbo[*n*]helicenes

The fusion of two or more helicene moieties together into one polycyclic framework leads to the formation of multiple helicenes, which have received considerable attention in recent years [133–135]. In the following text, they are categorized by the number of their helicene moieties (Figure 1.10).

The simplest double helicene **43a** was first prepared by Martin in 1983 [131]. More recently, Sarkar synthesized a bis(butyloxycarbonyl)double[5]helicene **43b** in an excellent 87% yield [136]. The contorted quadruple [5]helicene **43c** was obtained by Nuckolls in a sixfold photocyclization of the olefin precursor [140]. Watanabe prepared a triple [5]helicene **43d** in moderate yield through the double [2] + [1] + [2] strategy from 1,3,5-tris[2-(2-naphthyl)vinyl]benzene [141]. The first documented double [6]helicene **44a** was synthesized in a moderate yield by Laarhoven and Cuppen already in 1971 [142, 143]. Other double [6]helicenes (e.g. **44b**) were synthesized in even higher yields [77, 144, 145]. The double [7]helicene was obtained by Martin in 1974 [77]. Recently, Durola reported on the synthesis of *meso* double [7]helicene **45** in 83% yield from maleate-bridged chrysene trimer [137]. Wang afforded perylene diimide-embedded double [8]helicene **46a,** which represents the highest photochemically prepared double carbohelicene reported thus far [138]. Perylene diimide (PDI) is an increasingly popular functionality in

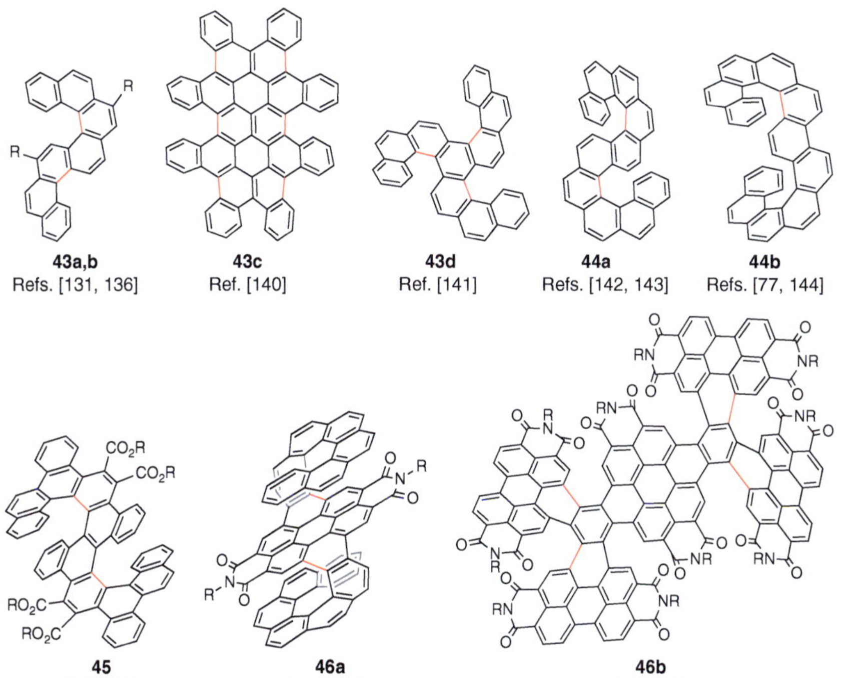

Figure 1.10 Representative examples of multiple carbo[*n*]helicenes. **43a**: Ref. [131] R = H; **b**: Ref. [136] R = CO_2Bu; **45**: Ref. [137] R = C_8H_{17}; **46a**: Ref. [138] R = $CH(C_5H_{11})_2$; **b**: Ref. [139] R = $CH(C_5H_{11})_2$.

helicene chemistry, as demonstrated by Nuckolls, who prepared helicene-based molecules with two PDI units [146–148]. These helicene-based structures exhibit the largest electronic circular dichroism (ECD) in the visible range of any molecule [148]. Recently, triperylene hexaimides with a three-bladed propeller-shaped π backbone [149] and pentaperylene decaimide dual-core sixfold [5]helicene **46b** were prepared by Wang in high yields using quadruple photocyclization [139].

1.3.5 Eliminative Photocyclizations of Carbo[*n*]helicenes

In addition to oxidative photocyclization, an eliminative approach can also be used to prepare carbo[*n*]helicenes (see Figure 1.1). In these cases either methanol or hydrogen halides, such as HF, HCl, HBr, and HI, were eliminated. The loss of TfOH was also observed.

The triple [5]helicene **47** (Figure 1.11) was synthesized in moderate yield by Watanabe through an acid-promoted double [2] + [1] + [2] photochemical strategy with the elimination of methanol [141]. The PDI-[6]helicene hybrid **48** was prepared in 94% yield by Nuckolls using the same method [150]. A photocyclodehydrofluorination reaction where pentafluorostilbene precursor gave a modest yield

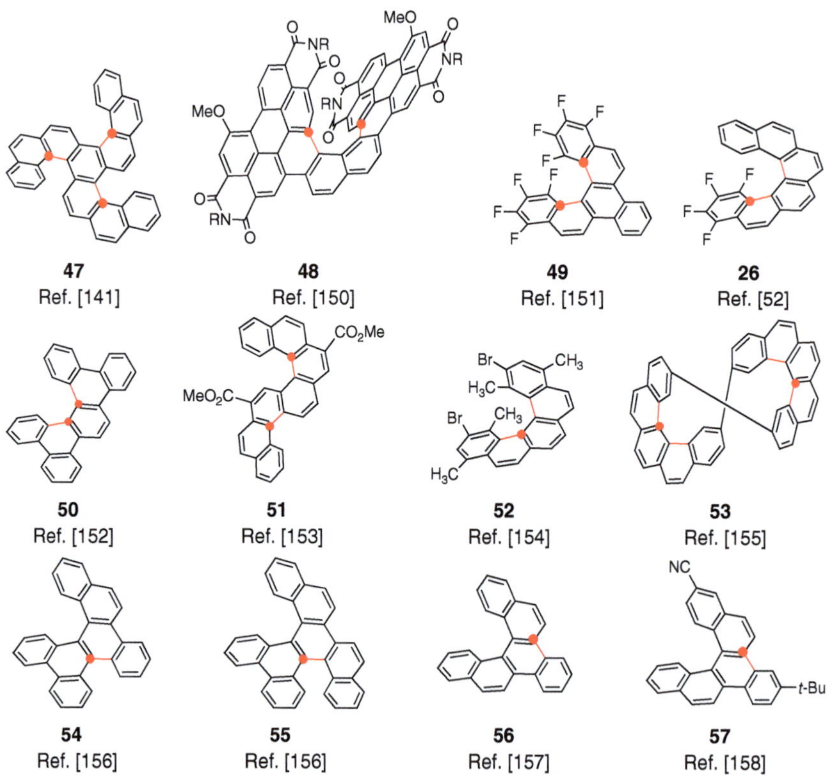

Figure 1.11 Carbo[*n*]helicenes formed by eliminative photocyclization. The red dot indicates an atom with a leaving group (i.e. OCH_3, F, Cl, Br, I, TfO); **48**: Ref. [150] R = $CH(C_5H_{11})_2$.

of fluorinated benzo[*i*][5]helicene **49** was reported by Twieg and Li [151]. Církva found that pentafluorophenyl derivative can provide the fluorinated [6]helicene **26** by dehydrofluorination reaction, however, in low yield [52]. Morin worked on photochemical cyclodehydrochlorination, yielding dibenzo[5]helicene **50** in 79% [152].

The eliminations of HBr and HI are far less common. The former was employed in the preparation of double helicene derivative **51** [153], while the latter was used to prepare [5]helicene derivative **52** [154]. Thulin and Wennerström prepared *propellicene*, bis-2,13-[5]helicenylene **53**, in 70% yield from diiodo derivative of paracyclophane-tetraene [155]. It was shown that the strain in *propellicene* can sometimes help to avoid the overannulation to benzo[*ghi*]perylene. Alabugin reported the preparation of fused helicenes from the alkynyl precursors using a photocyclization–dehydroiodination sequence. This new approach led to dibenzo[5]helicene **54** [156] and double helicene **55** [156] in high yields. Yorimitsu achieved a synthesis of benzo[5]helicene **56** in 59% yield by photoinduced cyclization with a loss of TfOH [157]. The derivative **57** was prepared analogously [158].

1.4 Photochemical Preparation of Heterohelicenes

1.4.1 Azahelicenes

Along with parent carbo[*n*]helicenes, the most studied and probably the most abundant group of helicenes, which are easily accessible via a photochemical approach, are *azahelicenes*. They contain one or more nitrogen atoms, usually as a part of *pyridine* (*quinoline*) or *pyrrole* (*carbazole*) moiety, while keeping the aromaticity along the helical backbone untouched. A special case, photochemically accessible azahelicenes containing quaternary nitrogen atoms (*azoniahelicenes*), is also discussed in this part. Principles of photochemistry using Mallory's or Katz's approach and preparation of starting material remain similar to general procedures described previously and will not be described herein unless stated otherwise. In general, iodine might be a less satisfactory oxidant than oxygen due to possible charge-transfer complex formation with azastilbenes [159, 160], which act as competitive light absorbers or singlet quenchers, but it is still frequently used.

Whereas the tendency to form packed (helical) or less strained (benzoanthracene) structures during photocyclization can be reasonably explained (based on free-valence atom indexes of the Hückel model [14, 161], Wiberg bond indexes [162, 163], or influence of solvent polarity [162]; see details in Section 1.3), the selectivity of a photoreaction in the context of nitrogen atom position remains unclear. The regioselectivity of photodehydrocyclization of 3-pyridyl stilbenes is markedly higher than in the case of *meta*-substituted carbostilbenes, providing 2-azaphenanthrenes [164, 165] or eventually 3-aza[*n*]helicenes in greater yields than the respective products of ring closure in the other available position (cf. **66a** vs. **66b** [166] in Figure 1.14). However, in the case of diazastilbenes, this observation based on molecular orbitals calculations [164, 167, 168] can be suppressed by

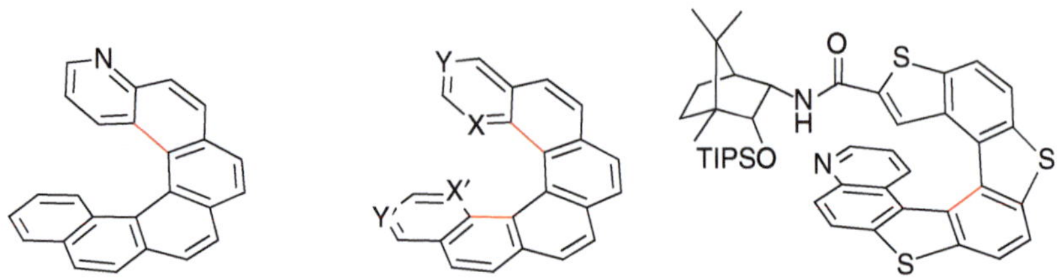

rac-**58** (18%)

59a–c

a: X = X′ = N, Y = Y′ = CH (<2%)
b: X = Y′ = N, X′ = Y = CH (25%)
c: Y = Y′ = N, X = X′ = CH (11%)

(*P*)/(*M*)-**60** (73%)

(*d.r.* 37:63)

Figure 1.12 First photochemical approaches to azahelicenes **58–60**.

a molecular design in which the electrostatic repulsion of lone electron pairs at nitrogen atoms predominates. Moreover, the observation of Carrona when synthesizing **62a,c** [163] or **63a,b** [169] (Figure 1.13) was opposite to the previously mentioned with the predominant generation of 1-aza[5]helicenes. Similarly, it insufficiently explains the ratio of regioisomers formed in the case of **59** [170] (Figure 1.12), where the higher-yielding isomer was 1,14-diazahelicene **59b** rather than 3,14-diazaisomer **59c**.

The photochemical approach to aza[*n*]helicenes can be compartmentalized into two main subgroups. While the first group uses the light-induced closure of stilbene-like moiety with C=C double bond substituted with various aromatic N-heterocycles (Figures 1.12–1.17; some examples from Figure 1.18), the second one utilizes photocyclization of arylimines or hydroxamic acids (Figure 1.19).

The first photochemical approach to 4-aza[6]helicene (**58**, Figure 1.12) was published by Martin in 1969 [171], while the first diazahelicene was not photochemically synthesized until 1994 [170] when Staab reported on the synthesis of **59a**. Unfortunately, instead of the desired 1,16-diaza[6]helicene (**59a**) when applying Katz's conditions [16], he isolated 1,14- (**59b**) and 3,14-diaza (**59c**) derivatives in 25 and 11% yields, respectively, with less than 2% yield of **59a**. To overcome the lack of regioselectivity, Staab prepared **59a** by intramolecular coupling published therein. The hybrid aza-thiahelicenes prepared by Tanaka in 1997 belong to one of the first asymmetric photochemical approaches [36]. The chiral auxiliary derived from D-camphor allowed the preparation of a diastereomeric mixture of **60** separable via standard chromatography and opened a way to isolate other pure enantiomers of such helicene hybrids.

It was Carrona who started extensive research on the photochemical preparation of aza[5]helicenes and studied their optical properties. In 2005, he reported on a series of various azahelicenes **61a–i** [172] (Figure 1.13) using Pyrex-filtered irradiation by visible light and neither iodine nor propylene oxide as an oxidant or HI scavenger. Thanks to the modified reaction conditions, he was able to reach excellent yields of up to 98% in some cases. Other [5]helicene derivatives (**62–64**) were

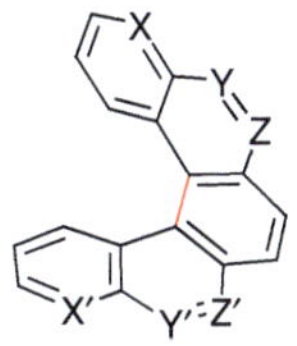

rac-**61a–i** (40–98%)

a: X = N, Y = Z = X′ = Y′ = Z′ = CH
b: Y = N, X = Z = X′ = Y′ = Z′ = CH
c: Z = N, X = Y = X′ = Y′ = Z′ = CH
d: X = X′ = N, Y = Z = Y′ = Z′ = CH
e: Y = Y′ = N, X = Z = X′ = Z′ = CH
f: Z = Z′ = N, X = Y = X′ = Y′ = CH
g: X = Z′ = N, Y = Z = X′ = Y′ = CH
h: X = Y′ = N, Y = Z = X′ = Z′ = CH
i: X = Z′ = N, Y = Z = X′ = Y′ = CH

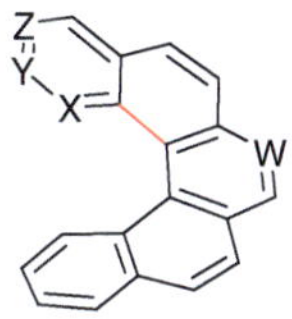

rac-**62a–d** (11–87%)

a: X = N, W = Y = Z = CH
b: Y = N, W = X = Z = CH
c: Z = N, W = X = Y = CH
d: W = N, X = Y = Z = CH

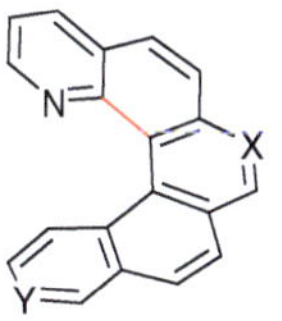

rac-**63a,b**

a: X = N, Y = CH (62%)
b: Y = N, X = CH (15%)

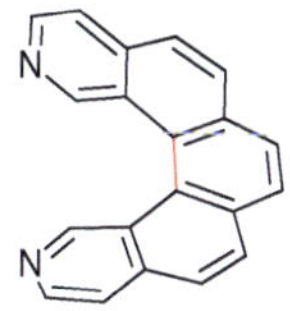

rac-**64** (27%)

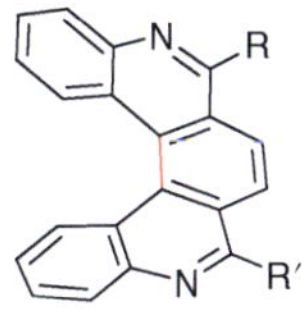

rac-**65a–c**

a: R = R′ = Cl (57%)
b: R = R′ = OMe (46%)
c: R = Cl, R′ = OMe (63%)

Figure 1.13 Photochemically prepared aza[5]helicenes **61–65.**

obtained after UV or vis light irradiation [162, 163, 169]. Among all reported structures, the author surprisingly observed no benzo[*ghi*]perylene formation except for the attempts to prepare 2-aza[5]helicene regardless of the starting stilbene used (for details see Section 1.6).

Photochemically prepared diazahelicenes with an appropriate position of nitrogen atoms can be alkylated to corresponding *viologens*, as shown by Clennan [173, 174] and Brichacek and Patterson, who studied gold heli-viologen [175].

Also, Dehaen, who introduced diversely functionalized diaza[5]helicenes **65**, (Figure 1.13) [176], observed the tendency of methoxy derivative **65b** to form an azabenzo[*ghi*]perylene after prolonged irradiation, unlike the case of 6,9-dichloro-5,10-diaza[5]helicene **65a**. This helicene was transformed by Buchwald–Hartwig amination with a chiral amine, providing (*P*,*S*,*S*)/(*M*,*S*,*S*)-diastereomers separable via standard chromatographic methods. In 2013, the author extended his work to diazadithia[7]helicene derivatives [177]. Based on his synthetic strategy, Dias and Skabara later published a synthesis of phenoxazine-functionalized [5]helicenes as novel thermally activated delayed fluorescence (TADF) materials for use in functional devices [178].

2-Aza- [179] and 5-aza[6]helicene [180] (**67** and **68**, Figure 1.14) were prepared photochemically by Abbate using visible light. Ben Hassine [166] used UV irradiation to prepare 3-aza[6]helicene (**66a**), while also managing to isolate its 1-aza derivative **66b** as a minor product. The chiral aza[6]helicenes (*P*)/(*M*)-**69a**, prepared and separated to optical antipodes by Crassous [41], served as a photoinitiator in photoinduced polymerization of (meth)acrylates. This is a remarkable example

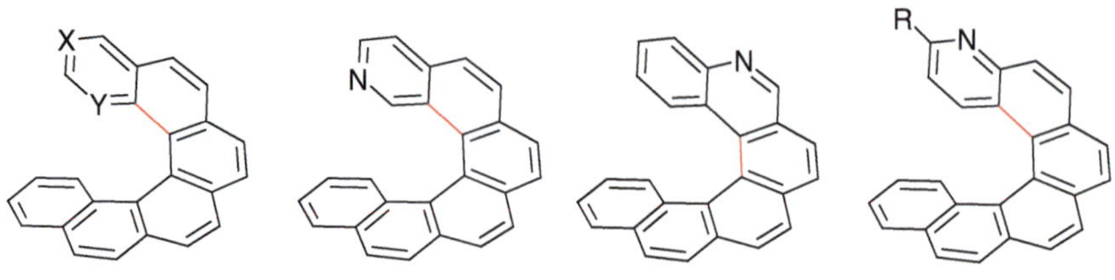

rac-**66a,b** *rac*-**67** (10%) *rac*-**68** (53%) (*P*)/(*M*)-**69a,b**

a: X = N, Y = CH (50%)
b: Y = N, X = CH (7%)

a: R = Me (53%)
b: R = Br (45%)

Figure 1.14 Photochemically prepared aza[6]helicenes **66–69.**

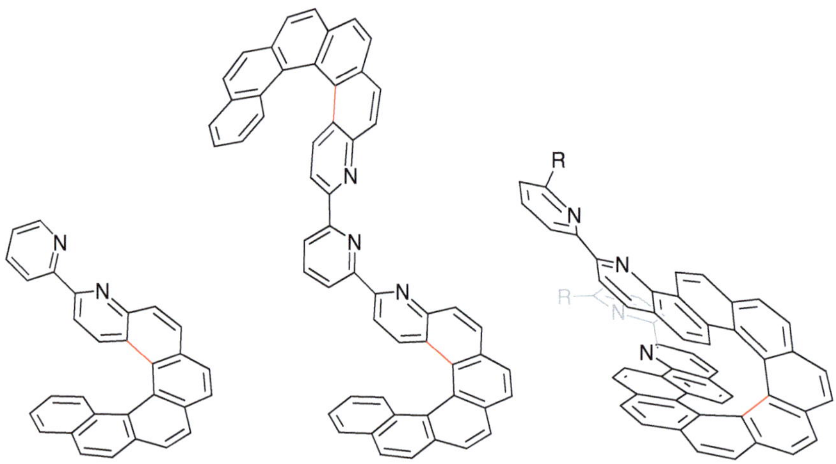

rac-**70** (90%) *rac*-**71** (45%) *rac*-**72a,b**
a: R = H (35%)
b: R = Br (76%)

Figure 1.15 Helicopodands **72** and helicenes adopting bipy- **70** and terpy-structures **71**.

of chiral properties introduced to polymer films by the photoinitiator through photopolymerization processes. Recently, Srebro-Hooper and Crassous published a synthesis of enantiopure bis-(4-aza[6]helicene-3-yl) adopting a 2,2′-bipy structure suitable for transition metal complexation [181]. It was accessible via Negishi coupling of 3-bromo-4-aza[6]helicene **69b**. A reversible protonation resulted in the tuning of the photophysical properties of the bipyridyl derivative. Other helicenes **70–72** adopting bipy- [182] and terpy-structures [183] and *helicopodands* [184] that were obtained photochemically are depicted in Figure 1.15. Their protonation and transition metal complexes were also studied [182, 183, 185–187].

The photochemically prepared diaza[7]helicenes with protected –OH groups served for the synthesis of polyaromatics bearing pyridon moieties at the terminal rings [188]. It was shown that some of such racemic derivatives self-assembled via

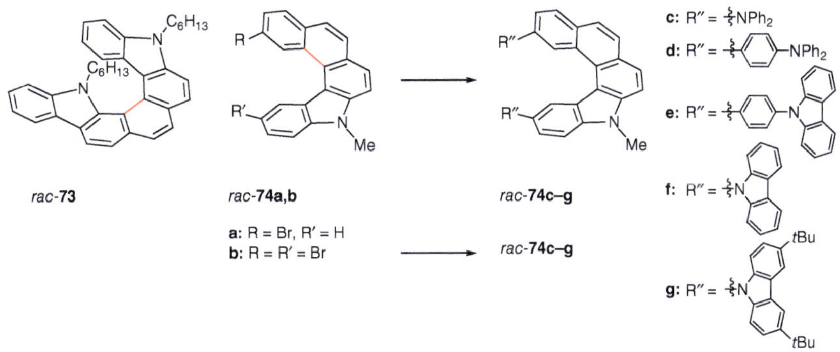

Figure 1.16 Helicenes **73-74** with carbazole subunits for OLED applications.

hydrogen bonding enantiospecifically in solution and diastereoselectively in solid state.

Pyrrole (carbazole) moiety represents another frequently used motif in azahelicene chemistry. Carbazole belongs among popular functional units in conjugated systems owing to its planarity and easy N–H bond substitution for other functional groups. Although the first pyrrolo[5]helicene synthesized by Meisenheimer and Witte [189] as early as 1903 is among the oldest helical structures ever, the photochemical approach leading to such helicenes was not published until 2012 when Liu reported a diazahelicene **73** (Figure 1.16) as a deep-blue dopant in organic light-emitting diode (OLED) [190]. In 2013, Ben Hassine prepared pentacyclic **74a** [191], taking advantage of a possible further transformation due to the presence of reactive bromine at the terminal benzene core. However, the reactivity was not exploited until 2016, when Chow and coworkers prepared a series of helical diamines **74c–g** [192], which were synthesized from photochemically prepared dibromo derivative **74b**. In this study, diamino derivatives were studied as materials suitable for use in OLED devices, as they exhibited a favorable range of photophysical and redox properties as well as thermal stability.

It was primarily Bedekar who succeeded in the preparation of various aza[*n*]helicenes ($n > 5$) with carbazole moiety by double photocyclodehydrogenation reaction from both symmetric (**75a–d**, **78**, Figure 1.17) [193] and asymmetric (**75e–i**) [194] bis-stilbene precursors including synthesis of bi-azahelicenes **76** and **77** [195]. In the same fashion as mentioned earlier (cf. Refs. [176, 177]), a carbazole–quinoline hybrid was prepared photochemically by Dehaen [196]. Among others, Liu prepared **79a** [197] and **79b** [198].

A specific group of nitrogen-containing helicenes is azoniahelicenes. The photochemical approach to these compounds dates back to the late 1980s and is mostly connected to the work of Arai, who unsuccessfully attempted a synthesis of the first quinolizinium-containing [5]helicene [199]. In fact, he identified the corresponding azoniabenzo[*ghi*]perylene in the reaction mixture after additional photocyclization, as discussed later.

Soon after, Arai et al. published a series of azoniahelicene structures endowed with quaternary nitrogen either on an inner (**81** [200, 201]) or outer (**80** [202], **82**

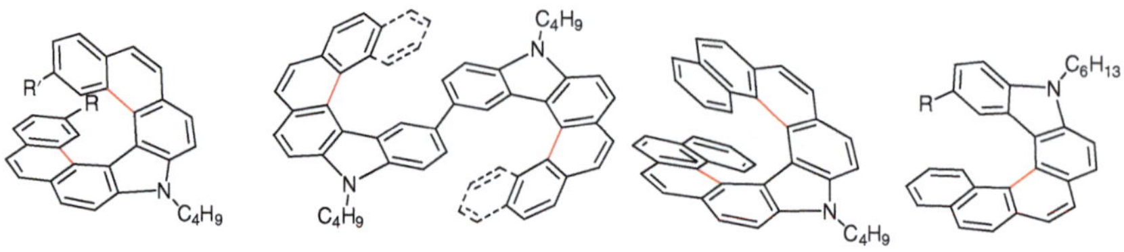

rac-**75a–i** *rac*-**76**, *rac*-**77** *rac*-**78** *rac*-**79a,b**

a: R = R′ = H
b: R = R′ = Me
c: R = R′ = OMe
d: R = R′ = F
e: R = H; R′ = Me
f: R = H; R′ = Cl
g: R = H; R′ = OMe
h: R = Me; R′ = Cl
i: R = OMe; R′ = Cl

a: R = H
b: R = Br

Figure 1.17 Carbazole containing azahelicenes **75–79** synthesized photochemically.

[203], **83–86** [204–206]) helix edge obtained upon photochemically induced quaternization or cyclodehydrogenation (Figure 1.18). The latter compounds were synthesized using stilbene-like precursors, which were typically prepared by dehydration of secondary alcohols followed by additional Knoevenagel condensation of suitably substituted quinolizinium derivative with aromatic aldehydes.

The phenomenon of the undesired formation of benzo[*ghi*]perylenes was observed only in the case of unsubstituted 6a-azonia[5]helicene [199] and partially in the case of **82a** [203] (where perylene was the minor product). Employing bulky substituents (**82b,c,f,g**) or thiophene(s) as a terminal ring(s) [204] (for example, in **83a–f**) avoids the formation of such by-products. When using monosubstituted precursors at one terminal ring, a mixture of [5]helicene/benzoperylene can be observed (e.g. **82e**), unlike in the case of kinetically driven helicene **82d** formation, where the helicene salt was isolated as a single product. Therefore, an efficient way to unsubstituted azonia[5]helicene or mono-methyl-substituted (e.g. **82h,i**) azonia[5]helicenes can be achieved by photocyclization of sterically demanding bromides (to form **82d,f,g**) and their subsequent debromination under mild conditions (see Figure 1.18).

Photochemically induced quaternization was used for the formation of compound **81** where the quaternary nitrogen is placed at the inner helix edge. To improve the yields, the Pyrex-filtered light was passed through aqueous nickel sulfate solution filter ($280 < \lambda < 360$ nm and $\lambda > 430$ nm) to rule out wavelengths corresponding to the absorption region of the product (around 410 nm) [207]. From the fundament of this chemical transformation, no oxidant is necessary. The stilbazoles as precursors of such azoniahelicenes are commonly synthesized by the condensation of 2-methylpyridine/quinoline with *o*-chlorobenzaldehyde or other 2-chloro aromatic aldehydes.

The results obtained from the preparation of azoniahelicenes showed that the N-heterocyclic subunit can be formed directly by the photochemical step. However, this is not the case of azahelicenes with pyridine or pyrrole subunits that are prepared almost exclusively from stilbene-like molecules with C=C double bond. Despite this, several examples can be found in the literature. In 1997, Howarth was the first who attempted to synthesize 8,11-diaza[7]heptahelicene photochemically from corresponding bisimine [124]. Unfortunately, the low yield of reaction conducted in pyridine using a Pyrex filter and propylene oxide discouraged researchers

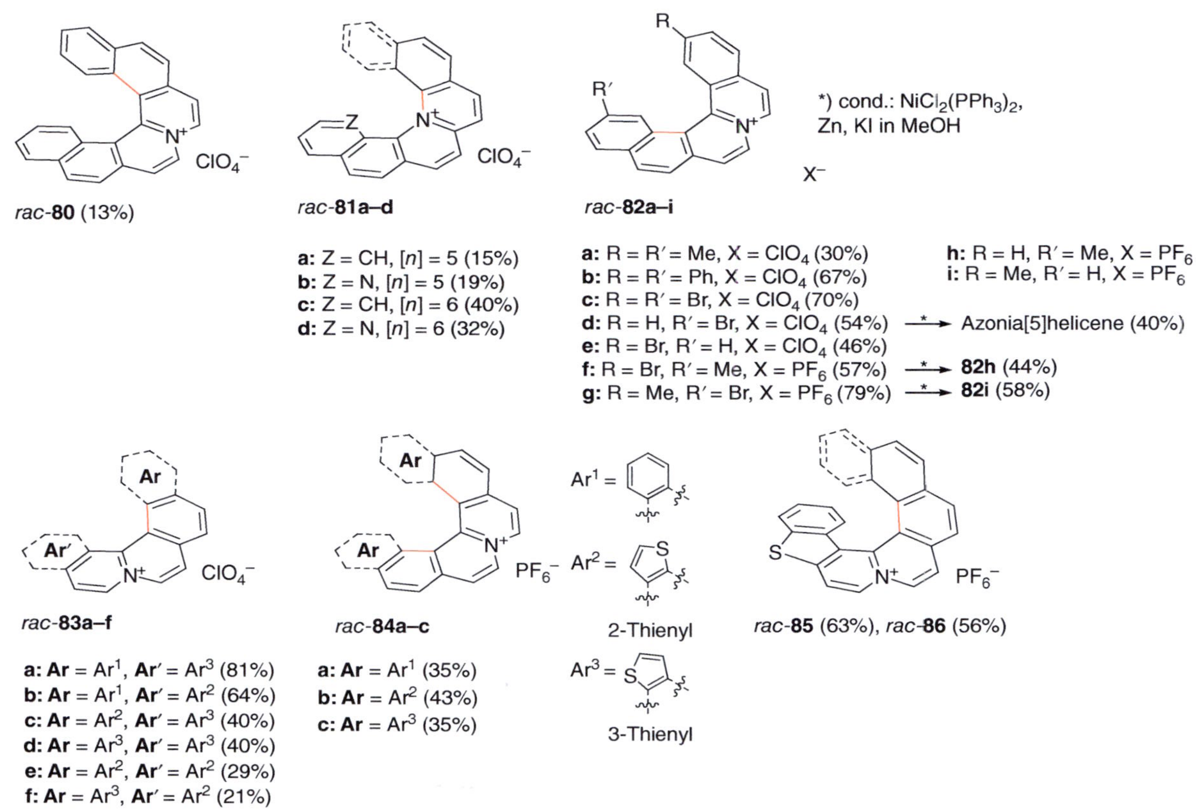

Figure 1.18 Photochemically synthesized azonia[*n*]helicenes **80–86**.

from further investigation of aromatic imines as precursors for azahelicenes. The poor reactivity of such imines is caused by the thermal instability of *Z*-conformers (short lifetime at room temp.) and unsuitable excited state (n, π^*) impeding cyclization from the (π, π^*) excited state [208–211]. Twenty years later, Murase came up with an idea to use hydroxamic acids as bidentate chelating ligands to circumvent this obstacle [50]. The boron complexes of *N*-phenylbenzohydroxamic acid **87** possessed a fixed *Z*-conformation diminishing the influence of the lone electron pair on the nitrogen atom, thus enabling successful photocyclization to *rac*-**88** upon N–O bond cleavage (Figure 1.19a). Recently, Alabugin, Církva, and Sýkora published Brønsted acid-promoted photocyclizations of arylimines (e.g. **89** and **91**) to various aza-PAHs, represented, among others, also by [5]helicenes **90** and **92** (Figure 1.19b) [212]. Unfortunately, higher aza[n]helicenes ($n \geq 6$) remained inaccessible using this protocol.

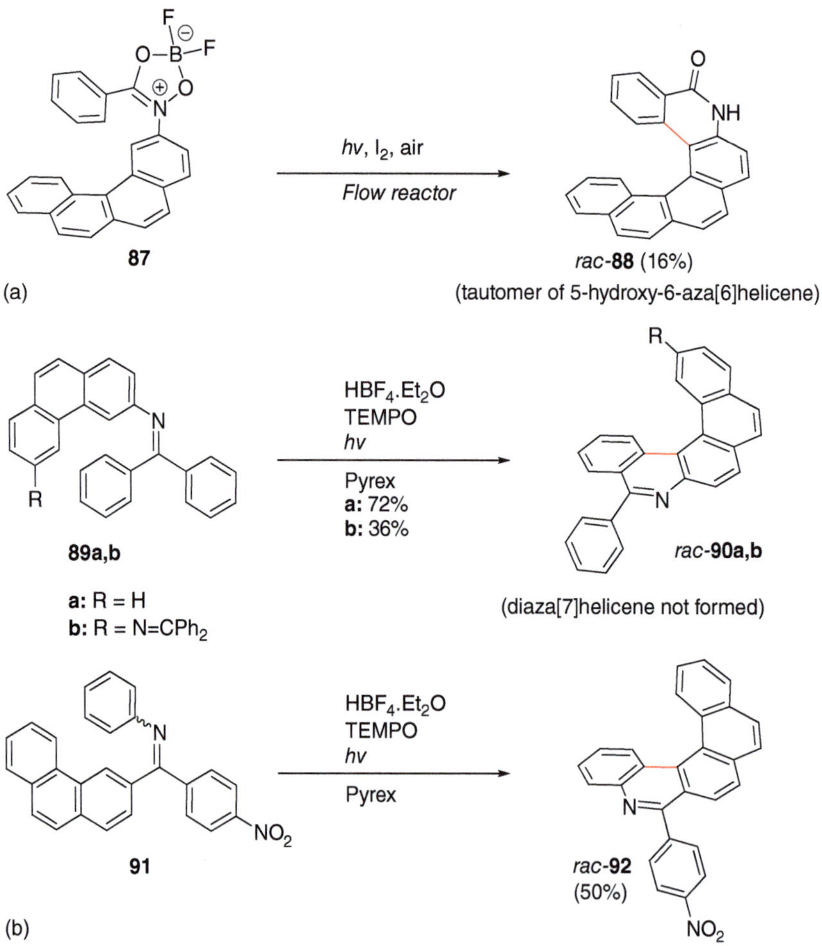

Figure 1.19 Photocyclization of (a) hydroxamic acids and (b) arylimines.

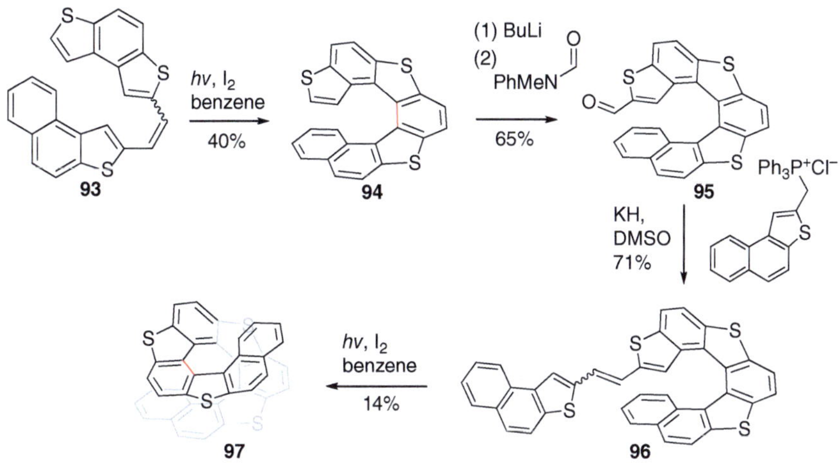

Figure 1.20 Preparation of tetrathia[11]helicene **97**.

1.4.2 Thiahelicenes

Thiahelicenes and other sulfur-containing derivatives make up a large part of helicene chemistry and therefore will be discussed separately. The first example of heterohelicene with thiophene replacing some of the benzenes in the original structure was published in 1968 by Wynberg and Groen, who synthesized dithia[6]helicene and trithia[7]helicene (Figure 1.20) [213]. The presence of five-membered rings in their structure has a significant impact on the shape of the molecule. The helix is more open, and, as a result, single enantiomers of these compounds are generally more prone to racemization than their carbohelicene counterparts [214–216]. On the other hand, the thiophene ring provides the possibility of easy derivatization of the thiahelicene backbone, which was used for the synthesis of longer derivatives by the same authors in 1971 [214, 215]. The main reaction in this sequence was the photocyclodehydrogenation of the asymmetrically substituted 1,2-benzothienylethene **93**, which formed helical compound **94** in 40% yield. The proton in the alpha position relative to the terminal thiophene ring was deprotonated in the presence of butyllithium, and the resulting carbanion was reacted with *N*-methyl phenyl formamide to give **95** in 65% yield. Subsequent Wittig reaction gave stilbene-type precursor **96**, which was irradiated by UV light for 12 hours in the benzene solution. The low yield of the reaction was caused by the fact that the reaction did not proceed to full conversion.

Long thiahelicenes with an alternating system of benzene and thiophene rings were also prepared with a similar approach (Figure 1.21) [217–219]. The starting material for the preparation of thia[5]- and [7]helicene was aldehyde **98**, which was used in the Wittig reaction to provide stilbene precursors **99** and **101,** respectively. These compounds then underwent the photocyclization reaction by 350 nm light to give tetrathia[7]helicene **100** and trithia[5]helicene **102**. Similarly, **102** was transformed to stilbenes **103** and **105** via formylation–Wittig reaction sequence, which were subsequently used as precursors for the preparation of pentathia[9]helicene

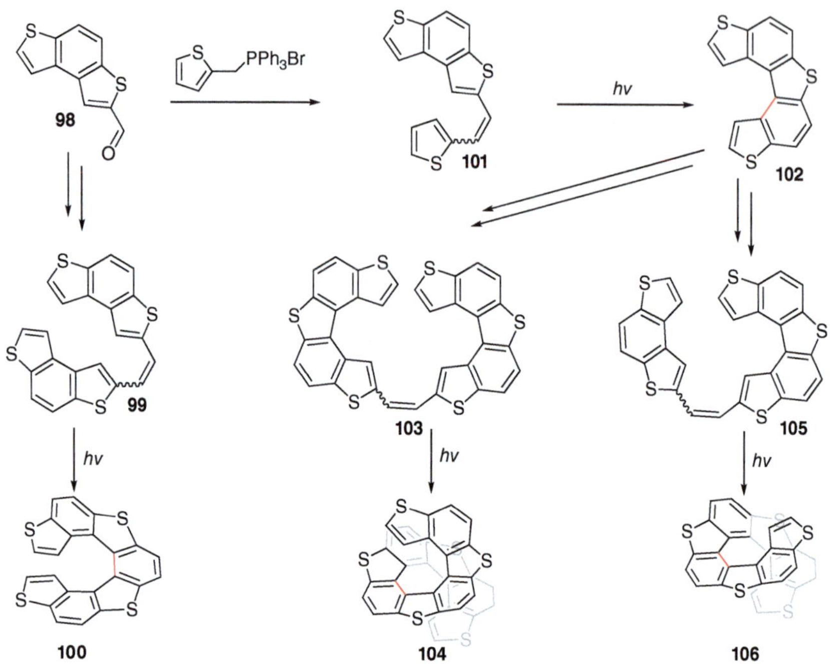

Figure 1.21 Preparation of alternate thia[*n*]helicenes **100**, **102**, **104**, and **106**.

106 and hexathia[11]helicene **104**, respectively. This approach was used for another series of substituted alternating thiahelicenes [220] and also for the preparation of even longer hexathia[13]helicene [221].

Other synthetic pathways for photochemical preparation of thiahelicenes have also been published. For example, 1,2-dithienylbenzenes can undergo the same cyclization procedure to yield naphthodithiophenes. This behavior was used for the synthesis of π-extended thiahelicenes, as shown in Figure 1.22 [222, 223]. Compound **107** underwent a threefold ring closure, providing thia[7]helicene **108** in 60% yield. The reaction, however, took 3 days of irradiation to observe full conversion, and the authors concluded that the precursor **107** can be more conveniently cyclized via Scholl reaction, providing the same product in higher yield within just 90 minutes. The same photochemical procedure was used in stepwise preparation of asymmetrically substituted thiahelicenes [224].

The aforementioned syntheses always produced thiahelicenes with sulfur atoms facing outward from the helical cavity as a result of the used precursors and their reactivity. The position of the sulfur can, however, be changed by using differently substituted thiophene precursors (Figure 1.23). Ben Hassine and coworkers observed that bisstilbene **109**, which was prepared by the double Wittig reaction, underwent the subsequent ring closure regioselectively to position 4 of the thiophene substituent [225]. Product **110** was isolated in 44% yield. On the other hand, other authors published a procedure where a similar precursor **111** differing only in the presence of two –COOH groups provided a mixture of two products in high

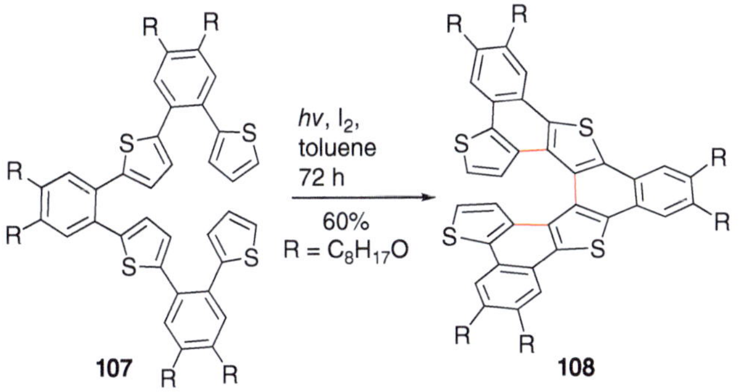

Figure 1.22 Preparation of tetrathia[7]helicene **108** via photocyclization of 1,2-diarylbenzenes.

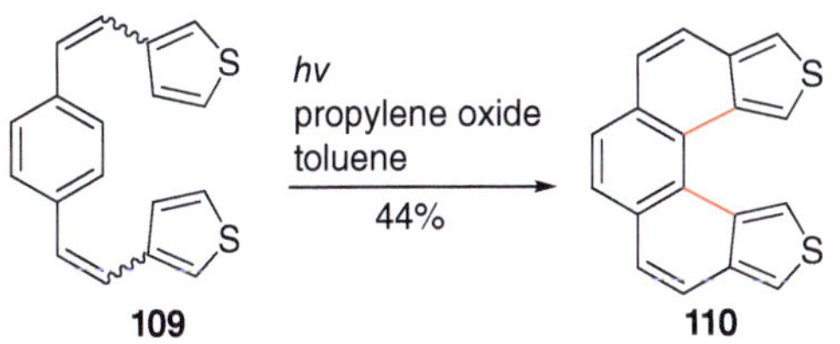

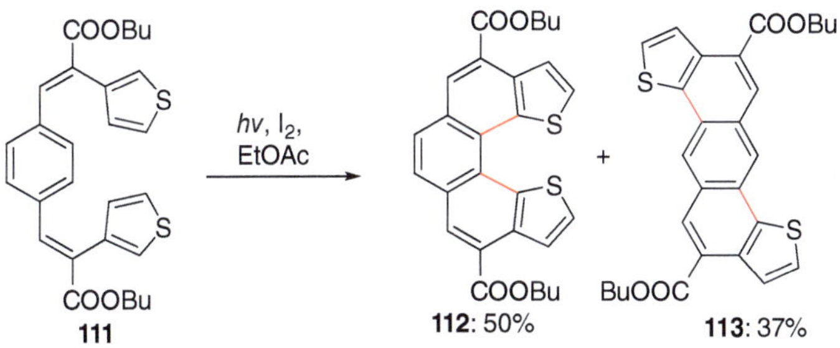

Figure 1.23 Examples of different regioselectivity in photocyclization of precursors **109** and **111**.

overall yield [136]. One was the desired dithiahelicene **112,** where both cyclizations proceeded selectively to position 2, and the other product (anthradithiophene **113**) with one thiophene ring reacted in position 2 and the other in position 5. These compounds were inseparable, and both regioisomers were distinguished by the means of NMR.

Aside from various other substituents that can be present during the photocyclization step of helicene preparation, chiral auxiliaries have been extensively studied for their possible influence on helicene formation. One of the first examples was published by Tanaka [226, 227], who studied the effect of chiral *exo*- and *endo*-bornane derivatives on the diastereoselectivity of the photocyclization reaction (Figure 1.24). The *exo* derivative **114a** (R = H) provided a very low excess of one diastereomer (10% *de*), although the diastereomers themselves were easily

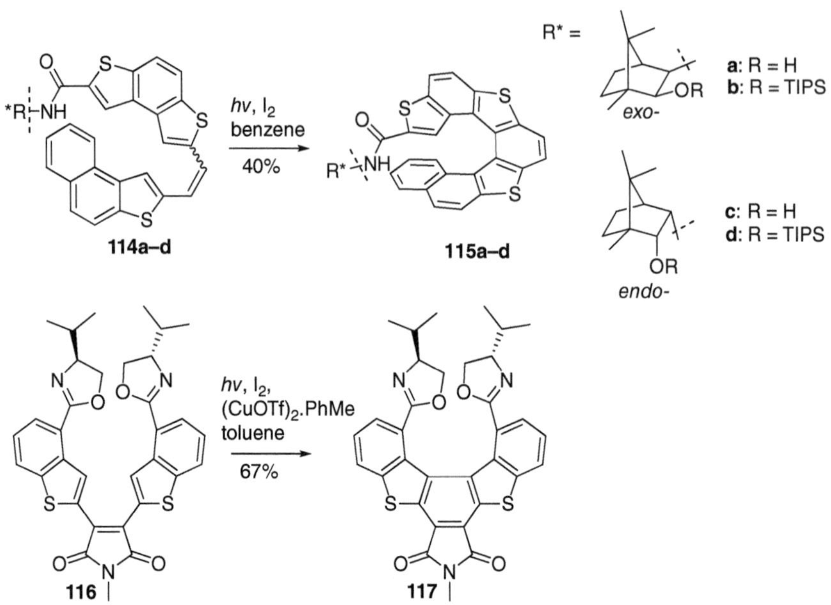

Figure 1.24 Chiral auxiliaries in photocyclization of thiahelicene precursors **114** and **116**.

separable by chromatography on silica gel. An improvement was observed upon triisopropylsilylation of the free hydroxyl (**114b**), which increased the diastereomeric excess to 36% *de*. The *endo* derivative **114c** showed no diastereomeric excess when free hydroxyl was used; however, upon triisopropylsilylation of the hydroxyl (**114d**), the diastereomers were observed in 75 : 25 ratio. The same approach was taken by Dehaen, who studied chiral oxazolines as auxiliaries [228]. The very presence of two oxazolines in the most sterically hindered positions of the precursor **116** provided the (*P,S,S*)-**117** with 32% *de*. The addition of copper salt $(CuOTf)_2PhMe$, forming a complex of suitable geometry with the starting material, increased the diastereomeric excess to 100% *de*.

Molecules such as **119** and **121** (Figure 1.25) are made mainly of fused thiophenes and are members of the family of thiahelicenes as well. Simpler hexathia[7]helicene

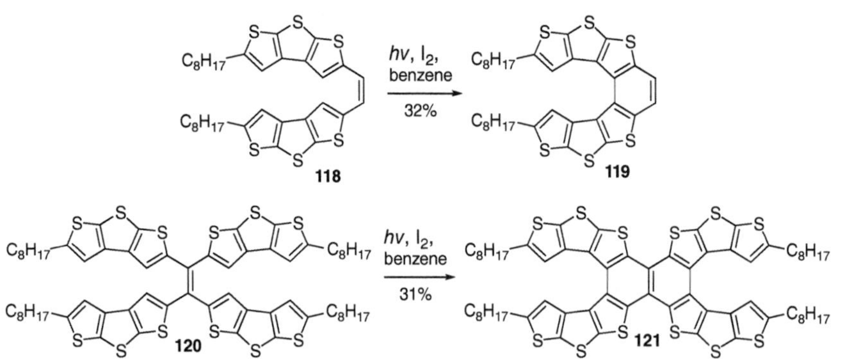

Figure 1.25 Preparation of thiahelicenes **119** and **121**.

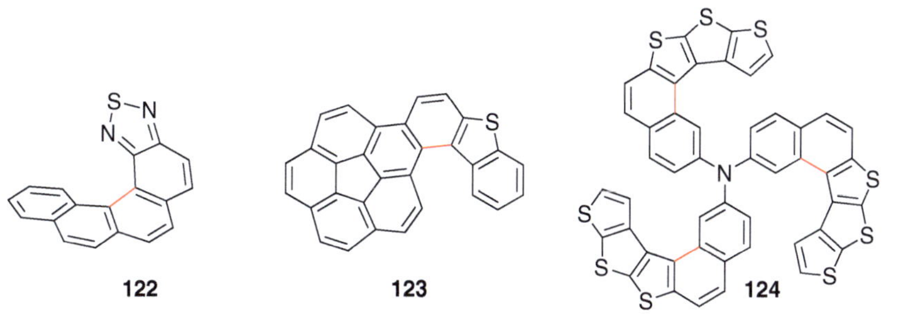

Figure 1.26 Structures of several thiahelicene-based derivatives.

Figure 1.27 Structures of other chalcogen-containing helicene derivatives.

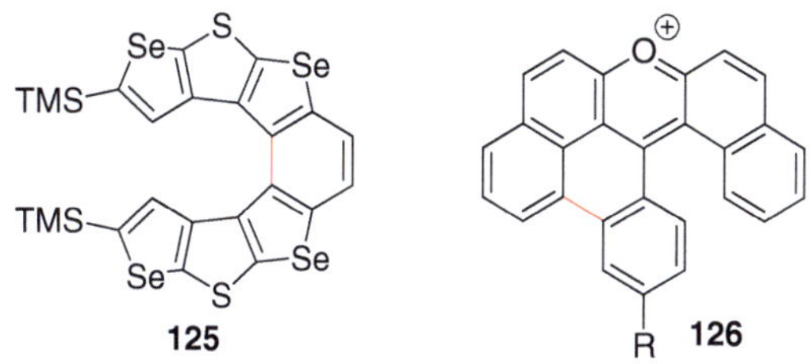

was prepared via photodehydrogenation reaction in benzene in 32% yield [229]. A similar reaction sequence started with a tetraaryl substituted ethylene **120** (Figure 1.25), which served as a suitable precursor for the preparation of bishelicene **121** [230]. The racemic **121** was obtained in 31% yield, while its *meso* form was isolated in 30% yield.

The aforementioned transformations, despite being based on the same principle, can be easily used for the introduction of a variety of different functionalities into the helicene backbone (Figure 1.26). For example, thia-aza [36] **60**, thia-azonia **83–86** [204, 205] (mentioned in Section 1.4.1), or thiadiazole (**122**) [231] helicenes were prepared in this manner. This approach was also used for the preparation of thiahelicene derived from large aromatic corannulene (**123**) [232], as well as for the synthesis of propeller-shaped *trishelicene* based on triarylamine (**124**) [233].

Other chalcogens were also successfully introduced into helicene structure via photochemistry. For example, selenium derivatives (e.g. **125**, Figure 1.27) containing alternating selenophene and thiophene rings were prepared [234]. The aromatic *furan* or *pyran* rings containing helicenes are well-established and well-studied systems (for an overview see Ref. [235] and Refs. [12] and [13] therein); however, non-photochemical approaches prevail in their preparations. A solitary example of photochemical photocyclization of xanthylium salt (*pyrylium* ion) under acidic conditions was disclosed in the 1930s, which led to pentacyclic oxahelicene **126** [236–238].

1.4.3 Phosphahelicenes and Other Heterohelicenes

Phosphorus-containing helicene derivatives are valuable for their performance in different areas, with enantioselective catalysis being probably the most important

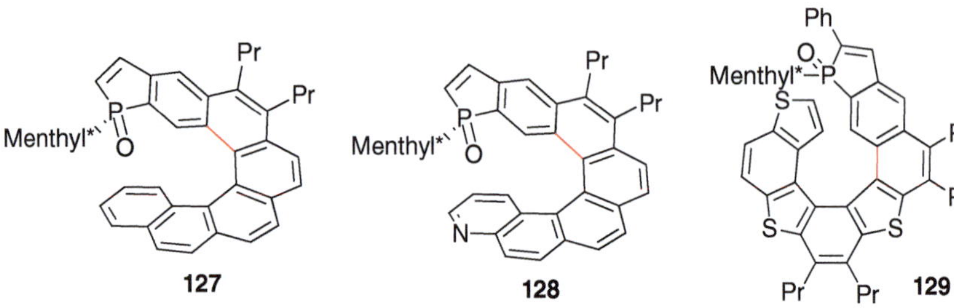

Figure 1.28 Structures of several phosphahelicene derivatives.

[239, 240]. The most significant contribution in this area of helicene chemistry was made by Marinetti and her coworkers, who studied the synthesis and applications of helicenes containing five-membered phosphole rings. These compounds can be prepared by a multitude of different approaches including photochemistry. The first photochemical approach was published in 2012 – the photocyclization of a suitable stilbene precursor provided phosphole-containing helicene **127** (Figure 1.28) [241]. The authors used the chirality on phosphorus, along with the L-menthyl group present in the structure, to form diastereomers easily separable by chromatography on silica gel. Subsequent formation of dimers of the product via [2 + 2] cycloaddition was found to significantly lower yields upon prolonged irradiation. This behavior was later found to be typical for the majority of these compounds and is discussed separately in Section 1.6. Phospholes based on other helicenes were later prepared, including different derivatives containing nitrogen (**128**) [242] or sulfur (**129**) [243]. Other helical compounds containing different functionalities on phosphorus atoms were also described in numerous instances [244–246].

Another way of utilizing photochemistry in the preparation of heterohelicenes involves the light-promoted synthesis of suitable building blocks prior to the non-photochemical formation of the helical backbone. Among various helical molecules accessible by this approach, metalla[*n*]helicenes [247, 248] and azaborahelicenes [249, 250] serve as typical examples.

1.5 Photochemical Preparation of Helicene-Like Molecules

In this part, only those screw-shaped molecules accomplished via photochemistry and containing non-aromatic rings will be discussed. The "helicene-world" is rich in systems possessing helical topology (known as *helicene-like molecules*, *helicenoids*, or *helicene-analogs*), and light might be used in different ways for their preparation, either for the generation of the catalytically active metal species promoting [2 + 2 + 2]-cyclotrimerization (e.g. in Refs. [251–256]) or simply for the preparation of molecular building blocks prior to the final non-photochemical cyclization (e.g. dioxepine[7]helicene-like [235]). Dihydrohelicenes may be formally included among these molecules too.

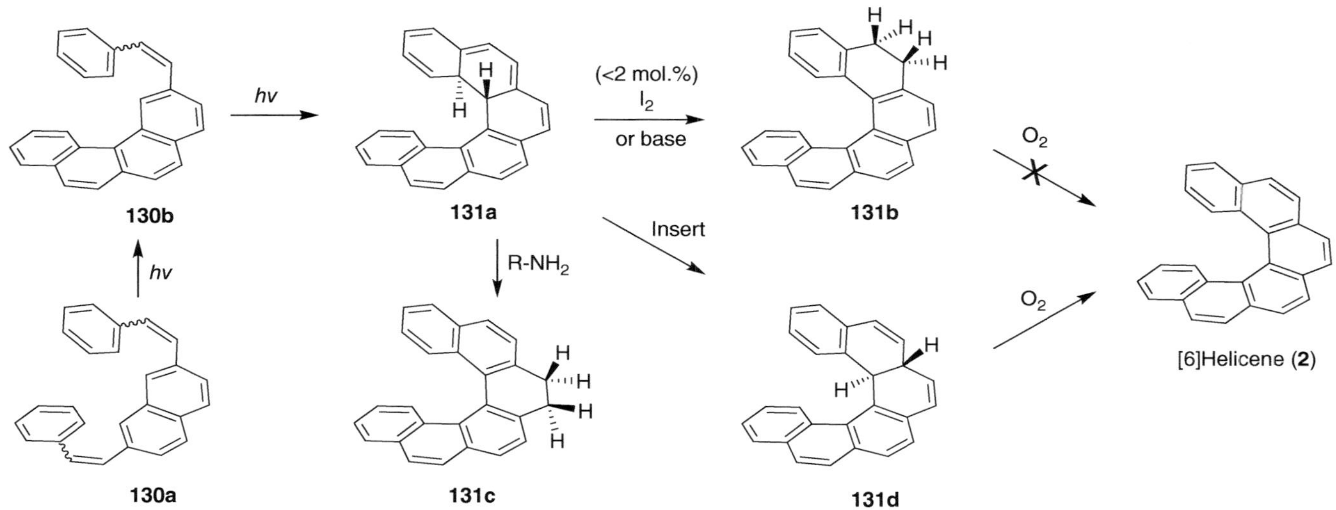

Figure 1.29 Photocyclizations leading to dihydro[6]helicenes**131**.

The irradiation of 2,7-distyrylnaphthalene **130a** or 2-styrylbenzo[*c*]phenanthrene **130b** (Figure 1.29) under oxygen-free conditions and in the presence of a small amount of iodine (0.5–2 mol %) provided 5,6-dihydro[6]helicene (**131b**) in 60% yield as a consequence of H-shifts from the primarily formed cyclization product **131a** [257]. Interestingly, the substance **131b** was not easily oxidized to [6]helicene. Under anaerobic conditions, *trans*-6a,16d-dihydro[6]helicene **131d** was formed from **131a** by a suprafacial 1,5-H shift of one hydrogen [258]. Dihydrohelicene

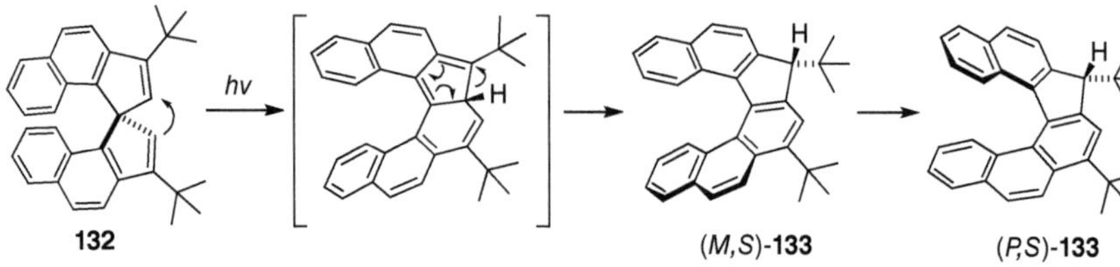

Figure 1.30 Other photochemical reactions to helicene analogs.

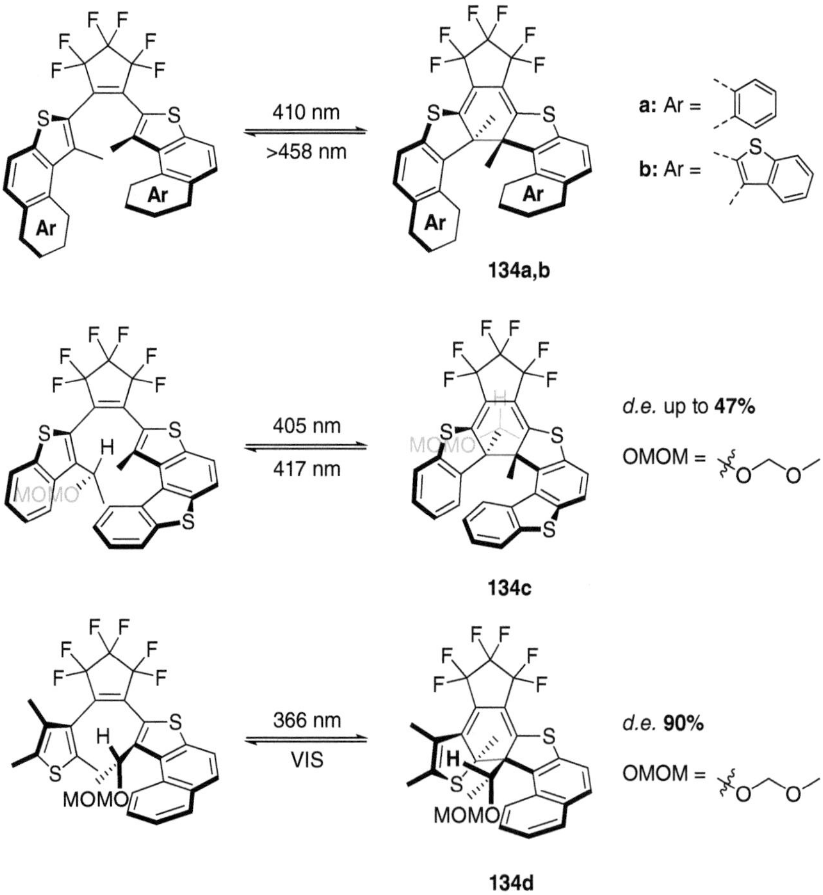

Figure 1.31 Photostationary states resulting in helicene-like structures.

131d was shown to be photolabile and easily oxidized to [6]helicene. Irradiation of 2-styrylbenzo[*c*]phenanthrene **130b** in alkylamines or basic alcoholic solutions resulted in the formation of a mixture of 5,6-dihydro- (**131b**) and 7,8-dihydro- (**131c**) [6]helicenes [259, 260]. In alkylamines, **131c** was the favored compound, whereas in basic alcoholic solutions compound **131b** predominated.

Shingu studied the stereospecific photochemical rearrangement of optically active *spiro*-compound **132** (Figure 1.30) that, after suprafacial 1,5-migration of the vinyl group, afforded 7,9-di-*tert*-butyl-9*H*-benzo[6,7]-indeno[2,1-*c*]phenanthrene (**133**), a helicene-like molecule [261]. The primary (*M,S*)-product epimerized to a more stable (*P,S*)-isomer.

Using photochemistry in the ring-closing reactions of other helicene-like molecules is rare and limited essentially to the preparation of *helicenoids* (e.g. **134a–d**, Figure 1.31) in their photostationary state (PSS). The photochromism of these systems was studied by Branda [262] and later on by Yokoyama et al. [263, 264]. Thanks to the allylic strain of the stereogenic carbon located at the periphery of the hexatriene moiety, Yokoyama prepared **134c,d** enantioselectively from their open forms [265]. Using the same strategy, Diederich et al. also succeeded in the preparation of other helicene-DAE (diarylethene) systems in PSS [266].

1.6 Photochemical Transformations of Helicenes

This part of the chapter discusses reactions where light is used to propel transformations of helicene skeleton, rather than for its preparation. The past couple of years has seen a surge in the development of photochemical methods for transformations of different substrates, and these methodologies are slowly finding their way into helicene chemistry as well.

The first transformations of this type, however, were observed as early as the late 1960s [44, 62, 267]. One of the typical by-products in photochemical preparations of [5]helicenes is their overannulation forming the corresponding benzo[*ghi*]perylenes (Figure 1.32a). The starting stilbene-type molecule **135** initially forms the desired [5]helicene **1**, but under the reaction conditions, another photocyclodehydrogenation takes place to form **136**. In most cases, this process is undesired, and significant effort was put into avoiding this reaction. For example, some bromo- [63], cyano- [86], or phenylene- [132] substituted [5]helicenes do not undergo such cyclization, because the orbitals involved in the cyclization process are destabilized by the presence of the substituents. In other cases, however, different benzoperylenes can be prepared in good yields, as shown in Figures 1.32b,c. For instance, dimethoxy benzoperylene **138** was isolated in 32% yield after irradiation of the bis(styryl)benzene **137** [86]. Similarly, π-extended derivative **140** was obtained in 89% yield from stilbene **139** [268].

A similar reaction can be conducted with certain aza- and azonia[5]helicenes as well. 2-Aza[5]helicene **62b** always undergoes overannulation to form the 7-azabenzo[*ghi*]perylene **141**, irrespective of the starting stilbene precursor (Figure 1.33a) [172, 269]. Azonium stilbene precursors can be irradiated to give the

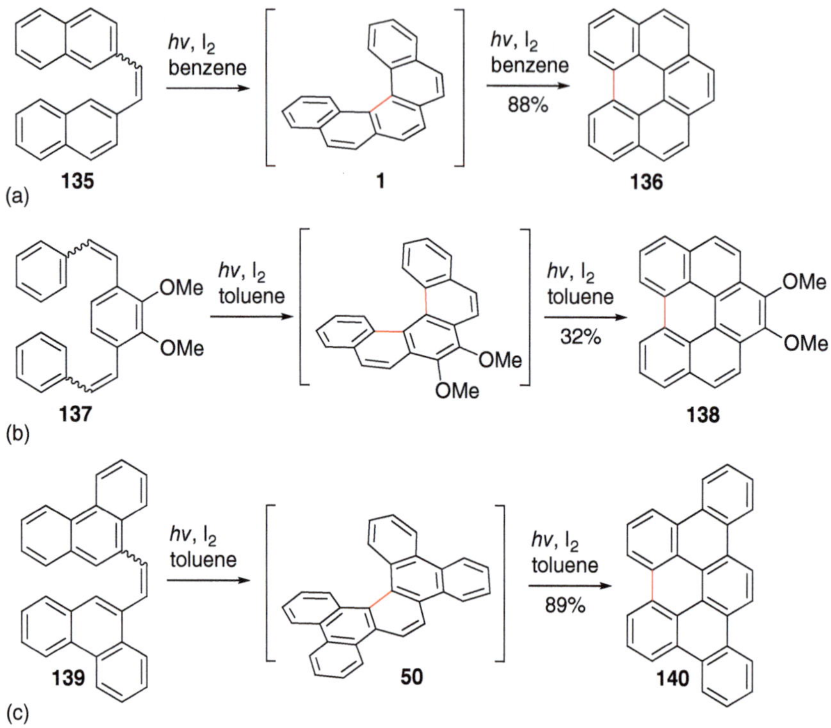

Figure 1.32 Preparation of (a) unsubstituted and (b, c) substituted benzo[*ghi*]perylenes from [5]helicenes.

respective azonia[5]helicenes, which subsequently undergo another ring closure to produce 2a-azoniabenzo[*ghi*]perylene perchlorate **142** in low yield.

Photochemical transformations are not limited to [5]helicenes, as has been shown recently in several cases. [2 + 2]-Photocycloadduct **144** was formed from a phosphole-containing helicene derivative **143** (Figure 1.34a) [245]. The process is regioselective and takes place under both UV and X-ray irradiation, both in crystal and solution. This behavior of phosphahelicenes is rather general and has been observed on multiple occasions [241, 242]. A similar example was published by Církva and coworkers, who isolated cyclobutane **145** substituted by two benzo[*c*]phenanthrenes as a major product in an attempted synthesis of 1,2,3,4-tetrafluoro[6]helicene **26** (Figure 1.34b) [52]. One example of [4 + 2]-photocycloaddition was published by Storch and Sýkora, who irradiated 9,10-diketo[7]helicene **146** (Figure 1.34c) with UV light in the presence of 2,3-dimethylbut-2-ene to obtain helicene-based 1,4-dioxane **147** in good yield [270].

Storch and König studied both reductive and oxidative photochemical transformations of helicene derivatives [42]. They showed that bromohelicenes can generate helicenyl radicals upon irradiation with blue light (Figure 1.35a, transformations of **32a** are shown as a representative example). This species can be subsequently trapped with a suitable trapping agent, ranging from various heterocycles (pyrrole,

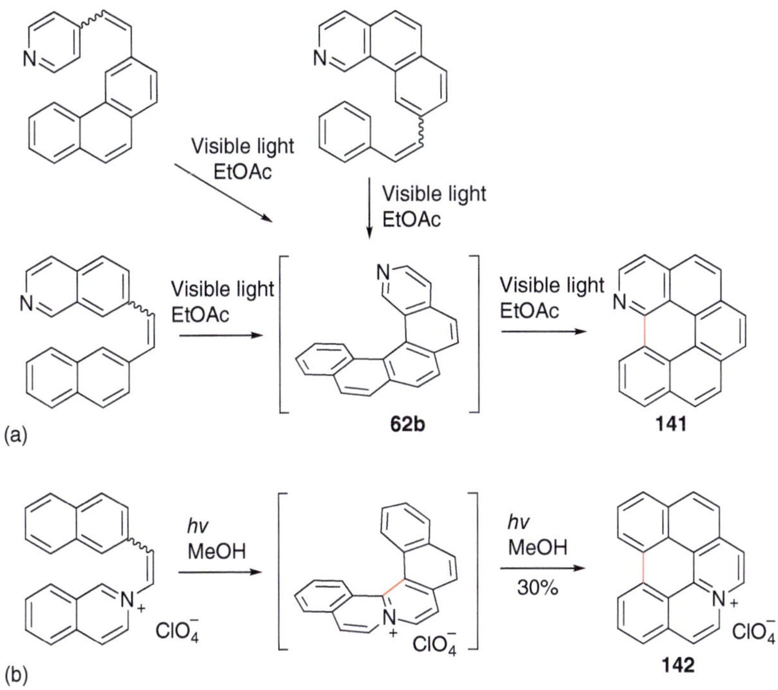

Figure 1.33 Preparation of (a) aza- and (b) azoniabenzo[*ghi*]perylenes.

thiophene, furan, indole) to heteroatomic compounds, such as triethyl phosphite, dimethyl disulfide, or bis(pinacolato)diboron to form helicenes **148a–h**. Three different examples of oxidative transformation of 2-methoxy[6]helicene **149** were also shown, yielding products **150a–c** with –Br, –CN, and –PO(OEt)$_2$ groups regioselectively in position 6 (Figure 1.35b). The same authors also published a simple oxidation of twisted arylamines into respective *ortho*-diketones (**146**, **151–154**) by oxygen [270]. The arylamine sensitizes the formation of singlet oxygen, which then oxidizes the amine-substituted double bond via [2 + 2]-cycloaddition. The prepared diketones are shown in Figure 1.35c and can be easily transformed into π-extended nitrogen-containing heterohelicenes [270].

1.7 Conclusions

The photochemical approach to carbo[*n*]helicenes and their derivatives provides a wide range of helical molecules differing in their length or substitution patterns and conceptually represents the most explored area in this field. The current state of knowledge is, however, starting to reach its limit, especially in the photosynthesis of long ($n \geq 10$) or multiple carbo[*n*]helicenes. The limiting factors are mainly the growing complexity of the starting materials and, consequently, their decreasing solubility. Additionally, an increasing number of bonds formed in a single

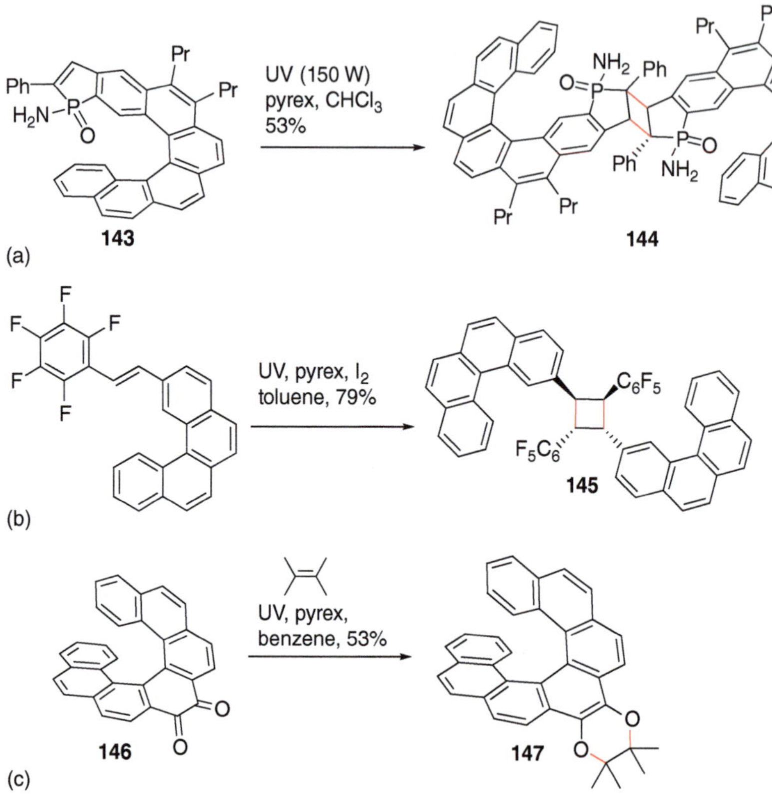

Figure 1.34 (a, b) [2+2]- and (c) [4+2]-Photocycloadditions of helicene molecules.

transformation typically lowers the yield of the reaction, limiting the applicability of the photochemical approach in the synthesis of complex helical structures.

Unlike carbohelicenes, the photochemical approach to hetero[*n*]helicenes is somewhat underdeveloped, despite the variability that these structures can offer. Thus, photosynthesis of long derivatives or helicenes with a higher number of heteroatoms in their backbone remains unexplored so far. Also, the preparation of helical structures with other heteroatoms (e.g. boron, silicon, etc.) that are commonly accessible with non-photochemical approaches remains elusive for photochemists.

The light-mediated transformations of helicenes represent significant interest in the recently emerging part of photochemistry. The results achieved in this area so far indicate an unexplored and potentially fast-growing field of photochemical research. The increasingly common LED-type photoreactors can play an important role in this field, as their development represents a major step forward in terms of their photoefficiency and availability of specific wavelengths.

Although the photochemical approach provides racemic helicenes only, it successfully competes with modern non-photochemical asymmetric syntheses requiring multistep reaction sequences or the use of economically demanding reagents/catalysts.

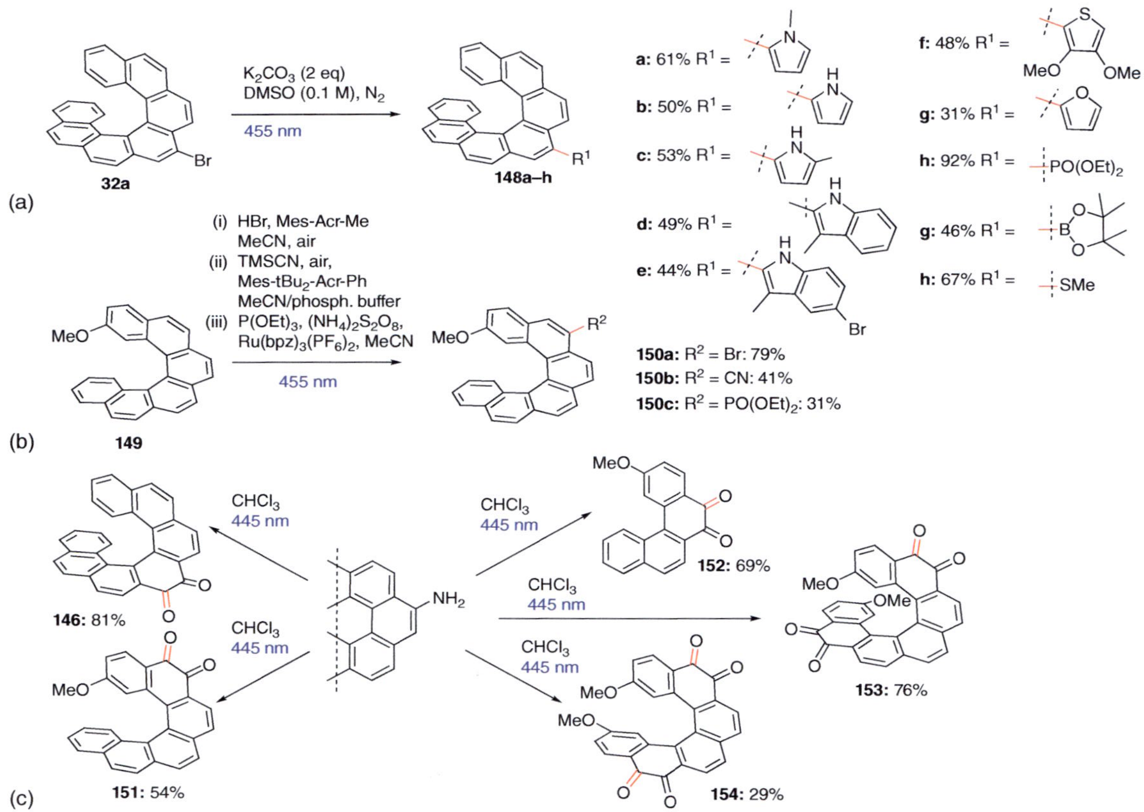

Figure 1.35 Photochemical (a) substitutions and (b, c) oxidative transformations of helicene backbone.

References

1 Moss, G.P., Smith, P.A.S., and Tavernier, D. (1995). Glossary of class names of organic compounds and reactive intermediates based on structure (IUPAC recommendations 1995). *Pure Appl. Chem.* 67 (8–9): 1307–1375.

2 Bromberg, A. and Muszkat, K.A. (1969). Oxidation of 4a,4b-dihydrophenanthrenes. I. Kinetics of the thermal reaction of 9,10-cyclopentano-4a,4b-dihydrophenanthrene with oxygen. *J. Am. Chem. Soc.* 91 (11): 2860–2866.

3 Knittel-Wismonsky, T., Fischer, G., and Fischer, E. (1972). Photochemical ring opening and true photo-oxidation in the 4A,4B-dihydrophenanthrenes. *Tetrahedron Lett.* 13 (28): 2853–2856.

4 Buckles, R.E. (1955). Illumination of *cis*- and *trans*-stilbenes in dilute solutions. *J. Am. Chem. Soc.* 77 (4): 1040–1041.

5 Mallory, F.B., Wood, C.S., and Gordon, J.T. (1964). Photochemistry of stilbenes. III. Some aspects of the mechanism of photocyclization to phenanthrenes. *J. Am. Chem. Soc.* 86 (15): 3094–3102.

6 Wood, C.S. and Mallory, F.B. (1964). Photochemistry of stilbenes. IV. The preparation of substituted phenanthrenes. *J. Org. Chem.* 29 (11): 3373–3377.

7 Cuppen, T.J.H.M. and Laarhoven, W.H. (1972). Photodehydrocyclizations of stilbene-like compounds. VI. Chemical evidence of an excited state mechanism. *J. Am. Chem. Soc.* 94 (16): 5914–5915.

8 Jørgensen, K.B. (2010). Photochemical oxidative cyclisation of stilbenes and stilbenoids – the Mallory-reaction. *Molecules* 15 (6): 4334–4358.

9 Giles, R.G.F. and Sargent, M.V. (1974). Photochemical synthesis of phenanthrenes from 2-methoxystilbenes. *J. Chem. Soc. Perkin Trans.* 1: 2447–2450.

10 Olsen, R.J. and Pruett, S.R. (1985). Photocyclization of *o*-halostilbenes. *J. Org. Chem.* 50 (26): 5457–5460.

11 Mallory, F.B., Rudolph, M.J., and Oh, S.M. (1989). Photochemistry of stilbenes. 8. Eliminative photocyclization of *o*-methoxystilbenes. *J. Org. Chem.* 54 (19): 4619–4626.

12 Zertani, R. and Meier, H. (1986). Photochemistry of 1,3-distyrylbenzene. A new route to syn-[2.2](1,3)cyclophanes. *Chem. Ber.* 119 (5): 1704–1715.

13 Mallory, F.B. and Mallory, C.W. (1984). Photocyclization of stilbenes and related molecules. *Org. React.* 30: 1–456.

14 Laarhoven, W.H. (1983). Photochemical cyclizations and intramolecular cycloadditions of conjugated arylolefins. Part I: photocyclization with dehydrogenation. *Recl. des Trav. Chim. des Pays-Bas* 102 (4): 185–204.

15 Sudhakar, A. and Katz, T.J. (1986). Directive effect of bromine on stilbene photocyclizations. An improved synthesis of [7]helicene. *Tetrahedron Lett.* 27 (20): 2231–2234.

16 Liu, L., Yang, B., Katz, T.J., and Poindexter, M.K. (1991). Improved methodology for photocyclization reactions. *J. Org. Chem.* 56 (12): 3769–3775.

17 Moradpour, A., Kagan, H., Baes, M. et al. (1975). Photochemistry with circularly polarized light – III. *Tetrahedron* 31 (17): 2139–2143.

18 Buchardt, O. (1974). Photochemistry with circularly polarized light. *Angew. Chem. Int. Ed. Engl.* 13 (3): 179–185.
19 Bernstein, W.J., Calvin, M., and Buchardt, O. (1973). Absolute asymmetric synthesis. III. Hindered rotation about aryl-ethylene bonds in the excited states of diaryl ethylenes. Structural effects on the asymmetric synthesis of 2- and 4-substituted hexahelicenes. *J. Am. Chem. Soc.* 95 (2): 527–532.
20 Bernstein, W.J., Calvin, M., and Buchardt, O. (1972). Absolute asymmetric synthesis. I. Mechanism of the photochemical synthesis of nonracemic helicenes with circularly polarized light. Wavelength dependence of the optical yield of octahelicene. *J. Am. Chem. Soc.* 94 (2): 494–498.
21 Goedicke, C. and Stegemeyer, H. (1972). Optical activity and stereochemistry of 4a,4b-dihydropentahelicene. *Chem. Phys. Lett.* 17 (4): 492–494.
22 Moradpour, A., Nicoud, J.F., Balavoine, G. et al. (1971). Photochemistry with circularly polarized light. The synthesis of optically active hexahelicene. *J. Am. Chem. Soc.* 93 (9): 2353–2354.
23 Kagan, H., Moradpour, A., Nicoud, J.F.F. et al. (1971). Photochemistry with circularly polarised light. II. Asymmetric synthesis of octa and nonahelicene. *Tetrahedron Lett.* 12 (27): 2479–2482.
24 Laarhoven, W.H. and Cuppen, T.J.H.M. (1977). Chiral solvent-induced asymmetric synthesis; photosynthesis of optically enriched hexahelicene. *J. Chem. Soc. Chem. Commun.* (2): 47–47.
25 Laarhoven, W.H. and Cuppen, T.J.H.M. (1978). Chiral solvent-induced asymmetric synthesis. Part 2. Photosynthesis of optically enriched hexahelicenes. *J. Chem. Soc. Perkin Trans.* 2 (4): 315–318.
26 Gingras, M., Félix, G., and Peresutti, R. (2013). One hundred years of helicene chemistry. Part 2. Stereoselective syntheses and chiral separations of carbohelicenes. *Chem. Soc. Rev.* 42 (3): 1007–1050.
27 Nakazaki, M., Yamamoto, K., Fujiwara, K., and Maeda, M. (1979). Mechanically directed absolute asymmetric syntheses of helicenes in a twisted nematic mesophase. *J. Chem. Soc. Chem. Commun.* (23): 1086–1087.
28 Hibert, M. and Solladie, G. (1980). Contribution of pitch and solute-solvent interactions to the photoasymmetric synthesis of hexahelicene in cholesteric liquid crystals. *J. Org. Chem.* 45 (26): 5393–5394.
29 Cochez, Y., Jespers, J., Libert, V. et al. (1975). Helicenes. chemically induced asymmetric photosyntheses of a helicene skeleton: 2- and 3-menthyloxycarbonyl [6] helicene. *Bull. des Sociétés Chim. Belges* 84 (11): 1033–1036.
30 Cochez, Y., Martin, R.H., and Jespers, J. (1977). Helicenes: chemically induced asymmetric photosyntheses of helicenes skeletons. *Isr. J. Chem.* 15 (1–2): 29–32.
31 Vanest, J.-M. and Martin, R.H. (1979). Helicenes: a striking temperature dependence in a chemically induced asymmetric photosynthesis. *Recl. des Trav. Chim. des Pays-Bas* 98 (3): 113–113.
32 Martin, R.H. and Libert, V. (1980). Helicenes. The use of resolved hexahelicene-2-carboxylic acid as a common precursor for the photochemical synthesis of optically pure octa-, nona-, deca-, undeca-, and trideca-helicenes.

Thermal racemization of deca- and undeca-helicenes. *J. Chem. Res.* 11 (30): 130–131.

33 Sudhakar, A., Katz, T.J., and Yang, B.W. (1986). Synthesis of a helical metallocene oligomer. *J. Am. Chem. Soc.* 108 (10): 2790–2791.

34 Katz, T.J., Sudhakar, A., Teasley, M.F. et al. (1993). Synthesis and properties of optically active helical metallocene oligomers. *J. Am. Chem. Soc.* 115 (8): 3182–3198.

35 Gilbert, A.M., Katz, T.J., Geiger, W.E. et al. (1993). Synthesis and properties of an optically active helical bis-cobaltocenium ion. *J. Am. Chem. Soc.* 115 (8): 3199–3211.

36 Osuga, H., Suzuki, H., and Tanaka, K. (1997). Practical synthesis of optically pure bifunctionalized heterohelicenes. *Bull. Chem. Soc. Jpn.* 70 (4): 891–897.

37 Pearson, M.S.M. and Carbery, D.R. (2009). Studies toward the photochemical synthesis of functionalized [5]- and [6]carbohelicenes. *J. Org. Chem.* 74 (15): 5320–5325.

38 Likhtenshtein, G. (2009). Stilbenes preparation and analysis. In: *Stilbenes: Applications in Chemistry, Life Sciences and Materials Science*, 1–41. Weinheim, Germany: Wiley-VCH Verlag GmbH & Co. KGaA.

39 Giacomini, E., Rupiani, S., Guidotti, L. et al. (2016). The use of stilbene scaffold in medicinal chemistry and multi-target drug design. *Curr. Med. Chem.* 23 (23): 2439–2489.

40 Klán, P. and Wirz, J. (2009). *Photochemistry of Organic Compounds: From Concepts to Practice, Postgraduate Chemistry Series.* Chichester, UK: Wiley.

41 Al Mousawi, A., Schmitt, M., Dumur, F. et al. (2018). Visible light chiral photoinitiator for radical polymerization and synthesis of polymeric films with strong chiroptical activity. *Macromolecules* 51 (15): 5628–5637.

42 Jakubec, M., Ghosh, I., Storch, J., and König, B. (2020). Photochemical functionalization of helicenes. *Chem. – A Eur. J.* 26 (2): 543–547.

43 Jakubec, M. and Storch, J. (2020). Recent advances in functionalizations of helicene backbone. *J. Org. Chem.* 85 (21): 13415–13428.

44 Scholz, M., Mühlstädt, M., and Dietz, F. (1967). Chemie angeregter zustände. I. Mitt. Die richtung der photocyclisierung naphthalinsubstituierter äthylene. *Tetrahedron Lett.* 8 (7): 665–668.

45 Flammang-Barbieux, M., Nasielski, J., and Martin, R.H. (1967). Synthesis of heptahelicene (1) benzo[c]phenanthro[4,3-g]phenanthrene. *Tetrahedron Lett.* 8 (8): 743–744.

46 Noël, T. (2017). *Photochemical Processes in Continuous-Flow Reactors. From Engineering Principles to Chemical Applications.* Word Scientific.

47 Hernandez-Perez, A.C., Vlassova, A., and Collins, S.K. (2012). Toward a visible light mediated photocyclization: Cu-based sensitizers for the synthesis of [5]helicene. *Org. Lett.* 14 (12): 2988–2991.

48 Bédard, A.-C., Vlassova, A., Hernandez-Perez, A.C. et al. (2013). Synthesis, crystal structure and photophysical properties of pyrene-helicene hybrids. *Chem. – A Eur. J.* 19 (48): 16295–16302.

49 Lefebvre, Q., Jentsch, M., and Rueping, M. (2013). Continuous flow photocyclization of stilbenes – scalable synthesis of functionalized phenanthrenes and helicenes. *Beilstein J. Org. Chem.* 9: 1883–1890.

50 Murase, T., Suto, T., and Suzuki, H. (2017). Azahelicenes from the oxidative photocyclization of boron hydroxamate complexes. *Chem. – An Asian J.* 12 (7): 726–729.

51 Jakubec, M., Beránek, T., Jakubík, P. et al. (2018). 2-Bromo[6]helicene as a key intermediate for [6]helicene functionalization. *J. Org. Chem.* 83 (7): 3607–3616.

52 Církva, V., Jakubík, P., Strašák, T. et al. (2019). Preparation and physicochemical properties of [6]helicenes fluorinated at terminal rings. *J. Org. Chem.* 84 (4): 1980–1993.

53 Storch, J., Církva, V., Bernard, M., and Vokál, J. (2013). Process for preparing [6]helicenes by photocyclization. CZ303997, issued 26 June 2013.

54 Církva, V. and Relich, S. (2011). Microwave photochemistry and photocatalysis. Part 1: principles and overview. *Curr. Org. Chem.* 15 (2): 248–264.

55 Martin, R.H. (1974). The helicenes. *Angew. Chem. Int. Ed. Engl.* 13 (10): 649–660.

56 Laarhoven, W.H. (1983). Photochemical cyclizations and intramolecular cycloadditions of conjugated arylolefins: Part 2: photocyclizations without dehydrogenation and photocycloadditions. *Recl. des Trav. Chim. des Pays-Bas* 102 (5): 241–254.

57 Laarhoven, W.H. and Prinsen, W.J.C. (1984). Carbohelicenes and heterohelicenes. *Top. Curr. Chem.* 125: 63–130.

58 Shen, Y. and Chen, C.-F. (2012). Helicenes: synthesis and applications. *Chem. Rev.* 112 (3): 1463–1535.

59 Gingras, M. (2013). One hundred years of helicene chemistry. Part 1: non-stereoselective syntheses of carbohelicenes. *Chem. Soc. Rev.* 42 (3): 968–1006.

60 Hoffmann, N. (2014). Photochemical reactions applied to the synthesis of helicenes and helicene-like compounds. *J. Photochem. Photobiol. C Photochem. Rev.* 19 (1): 1–19.

61 Chen, C.-F. and Shen, Y. (2017). *Helicene Chemistry*. Berlin, Heidelberg: Springer.

62 Dietz, F. and Scholz, M. (1968). Chemie angeregter zustände. IV. Die photocyclisierung der drei isomeren distyrylbenzole. *Tetrahedron* 24 (24): 6845–6849.

63 Liu, L. and Katz, T.J. (1991). Bromine auxiliaries in photosynthesis of [5]helicenes. *Tetrahedron Lett.* 32 (47): 6831–6834.

64 Martin, R.H., Marchant, M.J., and Baes, M. (1971). Syntheses in the field of polycyclic aromatic compounds. XXXI. Rapid syntheses of hexa and heptahelicene. *Helv. Chim. Acta* 54 (1): 358–360.

65 Martin, R.H.H., Flammang-Barbieux, M., Cosyn, J.P.P., and Gelbcke, M. (1968). 1-Synthesis of octa- and nonahelicenes. 2-New syntheses of hexa- and heptahelicenes. 3-Optical rotation and O.R.D. of heptahelicene. *Tetrahedron Lett.* 9 (31): 3507–3510.

66 Laarhoven, W.H., Cuppen, T.H.J.H.M., and Nivard, R.J.F. (1970). Photodehydrocyclizations in stilbene-like compounds – III. *Tetrahedron* 26 (20): 4865–4881.

67 El Abed, R., Ben Hassine, B., Genêt, J.-P. et al. (2004). An alternative procedure for the synthesis of [5]- and [7]carbohelicenes. *Eur. J. Org. Chem.* 2004 (7): 1517–1522.

68 Martin, R.H. and Cosyn, J.P. (1971). New synthesis of octahelicene involving non-interconvertible dl – intermediates. *Synth. Commun.* 1 (4): 257–265.

69 Mori, K., Murase, T., and Fujita, M. (2015). One-step synthesis of [16]helicene. *Angew. Chem. Int. Ed.* 54 (23): 6847–6851.

70 Martin, R.H. and Baes, M. (1975). Helicenes. Photosyntheses of [11], [12] and [14]helicene. *Tetrahedron* 31 (17): 2135–2137.

71 Martin, R.H., Morren, G., and Schurter, J.J. (1969). [13]Helicene and [13]helicene-10,21-d_2. *Tetrahedron Lett.* 10 (42): 3683–3688.

72 Roose, J., Achermann, S., Dumele, O., and Diederich, F. (2013). Electronically connected [*n*]helicenes: synthesis and chiroptical properties of enantiomerically pure (*E*)-1,2-di([6]helicen-2-yl)ethenes. *Eur. J. Org. Chem.* 2013 (16): 3223–3231.

73 Laarhoven, W.H., Cuppen, T.J.H.M., and Nivard, R.J.F. (1968). Photodehydrocyclizations in stilbene-like compounds. *Recl. des Trav. Chim. des Pays-Bas* 87 (6): 687–698.

74 Jungmann, H., Güsten, H., and Schulte-Frohlinde, D. (1968). Die photochemische Cyclisierung monosubstituierter Stilbene. *Chem. Ber.* 101 (8): 2690–2696.

75 Laarhoven, W.H. and Veldhuis, R.G.M. (1972). Conformational studies on hexahelicenes – II. *Tetrahedron* 28 (6): 1811–1822.

76 Laarhoven, W.H. and Cuppen, T.J.H.M. (1974). Conformational studies on hexahelicenes – IV. *Tetrahedron* 30 (9): 1101–1106.

77 Martin, R.H., Eyndels, C., and Defay, N. (1974). Double helicenes: diphenanthro[4,3-a; 3′,4′-o]picene and benzo[s]diphenanthro[4,3-a; 3′,4′-o]picene. *Tetrahedron* 30 (18): 3339–3342.

78 Gorsane, M., Defay, N., and Martin, R.H. (2010). Helicenes: Synthèse Photochimique et Étude RMN ^{19}F, ^{13}C et ^{1}H de Fluoro-1-Hexahelicenes et Fluoro-1-Heptahelicenes. *Bull. des Sociétés Chim. Belges* 94 (3): 215–231.

79 Mallory, F.B. and Mallory, C.W. (1983). Photocyclization of stilbenes. VII. Unusual fluorine atom rearrangement in the photocyclization of 1-fluoro[5]helicenes. *J. Org. Chem.* 48 (4): 526–532.

80 Gorsane, M. and Martin, R.H. (1985). Helicenes: synthèse photochimique de fluoro-1-pentahelicenes. *Bull. des Sociétés Chim. Belges* 94 (3): 205–214.

81 Ravat, P., Hinkelmann, R., Steinebrunner, D. et al. (2017). Configurational stability of [5]helicenes. *Org. Lett.* 19 (14): 3707–3710.

82 Puls, C., Stolle, A., De Meijere, A., and de Meijere, A. (1993). Preparation and properties of new methano-bridged dibenzo[*c,g*]phenanthrenes. *Chem. Ber.* 126 (7): 1635–1641.

83 El Abed, R., Ben Hassine, B., Genêt, J.-P. et al. (2004). Synthesis of a configurationally locked [5]helicene derivative. *Synthesis (Stuttg).* 2004 (15): 2513–2516.

84 Stammel, C., Fröhlich, R., Wolff, C. et al. (1999). Synthesis and X-ray analysis of new [5]helicenes – HMO calculations on the photocyclization of the stilbene precursors. *Eur. J. Org. Chem.* 1999 (7): 1709–1718.

85 Terfort, A., Görls, H., and Brunner, H. (1997). The first helical-chiral phosphane ligands: rac-[5]- and rac-[6]-heliphos. *Synthesis (Stuttg).* 1997 (1): 79–86.

86 Ito, N., Hirose, T., and Matsuda, K. (2014). Facile photochemical synthesis of 5,10-disubstituted [5]helicenes by removing molecular orbital degeneracy. *Org. Lett.* 16 (9): 2502–2505.

87 Kubo, H., Hirose, T., and Matsuda, K. (2017). Control over the emission properties of [5]helicenes based on the symmetry and energy levels of their molecular orbitals. *Org. Lett.* 19 (7): 1776–1779.

88 Robert, A., Naulet, G., Bock, H. et al. (2019). Cyclobishelicenes: shape-persistent figure-eight aromatic molecules with promising chiroptical properties. *Chem. – A Eur. J.* 25 (63): 14364–14369.

89 Belarmino Cabral, M.G., Pereira de Oliveira Santos, D.M., Cristiano, R. et al. (2017). From 1,4-phenylenebis(phenylmaleate) to a room-temperature liquid-crystalline benzo[*ghi*]perylene diimide. *ChemPlusChem* 82 (3): 342–346.

90 Wang, R., Shi, K., Cai, K. et al. (2016). Syntheses of polycyclic aromatic diimides via intramolecular cyclization of maleic acid derivatives. *New J. Chem.* 40 (1): 113–121.

91 Robert, A., Dechambenoit, P., Hillard, E.A. et al. (2017). Non-planar oligoarylene macrocycles from biphenyl. *Chem. Commun.* 53 (84): 11540–11543.

92 Naulet, G., Sturm, L., Robert, A. et al. (2018). Cyclic tris-[5]helicenes with single and triple twisted Möbius topologies and Möbius aromaticity. *Chem. Sci.* 9 (48): 8930–8936.

93 Raouafi, S., Aloui, F., and Hafedh, N. (2017). Synthesis, characterization, and photophysical properties of a new pentacyclic helicene. *C. R. Chim.* 20 (11–12): 1047–1052.

94 Jia, X., Nitsch, J., Ji, L. et al. (2019). Triarylborane-based helical donor-acceptor compounds: synthesis, photophysical, and electronic properties. *Chem. – A Eur. J.* 25 (46): 10845–10857.

95 Sudhakar, A. and Katz, T.J. (1986). Asymmetric synthesis of helical metallocenes. *J. Am. Chem. Soc.* 108 (1): 179–181.

96 Mallory, F.B., Regan, C.K., Bohen, J.M. et al. (2015). Discovery of deep-seated skeletal rearrangements in the photocyclizations of some tert-butyl-substituted 1,2-diarylethylenes. *J. Org. Chem.* 80 (1): 8–17.

97 Frimer, A.A., Kinder, J.D., Youngs, W.J., and Meador, M.A.B. (1995). Reinvestigation of the photocyclization of 1,4-phenylenebis[phenylmaleic anhydride]: preparation and structure of [5]helicene 5,6:9,10-dianhydride. *J. Org. Chem.* 60 (6): 1658–1664.

98 Yang, B., Liu, L., Katz, T.J. et al. (1991). Electron delocalization in helical quinone anion radicals. *J. Am. Chem. Soc.* 113 (23): 8993–8994.

99 Schwertel, M., Hillmann, S., and Meier, H. (2013). Synthesis of highly substituted hexahelicenes. *Helv. Chim. Acta* 96 (11): 2020–2032.

100 Meier, H., Schwertel, M., and Kolshorn, H. (2013). Hexahelicenophanes. *Helv. Chim. Acta* 96 (11): 2009–2019.

101 Aloui, F., El Abed, R., Guerfel, T., and Ben Hassine, B. (2006). Synthesis and X-ray analysis of a new [6]helicene. *Synth. Commun.* 36 (11): 1557–1567.

102 Schwertel, M., Hillmann, S., and Meier, H. (2013). Preparation of 1,2,5,6,9,10-hexaalkoxyhexahelicenes. *Z. Naturforsch. B* 68 (12): 1347–1355.

103 Kogiso, T., Yamamoto, K., Suemune, H., and Usui, K. (2012). Synthesis and characterization of 1,8-naphthalimide with [6]helicene skeleton. *Org. Biomol. Chem.* 10 (15): 2934–2936.

104 Chen, S., Ge, Z., Jia, Q. et al. (2019). The preparation of enantiopure [6]- and [7]helicenes from binaphthanol. *Chem. – An Asian J.* 14 (9): 1462–1466.

105 Murase, T., Matsuda, C., Adachi, K. et al. (2018). Triple photochemical domino reaction of a tetrafluorostilbene terminating in double fluorine atom transfer. *Commun. Chem.* 1 (1): 1–6.

106 Laarhoven, W.H. and Veldhuis, R.G.M. (1972). Conformational studies on hexahelicenes – III. *Tetrahedron* 28 (6): 1823–1827.

107 Tribout, J., Martin, R.H., Doyle, M., and Wynberg, H. (1972). Chemical assignment of absolute configurations in the helicene and heterohelicene series. Part XXXIV 1. Hexahelicene 2. Benzo[d]naphtho[1,2-d′]benzo[1,2-b; 4,3-b′]dithiophene. *Tetrahedron Lett.* 13 (28): 2839–2842.

108 Brown, J.M., Field, I.P., and Sidebottom, P.J. (1981). Structural specificity in asymmetric charge-transfer complexation of helicenes. *Tetrahedron Lett.* 22 (48): 4867–4870.

109 Wachsmann, C., Weber, E., Czugler, M., and Seichter, W. (2003). New functional hexahelicenes – synthesis, chiroptical properties, X-ray crystal structures, and comparative data bank analysis of hexahelicenes. *Eur. J. Org. Chem.* (15): 2863–2876.

110 Ben Hassine, B., Gorsane, M., Geerts-Evrard, F. et al. (1986). Atrolactic synthesis in the evaluation of the efficiency of inducers of asymmetric synthesis. *Bull. Soc. Chim. Belg.* 95 (7): 547–556.

111 Hellou, N., Mace, A., Martin, C. et al. (2018). Synthesis of carbo[6]helicene derivatives grafted with amino or aminoester substituents from enantiopure [6]helicenyl boronates. *J. Org. Chem.* 83 (1): 484–490.

112 Aloui, F., El Abed, R., Marinetti, A., and Ben Hassine, B. (2007). Synthesis and characterization of new hexahelicene derivatives. *Tetrahedron Lett.* 48 (11): 2017–2020.

113 Aloui, F., El Abed, R., Marinetti, A., and Ben Hassine, B. (2009). A new approach to 3,14-dihydroxyhexahelicene: resolution and attribution of the absolute configuration. *C. R. Chim.* 12 (1–2): 284–290.

114 Ben Braiek, M., Aloui, F., and Ben Hassine, B. (2016). Synthesis, enantiomeric resolution and optical properties of 8-cyanohexahelicene. *Tetrahedron Lett.* 57 (38): 4273–4276.

115 Talele, H.R., Chaudhary, A.R., Patel, P.R., and Bedekar, A.V. (2011). Expeditious synthesis of helicenes using an improved protocol of photocyclodehydrogenation of stilbenes. *Arkivoc* 2011 (9): 15–37.

116 Ben Braiek, M., Aloui, F., and Ben Hassine, B. (2013). Synthesis and resolution of 2-hydroxyhexahelicene. *Tetrahedron Lett.* 54 (5): 424–426.
117 Aloui, F., Moussa, S., and Ben Hassine, B. (2012). Synthesis and characterization of a new hexacyclic helicene. *Tetrahedron Lett.* 53 (26): 3216–3219.
118 El Abed, R., Aloui, F., Genêt, J.-P. et al. (2007). Synthesis and resolution of 2-(diphenylphosphino)heptahelicene. *J. Organomet. Chem.* 692 (5): 1156–1160.
119 Guedouar, H., Aloui, F., Moussa, S. et al. (2014). Synthesis and characterization of new heptacyclic helicenes. *Tetrahedron Lett.* 55 (45): 6167–6170.
120 Ben Braiek, M., Aloui, F., Moussa, S., and Ben Hassine, B. (2015). Synthesis and characterization of new helically chiral heptacyclic systems. *Tetrahedron Lett.* 56 (47): 6580–6584.
121 Gingras, M. and Collet, C. (2005). Functionalized heptahelicene bidentate ligands and chiral building blocks. *Synlett* (15): 2337–2341.
122 Joly, M., Defay, N., Martin, R.H. et al. (1977). Hélicènes pontés: ethano-3,15- et oxa-2-propano-3,15-[7]hélicène. Synthèse, spectrographie ^{1}H-RMN. et étude par diffraction aux rayons X. *Helv. Chim. Acta* 60 (2): 537–560.
123 Agou, T., Kohara, M., Tamura, Y. et al. (2020). Helicenes fused with hexafluorocyclopentene (HFCP): synthesis, structure, and properties. *Eur. J. Org. Chem.* 2020 (12): 1871–1880.
124 Howarth, J. and Finnegan, J. (1997). Synthesis of 9,10-dimethyl[7]helicene and 8,11-diaza[7]helicene. *Synth. Commun.* 27 (20): 3663–3668.
125 Milton, M., Schuster, N.J., Paley, D.W. et al. (2019). Defying strain in the synthesis of an electroactive bilayer helicene. *Chem. Sci.* 10 (4): 1029–1034.
126 Tinnemans, A.H.A. and Laarhoven, W.H. (1974). Photodehydrocyclizations in stilbene-like compounds. X. Rearrangements in the photocyclization of 4,5-diphenyltriphenylene and 4,5-diphenylphenanthrene. *J. Am. Chem. Soc.* 96 (14): 4617–4622.
127 Matsuura, K., Nishida, J., Ito, T. et al. (2019). Synthesis and properties of π-extended fluoranthene derivatives from 1,2-diarylacenaphthylene derivatives. *Tetrahedron* 75 (2): 278–285.
128 Fujikawa, T., Preda, D.V., Segawa, Y. et al. (2016). Corannulene-helicene hybrids: chiral π-systems comprising both bowl and helical motifs. *Org. Lett.* 18 (16): 3992–3995.
129 Wang, C.-Z., Kihara, R., Feng, X. et al. (2017). Synthesis, structure and photophysical properties of pyrene-based [5]helicenes: an experimental and theoretical study. *ChemistrySelect* 2 (4): 1436–1441.
130 Laarhoven, W.H. and Nivard, R.J.F. (1976). Conformational studies on helicenes. V. Alteration of the conformation of helicenes by annelation of benzo groups. *Tetrahedron* 32 (20): 2445–2450.
131 Brison, J., De Bakker, C., Defay, N. et al. (1983). Synthèse Photochimique D'Hydrocarbures Polycycliques Aromatiques et Étude en RMN-^{1}H Des Protons de Baie. Effets de Solvant Specifiques et Effets Nucleaires Overhauser. *Bull. Soc. Chim. Belg.* 92 (10): 901–912.
132 Tinnemans, A.H.A.A., Laarhoven, W.H., Sharafi-Ozeri, S., and Muszkat, K.A. (2010). Photodehydrocyclizations of stilbene-like compounds XV: electronic

overlap population as a reactivity measure in photocyclizations of pentahelicenes. *Recl. Trav. Chim. Pays-Bas* 94 (11): 239–243.

133 Lin, W.-B., Li, M., Fang, L., and Chen, C.-F. (2018). Recent progress on multidimensional construction of helicenes. *Chin. Chem. Lett.* 29 (1): 40–46.

134 Li, C., Yang, Y., and Miao, Q. (2018). Recent progress in chemistry of multiple helicenes. *Chem. Asian J.* 13 (8): 884–894.

135 Kato, K., Segawa, Y., and Itami, K. (2019). Symmetric multiple carbohelicenes. *Synlett* 30 (4): 370–377.

136 Sarkar, P., Das, B.K., Chakraborty, D., and Muthamma, K. (2019). Carbohelicenes and thiahelicene from phthalaldehydes through Perkin approach. *J. Mol. Struct.* 1195: 309–314.

137 Ferreira, M., Naulet, G., Gallardo, H. et al. (2017). A naphtho-fused double [7]helicene from a maleate-bridged chrysene trimer. *Angew. Chem. Int. Ed.* 56 (12): 3379–3382.

138 Liu, B., Böckmann, M., Jiang, W. et al. (2020). Perylene diimide-embedded double [8]helicenes. *J. Am. Chem. Soc.* 142 (15): 7092–7099.

139 Liu, G., Koch, T., Li, Y. et al. (2019). Nanographene imides featuring dual-core sixfold [5]helicenes. *Angew. Chem. Int. Ed.* 58 (1): 178–183.

140 Xiao, S., Kang, S.-J.J., Wu, Y. et al. (2013). Supersized contorted aromatics. *Chem. Sci.* 4 (5): 2018.

141 Saito, H., Uchida, A., and Watanabe, S. (2017). Synthesis of a three-bladed propeller-shaped triple [5]helicene. *J. Org. Chem.* 82 (11): 5663–5668.

142 Laarhoven, W.H. and Cuppen, T.J.H.M. (1971). Synthesis of a double helicene rac. and meso diphenanthro[3,4-c;3′4-1]chrysene. *Tetrahedron Lett.* 12 (2): 163–164.

143 Laarhoven, W.H. and Cuppen, T.H.J.M. (1973). Photodehydrocyclizations of stilbene-like compounds VII: synthesis and properties of the double helicene, diphenanthro[3,4-c ;3′,4′-l]chrysene. *Recl. des Trav. Chim. des Pays-Bas* 92 (5): 553–562.

144 Laarhoven, W.H., Cuppen, T.J.H.M., and Nivard, R.J.F. (1974). Photodehydrocyclizations of stilbene-like compounds. XI. Synthesis and racemization of the double helicene diphenanthro[4.3-a;3′,4′-o]picene. *Tetrahedron* 30 (18): 3343–3347.

145 Laarhoven, W.H. and de Jong, M.H. (1973). Photodehydrocyclizations of stilbene-like compounds VIII: synthesis of hexaheliceno[3,4-c]hexahelicene. *Recl. des Trav. Chim. des Pays-Bas* 92 (6): 651–657.

146 Schuster, N.J., Paley, D.W., Jockusch, S. et al. (2016). Electron delocalization in perylene diimide helicenes. *Angew. Chem. Int. Ed.* 55 (43): 13519–13523.

147 Schuster, N.J., Hernández Sánchez, R., Bukharina, D. et al. (2018). A helicene nanoribbon with greatly amplified chirality. *J. Am. Chem. Soc.* 140 (20): 6235–6239.

148 Schuster, N.J., Joyce, L.A., Paley, D.W. et al. (2020). The structural origins of intense circular dichroism in a waggling helicene nanoribbon. *J. Am. Chem. Soc.* 142 (15): 7066–7074.

149 Meng, D., Fu, H., Xiao, C. et al. (2016). Three-bladed rylene propellers with three-dimensional network assembly for organic electronics. *J. Am. Chem. Soc.* 138 (32): 10184–10190.

150 Khokhlov, K., Schuster, N.J., Ng, F., and Nuckolls, C. (2018). Functionalized helical building blocks for nanoelectronics. *Org. Lett.* 20 (7): 1991–1994.

151 Li, Z. and Twieg, R.J. (2015). Photocyclodehydrofluorination. *Chem. – A Eur. J.* 21 (44): 15534–15539.

152 Daigle, M., Picard-Lafond, A., Soligo, E., and Morin, J.-F.F. (2016). Regioselective synthesis of nanographenes by photochemical cyclodehydrochlorination. *Angew. Chem. Int. Ed.* 55 (6): 2042–2047.

153 Brison, J. and Martin, R.H. (1983). Synthèse, Par la Méthode de Hewett, Des Benzo [a]Naphto[2,1-j]Anthracéne, Naphto[2,1-c]Chrysene, Benzo[a]Picéne et 8,16-Diméthoxycarbonylbenzo[c]Naphto[2,1-1]Chrysene. *Bull. des Sociétés Chim. Belges* 92 (10): 893–899.

154 Yamamoto, K., Ikeda, T., Kitsuki, T. et al. (1990). Synthesis and chiral recognition of optically active crown ethers incorporating a helicene moiety as the chiral centre. *J. Chem. Soc. Perkin Trans. 1* 110 (2): 271–276.

155 Thulin, B., Wennerström, O., Nielsen, B.J. et al. (1976). Propellicene or Bi-2,13-pentahelicenylene. *Acta Chem. Scand.* 30B (7): 688–690.

156 Mohamed, R.K., Mondal, S., Guerrera, J.V. et al. (2016). Alkynes as linchpins for the additive annulation of biphenyls: convergent construction of functionalized fused helicenes. *Angew. Chem. Int. Ed.* 55 (39): 12054–12058.

157 Kurata, Y., Otsuka, S., Fukui, N. et al. (2017). Aromatic metamorphosis of dibenzofurans into triphenylenes starting with nickel-catalyzed ring-opening C-O arylation. *Org. Lett.* 19 (6): 1274–1277.

158 Hori, M., Guo, J.D., Yanagi, T. et al. (2018). Sigmatropic rearrangements of hypervalent-iodine-tethered intermediates for the synthesis of biaryls. *Angew. Chem. Int. Ed.* 57 (17): 4663–4667.

159 Perkampus, H.-H. and Bluhm, T. (1972). Zur photochemie der styryldiazine. *Tetrahedron* 28 (7): 2099–2110.

160 Fehn, H. and Perkampus, H.-H. (1978). Zur photokinetik der styryldiazine. *Tetrahedron* 34 (13): 1971–1977.

161 Muszkat, K.A. and Sharafi-Ozeri, S. (1976). Effects of electrostatic interactions on the photocyclization reactivity of diaza-1,2-diarylethylenes. *Chem. Phys. Lett.* 42 (1): 99–102.

162 Caronna, T., Fontana, F., Longhi, G. et al. (2009). 2,13-Diaza[5]helicene: synthesis, theoretical calculations and spectroscopic properties. *Arkivoc* (8): 145–155.

163 Abbate, S., Bazzini, C., Caronna, T. et al. (2006). Monoaza[5]helicenes. Part 2: synthesis, characterisation and theoretical calculations. *Tetrahedron* 62 (1): 139–148.

164 Loader, C.E. and Timmons, C.J. (1966). Studies in photochemistry. Part II. The photocyclisation of stilbazoles to azaphenanthrenes. *J. Chem. Soc. C Org.*: 1078–1081.

165 Galiazzo, G., Bortolus, P., and Cauzzo, G. (1966). Two-ways photocyclization of 3-styrylpyridine. *Tetrahedron Lett.* 7 (31): 3717–3721.
166 Aloui, F., El Abed, R., and Hassine, B.B. (2008). Synthesis of a new N-containing hexahelicene. *Tetrahedron Lett.* 49 (9): 1455–1457.
167 Bortolus, P., Cauzzo, G., Mazzucato, U., and Galiazzo, G. (1969). The photocyclization of styrylpyridines to azaphenanthrenes and their geometrical photoisomerization. *Z. Phys. Chem.* 63 (1_4): 29–38.
168 Muszkat, K.A. and Sharafi-Ozeri, S. (1976). Electronic overlap population as reactivity measure. The photodimerization of acenaphthylene. *Chem. Phys. Lett.* 38 (2): 346–348.
169 Bazzini, C., Caronna, T., Fontana, F. et al. (2008). Synthesis, crystal structure and crystal packing of diaza[5]helicenes. *New J. Chem.* 32 (10): 1710–1717.
170 Staab, H.A., Diehm, M., and Krieger, C. (1994). Synthesis, structure and basicity of 1,16-diaza[6]helicene. *Tetrahedron Lett.* 35 (45): 8357–8360.
171 Martin, R.H. and Deblecker, M. (1969). Synthesis of 4-azahexahelicene (benzo[c]phénanthro[1,2-f]quinoline). *Tetrahedron Lett.* 10 (41): 3597–3598.
172 Bazzini, C., Brovelli, S., Caronna, T. et al. (2005). Synthesis and characterization of some aza[5]helicenes. *Eur. J. Org. Chem.* 2005 (7): 1247–1257.
173 Zhang, X., Clennan, E.L., Arulsamy, N. et al. (2016). Synthesis, structure, and photochemical behavior of [5]heli-viologen isomers. *J. Org. Chem.* 81 (13): 5474–5486.
174 Clennan, E.L., Zhang, X., and Petek, T. (2017). Synthesis, reactivity, and sulfide quenching of helical viologens. *Phosphorus Sulfur Silicon Relat. Elem.* 192 (2): 222–226.
175 Patel, E.N., Arthur, R.B., Nicholas, A.D. et al. (2019). Synthesis, structure and photophysical properties of a 2D network with gold dicyanide donors coordinated to aza[5]helicene viologen acceptors. *Dalton Trans.* 48 (27): 10288–10297.
176 Waghray, D., Zhang, J., Jacobs, J. et al. (2012). Synthesis and structural elucidation of diversely functionalized 5,10-diaza[5]helicenes. *J. Org. Chem.* 77 (22): 10176–10183.
177 Waghray, D., Cloet, A., Van Hecke, K. et al. (2013). Diazadithia[7]helicenes: synthetic exploration, solid-state structure, and properties. *Chem. – A Eur. J.* 19 (36): 12077–12085.
178 Klimash, A., Pander, P., Klooster, W.T. et al. (2018). Intermolecular interactions in molecular crystals and their effect on thermally activated delayed fluorescence of helicene-based emitters. *J. Mater. Chem. C Mater. Opt. Electron. Devices* 6 (39): 10557–10568.
179 Abbate, S., Lebon, F., Longhi, G. et al. (2009). Experimental and calculated vibrational and electronic circular dichroism spectra of 2-Br-hexahelicene. *Phys. Chem. Chem. Phys.* 11 (40): 9039–9043.
180 Abbate, S., Longhi, G., Lebon, F. et al. (2014). Helical sense-responsive and substituent-sensitive features in vibrational and electronic circular dichroism, in circularly polarized luminescence, and in Raman spectra of some simple optically active hexahelicenes. *J. Phys. Chem. C* 118 (3): 1682–1695.

181 Isla, H., Saleh, N., Ou-Yang, J.-K. et al. (2019). Bis-4-aza[6]helicene: a bis-helicenic 2,2′-bipyridine with chemically triggered chiroptical switching activity. *J. Org. Chem.* 84 (9): 5383–5393.

182 Saleh, N., Moore, B., Srebro, M. et al. (2015). Acid/base-triggered switching of circularly polarized luminescence and electronic circular dichroism in organic and organometallic helicenes. *Chem. – A Eur. J.* 21 (4): 1673–1681.

183 Isla, H., Srebro-Hooper, M., Jean, M. et al. (2016). Conformational changes and chiroptical switching of enantiopure bis-helicenic terpyridine upon Zn^{2+} binding. *Chem. Commun.* 52 (35): 5932–5935.

184 Deshayes, K., Broene, R.D., Chao, I. et al. (1991). Synthesis of the helicopodands: novel shapes for chiral clefts. *J. Org. Chem.* 56 (24): 6787–6795.

185 Saleh, N., Srebro, M., Reynaldo, T. et al. (2015). *enantio*-Enriched CPL-active helicene–bipyridine–rhenium complexes. *Chem. Commun.* 51 (18): 3754–3757.

186 Saleh, N., Vanthuyne, N., Bonvoisin, J. et al. (2018). Redox-triggered chiroptical switching activity of ruthenium(III)-bis-(β-diketonato) complexes bearing a bipyridine-helicene ligand. *Chirality* 30 (5): 592–601.

187 Galland, M., Riobé, F., Ouyang, J. et al. (2019). Helicenic complexes of lanthanides: influence of the f-element on the intersystem crossing efficiency and competition between luminescence and oxygen sensitization. *Eur. J. Inorg. Chem.* 2019 (1): 118–125.

188 Murguly, E., McDonald, R., and Branda, N.R. (2000). Chiral discrimination in hydrogen-bonded [7]helicenes. *Org. Lett.* 2 (20): 3169–3172.

189 Meisenheimer, J. and Witte, K. (1903). Reduction von 2-nitronaphtalin. *Ber. Dtsch. Chem. Ges.* 36 (4): 4153–4164.

190 Shi, L., Liu, Z., Dong, G. et al. (2012). Synthesis, structure, properties, and application of a carbazole-based diaza[7]helicene in a deep-blue-emitting OLED. *Chem. – A Eur. J.* 18 (26): 8092–8099.

191 Ben Braiek, M., Aloui, F., Moussa, S. et al. (2013). Synthesis, X-ray analysis and photophysical properties of a new N-containing pentacyclic helicene. *Tetrahedron Lett.* 54 (40): 5421–5425.

192 Jhulki, S., Mishra, A.K., Chow, T.J., and Moorthy, J.N. (2016). Helicenes as all-in-one organic materials for application in OLEDs: synthesis and diverse applications of carbo- and aza[5]helical diamines. *Chem. – A Eur. J.* 22 (27): 9375–9386.

193 Upadhyay, G.M., Talele, H.R., Sahoo, S., and Bedekar, A.V. (2014). Synthesis of carbazole derived aza[7]helicenes. *Tetrahedron Lett.* 55 (39): 5394–5399.

194 Rajan, B. and Bedekar, A.V. (2017). Synthesis of unsymmetrical aza[7]helicenes by one-pot Wittig–Heck and photodehydrogenation sequence. *Synlett* 28 (17): 2262–2266.

195 Upadhyay, G.M. and Bedekar, A.V. (2015). Synthesis and photophysical properties of bi-aza[5]helicene and bi-aza[6]helicene. *Tetrahedron* 71 (34): 5644–5649.

196 Bucinskas, A., Waghray, D., Bagdziunas, G. et al. (2015). Synthesis, functionalization, and optical properties of chiral carbazole-based diaza[6]helicenes. *J. Org. Chem.* 80 (5): 2521–2528.

197 Hua, W., Liu, Z., Duan, L. et al. (2015). Deep-blue electroluminescence from nondoped and doped organic light-emitting diodes (OLEDs) based on a new monoaza[6]helicene. *RSC Adv.* 5 (1): 75–84.

198 Luo, X.-Y., Liu, Z., Zhang, B.-J. et al. (2018). Substituent effects on selective syntheses of fused-ring heterocycles based on carbazole and naphthalene by photocyclization: synthesis, crystal structures, and properties. *ChemistrySelect* 3 (12): 3426–3432.

199 Arai, S., Takeuchi, T., Ishikawa, M. et al. (1987). Syntheses of condensed polycyclic azonia aromatic compounds by photocyclization. *J. Chem. Soc. Perkin Trans.* 1: 481.

200 Arai, S., Ishikura, M., Sato, K., and Yamagishi, T. (1995). Synthesis of new azonia-helicenes with a quaternary nitrogen at the inner helix skeleton. *J. Heterocyclic Chem.* 32 (3): 1081–1083.

201 Arai, S., Ishikura, M., and Yamagishi, T. (1998). Synthesis of polycyclic azonia-aromatic compounds by photo-induced intramolecular quaternization: azonia derivatives of benzo[*c*]phenanthrene, [5]helicene and [6]helicene. *J. Chem. Soc. Perkin Trans.* 1 (9): 1561–1568.

202 Arai, S., Yafune, T., Ōkubo, M., and Hida, M. (1989). Synthesis of azonia derivative of hexahelicene. *Tetrahedron Lett.* 30 (51): 7217–7218.

203 Sato, K., Nakajima, K., Arai, S., and Yamagishi, T. (1996). Photocyclization of styrylbenzo[a]quinolizinium salts. Part 2. Efficient synthesis of 6a-azonia[5]helicene salts utilizing steric interactions of substituents. *Liebigs Ann.* (4): 439–446.

204 Sato, K., Yamagishi, T., and Arai, S. (2000). Synthesis of novel azonia[5]helicenes containing terminal thiophene rings. *J. Heterocyclic Chem.* 37 (4): 1009–1014.

205 Sato, K., Okazaki, S., Yamagishi, T., and Arai, S. (2004). The synthesis of azoniadithia[6]helicenes. *J. Heterocyclic Chem.* 41 (3): 443–447.

206 Sato, K., Katayama, Y., Yamagishi, T., and Arai, S. (2006). The synthesis of new azoniathiahelicenes. *J. Heterocyclic Chem.* 43 (1): 177–181.

207 Murov, S.L. (1973). *Handbook of Photochemistry*. New York: Marcel Dekker.

208 Anderson, D.G. and Wettermark, G. (1965). Photoinduced isomerizations in anils. *J. Am. Chem. Soc.* 87 (7): 1433–1438.

209 Wettermark, G., Wallström, E., Lindblom, T. et al. (1968). The *cis-trans* isomerization of Schiff bases. *Acta Chem. Scand.* 22 (2): 675–680.

210 Grellmann, K.H. and Tauer, E. (1973). Reaction pathways for the photochemical conversion of ortho-substituted benzylideneanilines to azoles. *J. Am. Chem. Soc.* 95 (10): 3104–3108.

211 Geibel, K., Staudinger, B., Grellmann, K.H., and Wendt, H. (1974). Investigations of the *cis-trans* isomerization of benzylidene aniline. III U.v. and n.m.r. spectra of *cis*-benzylidene anilines. *J. Photochem.* 3 (1): 241–246.

212 Kos, M., Žádný, J., Storch, J. et al. (2020). Oxidative photocyclization of aromatic schiff bases in synthesis of phenanthridines and other aza-PAHs. *Int. J. Mol. Sci.* 21 (16): 5868.

213 Wynberg, H. and Groen, M.B. (1968). Synthesis, resolution, and optical rotatory dispersion of a hexa- and a heptaheterohelicene. *J. Am. Chem. Soc.* 90 (19): 5339–5341.

214 Wynberg, H., Groen, M.B., and Schadenberg, H. (1971). Synthesis and resolution of some heterohelicenes. *J. Org. Chem.* 36 (19): 2797–2809.

215 Wynberg, H. (1971). Some observations on the chemical, photochemical, and spectral properties of thiophenes. *Acc. Chem. Res.* 4 (2): 65–73.

216 Ravat, P. (2021). Carbo[*n*]helicenes restricted to enantiomerize: an insight into the design process of configurationally stable functional chiral PAHs. *Chem. – A Eur. J.* 27 (12): 3957–3967. https://doi.org/10.1002/chem.202004488.

217 Caronna, T., Sinisi, R., Catellani, M. et al. (2001). Photochemical synthesis and optical properties of high membered thiohelicenes. *Synth. Met.* 119 (1–3): 79–80.

218 Caronna, T., Catellani, M., Luzzati, S. et al. (2001). Molecular crystal architecture and optical properties of a thiohelicenes series containing 5, 7, 9, and 11 rings prepared via photochemical synthesis. *Chem. Mater.* 13 (11): 3906–3914.

219 Yamamoto, A., Matsui, Y., Ohta, E. et al. (2016). Formation of a trithia[5]helicene in an unexpected photoreaction of a methyl-substituted bis(dithienylethenyl)thiophene through a double sequence of 6π-electrocyclization/aromatization (dehydrogenation/demethylation). *J. Photochem. Photobiol. A Chem.* 331: 48–55.

220 Lehman, P. and Wynberg, H. (1974). The synthesis of a series of regularly annelated 2-methylheterohelicenes. *Aust. J. Chem.* 27 (2): 315.

221 Larsen, J., Bechgaard, K., Bechgaard, K. et al. (1996). Thiaheterohelicenes. 1. Synthesis of unsubstituted thia[5]-, [9]- and [13]heterohelicenes. *Acta Chem. Scand.* 50 (1): 71–76.

222 Waghray, D., Nulens, W., and Dehaen, W. (2011). Efficient synthesis of benzo fused tetrathia[7]helicenes. *Org. Lett.* 13 (20): 5516–5519.

223 Waghray, D., de Vet, C., Karypidou, K., and Dehaen, W. (2013). Oxidative transformation to naphthodithiophene and thia[7]helicenes by intramolecular scholl reaction of substituted 1,2-bis(2-thienyl)benzene precursors. *J. Org. Chem.* 78 (22): 11147–11154.

224 Waghray, D. and Dehaen, W. (2013). A fragment based approach toward thia[*n*]helicenes. *Org. Lett.* 15 (12): 2910–2913.

225 Moussa, S., Aloui, F., and Ben Hassine, B. (2011). Synthesis and optoelectronic properties of some new thiahelicenes. *Synth. Commun.* 41 (7): 1006–1016.

226 Tanaka, K., Osuga, H., Suzuki, H., and Kishida, H. (1992). An efficient synthetic method for optically pure heterohelicenes. *Tetrahedron Lett.* 33 (32): 4599–4602.

227 Tanaka, K., Osuga, H., and Suzuki, H. (1993). Diastereocontrolled synthesis of optically pure functionalized heterohelicenes. *Tetrahedron Asymmetry* 4 (8): 1843–1856.

228 Waghray, D., Bagdziunas, G., Jacobs, J. et al. (2015). Diastereoselective strategies towards thia[*n*]helicenes. *Chem. – A Eur. J.* 21 (51): 18791–18798.

229 Wang, Z., Shi, J., Li, C. et al. (2010). The synthesis of octyl-substituted hexathia[7]heterohelicene based on dithieno[2,3-*b*:3,2-*d*]thiophene via irradiation. *Lett. Org. Chem.* 7 (1): 85–89.

230 Liu, X., Yu, P., Xu, L. et al. (2013). Synthesis for the mesomer and racemate of thiophene-based double helicene under irradiation. *J. Org. Chem.* 78 (12): 6316–6321.

231 Biet, T., Martin, K., Hankache, J. et al. (2017). Triggering emission with the helical turn in thiadiazole-helicenes. *Chem. – A Eur. J.* 23 (2): 437–446.

232 Rajeshkumar, V. and Stuparu, M.C. (2016). A photochemical approach to aromatic extension of the corannulene nucleus. *Chem. Commun. (Cambridge, UK)* 52 (64): 9957–9960.

233 Li, C., Zhang, Y., Zhang, S. et al. (2014). From *N,N*-diphenyl-*N*-naphtho[2,1-*b*]thieno[2,3-*b*:3′,2′-*d*]dithiophene-5-yl-amine to propeller-shaped *N,N,N*-tri(naphtho[2,1-*b*]thieno[2,3-*b*:3′,2′-*d*]dithiophene-5-yl)-amine: syntheses and structures. *Tetrahedron* 70 (25): 3909–3914.

234 Xu, W., Wu, L., Fang, M. et al. (2017). Diseleno[2,3-*b*:3′,2′-*d*]selenophene and diseleno[2,3-*b*:3′,2′-*d*] thiophene: building blocks for the construction of [7]helicenes. *J. Org. Chem.* 82 (20): 11192–11197.

235 Gupta, R., Cabreros, T.A., Muller, G., and Bedekar, A.V. (2018). Enantiomerically pure 5,13-dicyano-9-oxa[7]helicene: synthesis and study. *Eur. J. Org. Chem.* 2018 (39): 5397–5405.

236 Dilthey, W. and Quint, F. (1936). Dehydrenium-Farbstoffe. Eine neue Klasse von Carbeniumsalzen (XXV. Mitteil. über Pyrenium-Verbindungen). *Ber. Dtsch. Chem. Ges. (A and B Ser.)* 69 (7): 1575–1591.

237 Dilthey, W., Quint, F., and Stephen, H. (1939). Die Wirkung der Auxochrome in m- und p-Stellung bei Dehydreniumfarbstoffen. [Dehydrenium, III. Mitteilung]. [Pyreniumsalze, XXXIII]. *J. für Prakt. Chemie* 152 (3–6): 99–113.

238 Dilthey, W., Quint, F., and Heinen, J. (1939). Dehydreniumfarbstoffe, II. [Pyrenium, XXXII]. *J. Prakt. Chem.* 152 (3–6): 49–98.

239 Aillard, P., Voituriez, A., and Marinetti, A. (2014). Helicene-like chiral auxiliaries in asymmetric catalysis. *Dalton Trans.* 43 (41): 15263–15278.

240 Dhbaibi, K., Favereau, L., and Crassous, J. (2019). Enantioenriched helicenes and helicenoids containing main-group elements (B, Si, N, P). *Chem. Rev.* 119 (14): 8846–8953.

241 Yavari, K., Moussa, S., Ben Hassine, B.B. et al. (2012). 1*H*-Phosphindoles as structural units in the synthesis of chiral helicenes. *Angew. Chem. Int. Ed.* 51 (27): 6748–6752.

242 Yavari, K., Retailleau, P., Voituriez, A., and Marinetti, A. (2013). Heterohelicenes with embedded P-chiral 1*H*-phosphindole or dibenzophosphole units: diastereoselective photochemical synthesis and structural characterization. *Chem. – A Eur. J.* 19 (30): 9939–9947.

243 Aillard, P., Voituriez, A., Dova, D. et al. (2014). Phosphathiahelicenes: synthesis and uses in enantioselective gold catalysis. *Chem. – A Eur. J.* 20 (39): 12373–12376.

244 Febvay, J., Demmer, C.S., Retailleau, P. et al. (2019). Phosphahelicenes with (thio)phosphinic acid and ester functions by the oxidative photocyclisation approach. *Chem. – A Eur. J.* 25 (68): 15609–15614.

245 Demmer, C.S., Aillard, P., Febvay, J. et al. (2017). Photochemical [2+2] cyclization of helical phosphinamides in solution and in the solid state. *ChemPhotoChem* 1 (12): 535–538.

246 Yavari, K., Aillard, P., Zhang, Y. et al. (2014). Helicenes with embedded phosphole units in enantioselective gold catalysis. *Angew. Chem. Int. Ed.* 53 (3): 861–865.

247 Anger, E., Rudolph, M., Norel, L. et al. (2011). Multifunctional and reactive enantiopure organometallic helicenes: tuning chiroptical properties by structural variations of mono- and bis(platinahelicene)s. *Chem. – A Eur. J.* 17 (50): 14178–14198.

248 Yan, Z., Luo, X., Liu, W. et al. (2019). Configurationally stable platinahelicene enantiomers for efficient circularly polarized phosphorescent organic light-emitting diodes. *Chem. – A Eur. J.* 25 (22): 5672–5676.

249 Shen, C., Srebro-Hooper, M., Jean, M. et al. (2017). Synthesis and chiroptical properties of hexa-, octa-, and deca-azaborahelicenes: influence of helicene size and of the number of boron atoms. *Chem. – A Eur. J.* 23 (2): 407–418.

250 Domínguez, Z., López-Rodríguez, R., Álvarez, E. et al. (2018). Azabora[5]helicene charge-transfer dyes show efficient and spectrally variable circularly polarized luminescence. *Chem. – A Eur. J.* 24 (48): 12660–12668.

251 Andronova, A., Szydlo, F., Teplý, F. et al. (2009). The quest for alternative routes to racemic and nonracemic azahelicene derivatives. *Collect. Czechoslov. Chem. Commun.* 74 (1): 189–215.

252 Krausová, Z., Sehnal, P., Teplý, F. et al. (2007). Synthesis of methoxy substituted centrally chiral triynes as precursors of functionalised nonracemic helicene-like compounds. *Collect. Czechoslov. Chem. Commun.* 72 (11): 1499–1522.

253 Sehnal, P., Stara, I.G., Saman, D. et al. (2009). An organometallic route to long helicenes. *Proc. Natl. Acad. Sci.* 106 (32): 13169–13174.

254 Sehnal, P., Krausová, Z., Teplý, F. et al. (2008). On the origin of diastereoselectivity in [2 + 2 + 2] cycloisomerization of chiral triynes: controlling helicity of helicene-like compounds by thermodynamic factors. *J. Org. Chem.* 73 (6): 2074–2082.

255 Stará, I.G., Alexandrová, Z., Teplý, F. et al. (2005). Asymmetric synthesis of [7]helicene-like molecules. *Org. Lett.* 7 (13): 2547–2550.

256 Žádný, J., Jančařík, A., Andronova, A. et al. (2012). A general approach to optically pure [5]-, [6]-, and [7]heterohelicenes. *Angew. Chem. Int. Ed.* 51 (24): 5857–5861.

257 Borkent, J.H., Diesveld, J.W., and Laarhoven, W.H. (2010). Conformational studies on helicenes VII. Synthesis and conformation of 5,6-dihydrohexahelicene. *Recl. des Trav. Chim. des Pays-Bas* 100 (3): 114–118.

258 Laarhoven, W.H., Cuppen, T.J.H.M., and Brinkhof, H.H.K. (1982). The structure of the dihydrohexahelicene formed by the irradiation of styryl benzo[*c*]phenanthrene under anaerobic conditions. *Tetrahedron* 38 (21): 3179–3182.

259 Somers, J.B.M., Laarhoven, W.H., Couture, A., and Lablache-Combier, A. (1985). Base-induced proton tautomerism in the primary photocyclization product of stilbenes. *J. Am. Chem. Soc.* 107 (5): 1387–1394.

260 Prinsen, W.J.C. and Laarhoven, W.H. (1989). Formation of 5,6- and 7,8-dihydrohexahelicene: mechanistic details of the rearrangement of the primary photocyclization product of 2-styrylbenzo[*c*]phenanthrene in the presence of a base. *J. Org. Chem.* 54 (15): 3689–3694.

261 Imajo, S., Shingu, K., Kuritani, H., and Kato, A. (1981). Stereospecific, photochemical rearrangement of 1,1′-spirobiindene and 1,1′-spirobi[benzindene]s. *Tetrahedron Lett.* 22 (2): 107–110.

262 Norsten, T.B., Peters, A., McDonald, R. et al. (2001). Reversible [7]-thiahelicene formation using a 1,2-dithienylcyclopentene photochrome. *J. Am. Chem. Soc.* 123 (30): 7447–7448.

263 Okuyama, T., Tani, Y., Miyake, K., and Yokoyama, Y. (2007). Chiral helicenoid diarylethene with large change in specific optical rotation by photochromism. *J. Org. Chem.* 72 (5): 1634–1638.

264 Tani, Y., Ubukata, T., Yokoyama, Y., and Yokoyama, Y. (2007). Chiral helicenoid diarylethene with highly diastereoselective photocyclization. *J. Org. Chem.* 72 (5): 1639–1644.

265 Hoffmann, R.W. (1989). Allylic 1,3-strain as a controlling factor in stereoselective transformations. *Chem. Rev.* 89 (8): 1841–1860.

266 Milić, J.V., Schaack, C., Hellou, N. et al. (2018). Light-responsive pyrazine-based systems: probing aromatic diarylethene photocyclization. *J. Phys. Chem. C* 122 (33): 19100–19109.

267 Palewska, K., Meister, E.C., and Wild, U.P. (1989). Solid state photochemistry at 4.2 K: the photocyclization of [5] helicene to dihydrobenzoperylene in Shpolskii matrices. *J. Photochem. Photobiol. A Chem.* 50 (2): 239–248.

268 Fujino, S., Yamaji, M., Okamoto, H. et al. (2017). Systematic investigations on fused π-system compounds of seven benzene rings prepared by photocyclization of diphenanthrylethenes. *Photochem. Photobiol. Sci.* 16 (6): 925–934.

269 Gu, X., Wang, H., Roose, J. et al. (2015). A luminescent nitrogen-containing polycyclic aromatic hydrocarbon synthesized by photocyclodehydrogenation with unprecedented regioselectivity. *Chem. – A Eur. J.* 21 (49): 17973–17980.

270 Jakubec, M., Hansen-Troøyen, S., Císařová, I. et al. (2020). Photochemical oxidation specific to distorted aromatic amines providing *ortho* -diketones. *Org. Lett.* 22 (10): 3905–3910.

2

Synthesis of Helicenes by [2 + 2 + 2] Cycloisomerization of Alkynes and Related Systems

Irena G. Stará and Ivo Starý

Institute of Organic Chemistry and Biochemistry, Czech Academy of Sciences, Prague, Czech Republic

2.1 Introduction

The last few decades have seen remarkable developments in the chemistry of helicenes [1–32], stimulated by significant advances in their synthesis. At present, a whole portfolio of synthetic methodologies is available, from classical to state of the art, including traditional organic synthesis [33, 34], photochemical synthesis (see also Chapter 1 "*The Photochemical Approach to Helicenes*" by J. Storch and coworkers) [1–3, 35, 36], transition-metal-catalyzed synthesis [37–48], or organocatalysis [49, 50]. However, four of them clearly dominate: the well-established photocyclodehydrogenation of 1,2-diaryl olefins/arenes [51–54], the Diels–Alder cycloaddition of 1,4-benzoquinone to aromatic (bis)vinylethers [21, 55, 56], the recently developed transition-metal-catalyzed [2 + 2 + 2] cycloisomerization of π-electron systems (topics of this chapter) [57], and the oxidative cyclodehydrogenation of 1,2-diaryl arenes (Scholl reaction, popular today in the synthesis of chiral molecular nanocarbons as subgraphene architectures) [58–71].

Hand in hand with the development of new synthetic methods, the helicene family has diversified considerably and includes not only classical carbohelicenes or their heteroanalogs [31] but also helicene-like compounds [11, 72], cationic helicenes [73], partially hydrogenated helicenes [74], cyclometalated helicenes [29], or even new classes of compounds such as heliphenes [75, 76], helquats [77], multipole helicenes [30], extended helicenes [62, 63, 69, 78, 79], expanded helicenes [80], bilayer helicenes [81], and helicene nanoribbons [82, 83]. Thus, speaking about helicenes, nowadays we mean not only the parent carbohelicenes (or heteroanalogs), which rigorously consist of all-*ortho*-fused benzene rings (or heteroaromatic rings), but also their various counterparts, provided that they possess a related helically chiral scaffold. Therefore, in this chapter the term helicene will be used in the broadest sense and will go beyond the traditional definition.

The synthesis of helicenes by transition-metal-catalyzed [2 + 2 + 2] cycloisomerization of π-electron systems has recently been reviewed [57]. This approach to helicenes is gaining noticeable and growing attention because it complements

Helicenes: Synthesis, Properties, and Applications, First Edition.
Edited by Jeanne Crassous, Irena G. Stará, and Ivo Starý.

traditional methodologies such as photocyclodehydrogenation of 1,2-diaryl olefins or arenes, which in some respects surpasses in generality, robustness, high versatility, and helicity control. The methodology that was introduced into the synthesis of helicenes by Stará, Starý, and coworkers in 1998 [84] took advantage of the previous fundamental achievements of Vollhardt, who recognized considerable synthetic potential of transition-metal-catalyzed [2+2+2] alkyne cycloisomerization [85]. This method of synthesizing helicenes is gaining more and more attention, as its potential has already been demonstrated by other research groups, such as Vollhardt and coworkers [75, 76]; Tanaka and coworkers [86]; Teplý and coworkers [77]; Shibata and coworkers [87]; Carbery and coworkers [88]; Diederich, Gescheidt, and coworkers [89, 90]; Marinetti, Voituriez, Guinchard, and coworkers [91, 92]; Tsujihara, Kawano, and coworkers [93]; and Kotora and coworkers [94].

In the following text, the helical twist of all helicene molecules presented here is highlighted by thick lines (arbitrarily depicting (*P*) helicity) regardless of the fact that they are racemic (the prefix *rac-* is omitted in their numbers). When discussing nonracemic compounds, the absolute stereochemistry is indicated by thick lines/bold wedges/dashed wedges, and their number is supplemented by a prefix expressing the sign of optical rotation and the absolute stereochemistry of the individual chirality elements. If nonracemic helicenes are discussed but their helicity is unknown, a helical twist of the backbone is not indicated, and only a sign of optical rotation is added to the compound number.

2.2 Helicenes via [2 + 2 + 2] Cycloisomerization of π-Electron Systems

2.2.1 Intramolecular [2 + 2 + 2] Cycloisomerization

Springlike structure of helicenes introduces an inherent torsional strain into the molecule, because its full planarization is not possible due to steric hindrance. Accordingly, this extra energy must be compensated in the synthetic step in which the helical scaffold is formed. The transition-metal-catalyzed [2+2+2] cycloisomerization of π-electron systems fulfills this requirement as it is a highly exergonic process. A computational study of dibenzo(hetero)[5]helicenes **4–6** at the DFT level of theory shows that the cyclization of the respective triyne **1**, cyanodiyne **2**, and ynedinitrile **3** is energetically favored (Scheme 2.1) [95]. However, the energy released gradually decreases from carbohelicenes to pyridohelicenes and pyridazinohelicenes (indicating a significantly lower energy gain from the transformation of nitrile units than from alkyne units).

The mechanism of transition-metal-catalyzed [2+2+2] cycloisomerization of π-electron systems depends on the nature of the transition-metal catalyst and π-electron systems undergoing the reaction [17, 95, 96], but it is generally accepted that for late transition metals (Co^{I}, Ni^{0}, Rh^{I}) commonly used in helicene synthesis, the following organometallic intermediates play a key role in the simplified catalytic cycle: upon coordination of the two alkyne units of the starting triyne **A** to the

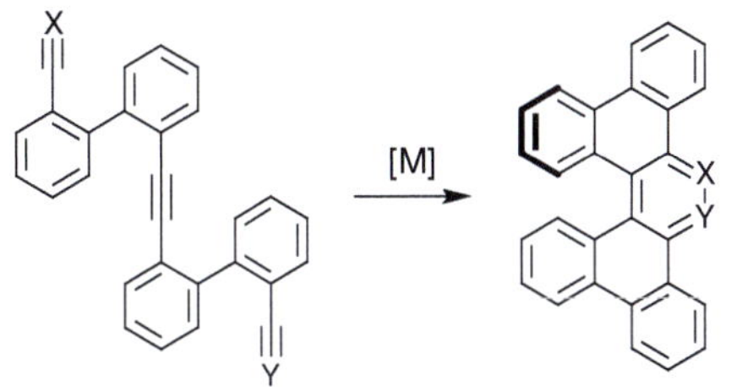

1–3 4–6

Reaction	X	Y	ΔG (kcal/mol)
1 → 4	CH	CH	−122.7
2 → 5	CH	N	−90.5
3 → 6	N	N	−37.8

Scheme 2.1 Free energy of [2 + 2 + 2] cycloisomerization of model triyne **1**, cyanodiyne **2**, and ynedinitrile **3** to dibenzo[5]helicene **4**, dibenzopyrido[5]helicene **5**, and dibenzopyridazino[5]helicene **6** as calculated by DFT (B3LYP/cc-pVTZ) *in vacuo* [95].

metal, an alkyne π-complex **B** is formed. It undergoes oxidative cyclization to deliver metallacyclopentadiene **C**. Coordination of the third alkyne unit provides metallacyclopentadiene alkyne π-complex **D**, which is a common intermediate for two alternative mechanistic pathways. The π-coordinated alkyne unit is either inserted into the metallacycle to give metallacycloheptatriene **E** or added to the metallacyclopentadiene unit in a [4 + 2] fashion to give metallanorbornadiene **F**. Both of these intermediates undergo reductive elimination, which leads to the helicene product **G** and the regenerated catalyst (Scheme 2.2). A more detailed recent discussion of the mechanism of [2 + 2 + 2] cycloisomerization of π-electron systems can be found in the literature [96].

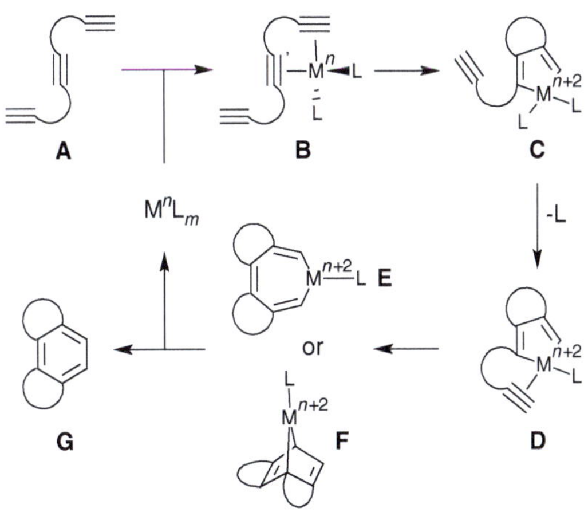

Scheme 2.2 A generally accepted mechanism of the intramolecular transition-metal-catalyzed [2 + 2 + 2] cycloisomerization of tryine to provide helicene.

Various transition metals such as Co^I, Ni^0, or Rh^I were found to be the most effective and therefore gained popularity in helicene synthesis. However, there is no universal catalytic system that exhibits high cyclization efficiency for most π-electron systems and provides a wide range of helicenes in high yield and possibly in high enantiomeric excess. Instead, catalyst/ligand screening is usually necessary to achieve good results (Table 2.1). The Co^I complexes such as $CpCo(CO)_2/PPh_3$ (commercially available) or CpCo(CO)(fum) (air stable and easy to make [98]) are usually catalysts of choice. Although a higher reaction temperature is required,

Table 2.1 The most common transition-metal catalysts used in the [2 + 2 + 2] cycloisomerization of π-electron systems to form (hetero)helicenes.

Catalyst	Notes (refs)	Reaction temperature[a)]	Use (refs)
Co^I (Co^0)			
$CpCo(CO)_2$	Commercially available	>100 °C	[72, 76, 84, 95]
$CpCo(CO)_2/PPh_3$		>100 °C	[84, 89, 97]
CpCo(CO)(fum)	[98]	>100 °C	[97, 99]
$CpCo(CH_2{=}CH_2)_2$	Jonas catalyst: [100–102]	rt	[103]
CpCo(cod)	[104, 105]	rt	[106]
$Co_2(CO)_8$	Commercially available	100 °C	[107]
Ni^0			
$Ni(cod)_2$	Commercially available	rt	[108]
$Ni(cod)_2$/cod (excess)		30 °C	[93]
$Ni(cod)_2/PR_3$	PR_3: PPh_3, PCy_3, dppe	rt	[90, 92, 93, 97, 109]
$Ni(CO)_2(PPh_3)_2$	Commercially available	>150 °C	[110]
$Ni(acac)_2$/EtMgCl/NHC[b)]	Commercially available	rt	[111]
$Ni(PPh_3)_2Br_2/Zn/Et_4NI$[b)]	Commercially available	45 °C	[112]
Rh^I			
$Rh(PPh_3)_3Cl$	Wilkinson's catalyst: commercially available	>100 °C	[77, 94, 113]
$[Rh(cod)_2]BF_4/PR_3$	PR_3: diphosphines	rt to 80 °C	[86, 114]
$[Rh(cod)_2]BAr^F{}_4/PR_3$	PR_3: diphosphines	rt	[115]
$[Rh(cod)_2]OTf/PR_3$	PR_3: diphosphines	80 °C	[116]
$Cp^*Rh(CH_2{=}CH_2)_2$	Commercially available	140 °C	[117]
$Rh_2(CH_2{=}CH_2)_4Cl_2$	Commercially available	140 °C	[117]
Ir^I			
$[Ir(cod)Cl]_2/PR_3$	Commercially available PR_3: diphosphines	120 °C	[80, 87]
Ru^{II}			
Cp*Ru(cod)Cl	Commercially available	100 °C	[77]

a) Reaction temperature typically used.
b) Ni^0L_n generated *in situ*.

these catalysts are robust and tolerate most of the functional groups. Advantageously, rings of different sizes can be closed during the cyclization. Moreover, irradiating the reaction mixture with a halogen lamp [84], conveniently performing cyclization in a microwave reactor (with the addition of an ionic liquid such as [bdmim]BF_4) [111] or in a flow reactor [117] can result in shorter reaction times and higher preparative yields. Ni^0 catalysts such as $Ni(cod)_2$ (commercially available) or complexes generated from it by adding appropriate ligands (usually monodentate phosphines) are significantly more reactive than the aforementioned Co^I catalysts. However, the Ni^0 complexes can rapidly decompose, and they perform best if rings being closed are no larger than six-membered ones. Ni^0 complexes generated *in situ* from the commercially available $Ni(acac)_2$ by treatment with EtMgCl (in the presence of phosphines or NHC ligands [111]) can be used instead of the unstable $Ni(cod)_2$. The application of chiral P, N-, or NHC ligands can lead to good control of helicity [111, 118]. $Ni(CO)_2(PPh_3)_2$ (air stable, commercially available) is active at much higher temperature, but it is robust and easy to handle [110]. Rh^I complexes are among the most effective catalysts of [2 + 2 + 2] cycloisomerization of π-electron systems. They can work well even at room or slightly elevated temperature (in particular cationic Rh^I species). Commercially available $[Rh(cod)_2]BF_4$ can be mixed with a variety of bidentate phosphines, including chiral diphosphines, which allow effective control of helicity [26]. The Rh^I catalysts seem to be less universal than their Co^I counterparts, but they work well in specific areas of the helicene synthesis. For instance, the simple Wilkinson's catalyst $Rh(PPh_3)_3Cl$ (air stable, commercially available) is advantageously utilized in the preparation of helquats [77] or helicene–spirobifluorene hybrids (with the addition of a silver salt) [94].

The intramolecular transition-metal-catalyzed [2 + 2 + 2] cycloisomerization of π-electron systems has several advantages over other methods for the helicene synthesis: (i) it is very universal and allows the preparation of a wide range of helicene derivatives using a uniform methodology from parent helicenes to complex helical architectures; (ii) it is compatible with the presence of various functional groups, heteroatoms, and moieties attached to or incorporated into the helicene backbone; (iii) it is possible to fine-tune the reaction conditions of the key cyclization step with respect to a number of catalysts and ligands available; (iv) with a few exceptions, it eliminates the problem of regioselectivity of the cyclization that forms the helicene scaffold; (v) unlike other synthetic methods, each intramolecular [2 + 2 + 2] cycloisomerization of π-electron systems leads to the formation of three new rings of the helicene scaffold in a single synthetic operation; as multiple cyclization of this kind is feasible, the helicene scaffold can grow rapidly; (vi) it allows scale-up so that helicenes can be prepared in multigram quantities; and (vii) last but not least, it allows enantio- and diastereoselective [2 + 2 + 2] cycloisomerization of π-electron systems to deliver highly enantio- and diastereoenriched helicenes by catalytic or stoichiometric asymmetric synthesis.

The use of transition-metal-catalyzed [2 + 2 + 2] cycloisomerization of π-electron systems significantly simplifies the synthesis of helicenes. It can be illustrated by the preparation of the racemic dibenzo[6]helicene derivative **13** from the easily accessible aromatic triyne **12** (Scheme 2.3) [118]. Although this example shows clearly the strength of the methodology, some problems may arise if properly di- or trisubstituted aromatics, which are used as starting building blocks (e.g. **7** or **8**), are not commercially available or difficult to synthesize. Aryl iodides are required for Sonogashira cross-coupling when nonactivated aromatics are subjected to the reaction. Furthermore, an undesired oxidative homocoupling of arylacetylene might occur at the expense of the formation of a required central 1,2-diarylacetylene subunit (e.g. **9**) via Sonogashira cross-coupling of the arylacetylene with aryl halide.

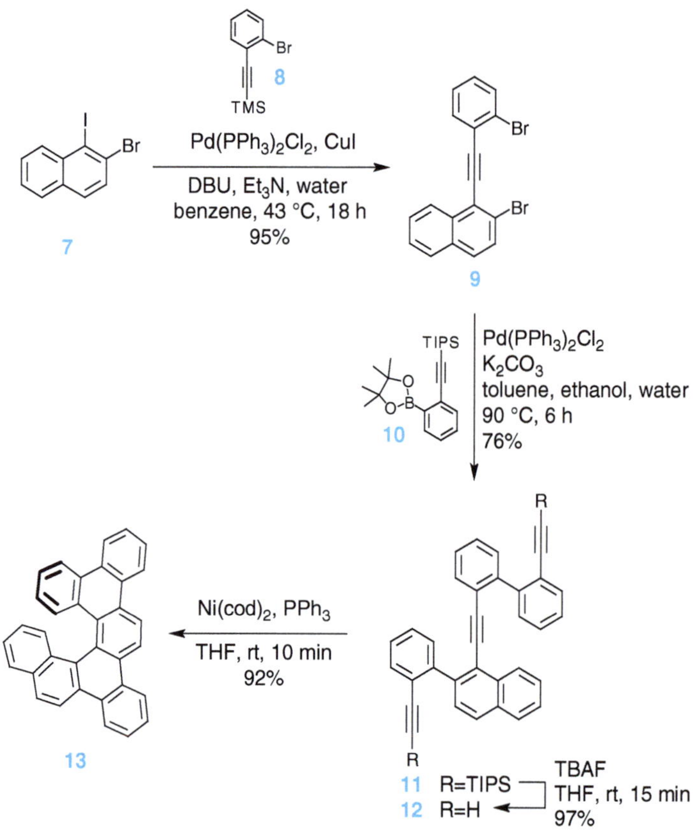

Scheme 2.3 An illustrative example of the synthesis of dibenzo[6]helicene **13** by intramolecular [2 + 2 + 2] cycloisomerization of triyne **12** under Ni^0 catalysis [118].

It is worth mentioning the possible dead ends of the key transition-metal-catalyzed [2 + 2 + 2] cycloisomerization of π-electron systems, of which there are in fact not many. The most abundant undesired reaction is simple polymerization of starting triynes or their analogs. Triynes (or oligoalkynes) with terminal unsubstituted alkyne units are more susceptible to this than triynes with all-alkylated/arylated alkyne units. In the case of Co^I-catalyzed cyclization, the cobaltacyclopentadiene intermediate primarily formed after oxidative cyclization (**C** in Scheme 2.2) can also undergo direct reductive elimination to give a poorly reactive Co^I cyclobutadiene complex [119] as a minor by-product, which may occasionally be isolated. In the case of Rh^I-catalyzed cyclization of triyne, an analogous Rh^I cyclobutadiene complex can also be formed, but it undergoes reverse conversion to isomeric rhodiacyclopentadiene, which yields the product of formal [2 + 1 + 2 + 1] cycloisomerization of triyne as a minor by-product (via a mechanism involving C≡C triple bond cleavage) [86].

2.2.2 Intermolecular [2 + 2 + 2] Cycloisomerization

Helicene scaffolds can also be formed using intermolecular alkyne [2 + 2 + 2] cycloisomerization. Although interesting examples of such a synthetic approach are given in the literature, there are not many of them. In this case, full regioselectivity of the ring closure may not be guaranteed if nonsymmetric components are to react. Moreover, intermolecular cyclization is entropically less favored compared with its intramolecular version.

Pioneered by Tanaka and coworkers, the intermolecular alkyne [2+2+2] cycloisomerization can be effectively used in the short synthesis of oxa[9]helicene derivatives such as **20** (Scheme 2.4) [120]. Starting from the functionalized naphthalene building block **14**, they prepared in three simple steps tetrayne **17**. Then,

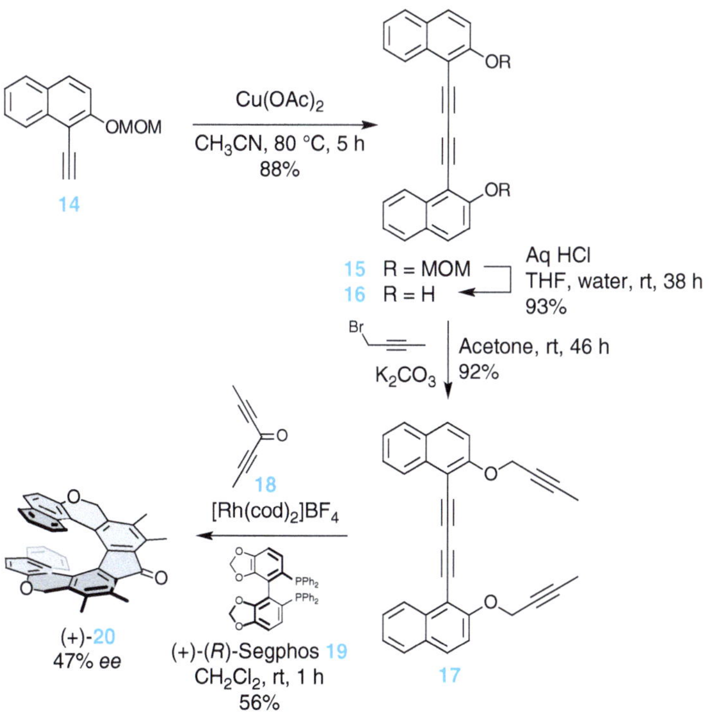

Scheme 2.4 Synthesis of the oxa[9]helicene derivative (+)-**20** by Rh^I-catalyzed intermolecular alkyne [2 + 2 + 2] cycloisomerization [120].

upon reaction with diyne **18** under Rh^I catalysis, the respective helical backbone of **20** was formed. It is worth noting that the five rings were closed in a single operation and a moderate enantiomeric excess of (+)-**20** was obtained using chiral diphosphine ligand (+)-(*R*)-Segphos **19** for rhodium. The same methodology was used in the enantioselective synthesis of parent carbohelicenes [121, 122] or related phosphahelicenes [121, 123] and silahelicenes [87, 115]. Cyclization yields usually range from moderate to good, typically for the preparation of [7]helicene derivatives, but may decrease for longer analogs such as [9]helicene derivatives. Further details are discussed in Section 2.5 and in Chapter 3 "*Enantioselective Synthesis of Helicenes*" by K. Tanaka.

Intriguing examples of intermolecular [2 + 2 + 2] cycloisomerization of formally alkyne-related aryne intermediates [124], which are generated *in situ* as highly reactive species, come from the synthesis of so-called multipole helicenes [30]. The methodology was pioneered by Pérez, Guitián, and coworkers, who performed Pd^0-catalyzed cycloisomerization of phenanthrenyl aryne to obtain hexabenzotriphenylene (three [5]helicenes centrally fused in a three-bladed propeller) [125]. With a time lag of almost two decades, Kamikawa and coworkers [126] and Gingras and coworkers [127] independently synthesized a more complex sextuple helicene **24** comprising six fused [5]helicene units and whose most thermodynamically stable (*M**,*P**,*M**,*P**,*M**,*P**) diastereomer has D_3 symmetry (Scheme 2.5). Although both approaches to **24** combine three [5]helicene derivatives such as **21** or **25** in a seemingly similar cyclotrimerization step, they can differ mechanistically. While **21** is thought to first provide a [5]helicenyl aryne intermediate **22**, which then undergoes true Pd^0-catalyzed intermolecular [2 + 2 + 2] cycloisomerization [128], dibromide **25** under Ni^0-mediated Yamamoto reaction conditions rather follows the sequence of three consecutive aryl–aryl couplings [129]. Relying on the latter concept, Gingras, Coquerel, and coworkers [130] and Ravat and coworkers [131] demonstrated the asymmetric synthesis of even higher sextuple helicenes (propeller-shaped nanocarbons) using either enantiopure (*M*)-7,8-dibromo[6]helicene or (*M*)-9,10-dibromo[7]helicene, respectively, to receive the most stable (*M*,*P*,*M*,*P*,*M*,*P*) diastereomer (out of a number of possible stereoisomers) in good yield and gram amounts.

Tsurusaki, Kamikawa, and coworkers recently applied Pd^0-catalyzed [2 + 2 + 2] cross-cyclotrimerization of two [5]helicenyl arynes (generated *in situ*) with an electron-deficient alkyne (dialkyl acetylenedicarboxylate) in the preparation of triple [5]helicenes such as **28** in moderate yield, which comprises three [5]helicene units (Scheme 2.6) [132]. This reaction with remarkable cross-selectivity can be performed enantioselectively when the chiral ligand (-)-(*S*)-QUINAP **27** is used. It is worth noting that the most stable heterochiral (*M*,*P*,*M*) diastereomer is preferably formed and that the absolute configuration of this product is controlled by the dynamic kinetic resolution of the chiral intermediary five-membered palladacycle (according to DFT calculations).

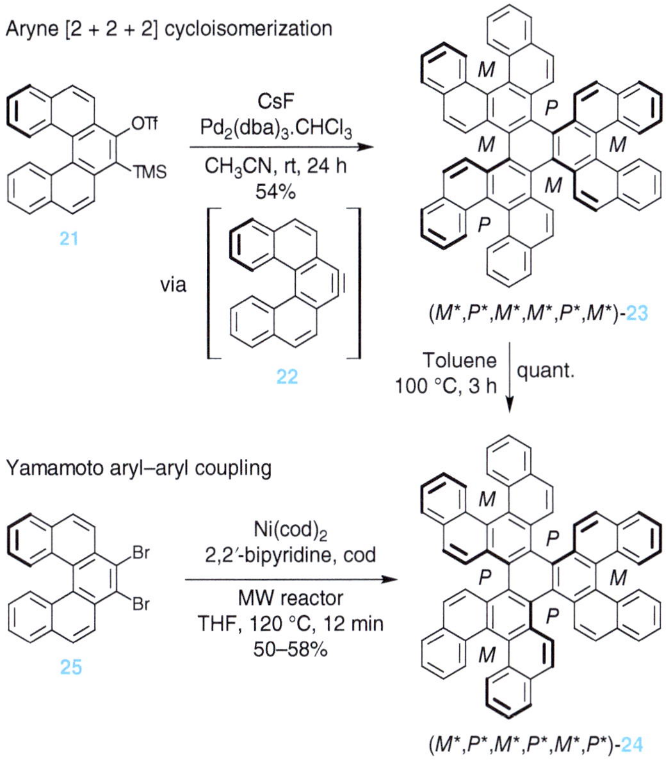

Scheme 2.5 Synthesis of sextuple [5]helicene **24** as its most thermodynamically stable (*M*,P*,M*,P*,M*,P**) diastereomer of D_3 symmetry [126, 127].

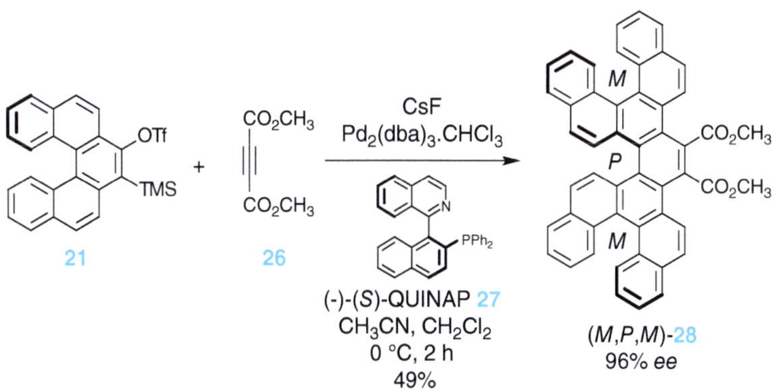

Scheme 2.6 Enantioselective synthesis of triple (*M,P,M*)-[5]helicene **28** via Pd^0-catalyzed [2 + 2 + 2] cross-cyclotrimerization [132].

2.3 Carbo- and Heterohelicenes

2.3.1 Carbohelicenes

The transition-metal-catalyzed intramolecular [2 + 2 + 2] cycloisomerization of aromatic triynes is definitely a suitable method for the construction of a variety of all-angularly condensed fully aromatic carbohelicenes. Actually, there are several complemental synthetic scenarios available with respect to the structure of triynes, as developed by Stará, Starý, and coworkers. Starting from appropriate energy-rich aromatic *cis,cis*-dienetriynes, atom economic isomerization leads directly to [5]-, [6]-, or [7]helicenes (**29** → **30**, Scheme 2.7a) [108]. The cyclization, catalyzed preferably by Ni^0 complexes, is very fast at room temperature (the reaction rate is rather limited only by mixing the reagents) and usually provides good yields but rarely quantitative. The reason for this and the main disadvantage of this method is the limited stability of fully unsaturated *cis,cis*-dienetriynes, which are prone to uncontrolled polymerization during synthesis or cyclization. It is worth noting that, unlike traditional photocyclodehydrogenation of 1,2-diaryl olefins in helicene synthesis, where the necessary *cis* configuration of olefins is continuously generated by UV irradiation, *cis,cis*-dienetriynes must be synthesized stereoselectively (no *trans*-to-*cis* olefin isomerization occurs during the non-photochemical cyclization).

The use of triynes containing saturated ethane-1,2-diyl tethers instead of unsaturated *cis*-ethene-1,2-diyl ones usually reduces these disadvantages discussed previously. Here, after smooth [2 + 2 + 2] cycloisomerization catalyzed by Ni^0 or Co^I complexes (**31** → **32**, Scheme 2.7b), the initially formed tetrahydrohelicenes can be oxidized with a tritylium salt (which is more effective than the commonly used oxidizing agent DDQ) to fully aromatic carbohelicenes (**32** → **33**, Scheme 2.7b) [133]. There is another way to aromatize tetrahydrohelicene intermediates when the ethane-1,2-diyl tether bears an oxygen substituent (AcO– or MOMO–). It can be removed by acid-assisted elimination, leading also to fully aromatic carbohelicenes (**34** → **35** → **36**, Scheme 2.7c) [134]. This aromatization method was advantageously used in the asymmetric synthesis of helicenes by Stará, Starý, and coworkers (for details, see Section 2.5.1, Scheme 2.21) [135] or functionalized helicenes by Diederich and coworkers [89, 90].

Finally, the incorporation of *ortho*-phenylene linkers into the aromatic triynes substantially simplifies their synthesis and solves completely the instability of the aforementioned fully unsaturated *cis,cis*-dienetriynes. Thus, a wide range of dibenzohelicenes can be prepared in a straightforward way by combining powerful synthetic tools such as a Pd^0/Cu^I-catalyzed Sonogashira coupling, Pd^0-catalyzed Suzuki–Miyaura coupling, and Ni^0- or Co^I-catalyzed [2 + 2 + 2] cycloisomerization as reported by Stará, Starý, and coworkers (**37** → **38**, Scheme 2.7d) [118]. This methodology was also employed successfully by Marinetti and coworkers [109]. It provides a general and non-photochemical access to helicenes (here dibenzohelicenes) with a significant potential to mimic or even substitute parent helicenes in their applications.

The synthetic methodology based on the transition-metal-catalyzed intramolecular [2 + 2 + 2] cycloisomerization of alkynes for the preparation of carbohelicenes

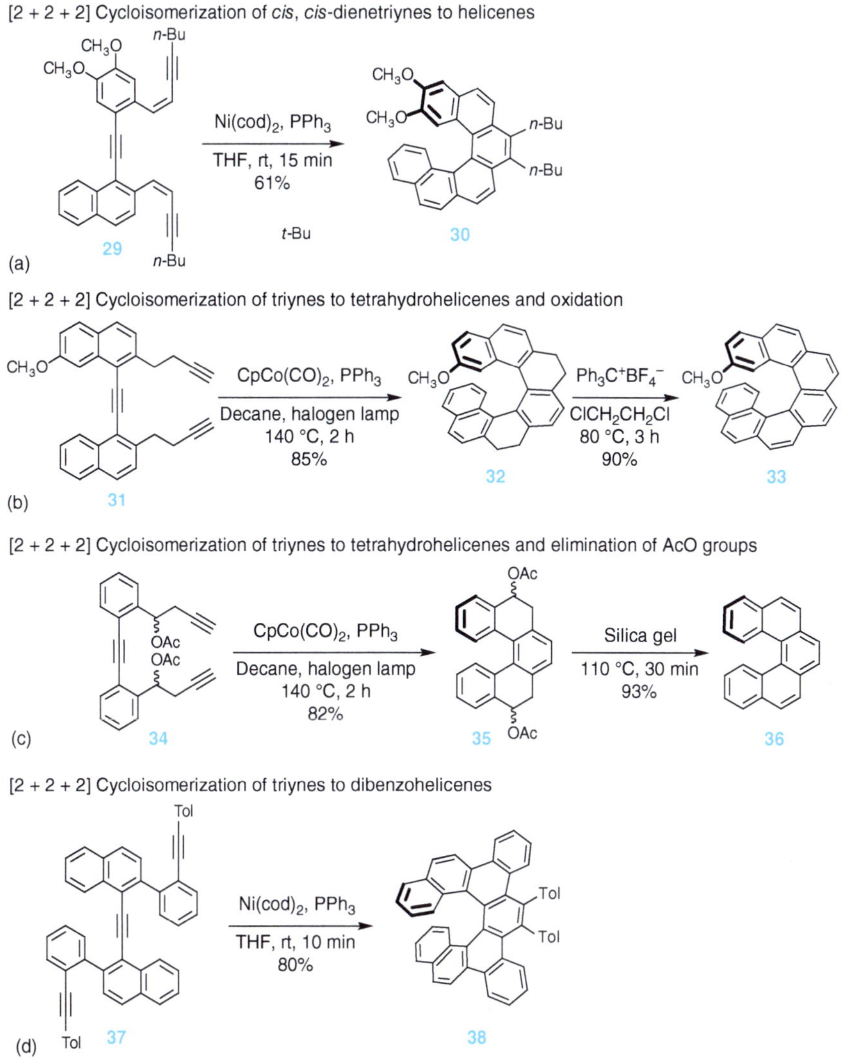

Scheme 2.7 Alternative synthetic approaches to helicenes based on Ni^0- or Co^I-catalyzed intramolecular [2 + 2 + 2] cycloisomerization of triynes to give fully aromatic carbohelicenes (a) [108], (b) [133], (c) [134] and (d) [118].

exhibits a high degree of modularity, as it can be adapted to the intermolecular version described by Tanaka and coworkers (Scheme 2.4) [120]. In this way, parent and functionalized (hetero)helicenes, either racemic or nonracemic, can be prepared, as reported by several authors (for details, see other sections of this chapter).

2.3.2 Azahelicenes

The introduction of the pyridine subunit(s) into the starting heteroaromatic triyne leads to the formation of the corresponding azahelicenes (optionally named

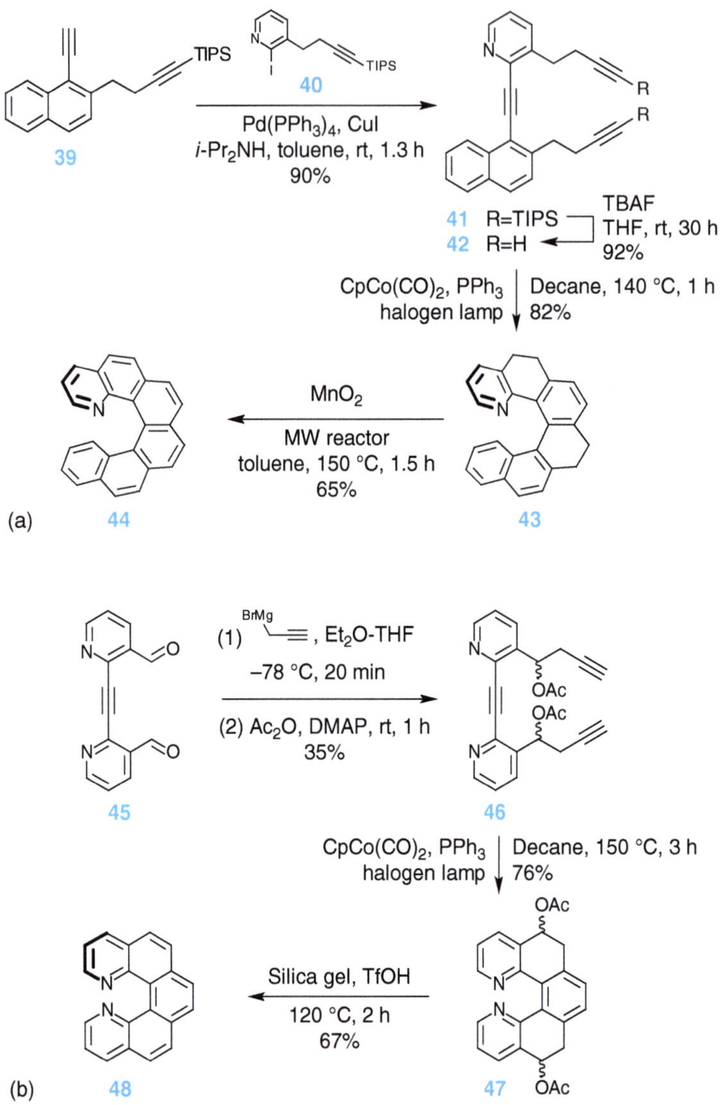

Scheme 2.8 Synthesis of racemic pyridohelicenes by Co^{I}-catalyzed [2 + 2 + 2] cycloisomerization of triynes finalized by aromatization via (a) oxidation [136] or (b) elimination of oxygen functional groups [134].

pyridohelicenes), since the presence of the heterocyclic subunit(s) does not prevent the cyclization by Ni^{0} or Co^{I} complexes as described by Stará, Starý, and coworkers (**44**, Scheme 2.8a) [136]. However, the final oxidation of the partially hydrogenated pyridohelicene intermediates (**43**) can be difficult. An alternative may be aromatization by eliminating the installed oxygen functional group with the aid of acids (**47** → **48**, Scheme 2.8b) [134]. The synthesis of dibenzopyrido[*n*]helicenes ($n = 5$, 6, 7) by the transition-metal-catalyzed [2 + 2 + 2] cycloisomerization of heteroaromatic triynes solves the problem with the final aromatization of pyridohelicenes.

These helically chiral heteroaromatics can be prepared usually within four to five steps in total yields ranging from 35 to 53% using a short sequence of reliable processes such as Sonogashira coupling (**49** + **50** → **51**), Suzuki–Miyaura coupling (**51** + **10** → **52**), desilylation (**52** → **53**), and [2+2+2] triyne cycloisomerization (**53** → **54**) according to the work of Stará, Starý, and coworkers (Scheme 2.9) [118]. Dibenzopyridohelicenes have an advantage over the parent pyridohelicenes in the simplicity of their non-photochemical preparation and, therefore, may have the potential to mimic or even replace them in intended applications.

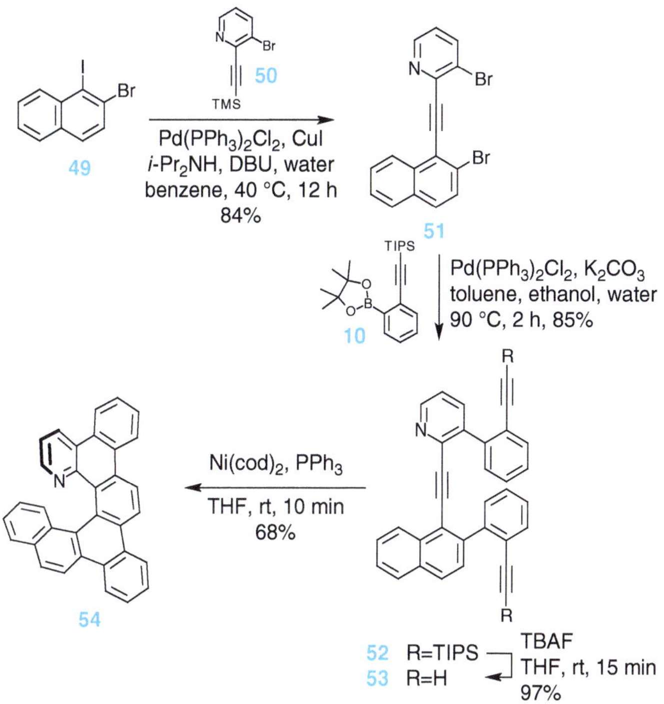

Scheme 2.9 Synthesis of racemic dibenzopyridohelicenes by Ni^0-catalyzed [2 + 2 + 2] cycloisomerization of heteroaromatic triynes [118].

Although it is well known that co-cycloisomerization of two alkynes and one nitrile provides pyridine derivatives, this synthetic methodology was surprisingly never used in the synthesis of pyridohelicenes until Stará, Starý, and coworkers developed a direct approach to pyridohelicenes and their analogs using Ni^0-, Co^I-, or Rh^I-mediated intramolecular [2+2+2] cycloisomerization of cyanodiynes (**55** → **56**, Scheme 2.10a) [137]. This allowed the folding of linear precursors into the corresponding helical backbones containing the newly formed pyridine unit in their central part. Along with racemic pyrido[n]helicenes (n = 5, 6, 7) and their derivatives, enantio- and diastereomerically pure oxapyrido[n]helicenes (n = 5, 6) were prepared by chiral substrate-controlled cyclization of the corresponding enantiopure cyanodiynes ((–)-(R)-**57** → (–)-(M,R)-**58**, Scheme 2.10b).

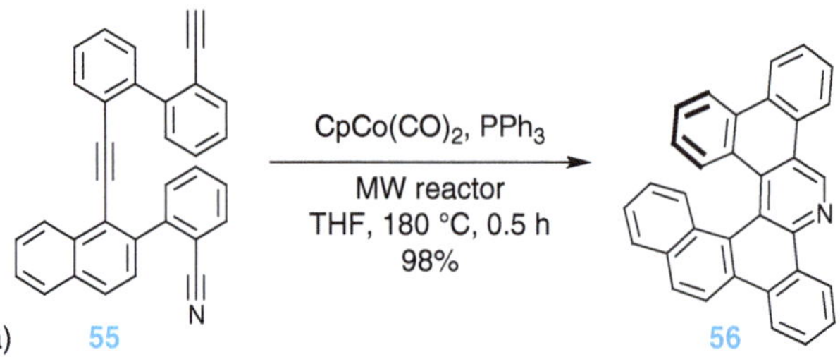

(a) 55 56

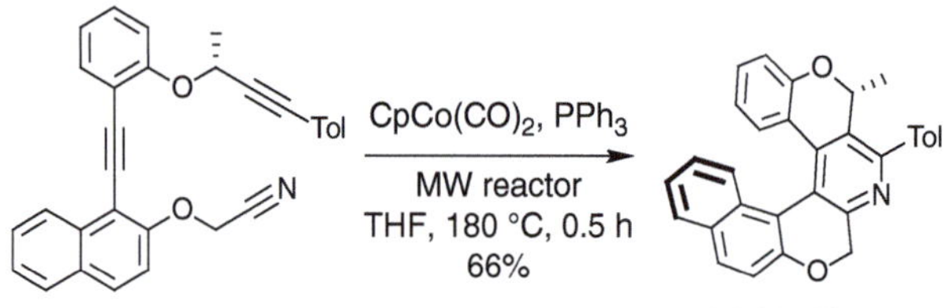

(b) (−)-(*R*)-57 (−)-(*M*,*R*)-58
dr = >99 : <1

Scheme 2.10 Synthesis of racemic (a) and optically pure (b) pyridohelicenes by Co^{I}-catalyzed [2 + 2 + 2] cycloisomerization of cyanodiynes [137].

Until independent reports by Stará, Starý, and coworkers [95] and Snyder and coworkers [138], the complementary [2+2+2] co-cycloisomerization of one alkyne and two nitrile units to obtain pyridazine derivatives was not known. Stará, Starý, and coworkers developed a cobalt-mediated/catalyzed [2+2+2] cycloisomerization of ynedinitriles to pyridazine helicenes in good to high yields (**59** → **60**, Scheme 2.11). As indicated by circumstantial evidence, a single electron transfer-triggered radical cyclization of ynedinitrile mediated by the $CpCo^{II}L_n$ species is proposed as a reaction mechanism. They applied this synthetic methodology to the preparation of a series of helical pyridazines including [5]-, [6]-, and [7]helicene derivatives. The discovery of this new reaction is still awaiting recognition by the chemical community as a useful tool for the preparation of complex pyridazines.

Stará, Starý, and coworkers reported the asymmetric synthesis of the C_2 symmetric enantio- and diastereopure oxa[5]- and oxa[6]helicene 2,2′-bipyridines [112]. Their versatile and straightforward synthesis employs either the double intramolecular [2+2+2] cycloisomerization of a centrally chiral dicyanotetrayne/triyne to form simultaneously the 2,2′-bipyridine unit as well as azaoxahelicene backbone ((−)-(*R*,*R*),(*R*,*R*)-**64** → (−)-(*M*,*R*,*R*),(*M*,*R*,*R*)-**65**) or a one-pot/sequential cascade of the intramolecular [2+2+2] cycloisomerization of a centrally chiral triyne followed by an intermolecular haloazaoxahelicene reductive Yamamoto-type homocoupling ((−)-(*R*,*R*)-**63** → (−)-(*M*,*R*,*R*),(*M*,*R*,*R*)-**65**), Scheme 2.12). Notably, they reached an effective central-to-helical-to-axial chirality transfer.

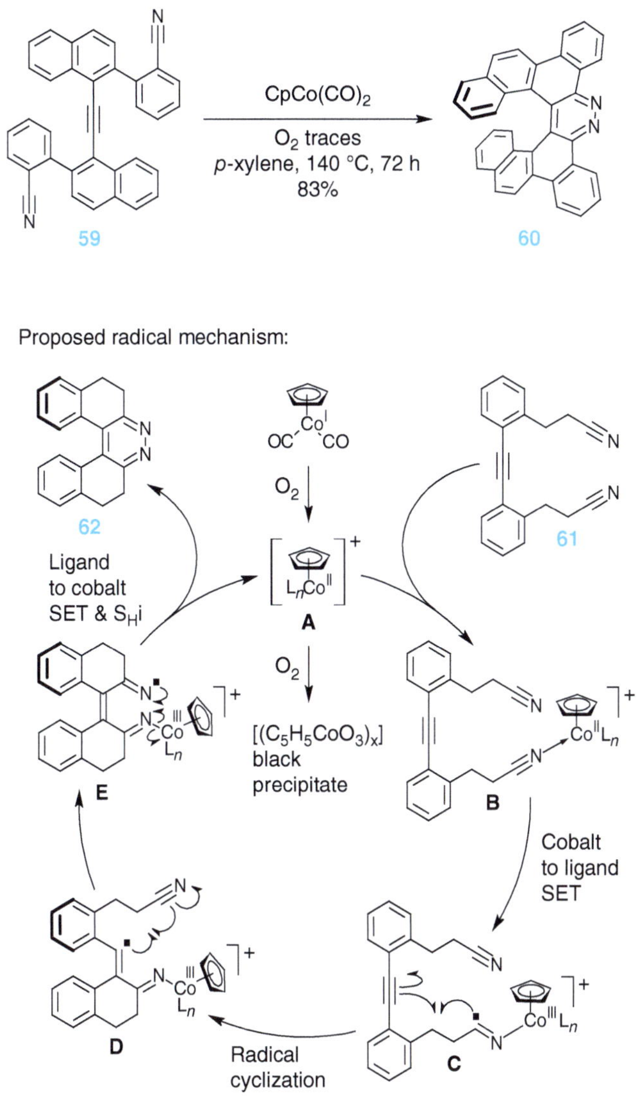

Scheme 2.11 Synthesis of pyridazinohelicenes by new Co^{I}-mediated [2 + 2 + 2] cycloisomerization of ynedinitriles and the proposed reaction mechanism [95].

There are not many examples where pyridohelicenes are formed from the corresponding triynes whose aliphatic tether(s) contain a nitrogen atom. Stará, Starý, and coworkers demonstrated this methodology by preparing a dihydropyridine analog of [5]helicene **67** by Co^{I}-mediated [2+2+2] cycloisomerization of triyne in moderate yield (Scheme 2.13) [139]. The same principle was later used by Carbery and coworkers who prepared racemic or optically pure azaoxa[6]helicenes by Rh^{I}-catalyzed [2+2+2] cycloisomerization of triynes in good yield (for other details, see Section 2.5.1) [88, 113].

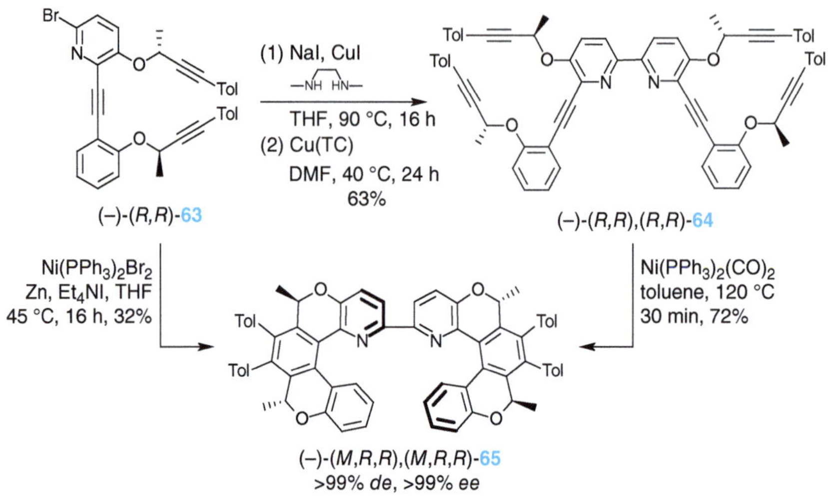

Scheme 2.12 Asymmetric synthesis of diastereo- and enantiopure oxahelicene 2,2′-bipyridines by Ni^0-mediated [2 + 2 + 2] cycloisomerization of heterotriyne/hexayne [112].

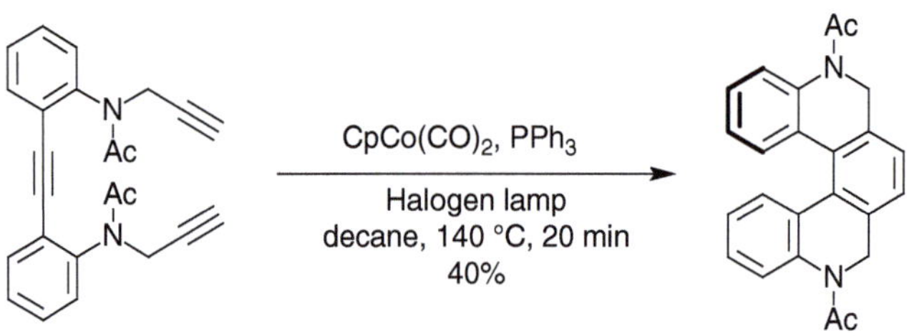

66 67

Scheme 2.13 Synthesis of dihydropyridine [5]helicene analog by Co^I-mediated [2 + 2 + 2] cycloisomerization of triyne [139].

2.3.3 Azoniahelicenes

Azoniahelicenes, i.e. cationic azahelicenes with a tetravalent positively charged nitrogen atom placed in a helical scaffold, constitute a specific class of heterohelicenes, which were traditionally synthesized by photocyclodehydrogenation of 1,2-aryl-heteroaryl olefins containing a quinolizinium subunit [10]. However, Teplý and coworkers changed the paradigm of their preparation by developing a specific class of azahelicene dications (diazoniahelicenes) and coining their name helquats, whose synthesis is based on transition-metal-catalyzed [2 + 2 + 2] cycloisomerization of pyridinium-type triynes [77, 140–143]. Helquats encompass structural features of both helicenes and viologens (e.g. paraquat), and their synthesis is very straightforward. It capitalizes on the facile quaternization of a symmetric or non-symmetric diazaarylacetylene precursor, such as **68**, to form the dicationic triyne precursor, such as **69**, which smoothly undergoes [2 + 2 + 2] cycloisomerization in the presence of Wilkinson's catalyst (preferably) or Cp*Ru(cod)Cl to form a helical backbone of, e.g. [5]helquat **70** (Scheme 2.14) [77].

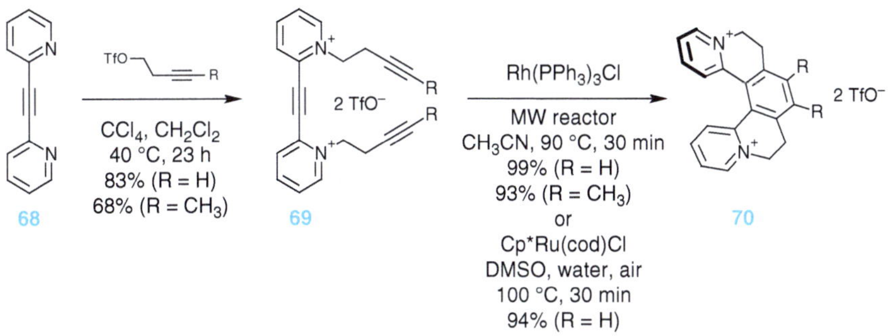

Scheme 2.14 Synthesis of helquats (diazoniahelicenes) by [2 + 2 + 2] cycloisomerization of pyridinium-type triynes under Rh^{I}- or Ru^{II}-catalysis [77].

In this way, various helquats can be synthesized in a racemic form, for example, [6]helquat **72** [77], benzo[6]helquat **74** [142], or [7]helquat **76** [77] (Scheme 2.15), some of them in a few steps on a multigram scale. Since two consecutive distinct pyridine-type nitrogen quaternizations are possible, different nonsymmetric [5]-, [6]-, and [7]helquats can be prepared from diazaarylacetylene precursors by rhodium-catalyzed [2+2+2] cycloisomerization of the corresponding dicationic triynes [140]. This pathway allows direct molecular editing of cationic helical scaffolds to vary the size of the inserted partially saturated cationic heterocycles. In fact, helquats were uniformly synthesized as partially hydrogenated cationic heteroaromatics, and no example of their conversion to fully aromatic systems exists to date. The methodology was also used to synthesize a tricationic helicene-like system with an imidazolium core unit, such as tricationic [9]helquat **78**, which relied on the double Rh^{I}-catalyzed [2+2+2] cycloisomerization of tricationic hexayne **77** [144]. It is the longest nitrogen-based cationic helicene-like system to date, containing nine contiguously *ortho*-annulated rings.

Although asymmetric synthesis of helquats was not yet reported, their racemates can be resolved into enantiomers on a preparative scale by crystallization of the corresponding diastereomeric dibenzoyltartrate salts [145–147]. Moreover, using preferential crystallization of slightly enantioenriched helquats, which requires highly enantioenriched seeds of both enantiomers, they can be resolved into individual enantiomers on a multigram scale by repeating the two-step cycle several times: adding alternately one and the other enantiomer as seeds before each crystallization while supplementing the system with the racemate to the original amount after each crystallization [147]. Capillary electrophoresis employing randomly sulfated α-, β-, and γ-cyclodextrins as chiral selectors allows fast and baseline resolution of a wide range of racemic helquats on an analytical scale [145, 148]. Using this separation technique, the complex pathway of racemization of a nonracemic and nonsymmetric [6]helquat derivative via an isolable saddle-shaped intermediate [6]saddlequat was followed [141].

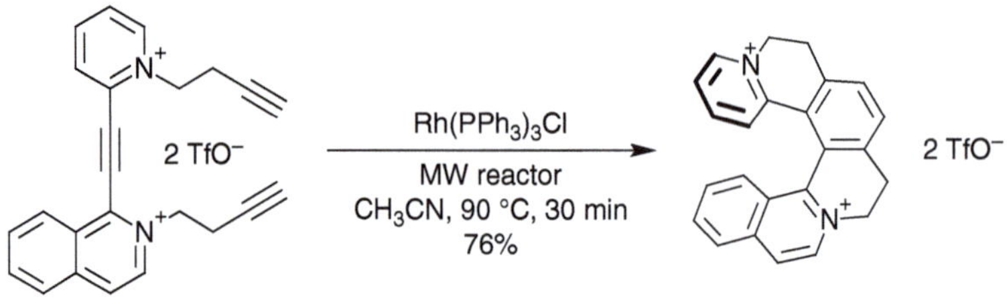

71 72

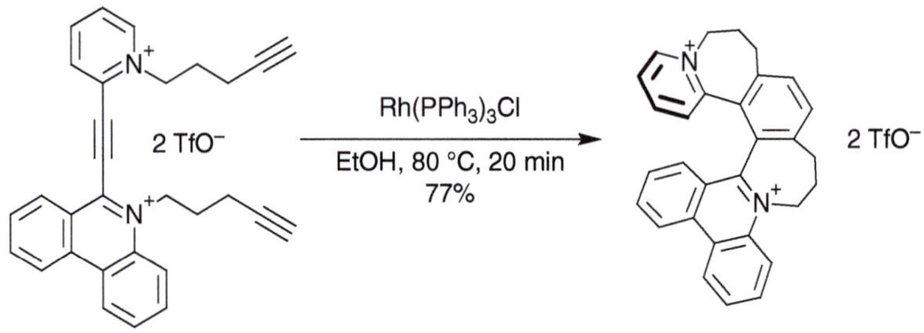

73 74

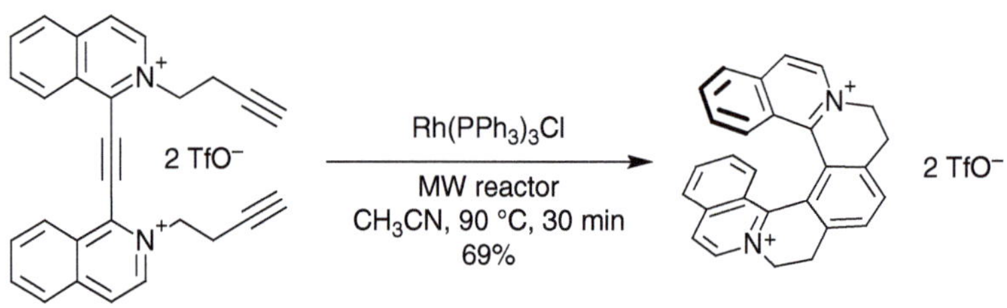

75 76

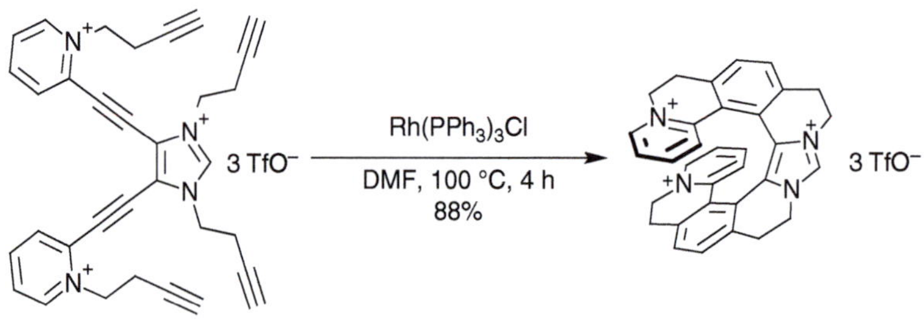

77 78

Scheme 2.15 Synthesis of various helquats by [2 + 2 + 2] cycloisomerization of pyridinium-type triynes under Rh^{I}-catalysis such as [6]helquat **72** [77], benzo[6]helquat **74** [142], [7]helquat **76** [77] and tricationic [9]helquat **78** [144].

It is worth noting that a general one-step conversion of helquats with the active methyl group(s) to cationic heterohelicene styryl-type dyes by Knoevenagel condensation with various aromatic aldehydes is possible [143] and that enantioselective synthesis of azoniahelicenes by alternative Rh^{I}-catalyzed C–H annulation with alkynes was recently developed [48].

2.3.4 Oxahelicenes

The robustness and high versatility of transition-metal-catalyzed [2 + 2 + 2] cycloisomerization of alkynes to give heteroanalogs of the parent carbohelicenes can be demonstrated by the synthesis of a variety of oxahelicenes and oxahelicene-like compounds. The insertion of the oxygen atom(s) into the helical scaffold does not usually hamper the transition-metal catalysis or chemically destabilize the resulting oxahelicenes. In addition, the presence of an ether moiety in the alkyne precursor of oxahelicene may lead to its simplified synthesis. This can be documented by the preparation of the longest axially extended helicene to date, which is oxa[19]helicene **159** (Scheme 2.27b; for details and another example of complex oxahelicene architecture, see Section 2.6).

Oxahelicenes synthesized in this manner typically contain six-membered 2*H*-pyran (eventually 2*H*-pyran-2-one) cycle(s) or seven-membered 2,7-dihydrooxepine cycle(s), which are *ortho*-fused to adjacent benzene rings. The corresponding alkyne [2 + 2 + 2] cycloisomerization, which may be intra- or intermolecular, is usually catalyzed by Co^{I}, Ni^{0}, or Rh^{I} complexes to afford oxahelicenes in generally good yield. Stará, Starý, and coworkers pioneered the synthesis of oxahelicenes by this methodology, when they reported the preparation of the oxa[6]helicene derivative **80** in good yield using Co^{I} catalysis (Scheme 2.16a) [84]. Importantly, the preparation of enantio- and diastereopure oxahelicenes by [2 + 2 + 2] cycloisomerization of centrally chiral triynes was extensively investigated by Stará, Starý, and coworkers in order to develop their practical asymmetric synthesis (cf. Scheme 2.23; for details, see Section 2.5.1) [97].

Similarly, Tanaka and coworkers paid systematic attention to the enantioselective [2 + 2 + 2] cycloisomerization of achiral alkynes (catalyzed by cationic Rh^{I} complexes in the presence of chiral diphosphine ligands) to obtain structurally diverse nonracemic oxahelicenes mostly in moderate to good yield and up to 99% enantiomeric excess [86, 114, 152–154] (see also Chapter 3 "*Enantioselective Synthesis of Helicenes*" by K. Tanaka). They published an efficient synthesis of oxa[7]helicene derivative (−)-**83** [86] (Scheme 2.16b) and oxa[9]helicene derivative (−)-**86** (Scheme 2.16c) [120], both in good yield and in a promising enantiomeric excess. While [2 + 2 + 2] cycloisomerization was intramolecular in the first case, the second double cyclization was intermolecular combining tetrayne **84** with diyne **85**.

Other examples of oxahelicenes synthesized by [2 + 2 + 2] cycloisomerization of alkynes should be briefly mentioned. The presence of oxygen in the helicene scaffold can be combined also with other heteroatoms as demonstrated by Voituriez, Marinetti, and coworkers, who synthesized 1*H*-phosphole-extended oxaphospha[6]helicene **98** in good yield (Scheme 2.18b; for details, see Section 2.3.5) [91]

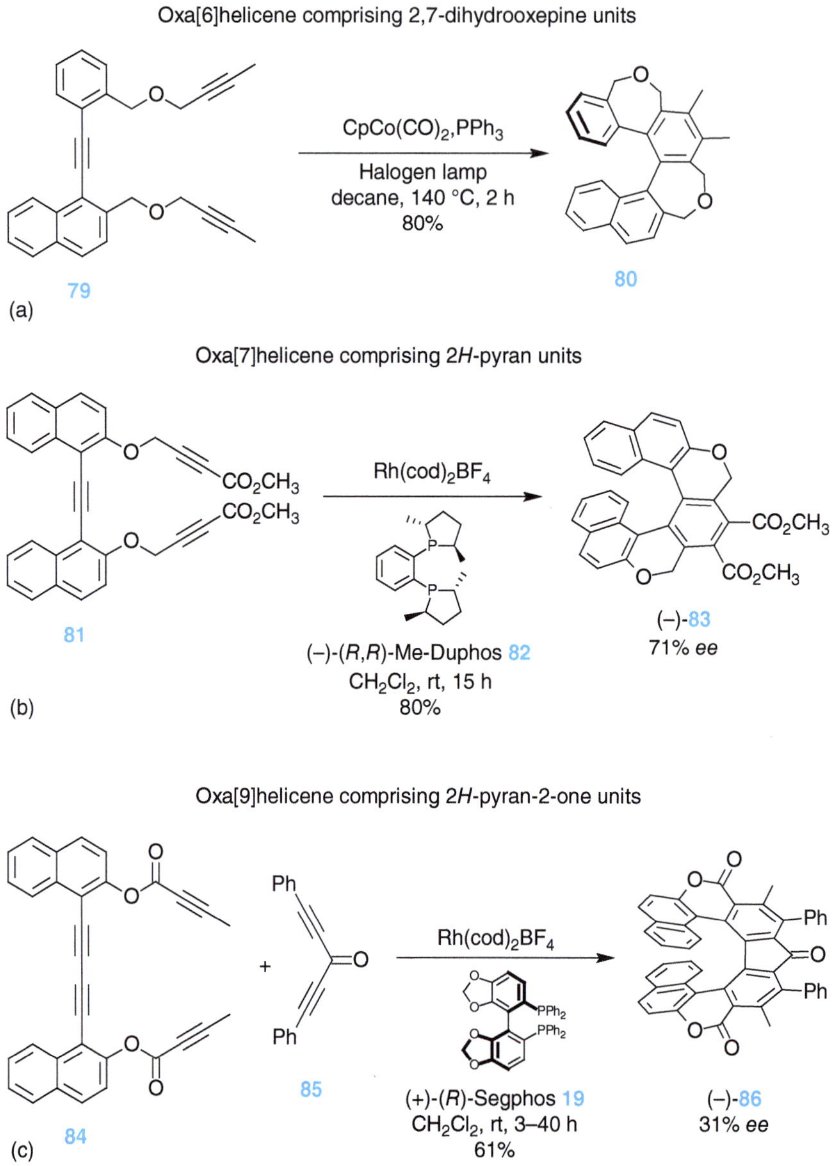

Scheme 2.16 Synthesis of various oxahelicenes, such as **80** [84] (a), (−)-**83** [86] (b), or (−)-**86** [120] (c), by Co^{I}- or Rh^{I}-catalyzed intramolecular [2 + 2 + 2] cycloisomerization of triynes or by intermolecular double [2 + 2 + 2] co-cycloisomerization of tetrayne and diyne.

and by Tanaka and coworkers, who prepared oxaphospha[7]- and [9]helicenes in moderate yield and low-to-good enantiomeric excess, placing a 1*H*-phosphole unit in the center of a helical scaffold [123]. There are also enantio- and diastereopure azaoxahelicenes of various structures reported by Stará, Starý, and coworkers [112] or Carbery and coworkers [113] (cf. Scheme 2.22; for details, see Section 2.5.1).

2.3.5 Sila-, Phospha-, and Thiahelicenes

The versatility of transition-metal-catalyzed [2+2+2] cycloisomerization of alkynes in the synthesis of heterohelicenes can be documented by the successful preparation of a series of sila-, phospha-, and thiahelicenes, which may carry additional functionalities. In general, the presence of Si, P, or S atom(s) in the starting alkyne or embedded in the formed helicene scaffold should be compatible with the Co^{I}, Ni^{0}, Rh^{I}, or Ir^{I} catalysts most commonly used to fold the precursor into the helix. The family of these already published heterohelicenes can be expected to grow in the near future as the presence of specific heteroatoms can significantly affect their physicochemical properties and lead to new applications.

Shibata and coworkers reported the first example of the synthesis of sila-helicenes, which contain two silole subunits in their helical scaffold, using transition-metal-catalyzed [2+2+2] co-cycloisomerization of tetraynes with diynes (Scheme 2.17a) [87]. In fact, dual catalysis was used in two separate steps: first, under Ir^{I}-catalysis and in the presence of the chiral ligand (−)-(*S*,*S*)-EtFerroTANE **89**, disilatetrayne **87** reacted intermolecularly with diyne **88** to form the axially chiral intermediate (−)-**90** in good yield and high enantiomeric excess (the second co-cycloisomerization was blocked under these conditions). Then, in the presence a stoichiometric amount of the Ni^{0} complex and achiral triphenylphosphine, highly diastereoselective intramolecular [2+2+2] cycloisomerization took place to deliver the final sila[7]helicene (+)-**91** in high yield and diastereomeric excess. Tanaka and coworkers succeeded in completing the enantioselective double Rh^{I}-catalyzed intramolecular [2+2+2] co-cycloisomerization of tetrayne **92** with siladiyne **93** in the presence of the chiral ligand (−)-(*S*)-Segphos **19** (Scheme 2.17b) [115]. The sila[7]helicene derivative (+)-**94** was received in high enantiomeric excess, but preparative yield was low.

Two complementary methodologies for the synthesis of phosphahelicenes by transition-metal-catalyzed [2+2+2] cycloisomerization of alkynes were developed by Tanaka and coworkers [121, 123, 155] and Marinetti, Voituriez, Guinchard, and coworkers [91, 92, 109]. While in the former case the methodology relies on enantioselective intermolecular double Rh^{I}-catalyzed [2+2+2] co-cycloisomerization of tetraynes such as **92** with dialkynyl phosphinates such as **95** in the presence of chiral ligand (−)-(*S*)-Segphos **19** to provide phospha[7]helicenes such as (+)-**96** in moderate yield and good enantiomeric excess (Scheme 2.18a) [121], the latter methodology is based on intramolecular Ni^{0}-catalyzed [2+2+2] cycloisomerization of triynes with the incorporated 1*H*-phosphindolone unit such as **97** to afford oxaphospha[6]helicenes such as **98** in good yield and low diastereomeric excess (Scheme 2.18b) [91].

The synthesis of thiahelicenes by transition-metal-catalyzed [2+2+2] cycloisomerization of alkynes with incorporated sulfur subunits was unknown until the first two examples by Stará, Starý, and coworkers (the first note in the literature [57]) [173] and by Guinchard, Voituriez, and coworkers [92] were recently reported. Dithia[5]helicene **100** was prepared from thiophenotriyne **99**, when in one operation two synthetic steps took place: Co^{I}-mediated cyclization followed by

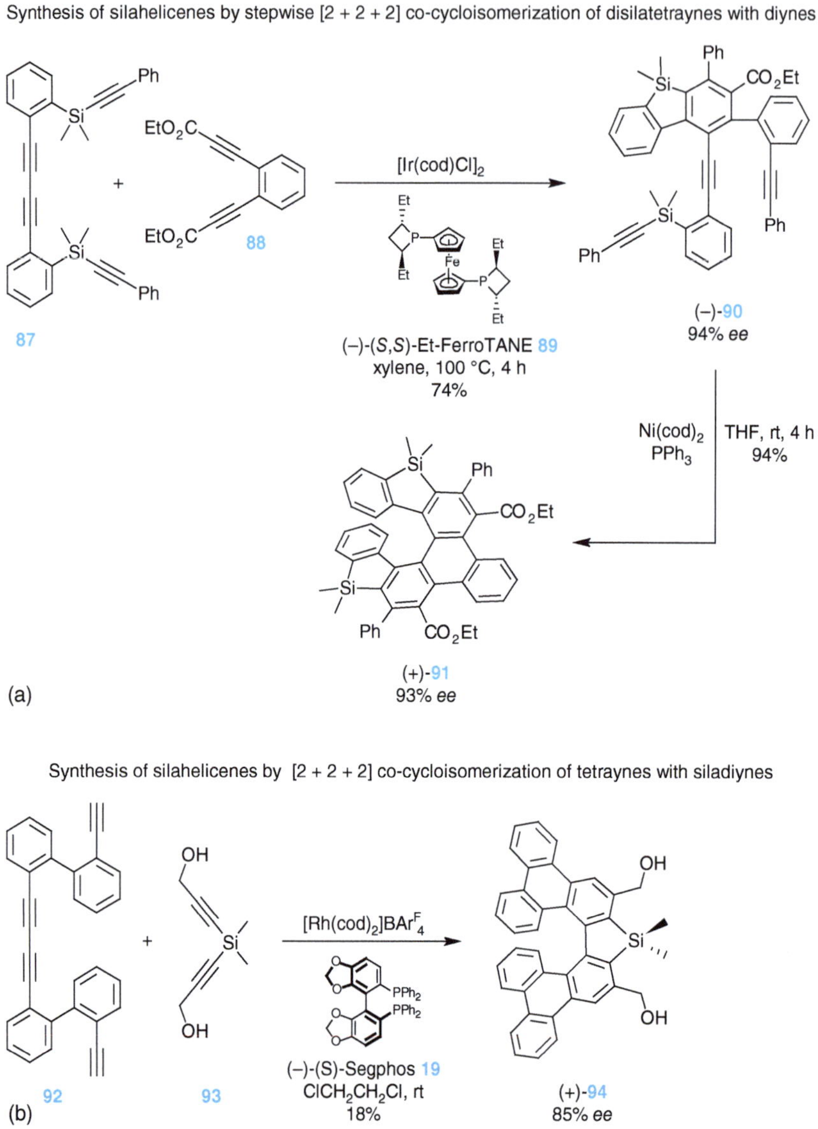

Scheme 2.17 Synthesis of nonracemic silahelicenes (+)-**91** and (+)-**94** by enantioselective stepwise/double [2 + 2 + 2] co-cycloisomerization of disilatetrayne **87** with diyne **88** (a) [87] or tetrayne **92** with siladiyne **93** (b) [115] under Ir^I/Ni^0- or Rh^I-catalysis.

spontaneous acetic acid elimination (Scheme 2.19a) [173]. The synthesis of the phosphathia[7]helicene derivative **102** from the enantiopure 1*H*-phospholothiophenotriyne (+)-(R_P)-**101** employing Ni^0 catalysis led to roughly equal mixture of two diastereomers, which were conveniently separated by column chromatography (Scheme 2.19b) [92]. The last example documents the advantages of the transition-metal-catalyzed [2+2+2] cycloisomerization approach to helicenes over the photocyclization method, as it provides products in high yield and allows easy scale-up to gram quantities.

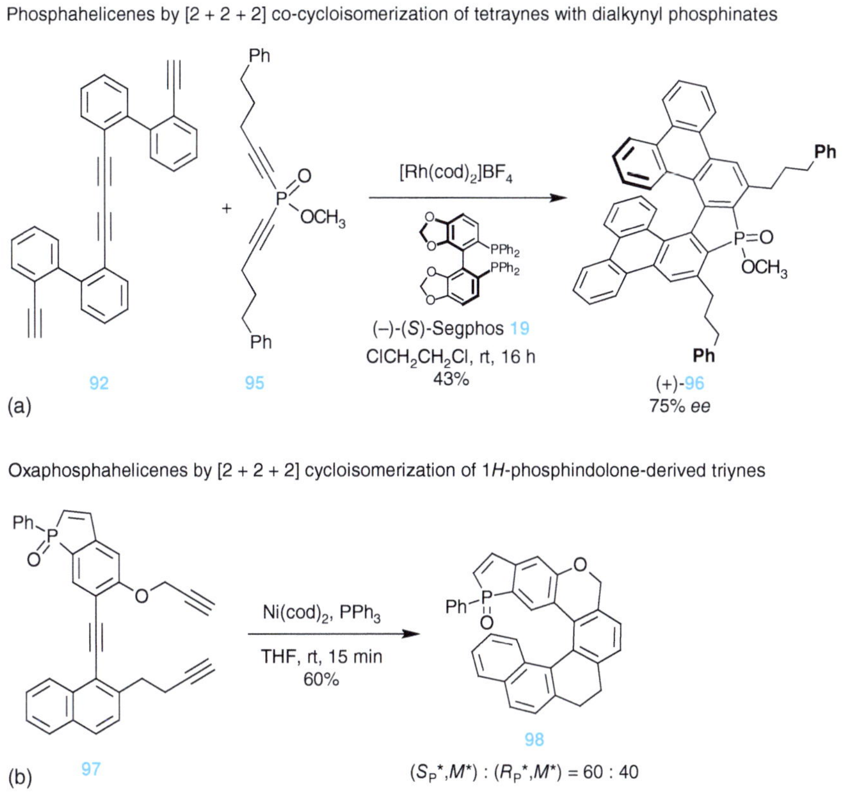

Scheme 2.18 Synthesis of phosphahelicenes by intermolecular/intramolecular [2 + 2 + 2] cycloisomerization of alkynes under Rh^{I}- (a) [121] or Ni^{0}-catalysis (b) [91].

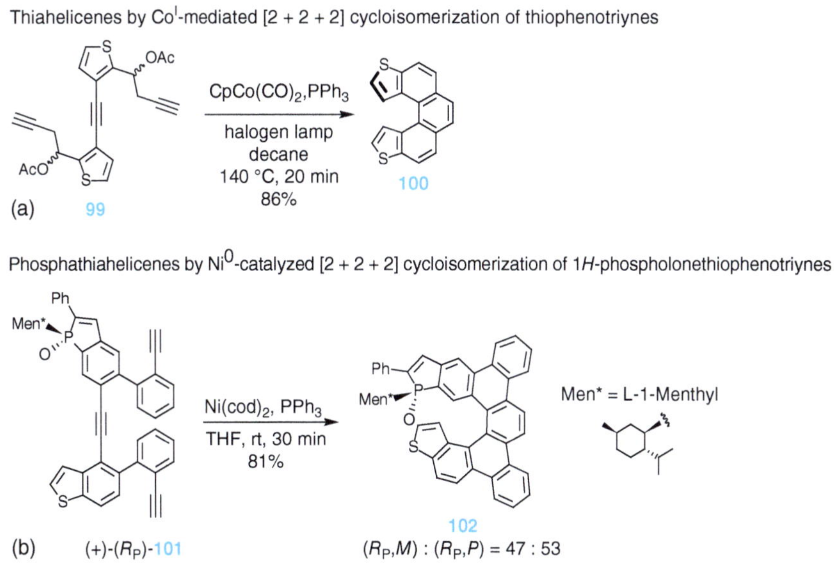

Scheme 2.19 Synthesis of thiahelicenes by intramolecular [2 + 2 + 2] cycloisomerization of alkynes under Co^{I}- (a) [57] or Ni^{0}-catalysis (b) [92].

2.4 Functionalized Helicenes

Although there are intriguing exceptions reported by Stará, Starý, and coworkers [156] or Lacour and coworkers (see also chapter 4 "*Cationic Azahelicenes*" by J. Lacour and coworkers) [157–160], post-synthetic functionalization of bare helicene scaffolds [32] usually suffers from poor regioselectivity control. Therefore, it is advantageous to start the synthesis of functionalized helicenes with already correctly substituted building blocks, provided that all synthetic steps are regioselective and tolerant to the respective functional groups. Actually, transition-metal-catalyzed [2 + 2 + 2] cycloisomerization of π-electron systems almost ideally meets these requirements, and diverse functional groups are tolerated, and no scrambling of their position occurs during the cyclization, namely, CH_3O–, AcO–, MOMO–, CH_3O_2C–, TMS–, R_2N– (R = Boc or Bn), O_2N–, Cl–, F–, and occasionally Br– or HO– (Scheme 2.20). Since there is sufficient room to tune the cyclization reaction conditions (screening Co^I, Ni^0, or Rh^I catalysts, various ligands, conventional/microwave/flow reactor heating), it is very likely that some of them will tolerate the functional groups present. Furthermore, as shown in other sections of this chapter, this synthetic methodology is also highly tolerant to various subunits incorporated into the helicene scaffold, such as pyridine, pyridazine, bipyridine, 1-alkylpyridin-1-ium, thiophene, 2*H*-pyran, 2,7-dihydrooxepine, 1-alkyl- or 1-acetyl-1,2-dihydropyridine, 1-*p*-toluenesulfonyl-2,7-dihydro-1*H*-azepine, 1-aryl- and 1-alkoxy-1*H*-1λ^5-phosphol-1-one, 1,1-dialkyl-1*H*-silole, cyclobutane, and cyclopentadienone (fluorenone).

2.5 Asymmetric Synthesis of Nonracemic Helicenes

2.5.1 Stoichiometric Diastereoselective Synthesis

Resolution of racemic helicenes by HPLC on commercially available chiral columns is the simplest way to obtain enantiopure helicenes on a (semi)preparative scale (from a few to hundreds of milligrams), provided they are sufficiently soluble in the respective eluents [27]. On the other hand, asymmetric synthesis represents a challenging alternative to this method and may be more practical in specific situations if larger amounts of enantiopure helicenes are required (from hundreds of milligrams to grams). The synthetic methodology based on [2 + 2 + 2] cycloisomerization of π-electron systems allows stereochemical control in the preparation of helicenes either through diastereoselective synthesis starting from commercially or easily available chiral building blocks or through enantioselective catalysis using chiral catalysts or ligands (for details, see Section 2.5.2 and Chapter 3 "Enantioselective Synthesis of Helicenes" by K. Tanaka).

Stará, Starý, and coworkers developed a general synthetic methodology for the preparation of uniformly enantiopure fully aromatic [5]-, [6]-, and [7]helicenes by diastereoselective synthesis (Scheme 2.21) [110, 135]. Here, Co^I-catalyzed [2 + 2 + 2] cycloisomerization of centrally chiral triynes is coupled with post-cyclization

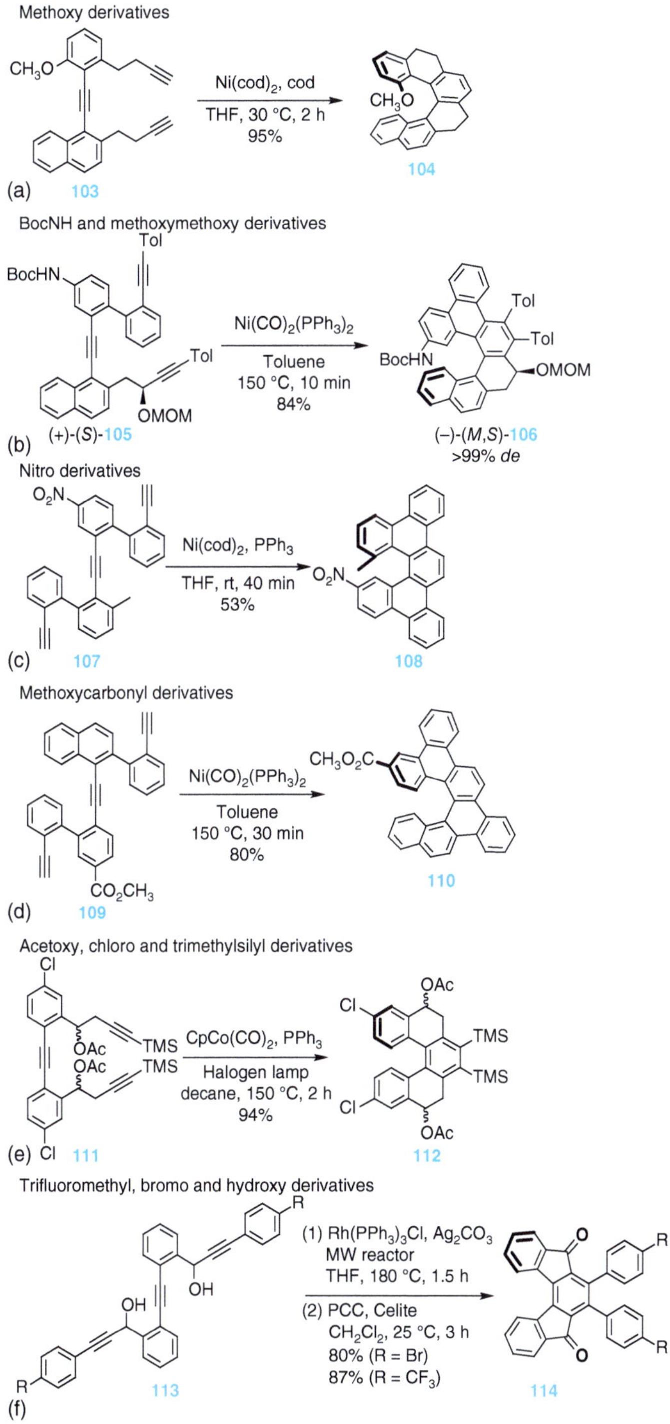

Scheme 2.20 Tolerance of transition-metal-catalyzed [2 + 2 + 2] cycloisomerization of π-electron systems to various functional groups in the synthesis of helicenes: (a) CH_3O– [93]; (b) BocNH– and MOMO– [110]; (c) O_2N– [118]; (d) CH_3O_2C– [161]; (e) AcO–, Cl–, and TMS– [134]; and (f) CF_3–, Br– and HO– [94].

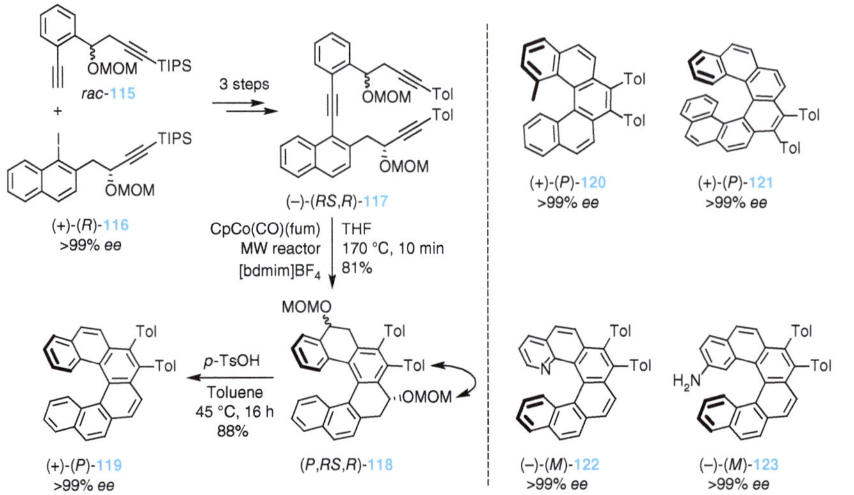

Scheme 2.21 Stoichiometric asymmetric synthesis of enantiopure fully aromatic helicenes (+)-(*P*)-**119–121**, (−)-(*M*)-**122** [135] and (−)-(*M*)-**123** [110] by diastereoselective Co^{I}-catalyzed [2 + 2 + 2] cycloisomerization of nonracemic chiral triynes.

thermodynamic equilibration of diastereomeric tetrahydrohelicene derivatives, the relative stability of which is ultimately controlled by a 1,3-allylic-type strain ((−)-(*RS*,*R*)-**117** → (*P*,*RS*,*R*)-**118**). That is, the molecule is forced to adopt a conformation, here helicity, which minimizes steric repulsion (indicated by an arrow) between the *p*-tolyl group on the newly formed central benzene ring and the adjacent methoxymethoxy substituent on the stereogenic center at the benzylic position in (*P*,*RS*,*R*)-**118**. Such an effect can lead to a significant difference in free energy between the corresponding diastereomers of opposite helicity, usually around 5 kcal mol^{-1}, which is more than enough to achieve a >99% diastereomeric excess [135]. Nevertheless, the cyclization itself may show some level of diastereoselectivity (under kinetic control), but may not determine the stereochemical outcome of the whole reaction as it is controlled by the subsequent asymmetric transformation of the first order. The stereogenic center in triyne (−)-(*RS*,*R*)-**117**, which bears the methoxymethoxy group in the propargylic position, serves as a traceless chiral auxiliary because it is removed in the final aromatization step ((*P*,*RS*,*R*)-**118** → (+)-(*P*)-**119**). Such an efficient point-to-helical chirality transfer exhibits remarkable independence from the structure of triyne (the stereogenic center in the homopropargylic position in triyne (−)-(*RS*,*R*)-**117** is stereochemically innocent), allowing the preparation of a variety of helicenes in enantiomer ratios uniformly of >99 : <1. Thus, the highest control of helicity is guaranteed, and the sense of the newly formed helix can be predicted: the *R* configuration of the stereogenic center in the propargylic position in triyne leads to *P* helicene and vice versa, its *S* configuration to *M* helicene. In addition, the stereochemical outcome of the reaction can also be easily predicted by a routine calculation at the DFT level of theory. Importantly, racemic diyne *rac*-**115** can be used, *vide supra*, and the key enantiopure alkyne (+)-(*R*)-**116** along with its congeners are available on a multigram scale in both enantiomeric forms using routine enantioselective biocatalysis (lipase-catalyzed resolution of racemic secondary alcohols optionally associated with their interconversion by Mitsunobu reaction) [135]. This synthetic methodology is suitable for the preparation of various enantiopure helicenes such as (+)-(*P*)-**120** and (+)-(*P*)-**121**, including functionalized derivatives such as pyrido[6]helicene and amino[6]helicene derivatives (−)-(*M*)-**122** and (−)-(*M*)-**123**, respectively.

Even simpler is stoichiometric asymmetric synthesis of enantio- and diastereopure oxa[5]-, oxa[6]-, and oxa[7]helicenes (and their more complex congeners) with the embedded 2*H*-pyran ring(s) developed by Stará, Starý, and coworkers [97], which represent easily accessible replacements for nonracemic parent helicenes (Scheme 2.22a). Also in this case, the stereochemical outcome of Co^{I}-catalyzed [2+2+2] cycloisomerization of centrally chiral triynes is governed by the post-cyclization thermodynamic equilibration, which is controlled by 1,3-allylic-type strain as evidenced, e.g. by the ultimately diastereoselective transformation of the centrally chiral triyne (−)-(*R*,*R*)-**126** to the oxa[7]helicene derivative (−)-(*M*,*R*,*R*)-**127** in good yield [97]. The effect of strain in this case is so strong that the difference in free energy between the corresponding diastereomers of opposite helicity is about 9 kcal mol^{-1} (regardless of the length of the oxahelicene backbone), provided that two stereogenic centers are present in the molecule. This means that already oxa[5]helicene derivatives, such as (−)-(*M*,*R*,*R*)-**128** [97], exist as a single diastereomer and enantiomer in the entire temperature range of their chemical stability, in contrast to the configurationally unstable parent [5]helicene **36** (Scheme 2.7c) (unless it is conformationally locked, such as (+)-(*P*)-**120** [135], Scheme 2.21). Actually, one stereogenic center is sufficient to see the respective oxahelicene derivative, for example, (−)-(*M*,*R*),(*M*,*R*)-**133** [112] as a single diastereomer and enantiomer regardless of temperature (Scheme 2.22b). Helicity can be easily predicted: if the stereogenic center(s) in the propargylic position of triyne has the *R* configuration, the resulting oxahelicene accepts *M* helicity, and, conversely, the *S* configuration leads to the *P* helicity [97]. This general rule can be easily supported computationally in particular cases. The synthesis of chiral triynes is straightforward and benefits from the commercial availability of both *R* and *S* enantiomers of but-3-yn-2-ol, a close precursor of, for example, (−)-(*S*)-**125**. Various oxahelicenes can be prepared using this synthetic methodology (e.g. (−)-(*M*,*R*,*R*)-**129**), including functionalized derivatives such as (−)-(*M*,*R*)-**128** [97] and (−)-(*M*,*R*)-**131** [149] or complex helical architectures such as (−)-(*M*,*R*)-**132** [99], (−)-(*M*,*R*),(*M*,*R*)-**133** [112] and (−)-(*M*,*R*,*R*)-**134** [117]. Multigram-scale synthesis of enantio- and diastereopure oxahelicenes is possible. The usefulness of this methodology was underlined by the scalable and resolution-free highly diastereoselective synthesis of the nonracemic azaoxa[6]helicene DMAP organocatalyst precursor (+)-(*P*,*S*)-**130** (after chromatography, the pure major diastereomer was isolated), reported by Carbery and coworkers [113].

A different concept of diastereoselective [2+2+2] cycloisomerization of triynes was innovatively applied to the Ni^{0}-catalyzed cyclization of the chiral P-menthyl substituted phosphindolones delivering oxaphospha[7]helicene analogs, such as

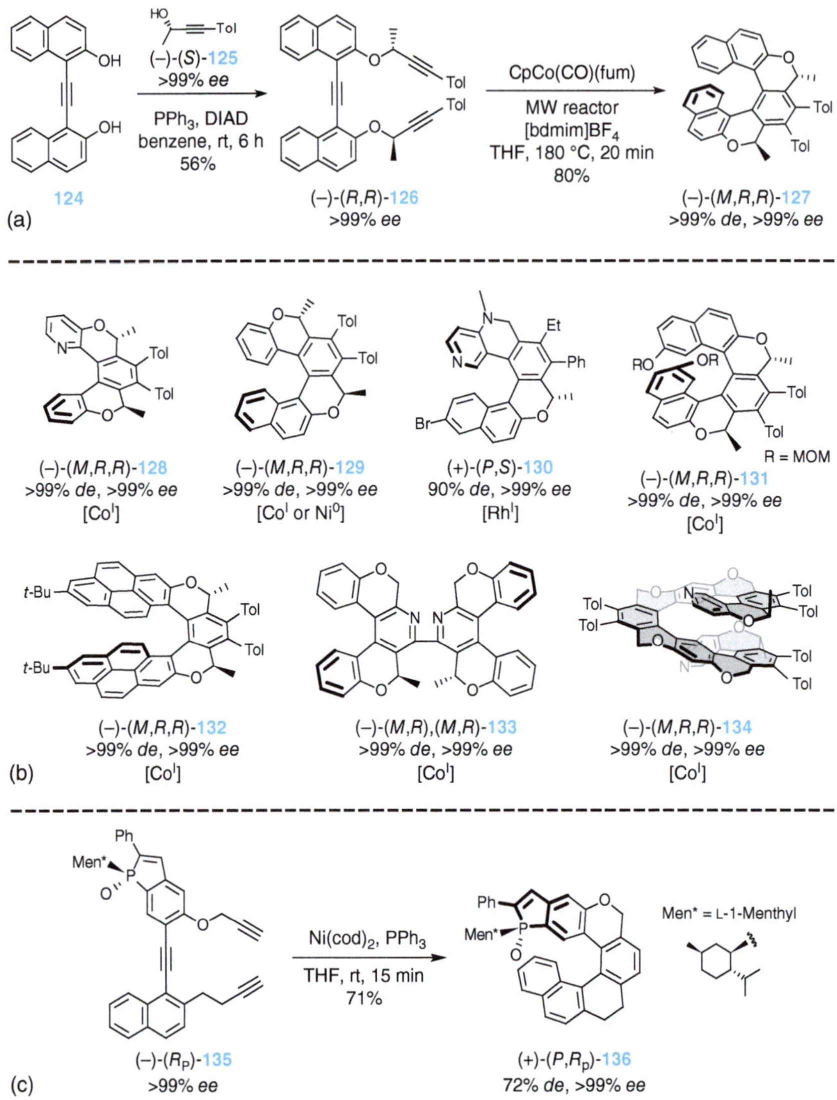

Scheme 2.22 Stoichiometric asymmetric synthesis of highly enantio- and diastereoenriched 2*H*-pyran oxahelicene derivatives (−)-(*M*,*R*,*R*)-**127**–**129** [97], (+)-(*P*,*S*)-**130** [113], (−)-(*M*,*R*,*R*)-**131** [149], (−)-(*M*,*R*,*R*)-**132** [99], (−)-(*M*,*R*),(*M*,*R*)-**133** [112], (−)-(*M*,*R*,*R*)-**134** [117] and (+)-(*P*,R_P)-**136** [91] by diastereoselective transition-metal-catalyzed [2 + 2 + 2] cycloisomerization of nonracemic chiral triynes or multiple alkynes (used metals in square brackets).

(+)-(*P*,R_P)-**136** [91], with promising *dr*'s depending on the phosphorus-centered chirality as reported by Voituriez, Marinetti, and coworkers (Scheme 2.22c).

Stará, Starý, and coworkers showed that oxahelicenes with an embedded seven-membered 2,7-dihydrooxepine ring(s) can also be prepared enantio- and diastereoenriched using the principles of the aforementioned diastereoselective synthesis of helicenes [72, 150, 151]. Although Co^{I}-catalyzed [2 + 2 + 2] cycloisomerization of a centrally chiral triyne was reported to provide, for example, the enantio- and diastereopure oxa[7]helicene derivative (+)-(*P*,*S*)-**138** in good yield (Scheme 2.23), stereochemical control of cyclization is usually less effective for such nonracemic 2,7-dihydrooxepine oxahelicenes than in the case of the nonracemic 2*H*-pyran oxahelicenes discussed earlier.

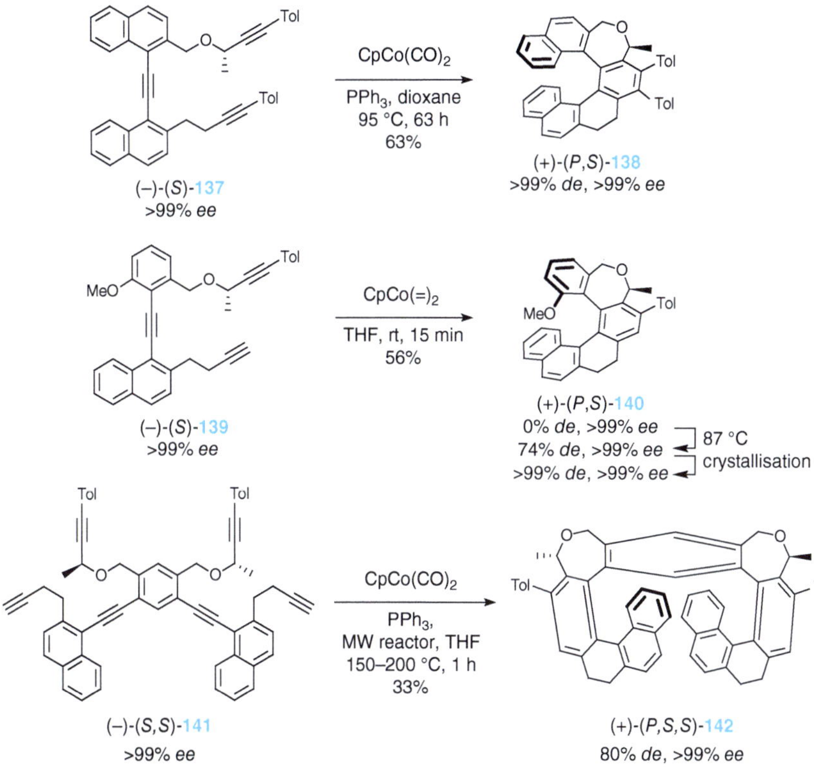

Scheme 2.23 Stoichiometric asymmetric synthesis of enantio- and diastereoenriched 2,7-dihydrooxepine oxahelicene derivatives (+)-(*P*,*S*)-**138** [72], (+)-(*P*,*S*)-**140** [150] and (+)-(*P*,*S*,*S*)-**142** [151] by diastereoselective Co^{I}-mediated [2 + 2 + 2] cycloisomerization of nonracemic chiral triynes or multiple alkynes.

In this respect, the case of the enantiopure triyne (−)-(*S*)-**139** and its controlled transformation to the functionalized oxa[6]helicene is illustrative: the use of the highly reactive Jonas catalyst [100–102] at room temperature resulted in the formation of both diastereomers (*M*,*S*)-**140** and (*P*,*S*)-**140** equally populated [150]. However, subsequent thermal equilibration led to a decent preference for (+)-(*P*,*S*)-**140**. This proves that the stereochemical outcome of the cyclization is, in fact, controlled by the post-cyclization equilibration. It is worth noting that the enantio- and diastereopure (+)-(*P*,*S*)-**140** can be obtained by crystallization of the diastereomerically enriched mixture. Using this synthetic methodology, even a more complex oxa[11]helicene derivative (+)-(*P*,*S*,*S*)-**142** can be prepared by double

[2+2+2] cycloisomerization of a centrally chiral hexayne (−)-(*S*,*S*)-**141** with good diastereoselectivity [151].

It is worth noting that a case was described where a strong 1,3-allylic-type strain between two carbon substituents in the molecule is not necessary to achieve complete diastereoselectivity in the Co^{I}-mediated [2+2+2] cycloisomerization of the enantiopure starting triyne. Stará, Starý, and coworkers reported asymmetric synthesis of an enantio- and diastereopure dihydroazepine-derived [5]helicene analog in moderate yield, in which the nitrogen atom can become part of a newly formed seven-membered ring (Scheme 2.24) [139].

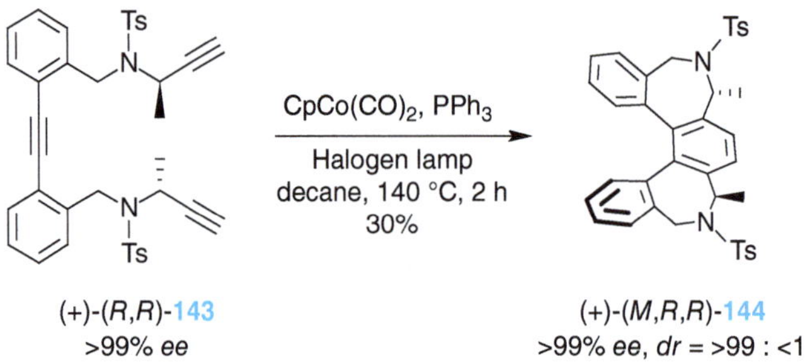

Scheme 2.24 Diastereoselective synthesis of a dihydroazepine-derived aza[5]helicene analog (+)-(*M*,*R*,*R*)-**144** by Co^{I}-mediated [2 + 2 + 2] cycloisomerization of enantiopure triyne (no strong 1,3-allylic-type strain is present in the cyclized molecule) [139].

2.5.2 Catalytic Enantioselective Synthesis

Transition-metal-catalyzed [2+2+2] cycloisomerization of alkynes in the synthesis of helicenes and their derivatives calls for the application of chiral ligands to metals in order to obtain the helical products nonracemic, ideally enantiopure. This idea was first explored by Stará, Starý, and coworkers shortly after they introduced this general methodology for the synthesis of helicenes [84]. Employing Ni^{0} catalysts along with the axially chiral monophosphine MOP [162], they received a tetrahydro[6]helicene derivative in 48% enantiomeric excess [74]. Since then, enantioselective Ni^{0}-catalyzed [2+2+2] cycloisomerization of triynes was further developed to achieve high enantioselectivity on model substrates: the use of Ni^{0} catalysis in combination with the axially chiral P-N ligand (+)-(*R*)-QUINAP led to the dibenzo[6]helicene derivative (+)-(*P*)-**146** in high 85% enantiomeric excess (>99% *ee* after crystallization) [118] and with the novel helically chiral NHC ligand precursor (−)-(*M*,*R*,*R*),(*M*,*R*,*R*)-**148** led to the dibenzo[7]helicene derivative (+)-(*P*)-**149** also in high 86% enantiomeric excess (Scheme 2.25) [111].

Only a few chiral Co^{I} complexes were so far investigated in the enantioselective [2+2+2] cycloisomerization of alkynes to give nonracemic tetrahydrohelicenes as reported by Heller, Stará, and coworkers [106]. In the case of a Co^{I} catalyst, the only ligand that is permanently bound to the cobalt atom during the multistep catalytic cycle is the Cp unit, which must be chiral in order to control helicity of the

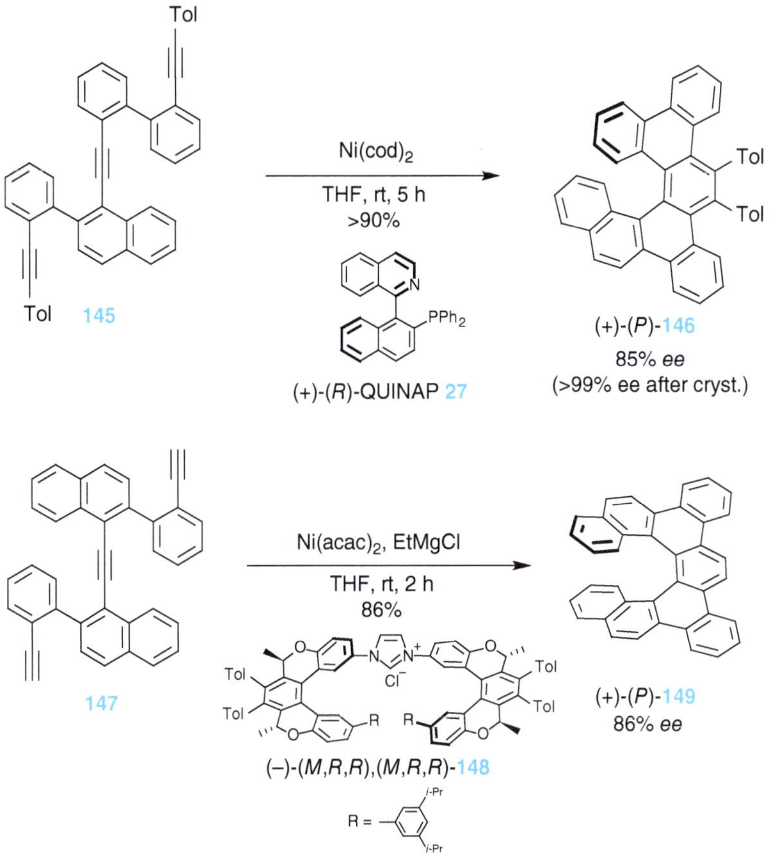

Scheme 2.25 Catalytic asymmetric synthesis of enantioenriched dibenzohelicenes (+)-(*P*)-**146** [118] and (+)-(*P*)-**149** [111] by enantioselective Ni^0-catalyzed [2 + 2 + 2] cycloisomerization of triynes.

product (e.g. (−)-(1*S*,2*S*,5*R*,S_p)-**151**, Scheme 2.26). However, the best enantiomeric excess thus obtained was so far only modest as evidenced by the formation of the tetrahydro[6]helicene derivative (+)-**152** in high yield but with 25% *ee*.

Importantly, the most efficient chirality transfer from the catalyst to the helicene product was observed in enantioselective [2+2+2] cycloisomerization of alkynes catalyzed by cationic Rh^I complexes, which was introduced and systematically explored by Tanaka and coworkers (see also Chapter 3 "Enantioselective Synthesis of Helicenes" by K. Tanaka). Although the preparative yield of intramolecular/intermolecular cyclization is moderate in some cases and a [2+1+2+1] cycloaddition by-product can also be formed [86, 116], this methodology benefits from the commercial availability of a number of chiral biaryl diphosphines and the enantiomeric excess of the helicene product can be well above 90%. To illustrate, there are several such examples: double intermolecular Rh^I-catalyzed [2+2+2] cycloisomerization of tetrayne and diyne resulted in highly enantioenriched helical

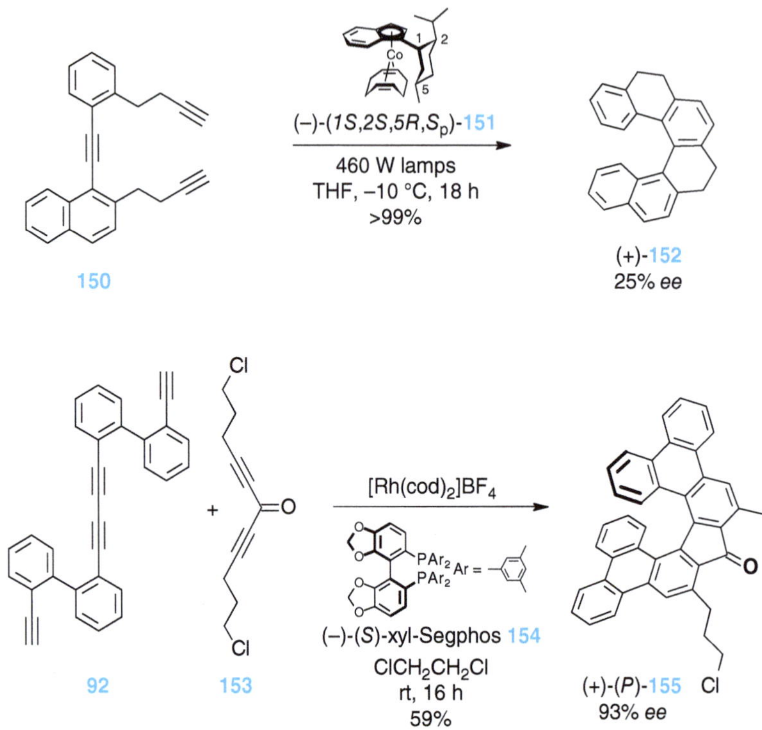

Scheme 2.26 Catalytic asymmetric synthesis of enantioenriched helicene derivatives (+)-**152** [106] and (+)-(*P*)-**155** [121] by enantioselective Co^{I}- or Rh^{I}-catalyzed [2 + 2 + 2] cycloisomerization of alkynes.

1,1′-bitriphenylenes (dibenzo[7]helicene analogs), such as (+)-(*P*)-**155**, in good yield and with 93% *ee* (Scheme 2.26) [121], or optically pure S-shaped double oxahelicene (−)-**169** was synthesized by enantioselective cyclization of hexayne **167** in good yield and with 99% *ee* [154] (for details, see Section 2.6, Scheme 2.29). Interestingly, Tanaka and coworkers described the preparation of the virtually enantiopure dibenzo[7]helicene derivative (+)-(*P*)-**38** by [2 + 2 + 2] cycloisomerization of triyne **37** (cf. Scheme 2.7) catalyzed by $[Rh(cod)_2]OTf$/(+)-(*R*)-BINAP [116]. However, the true enantiomeric excess of 64% was spontaneously increased during preparative TLC chromatography. Finally, the same laboratory demonstrated the enantioselective synthesis of all-angularly condensed [9]- and [11]helicene-like molecules through a double Rh^{I}-catalyzed intramolecular [2 + 2 + 2] cycloisomerization of hexaynes reaching up to 95% *ee*, although in modest yields [114, 152].

There is only one example of Ir^{I}-catalyzed enantioselective [2 + 2 + 2] co-cycloisomerization of alkynes to give a nonracemic helicene derivative, sila[7]helicene (+)-**91**, which was formed stepwise from disilatetrayne **87** and diyne **88** under Ir^{I}/Ni^{0}-catalysis [87] (for details, see Section 2.3.5, Scheme 2.17a). Conceptually, this asymmetric synthesis of (+)-**91** combines the principles of enantioselective (first step) and diastereoselective (second step) synthesis.

2.6 Advanced Helicene Architectures

Alkyne [2+2+2] cycloisomerization, whether performed inter- or intramolecularly, is a robust and versatile synthetic method that allows the preparation of advanced helically chiral molecular architectures. Naturally, extending helicenes axially (lengthen their scaffold) or laterally (widen their scaffold) is a challenging task. Although a number of methods for the synthesis of helicenes have been published (for references, see Section 2.1), only a few of them are suitable for the construction of the most complex helical nanocarbons. Because multiple ring closure is required to form a polycyclic backbone, the individual cyclization steps have to provide the highest possible yields, and the reaction conditions should tolerate polyaromatic/unsaturated starting materials and should optimally be intramolecular. To date, only a limited number of multiple processes have been found to be effective in this regard such as photocyclodehydrogenation of 1,2-diaryl olefins [78, 163] or 1,2-diaryl arenes [64, 82, 164–166], Scholl reaction (oxidative cyclodehydrogenation of 1,2-diaryl arenes) [58–61, 63–71, 79], palladium catalyzed C–H arylation (cyclodehydrohalogenation) [167, 168], photocyclodehydrochlorination [83] and transition-metal-catalyzed [2+2+2] cycloisomerization of oligoalkynes [75, 76, 117, 151].

The strength and efficiency of the last method for the construction of monodisperse axially extended helicenes, which contain more than 11 carbo/heterocycles fused to a helical scaffold, can be documented by two notable examples. Vollhardt and coworkers reported a landmark synthesis of [17]heliphene **157** (all-angularly condensed [17]helicene analog or angular [9]phenylene, Scheme 2.27a) [76]. They employed a triple Co^{I}-mediated [2+2+2] cycloisomerization of the aromatic nonayne **156** to form nine rings in a single operation, but the yield of the key multicyclization was very low (2–3.5%) probably owing to the simultaneous closure of six 4-membered strained rings and the presence of six terminal alkyne units in the starting material prone to polymerization. The limits of this method were further surmounted by Stará, Starý, and coworkers who synthesized the longest monodisperse helicene to date, oxa[19]helicene **159** containing 19 angularly/linearly annulated benzene/2*H*-pyran rings in its helical backbone (Scheme 2.27b) [117]. This record helicene was prepared using quadruple Co^{I}-mediated alkyne [2+2+2] cycloisomerization, which was carried out in a flow reactor. As a result, the starting aromatic dodecayne **158** (with most of the *p*-tolylated pendant alkyne units) was folded into a helical shape to form 12 C–C bonds and 12 rings in one synthetic operation in decent preparative yield (47%). Importantly, this synthetic methodology made it possible to prepare the optically pure oxa[19]helicene derivatives by asymmetric synthesis [117, 149]. By installing two stereogenic centers of known absolute configuration into the respective dodecayne precursor, such as (−)-(*R*,*R*)-**160**, [2+2+2] cycloisomerization proceeded with ultimate diastereoselectivity, which was controlled by a 1,3-allylic-type strain (for principles of diastereoselective synthesis of nonracemic helicenes, see Section 2.5.1). Although the preparative yield may be lower, functionalized derivatives of the enantio- and diastereopure longest helicenes, such as dichloro (−)-(*M*,*R*,*R*)-oxa[19]helicene **161**, could be synthesized in this way (Scheme 2.27c) [149].

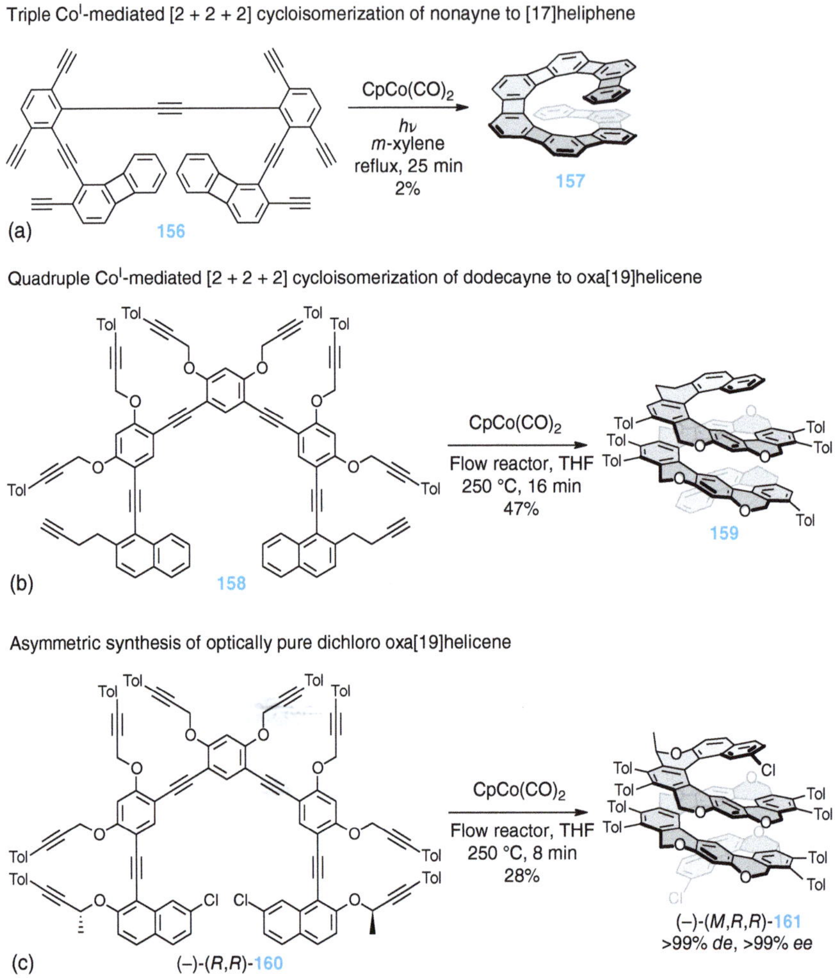

Scheme 2.27 Synthesis of the longest axially extended helicenes: (a) [17]heliphene **157** [76], (b) oxa[19]helicene **159** [117] and (c) enantio- and diastereopure dichloro oxa[19] helicene (−)-(*M*,*R*,*R*)-**161** [149] by multiple Co^{I}-mediated alkyne [2 + 2 + 2] cycloisomerization.

Chiral nanographenes as an emerging class of advanced aromatics [169] attract considerable attention because they represent structurally uniform pieces of graphene with broken mirror symmetry. This class of compounds is structurally diverse and includes also carbon-rich laterally extended helicenes that were synthesized using photocyclodehydrogenation of 1,2-diaryl olefins [62] or 1,2-diaryl arenes [82, 165], Scholl reaction [63, 69, 81] and transition-metal-catalyzed [2 + 2 + 2] cycloisomerization of alkynes [80, 99] to form the helical scaffold (see also Chapter 9 "*Helical Nanographenes: Synthetic and Chiroptical Achievements*" by A. G. Campaña and coworkers). Employing the last method, Stará, Starý, and coworkers described the efficient synthesis of the laterally extended chimerical pyrene [7]helicene derivative **163** in high yield (Scheme 2.28a) [99]. In the presence of the chiral ligand (+)-(*R*)-QUINAP, the intramolecular Ni^0-catalyzed [2+2+2] cycloisomerization of the pyrene-derived triyne **162** provided (+)-(*P*)-**163** in a moderate enantiomeric excess (57% *ee*) but in lower yield (20%). The analogous chimerical pyrene oxa[7]helicene (-)-(*M*,*R*,*R*)-**132** was obtained in an enantio- and diastereopure form using asymmetric synthesis, which capitalized on ultimately diastereoselective [2+2+2] cycloisomerization of centrally chiral

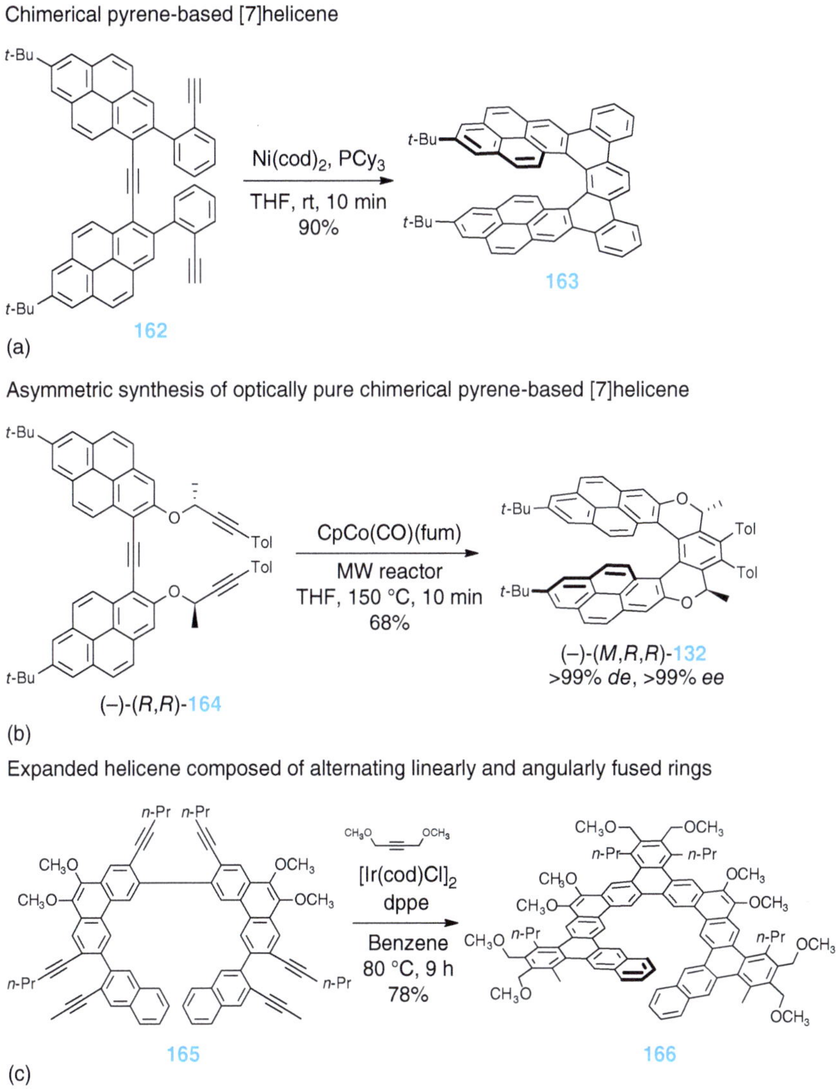

Scheme 2.28 Synthesis of laterally extended helicenes: (a) chimerical pyrene [7]helicene derivative **163** [99], (b) enantio- and diastereopure pyrene oxa[7]helicene (–)-(*M*,*R*,*R*)-**132** [99] and (c) expanded [13]helicene **166** [80] using multiple transition-metal-catalyzed/mediated alkyne [2+2+2] cycloisomerization.

triyne (−)-(*R*,*R*)-**164** (Scheme 2.28b) (for principles of diastereoselective synthesis of nonracemic helicenes, see Section 2.5.1). Tilley and coworkers reported a new class of the "expanded helicenes," which are composed of alternating linearly and angularly fused rings (Scheme 2.28c) [80]. Their synthesis was based on the threefold intermolecular Ir^{I}-catalyzed [2 + 2 + 2] co-cycloisomerization of hexaynes with alkynes, which can be illustrated by the conversion of hexayne **165** to the [13]helicene congener **166** (with a more loose helix than the parent all-*ortho*-fused helicenes) in good yield.

An intriguing oxabenzohelicene architecture was prepared by Tanaka and coworkers (Scheme 2.29) [154]. Using enantioselective double [2 + 2 + 2] cycloisomerization of achiral hexayne **167**, which was catalyzed by the cationic Rh^{I} complex in the presence of chiral diphosphine (+)-(*S*)-Difluorphos **168**, the S-shaped double oxahelicene molecule (−)-**169** was obtained in good yield and practically enantiopure. It consists of two oxa[6]helicene units of the same helicity, sharing a common central benzene ring. The molecule (−)-**169** is resistant to racemization up to 100 °C, because extra-condensed benzene rings increase its configurational stability, and exhibits enhanced chiroptical properties (ECD, CPL) with respect to its half counterpart.

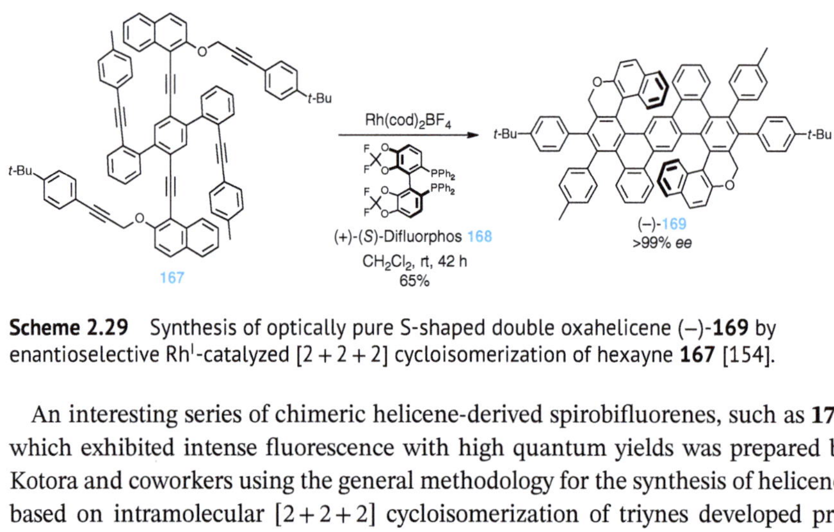

Scheme 2.29 Synthesis of optically pure S-shaped double oxahelicene (−)-**169** by enantioselective Rh^{I}-catalyzed [2 + 2 + 2] cycloisomerization of hexayne **167** [154].

An interesting series of chimeric helicene-derived spirobifluorenes, such as **173**, which exhibited intense fluorescence with high quantum yields was prepared by Kotora and coworkers using the general methodology for the synthesis of helicenes based on intramolecular [2 + 2 + 2] cycloisomerization of triynes developed previously by Stará, Starý, and coworkers, as thoroughly discussed in this chapter (Scheme 2.30) [94].

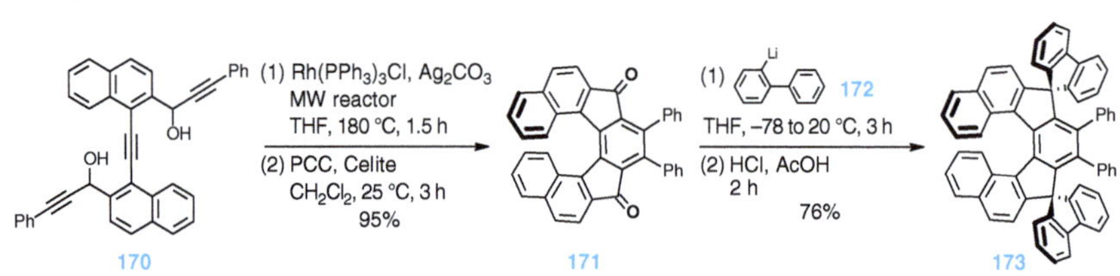

Scheme 2.30 Synthesis of chimeric helicene-derived spirobifluorene **173** based on Rh^{I}-catalyzed [2 + 2 + 2] cycloisomerization of aromatic triyne **170** [94].

2.7 Summary and Outlook

The constant development of a versatile synthetic methodology for the preparation of various helicene derivatives makes them available for the intended applications. The results discussed show that the intramolecular/intermolecular transition-metal-catalyzed [2+2+2] cycloisomerization of π-electron systems has several advantages over other methods for the helicene synthesis: (i) it is universal and allows the preparation of a wide range of helicene derivatives using a uniform methodology from parent helicenes to complex helical architectures; (ii) it is compatible with the presence of various functional groups, heteroatoms, and moieties attached to or incorporated into the helicene backbone; (iii) the reaction conditions of the key cyclization step can be fine-tuned with respect to a number of catalysts and ligands available; (iv) with a few exceptions, it eliminates the problem of regioselectivity of the cyclization that forms the helicene scaffold; (v) unlike other synthetic methods, each intramolecular [2+2+2] cycloisomerization of π-electron systems leads to the formation of three new rings of the helicene scaffold in a single synthetic operation; as multiple cyclization of this kind is feasible, the helicene scaffold can grow rapidly; (vi) it allows scale-up so that helicenes can be prepared in multigram quantities; and (vii) last but not least, enantio- or diastereoselective [2+2+2] cycloisomerization of π-electron systems can deliver highly enantioenriched or enantio/diastereopure helicenes by catalytic or stoichiometric asymmetric synthesis, respectively. It constitutes an alternative to the traditional photocyclodehydrogenation methodology and, in some respects, surpasses it.

The combination of a unique π-electron character with chirality makes helicenes highly attractive for various branches of science. Actually, there is an urgent need for the development of synthetic methodologies that are suitable for the construction of advanced chiral molecular nanocarbons of high complexity. Increasing attention is expected to be paid to the synthesis of tailor-made helical molecular nanocarbons in the context of spintronics (chirality induced spin selectivity) [170], circularly polarized luminescence [171], enantioselective catalysis [22, 24] and chiral single-molecule devices [172].

2.8 Abbreviations

Ac	Acetyl
acac	Acetylacetonate
aq	Aqueous
$BAr^F{}_4^-$ (or BARF)	Tetrakis[3,5-bis(trifluoromethyl)phenyl]borate
$[bdmim]BF_4$	1-Butyl-2,3-dimethyl-1*H*-imidazol-3-ium tetrafluoroborate
BINAP	(1,1′-Binaphthalene-2,2′-diyl)bis(diphenylphosphine)
Bn	Benzyl
Boc	*tert*-Butyloxycarbonyl
Celite®	Diatomaceous earth
cod	*cis*-1,5-Cyclooctadiene

Cp	Cyclopentadienyl
Cp*	Pentamethylcyclopentadienyl
CPL	Circularly polarized luminescence
Cu(TC)	Copper(I) thiophene-2-carboxylate
dba	Bis(dibenzylideneacetone)
DBU	1,8-Diazabicyclo[5.4.0]undec-7-ene
DDQ	Dichloro-5,6-dicyano-*p*-benzoquinone
de	Diastereomeric excess
DIAD	Diisopropyl azodicarboxylate
Difluorphos	1,1′-[2,2,2′,2′-Tetrafluoro[4,4′-bi-1,3-benzodioxole]-5,5′-diyl]bis[1,1-diphenylphosphine]
DMAP	4-Dimethylaminopyridine
DMF	Dimethylformamide
DMSO	Dimethyl sulfoxide
ECD	Electronic circular dichroism
ee	Enantiomeric excess
Et	Ethyl
Et-FerroTANE	1,1′-Bis-2,4-diethylphosphotano)ferrocene
fum	Dimethyl fumarate
HPLC	High performance liquid chromatography
i-Pr	Propan-2-yl (isopropyl)
Jonas catalyst	$CpCo(CH_2{=}CH_2)_2$
L	Ligand
Me-Duphos	1,2-Bis-2,5-dimethylphospholano]benzene
Men*	L-1-Menthyl or (1*R*,2*S*,5*R*)-5-methyl-2-(propan-2-yl)cyclohexyl
MOM	Methoxymethyl
MOP	2-(Diphenylphosphino)-2′-methoxy-1,1′-binaphthyl
MW	Microwave
n-Bu	Butan-1-yl
n-Pr	Propan-1-yl
PCC	Pyridinium chlorochromate
Ph	Phenyl
QUINAP	1-(2-Diphenylphosphino-1-naphthyl)isoquinoline
Segphos	5,5′-Bis(diphenylphosphino)-4,4′-bi-1,3-benzodioxole
SET	Single electron transfer
S_Hi	Substitution radical internal
TBAF	Tetrabutylammonium fluoride
Tf	Trifluoromethanesulfonyl
TfO	Trifluoromethanesulfonyloxy
THF	Tetrahydrofuran
TIPS	Triisopropylsilyl
TMS	Trimethylsilyl
Tol	*p*-Tolyl
Ts	*p*-Toluenesulfonyl

References

1 Wynberg, H. (1971). Some observations on the chemical, photochemical, and spectral properties of thiophenes. *Acc. Chem. Res.* 4 (2): 65–73.
2 Martin, R.H. (1974). The helicenes. *Angew. Chem. Int. Ed.* 13 (10): 649–660.
3 Laarhoven, W.H. and Prinsen, W.J.C. (1984). Carbohelicenes and heterohelicenes. In: *Topics in Current Chemistry: Stereochemistry*, 63–130. Berlin: Springer.
4 Meurer, K.P. and Vögtle, F. (1985). Helical molecules in organic chemistry: organic chemistry. In: *Topics in Current Chemistry*, 1–76. Berlin: Springer.
5 Oremek, G., Seiffert, U., and Janecka, A. (1987). Synthese und Eigenschaften von Helicenen. *Chem. Ztg.* 111 (2): 69–75.
6 Vögtle, F. (1992). *Fascinating Molecules in Organic Chemistry*. Chichester: Wiley.
7 Grimme, S., Harren, J., Sobanski, A., and Vögtle, F. (1998). Structure/chiroptics relationships of planar chiral and helical molecules. *Eur. J. Org. Chem.* 1998 (8): 1491–1509.
8 Hopf, H. (2000). *Classics in Hydrocarbon Chemistry: Syntheses, Concepts, Perspectives*. Weinheim: Wiley-VCH.
9 Katz, T.J. (2000). Syntheses of functionalized and aggregating helical conjugated molecules. *Angew. Chem. Int. Ed.* 39 (11): 1921–1923.
10 Sato, K. and Arai, S. (2002). Heterohelicenes containing nitrogen aromatics: Azahelicenes and azoniahelicenes. In: *Cyclophane Chemistry for the 21st Century* (ed. H. Takemura), 173–197. Trivandrum: Research Signpost.
11 Urbano, A. (2003). Recent developments in the synthesis of helicene-like molecules. *Angew. Chem. Int. Ed.* 42 (34): 3986–3989.
12 Collins, S.K. and Vachon, M.P. (2006). Unlocking the potential of thiaheterohelicenes: chemical synthesis as the key. *Org. Biomol. Chem.* 4 (13): 2518–2524.
13 Stará, I.G. and Starý, I. (2010). Phenanthrenes, helicenes, and other angular acenes. In: *Science of Synthesis, 45b: Category 6, Compounds with All-Carbon Functions* (ed. J.S. Siegel), 855–953. Stuttgart, Chapter 45.21: Thieme.
14 Starý, I. and Stará, I.G. (2009). Helicenes. In: *Strained Hydrocarbons. Beyond the van't Hoff and Le Bel Hypothesis* (ed. H. Dodziuk), 166–176. Weinheim, Chapter 4.3: Wiley-VCH.
15 Dumitrascu, F., Dumitrescu, D.G., and Aron, I. (2010). Azahelicenes and other similar tri- and tetracyclic helical molecules. *Arkivoc* (i): 1–32.
16 Shen, Y. and Chen, C.-F. (2012). Helicenes: synthesis and applications. *Chem. Rev.* 112 (3): 1463–1535.
17 Tanaka, K. (2013). Synthesis of helically chiral aromatic compounds via [2 + 2 + 2] cycloaddition. In: *Transition-Metal-Mediated Aromatic Ring Construction* (ed. K. Tanaka), 281–298. Hoboken, Chapter 10: Wiley.
18 Gingras, M. (2013). One hundred years of helicene chemistry. Part 1: non-stereoselective syntheses of carbohelicenes. *Chem. Soc. Rev.* 42 (3): 968–1006.
19 Gingras, M., Félix, G., and Peresutti, R. (2013). One hundred years of helicene chemistry. Part 2: stereoselective syntheses and chiral separations of carbohelicenes. *Chem. Soc. Rev.* 42 (3): 1007–1050.

20 Gingras, M. (2013). One hundred years of helicene chemistry. Part 3: applications and properties of carbohelicenes. *Chem. Soc. Rev.* 42 (3): 1051–1095.

21 Urbano, A. and Carreño, M.C. (2013). Enantioselective synthesis of helicenequinones and -bisquinones. *Org. Biomol. Chem.* 11 (5): 699–708.

22 Narcis, M.J. and Takenaka, N. (2014). Helical-chiral small molecules in asymmetric catalysis. *Eur. J. Org. Chem.* 2014 (1): 21–34.

23 Bosson, J., Gouin, J., and Lacour, J. (2014). Cationic triangulenes and helicenes: synthesis, chemical stability, optical properties and extended applications of these unusual dyes. *Chem. Soc. Rev.* 43 (8): 2824–2840.

24 Aillard, P., Voituriez, A., and Marinetti, A. (2014). Helicene-like chiral auxiliaries in asymmetric catalysis. *Dalton Trans.* 43 (41): 15263–15278.

25 Hoffmann, N. (2014). Photochemical reactions applied to the synthesis of helicenes and helicene-like compounds. *J. Photochem. Photobiol. C Photochem. Rev.* 19: 1–19.

26 Tanaka, K., Kimura, Y., and Murayama, K. (2015). Enantioselective helicene synthesis by rhodium-catalyzed [2 + 2 + 2] cycloadditions. *Bull. Chem. Soc. Jpn.* 88 (3): 375–385.

27 Starý, I. and Stará, I.G. (2017). The synthesis of azahelicenes. In: *Targets in Heterocyclic Systems. Chemistry and Properties*, vol. 21, 23–53. Italian Society of Chemistry.

28 Chen, C.-F. and Shen, Y. (2017). *Helicene Chemistry*. Berlin: Springer.

29 OuYang, J. and Crassous, J. (2018). Chiral multifunctional molecules based on organometallic helicenes: recent advances. *Coord. Chem. Rev.* 376: 533–547.

30 Li, C., Yang, Y., and Miao, Q. (2018). Recent progress in chemistry of multiple helicenes. *Chem. Asian J.* 13 (8): 884–894.

31 Dhbaibi, K., Favereau, L., and Crassous, J. (2019). Enantioenriched helicenes and helicenoids containing main-group elements (B, Si, N, P). *Chem. Rev.* 119 (14): 8846–8953.

32 Jakubec, M. and Storch, J. (2020). Recent advances in functionalizations of helicene backbone. *J. Org. Chem.* 85 (21): 13415–13428.

33 Newman, M.S. and Lednicer, D. (1956). The synthesis and resolution of hexahelicene. *J. Am. Chem. Soc.* 78 (18): 4765–4770.

34 Stará, I.G., Starý, I., Tichý, M. et al. (1994). Stereochemical dichotomy in the stevens rearrangement of axially twisted dihydroazepinium and dihydrothiepinium salts. A novel enantioselective synthesis of pentahelicene. *J. Am. Chem. Soc.* 116 (12): 5084–5088.

35 Flammang-Barbieux, M., Nasielski, J., and Martin, R.H. (1967). Synthesis of heptahelicene (1) benzo[*c*]phenanthro[4,3-g]phenanthrene. *Tetrahedron Lett.* 8 (8): 743–744.

36 Wynberg, H. and Groen, M.B. (1968). Synthesis, resolution, and optical rotatory dispersion of a hexa- and a heptaheterohelicene. *J. Am. Chem. Soc.* 90 (19): 5339–5341.

37 Mamane, V., Hannen, P., and Fürstner, A. (2004). Synthesis of phenanthrenes and polycyclic heteroarenes by transition-metal catalyzed cycloisomerization reactions. *Chem. A Eur. J.* 10 (18): 4556–4575.

38 Collins, S.K., Grandbois, A., Vachon, M.P., and Côté, J. (2006). Preparation of helicenes through olefin metathesis. *Angew. Chem. Int. Ed.* 45 (18): 2923–2926.

39 Kamikawa, K., Takemoto, I., Takemoto, S., and Matsuzaka, H. (2007). Synthesis of helicenes utilizing palladium-catalyzed double C–H arylation reaction. *J. Org. Chem.* 72 (19): 7406–7408.

40 Takenaka, N., Sarangthem, R.S., and Captain, B. (2008). Helical chiral pyridine N-oxides: a new family of asymmetric catalysts. *Angew. Chem. Int. Ed.* 47 (50): 9708–9710.

41 Grandbois, A. and Collins, S.K. (2008). Enantioselective synthesis of [7]helicene: dramatic effects of olefin additives and aromatic solvents in asymmetric olefin metathesis. *Chem. A Eur. J.* 14 (30): 9323–9329.

42 Storch, J., Sýkora, J., Čermák, J. et al. (2009). Synthesis of hexahelicene and 1-methoxyhexahelicene via cycloisomerization of biphenylyl-naphthalene derivatives. *J. Org. Chem.* 74 (8): 3090–3093.

43 Kaneko, E., Matsumoto, Y., and Kamikawa, K. (2013). Synthesis of azahelicene N-oxide by palladium-catalyzed direct C-H annulation of a pendant (*Z*)-bromovinyl side chain. *Chem. A Eur. J.* 19 (36): 11837–11841.

44 Weimar, M., Correa da Costa, R., Lee, F.-H., and Fuchter, M.J. (2013). A scalable and expedient route to 1-aza[6]helicene derivatives and its subsequent application to a chiral-relay asymmetric strategy. *Org. Lett.* 15 (7): 1706–1709.

45 Tanaka, M., Shibata, Y., Nakamura, K. et al. (2016). Gold-catalyzed enantioselective synthesis, crystal structure, and photophysical/chiroptical properties of aza[10]helicenes. *Chem. A Eur. J.* 22 (28): 9537–9541.

46 González-Fernández, E., Nicholls, L.D.M., Schaaf, L.D. et al. (2017). Enantioselective synthesis of [6]carbohelicenes. *J. Am. Chem. Soc.* 139 (4): 1428–1431.

47 Wang, Z., Jiang, L., Ji, J. et al. (2020). Construction of cationic azahelicenes: regioselective three-component annulation using in situ activation strategy. *Angew. Chem. Int. Ed.* 59 (52): 23532–23536.

48 Wang, Q., Zhang, W.-W., Zheng, C. et al. (2021). Enantioselective synthesis of azoniahelicenes by Rh-catalyzed C-H annulation with alkynes. *J. Am. Chem. Soc.* 143 (1): 114–120.

49 Kötzner, L., Webber, M.J., Martínez, A. et al. (2014). Asymmetric catalysis on the nanoscale: the organocatalytic approach to helicenes. *Angew. Chem. Int. Ed.* 53 (20): 5202–5205.

50 Jia, S., Li, S., Liu, Y. et al. (2019). Enantioselective control of both helical and axial stereogenic elements though an organocatalytic approach. *Angew. Chem. Int. Ed.* 58 (51): 18496–18501.

51 Liu, L. and Katz, T.J. (1991). Bromine auxiliaries in photosyntheses of [5]helicenes. *Tetrahedron Lett.* 32 (47): 6831–6834.

52 Liu, L., Yang, B., Katz, T.J., and Poindexter, M.K. (1991). Improved methodology for photocyclization reactions. *J. Org. Chem.* 56 (12): 3769–3775.

53 Storch, J., Církva, V., Bernard, M., and Vokál, J. (2013). Způsob výroby [6]helicenů fotocyklizací. Úřad průmyslového vlastnictví CZ PV 2012-245 A3, filed 11 April 2012 and issued 7 August 2013.

54 Bédard, A.-C., Vlassova, A., Hernandez-Perez, A.C. et al. (2013). Synthesis, crystal structure and photophysical properties of pyrene-helicene hybrids. *Chem. A Eur. J.* 19 (48): 16295–16302.

55 Willmore, N.D., Liu, L., and Katz, T.J. (1992). A diels-alder route to [5]- and [6]-helicenes. *Angew. Chem. Int. Ed.* 31 (8): 1093–1095.

56 Phillips, K.E.S., Katz, T.J., Jockusch, S. et al. (2001). Synthesis and properties of an aggregating heterocyclic helicene. *J. Am. Chem. Soc.* 123 (48): 11899–11907.

57 Stará, I.G. and Starý, I. (2020). Helically chiral aromatics: the synthesis of helicenes by [2 + 2 + 2] cycloisomerization of π-electron systems. *Acc. Chem. Res.* 53 (1): 144–158.

58 Fujikawa, T., Segawa, Y., and Itami, K. (2015). Synthesis, structures, and properties of π-extended double helicene: a combination of planar and nonplanar π-systems. *J. Am. Chem. Soc.* 137 (24): 7763–7768.

59 Fujikawa, T., Segawa, Y., and Itami, K. (2016). Synthesis and structural features of quadruple helicenes: highly distorted π systems enabled by accumulation of helical repulsions. *J. Am. Chem. Soc.* 138 (10): 3587–3595.

60 Hu, Y., Wang, X.-Y., Peng, P.-X. et al. (2017). Benzo-fused double [7]carbohelicene: synthesis, structures, and physicochemical properties. *Angew. Chem. Int. Ed.* 56 (12): 3374–3378.

61 Zhu, Y., Xia, Z., Cai, Z. et al. (2018). Synthesis and characterization of hexapole [7]helicene, a circularly twisted chiral nanographene. *J. Am. Chem. Soc.* 140 (12): 4222–4226.

62 Nakakuki, Y., Hirose, T., Sotome, H. et al. (2018). Hexa-*peri*-hexabenzo[7] helicene: homogeneously π-extended helicene as a primary substructure of helically twisted chiral graphenes. *J. Am. Chem. Soc.* 140 (12): 4317–4326.

63 Reger, D., Haines, P., Heinemann, F.W. et al. (2018). Oxa[7]superhelicene: a π-extended helical chromophore based on hexa-*peri*-hexabenzocoronenes. *Angew. Chem. Int. Ed.* 57 (20): 5938–5942.

64 Guo, X., Yuan, Z., Zhu, Y. et al. (2019). A nitrogen-doped hexapole [7]helicene *versus* its all-carbon analogue. *Angew. Chem. Int. Ed.* 58 (47): 16966–16972.

65 Zhu, Y., Guo, X., Li, Y., and Wang, J. (2019). Fusing of seven HBCs toward a green nanographene propeller. *J. Am. Chem. Soc.* 141 (13): 5511–5517.

66 Wang, Y., Yin, Z., Zhu, Y. et al. (2019). Hexapole [9]helicene. *Angew. Chem. Int. Ed.* 58 (2): 587–591.

67 Navakouski, M., Zhylitskaya, H., Chmielewski, P.J. et al. (2019). Stereocontrolled synthesis of chiral heteroaromatic propellers with small optical bandgaps. *Angew. Chem. Int. Ed.* 58 (15): 4929–4933.

68 Hu, Y., Paternò, G.M., Wang, X.-Y. et al. (2019). π-Extended pyrene-fused double [7]carbohelicene as a chiral polycyclic aromatic hydrocarbon. *J. Am. Chem. Soc.* 141 (32): 12797–12803.

69 Martin, M.M., Hampel, F., and Jux, N. (2020). A hexabenzocoronene-based helical nanographene. *Chem. A Eur. J.* 26 (45): 10210–10212.

70 Ma, S., Gu, J., Lin, C. et al. (2020). Supertwistacene: a helical graphene nanoribbon. *J. Am. Chem. Soc.* 142 (39): 16887–16893.

71 Dusold, C., Sharapa, D.I., Hampel, F., and Hirsch, A. (2021). π-Extended diaza[7]helicenes by hybridization of naphthalene diimides and hexa-*peri*-hexabenzocoronenes. *Chem. A Eur. J.* 27 (7): 2332–2341.
72 Stará, I.G., Alexandrová, Z., Teplý, F. et al. (2005). Asymmetric synthesis of [7]helicene-like molecules. *Org. Lett.* 7 (13): 2547–2550.
73 Herse, C., Bas, D., Krebs, F.C. et al. (2003). A highly configurationally stable [4]heterohelicenium cation. *Angew. Chem. Int. Ed.* 42 (27): 3162–3166.
74 Stará, I.G., Starý, I., Kollárovič, A. et al. (1999). Transition metal catalysed synthesis of tetrahydro derivatives of [5]-, [6]- and [7]helicene. *Tetrahedron Lett.* 40 (10): 1993–1996.
75 Han, S., Bond, A.D., Disch, R.L. et al. (2002). Total syntheses and structures of angular [6]- and [7]phenylene: the first helical phenylenes (heliphenes). *Angew. Chem. Int. Ed.* 41 (17): 3223–3227.
76 Han, S., Anderson, D.R., Bond, A.D. et al. (2002). Total syntheses of angular [7]-, [8]-, and [9]phenylene by triple cobalt-catalyzed cycloisomerization: remarkably flexible heliphenes. *Angew. Chem. Int. Ed.* 41 (17): 3227–3230.
77 Adriaenssens, L., Severa, L., Šálová, T. et al. (2009). Helquats: a facile, modular, scalable route to novel helical dications. *Chem. A Eur. J.* 15 (5): 1072–1076.
78 Mori, K., Murase, T., and Fujita, M. (2015). One-step synthesis of [16]helicene. *Angew. Chem. Int. Ed.* 54 (23): 6847–6851.
79 Cruz, C.M., Márquez, I.R., Castro-Fernández, S. et al. (2019). A triskelion-shaped saddle-helix hybrid nanographene. *Angew. Chem. Int. Ed.* 58 (24): 8068–8072.
80 Kiel, G.R., Patel, S.C., Smith, P.W. et al. (2017). Expanded helicenes: a general synthetic strategy and remarkable supramolecular and solid-state behavior. *J. Am. Chem. Soc.* 139 (51): 18456–18459.
81 Evans, P.J., Ouyang, J., Favereau, L. et al. (2018). Synthesis of a helical bilayer nanographene. *Angew. Chem. Int. Ed.* 57 (23): 6774–6779.
82 Schuster, N.J., Hernández Sánchez, R., Bukharina, D. et al. (2018). A helicene nanoribbon with greatly amplified chirality. *J. Am. Chem. Soc.* 140 (20): 6235–6239.
83 Daigle, M., Miao, D., Lucotti, A. et al. (2017). Helically coiled graphene nanoribbons. *Angew. Chem. Int. Ed.* 56 (22): 6213–6217.
84 Stará, I.G., Starý, I., Kollárovič, A. et al. (1998). A novel strategy for the synthesis of molecules with helical chirality. Intramolecular [2 + 2 + 2] cycloisomerization of triynes under cobalt catalysis. *J. Org. Chem.* 63 (12): 4046–4050.
85 Vollhardt, K.P.C. (1984). Cobalt-mediated [2 + 2 + 2]-cycloadditions: a maturing synthetic strategy. *Angew. Chem. Int. Ed.* 23 (8): 539–556.
86 Tanaka, K., Kamisawa, A., Suda, T. et al. (2007). Rh-catalyzed synthesis of helically chiral and ladder-type molecules via [2 + 2 + 2] and formal [2 + 1 + 2 + 1] cycloadditions involving C≡C triple bond cleavage. *J. Am. Chem. Soc.* 129 (40): 12078–12079.
87 Shibata, T., Uchiyama, T., Yoshinami, Y. et al. (2012). Highly enantioselective synthesis of silahelicenes using Ir-catalyzed [2 + 2 + 2] cycloaddition. *Chem. Commun.* 48 (9): 1311–1313.

88 Crittall, M.R., Rzepa, H.S., and Carbery, D.R. (2011). Design, synthesis, and evaluation of a helicenoidal DMAP Lewis base catalyst. *Org. Lett.* 13 (5): 1250–1253.

89 Roose, J., Achermann, S., Dumele, O., and Diederich, F. (2013). Electronically connected [n]helicenes: synthesis and chiroptical properties of enantiomerically pure (*E*)-1,2-di([6]helicen-2-yl)ethenes. *Eur. J. Org. Chem.* 2013 (16): 3223–3231.

90 Schweinfurth, D., Zalibera, M., Kathan, M. et al. (2014). Helicene quinones: redox-triggered chiroptical switching and chiral recognition of the semiquinone radical anion lithium salt by electron nuclear double resonance spectroscopy. *J. Am. Chem. Soc.* 136 (37): 13045–13052.

91 Aillard, P., Retailleau, P., Voituriez, A., and Marinetti, A. (2014). A [2+2+2] cyclization strategy for the synthesis of phosphorus embedding [6]helicene-like structures. *Chem. Commun.* 50 (17): 2199–2201.

92 Magné, V., Sanogo, Y., Demmer, C.S. et al. (2020). Chiral phosphathiahelicenes: improved synthetic approach and uses in enantioselective gold(I)-catalyzed [2+2] cycloadditions of N-homoallenyl tryptamines. *ACS Catal.* 10 (15): 8141–8148.

93 Tsujihara, T., Inada-Nozaki, N., Takehara, T. et al. (2016). Nickel-catalyzed construction of chiral 1-[6]helicenols and application in the synthesis of [6]helicene-based phosphinite ligands. *Eur. J. Org. Chem.* 2016 (29): 4948–4952.

94 Kaiser, R.P., Nečas, D., Cadart, T. et al. (2019). Straightforward synthesis and properties of highly fluorescent [5]- and [7]-helical dispiroindeno[2,1-*c*]fluorenes. *Angew. Chem. Int. Ed.* 58 (48): 17169–17174.

95 Chercheja, S., Klívar, J., Jančařík, A. et al. (2014). The use of cobalt-mediated cycloisomerisation of ynedinitriles in the synthesis of pyridazinohelicenes. *Chem. A Eur. J.* 20 (27): 8477–8482.

96 Yamamoto, K., Nagae, H., Tsurugi, H., and Mashima, K. (2016). Mechanistic understanding of alkyne cyclotrimerization on mononuclear and dinuclear scaffolds: [4+2] cycloaddition of the third alkyne onto metallacyclopentadienes and dimetallacyclopentadienes. *Dalton Trans.* 45 (43): 17072–17081.

97 Žádný, J., Jančařík, A., Andronova, A. et al. (2012). A general approach to optically pure [5]-, [6]-, and [7]heterohelicenes. *Angew. Chem. Int. Ed.* 51 (24): 5857–5861.

98 Geny, A., Agenet, N., Iannazzo, L. et al. (2009). Air-stable $\{(C_5H_5)Co\}$ catalysts for [2+2+2] cycloadditions. *Angew. Chem. Int. Ed.* 48 (10): 1810–1813.

99 Buchta, M., Rybáček, J., Jančařík, A. et al. (2015). Chimerical pyrene-based [7]helicenes as twisted polycondensed aromatics. *Chem. A Eur. J.* 21 (24): 8910–8917.

100 Jonas, K., Deffense, E., and Habermann, D. (1983). Synthesis and reactions of η^5-cyclopentadienylbis(ethene)cobalt. *Angew. Chem. Int. Ed.* 22 (9): 716–717.

101 Jonas, K., Deffense, E., and Habermann, D. (1983). Synthese und Reaktionen von η^5-Cyclopentadienylbis(ethen) Cobalt. *Angew. Chem. Int. Ed.* 22 (S9): 1005–1016.

102 Cammack, J.K., Jalisatgi, S., Matzger, A.J. et al. (1996). Room temperature CpCo-mediated cyclization of α,δ,ω-enediynes to rearranging strained tricylic dienes. Some observations of kinetic versus thermodynamic control. *J. Org. Chem.* 61 (14): 4798–4800.

103 Alexandrová, Z., Sehnal, P., Stará, I.G. et al. (2006). Modified synthesis of heptahelicene and its resolution into single enantiomers. *Collect. Czech. Chem. Commun.* 71 (8): 1256–1264.

104 Bönnemann, H., Brijoux, W., Brinkmann, R. et al. (1984). A correlation between ^{13}C and ^{59}Co NMR data and the catalytic activity of organocobalt complexes in the synthesis of pyridine derivatives. *J. Organomet. Chem.* 272 (2): 231–249.

105 Bönnemann, H., Bogdanović, B., Brinkmann, R. et al. (1983). "Organically solvated" magnesium for the synthesis of transition-metal complexes and catalysts. *Angew. Chem. Int. Ed.* 22 (9): 728–728.

106 Heller, B., Hapke, M., Fischer, C. et al. (2013). Chiral cobalt^{I} and nickel^{0} complexes in the synthesis of nonracemic helicenes through the enantioselective [2+2+2] cyclotrimerisation of alkynes. *J. Organomet. Chem.* 723: 98–102.

107 Teplý, F., Stará, I.G., Starý, I. et al. (2003). Synthesis of 3-hexahelicenol and its transformation to 3-hexahelicenylamines, diphenylphosphine, methyl carboxylate, and dimethylthiocarbamate. *J. Org. Chem.* 68 (13): 5193–5197.

108 Teplý, F., Stará, I.G., Starý, I. et al. (2002). Synthesis of [5]-, [6]-, and [7]helicene via Ni(0)- or Co(I)-catalyzed isomerization of aromatic *cis,cis*-dienetriynes. *J. Am. Chem. Soc.* 124 (31): 9175–9180.

109 Aillard, P., Retailleau, P., Voituriez, A., and Marinetti, A. (2015). Synthesis of new phosphahelicene scaffolds and development of gold(I)-catalyzed enantioselective allenene cyclizations. *Chem. A Eur. J.* 21 (34): 11989–11993.

110 Karras, M., Holec, J., Bednárová, L. et al. (2018). Asymmetric synthesis of nonracemic 2-amino[6]helicenes and their self-assembly into langmuir films. *J. Org. Chem.* 83 (10): 5523–5538.

111 Gay Sánchez, I., Šámal, M., Nejedlý, J. et al. (2017). Oxahelicene NHC ligands in the asymmetric synthesis of nonracemic helicenes. *Chem. Commun.* 53 (31): 4370–4373.

112 Klívar, J., Šámal, M., Jančařík, A. et al. (2018). Asymmetric synthesis of diastereo- and enantiopure bioxahelicene 2,2′-bipyridines. *Eur. J. Org. Chem.* 2018 (37): 5164–5178.

113 Crittall, M.R., Fairhurst, N.W.G., and Carbery, D.R. (2012). Point-to-helical chirality transfer for a scalable and resolution-free synthesis of a helicenoidal DMAP organocatalyst. *Chem. Commun.* 48 (91): 11181–11183.

114 Kimura, Y., Fukawa, N., Miyauchi, Y. et al. (2014). Enantioselective synthesis of [9]- and [11]helicene-like molecules: double intramolecular [2+2+2] cycloaddition. *Angew. Chem. Int. Ed.* 53 (32): 8480–8483.

115 Murayama, K., Oike, Y., Furumi, S. et al. (2015). Enantioselective synthesis, crystal structure, and photophysical properties of a 1,1′-bitriphenylene-based sila[7]helicene. *Eur. J. Org. Chem.* 2015 (7): 1409–1414.

116 Yamano, R., Shibata, Y., and Tanaka, K. (2018). Synthesis of single and double dibenzohelicenes by rhodium-catalyzed intramolecular [2+2+2] and [2+1+2+1] cycloaddition. *Chem. A Eur. J.* 24 (24): 6364–6370.

117 Nejedlý, J., Šámal, M., Rybáček, J. et al. (2017). Synthesis of long oxahelicenes by polycyclization in a flow reactor. *Angew. Chem. Int. Ed.* 56: 5839–5843.

118 Jančařík, A., Rybáček, J., Cocq, K. et al. (2013). Rapid access to dibenzohelicenes and their functionalized derivatives. *Angew. Chem. Int. Ed.* 52 (38): 9970–9975.

119 Ville, G.A., Vollhardt, K.P.C., and Winter, M.J. (1984). Thermally induced diastereoisomerization of (cyclobutadiene)cobalt complexes as a probe for the reversibility of their formation from complexed alkynes. *Organometallics* 3 (8): 1177–1187.

120 Tanaka, K., Fukawa, N., Suda, T., and Noguchi, K. (2009). One-step construction of five successive rings by rhodium-catalyzed intermolecular double [2 + 2 + 2] cycloaddition: enantioenriched [9]helicene-like molecules. *Angew. Chem. Int. Ed.* 48 (30): 5470–5473.

121 Sawada, Y., Furumi, S., Takai, A. et al. (2012). Rhodium-catalyzed enantioselective synthesis, crystal structures, and photophysical properties of helically chiral 1,1′-bitriphenylenes. *J. Am. Chem. Soc.* 134 (9): 4080–4083.

122 Murayama, K., Shibata, Y., Sugiyama, H. et al. (2017). Synthesis, structure, and photophysical/chiroptical properties of benzopicene-based π-conjugated molecules. *J. Org. Chem.* 82 (2): 1136–1144.

123 Fukawa, N., Osaka, T., Noguchi, K., and Tanaka, K. (2010). Asymmetric synthesis and photophysical properties of benzopyrano- or naphthopyrano-fused helical phosphafluorenes. *Org. Lett.* 12 (6): 1324–1327.

124 Pozo, I., Guitián, E., Pérez, D., and Peña, D. (2019). Synthesis of nanographenes, starphenes, and sterically congested polyarenes by aryne cyclotrimerization. *Acc. Chem. Res.* 52 (9): 2472–2481.

125 Peña, D., Pérez, D., Guitián, E., and Castedo, L. (1999). Synthesis of hexabenzotriphenylene and other strained polycyclic aromatic hydrocarbons by palladium-catalyzed cyclotrimerization of arynes. *Org. Lett.* 1 (10): 1555–1557.

126 Hosokawa, T., Takahashi, Y., Matsushima, T. et al. (2017). Synthesis, structures, and properties of hexapole helicenes: assembling six [5]helicene substructures into highly twisted aromatic systems. *J. Am. Chem. Soc.* 139 (51): 18512–18521.

127 Berezhnaia, V., Roy, M., Vanthuyne, N. et al. (2017). Chiral nanographene propeller embedding six enantiomerically stable [5]helicene units. *J. Am. Chem. Soc.* 139 (51): 18508–18511.

128 Hosokawa, T., Asada, T., and Kamikawa, K. (2020). Theoretical examination of the plausible reaction process for stereoselective synthesis of hexapole helicene via a palladium-catalyzed [2 + 2 + 2] cyclotrimerization of [5]helicenyl aryne. *J. Phys. Chem. A* 124 (4): 652–661.

129 Yamamoto, T., Wakabayashi, S., and Osakada, K. (1992). Mechanism of C-C coupling reactions of aromatic halides, promoted by $Ni(COD)_2$ in the presence of 2,2′-bipyridine and PPh_3, to give biaryls. *J. Organomet. Chem.* 428 (1): 223–237.

130 Roy, M., Berezhnaia, V., Villa, M. et al. (2020). Stereoselective syntheses, structures, and properties of extremely distorted chiral nanographenes embedding hextuple helicenes. *Angew. Chem. Int. Ed.* 59 (8): 3264–3271.

131 Zhang, F., Michail, E., Saal, F. et al. (2019). Stereospecific synthesis and photophysical properties of propeller-shaped $C_{90}H_{48}$ PAH. *Chem. A Eur. J.* 25 (71): 16241–16245.
132 Yubuta, A., Hosokawa, T., Gon, M. et al. (2020). Enantioselective synthesis of triple helicenes by cross-cyclotrimerization of a helicenyl aryne and alkynes via dynamic kinetic resolution. *J. Am. Chem. Soc.* 142 (22): 10025–10033.
133 Rybáček, J., Huerta-Angeles, G., Kollárovič, A. et al. (2011). Racemic and optically pure heptahelicene-2-carboxylic acid: its synthesis and self-assembly into nanowire-like aggregates. *Eur. J. Org. Chem.* 2011 (5): 853–860.
134 Songis, O., Míšek, J., Schmid, M.B. et al. (2010). A versatile synthesis of functionalized pentahelicenes. *J. Org. Chem.* 75 (20): 6889–6899.
135 Šámal, M., Chercheja, S., Rybáček, J. et al. (2015). An ultimate stereocontrol in asymmetric synthesis of optically pure fully aromatic helicenes. *J. Am. Chem. Soc.* 137 (26): 8469–8474.
136 Míšek, J., Teplý, F., Stará, I.G. et al. (2008). A straightforward route to helically chiral N-heteroaromatic compounds: practical synthesis of racemic 1,14-diaza[5]helicene and optically pure 1- and 2-aza[6]helicenes. *Angew. Chem. Int. Ed.* 47 (17): 3188–3191.
137 Klívar, J., Jančařík, A., Šaman, D. et al. (2016). [2 + 2 + 2] Cycloisomerisation of aromatic cyanodiynes in the synthesis of pyridohelicenes and their analogues. *Chem. A Eur. J.* 22 (40): 14401–14405.
138 Snyder, J.K., Cai, C., and Audet, M.A. (2014). N,N-bond formation in intramolecular cobalt-catalyzed [2 + 2 + 2] cyclizations of alkynyl-linked bis-nitriles, and the preparation of annulated pyridazines. *Heterocycles* 88 (1): 179–186.
139 Andronova, A., Szydlo, F., Teplý, F. et al. (2009). The quest for alternative routes to racemic and nonracemic azahelicene derivatives. *Collect. Czech. Chem. Commun.* 74 (1): 189–215.
140 Severa, L., Adriaenssens, L., Vávra, J. et al. (2010). Highly modular assembly of cationic helical scaffolds: rapid synthesis of diverse helquats via differential quaternization. *Tetrahedron* 66 (19): 3537–3552.
141 Adriaenssens, L., Severa, L., Koval, D. et al. (2011). [6]Saddlequat: a [6]helquat captured on its racemization pathway. *Chem. Sci.* 2 (12): 2314–2320.
142 Sonawane, M.R., Vávra, J., Šaman, D. et al. (2015). Synthesis of helquats based on phenanthridinium units: four-step procedure to novel extended helical dications. *Synthesis* 47 (22): 3479–3488.
143 Coe, B.J., Rusanova, D., Joshi, V.D. et al. (2016). Helquat dyes: helicene-like push-pull systems with large second-order nonlinear optical responses. *J. Org. Chem.* 81 (5): 1912–1920.
144 Čížková, M., Šaman, D., Koval, D. et al. (2014). Modular synthesis of helicene-like compounds based on the imidazolium motif. *Eur. J. Org. Chem.* 2014 (26): 5681–5685.
145 Severa, L., Koval, D., Novotná, P. et al. (2010). Resolution of a configurationally stable [5]helquat: enantiocomposition analysis of a helicene congener by capillary electrophoresis. *New J. Chem.* 34 (6): 1063–1067.

146 Vávra, J., Severa, L., Švec, P. et al. (2012). Preferential crystallization of a helicene-viologen hybrid – an efficient method to resolve [5]helquat enantiomers on a 20 g scale. *Eur. J. Org. Chem.* 2012 (3): 489–499.

147 Vávra, J., Severa, L., Císařová, I. et al. (2013). Search for conglomerate in set of [7]helquat salts: multigram resolution of helicene-viologen hybrid by preferential crystallization. *J. Org. Chem.* 78 (4): 1329–1342.

148 Koval, D., Severa, L., Adriaenssens, L. et al. (2011). Chiral analysis of helquats by capillary electrophoresis: resolution of helical N-heteroaromatic dications using randomly sulfated cyclodextrins. *Electrophoresis* 32 (19): 2683–2692.

149 Nejedlý, J., Šámal, M., Rybáček, J. et al. (2020). Synthesis of racemic, diastereopure, and enantiopure carba- or oxa[5]-, [6]-, [7]-, and -[19]helicene (di)thiol derivatives. *J. Org. Chem.* 85 (1): 248–276.

150 Sehnal, P., Krausová, Z., Teplý, F. et al. (2008). On the origin of diastereoselectivity in [2 + 2 + 2] cycloisomerization of chiral triynes: controlling helicity of helicene-like compounds by thermodynamic factors. *J. Org. Chem.* 73 (6): 2074–2082.

151 Sehnal, P., Stará, I.G., Šaman, D. et al. (2009). An organometallic route to long helicenes. *Proc. Natl. Acad. Sci.* 106 (32): 13169–13174.

152 Yamano, R., Hara, J., Murayama, K. et al. (2017). Rh-mediated enantioselective synthesis, crystal structures, and photophysical/chiroptical properties of phenanthrenol-based [9]helicene-like molecules. *Org. Lett.* 19 (1): 42–45.

153 Kimura, Y., Shibata, Y., Noguchi, K., and Tanaka, K. (2019). Enantioselective synthesis and epimerization behavior of a chiral S-shaped [11]helicene-like molecule having collision between terminal benzene rings. *Eur. J. Org. Chem.* 2019 (6): 1390–1396.

154 Kinoshita, S., Yamano, R., Shibata, Y. et al. (2020). Rhodium-catalyzed highly diastereo- and enantioselective synthesis of a configurationally stable s-shaped double helicene-like molecule. *Angew. Chem. Int. Ed.* 59 (27): 11020–11027.

155 Nishigaki, S., Murayama, K., Shibata, Y., and Tanaka, K. (2018). Rhodium-mediated enantioselective synthesis of a benzopicene-based phospha[9]helicene: the structure-property relationship of triphenylene- and benzopicene-based carbo- and phosphahelicenes. *Mater. Chem. Front.* 2 (3): 585–590.

156 Perzyna, A., Zotto, C.D., Durand, J.-O. et al. (2007). Reaction of isocyanate-functionalised silicon wafers with complex amino compounds. *Eur. J. Org. Chem.* 2007 (24): 4032–4037.

157 Torricelli, F., Bosson, J., Besnard, C. et al. (2013). Modular synthesis, orthogonal post-functionalization, absorption, and chiroptical properties of cationic [6]helicenes. *Angew. Chem. Int. Ed.* 52 (6): 1796–1800.

158 Delgado, I.H., Pascal, S., Wallabregue, A. et al. (2016). Functionalized cationic [4]helicenes with unique tuning of absorption, fluorescence and chiroptical properties up to the far-red range. *Chem. Sci.* 7 (7): 4685–4693.

159 Duwald, R., Pascal, S., Bosson, J. et al. (2017). Enantiospecific elongation of cationic helicenes by electrophilic functionalization at terminal ends. *Chem. A Eur. J.* 23 (55): 13596–13601.

160 Delgado, I.H., Pascal, S., Besnard, C. et al. (2018). C-Functionalized cationic diazaoxatriangulenes: late-stage synthesis and tuning of physicochemical properties. *Chem. A Eur. J.* 24 (40): 10186–10195.

161 Holec, J., Rybáček, J., Vacek, J. et al. (2019). Chirality-controlled self-assembly of amphiphilic dibenzo[6]helicenes into langmuir-blodgett thin films. *Chem. A Eur. J.* 25 (49): 11494–11502.

162 Uozumi, Y., Tanahashi, A., Lee, S.Y., and Hayashi, T. (1993). Synthesis of optically active 2-(diarylphosphino)-1, 1′-binaphthyls, efficient chiral monodentate phosphine ligands. *J. Org. Chem.* 58 (7): 1945–1948.

163 Saal, F., Zhang, F., Holzapfel, M. et al. (2020). [n]Helicene diimides (n = 5, 6, and 7): through-bond *versus* through-space conjugation. *J. Am. Chem. Soc.* 142 (51): 21298–21303.

164 Meng, D., Liu, G., Xiao, C. et al. (2019). Corannurylene pentapetalae. *J. Am. Chem. Soc.* 141 (13): 5402–5408.

165 Schuster, N.J., Joyce, L.A., Paley, D.W. et al. (2020). The structural origins of intense circular dichroism in a waggling helicene nanoribbon. *J. Am. Chem. Soc.* 142 (15): 7066–7074.

166 Liu, B., Böckmann, M., Jiang, W. et al. (2020). Perylene diimide-embedded double [8]helicenes. *J. Am. Chem. Soc.* 142 (15): 7092–7099.

167 Gu, X., Xu, X., Li, H. et al. (2015). Synthesis, molecular packing, and thin film transistors of dibenzo[*a*,*m*]rubicenes. *J. Am. Chem. Soc.* 137 (51): 16203–16208.

168 Kato, K., Segawa, Y., Scott, L.T., and Itami, K. (2018). A quintuple [6]helicene with a corannulene core as a C_5-symmetric propeller-shaped π-system. *Angew. Chem. Int. Ed.* 57 (5): 1337–1341.

169 Fernández-García, J.M., Evans, P.J., Filippone, S. et al. (2019). Chiral molecular carbon nanostructures. *Acc. Chem. Res.* 52 (6): 1565–1574.

170 Kiran, V., Mathew, S.P., Cohen, S.R. et al. (2016). Helicenes – a new class of organic spin filter. *Adv. Mater.* 28 (10): 1957–1962.

171 Mori, T. (2021). Chiroptical properties of symmetric double, triple, and multiple helicenes. *Chem. Rev.* 121 (4): 2373–2412.

172 Stetsovych, O., Mutombo, P., Švec, M. et al. (2018). Large converse piezoelectric effect measured on a single molecule on a metallic surface. *J. Am. Chem. Soc.* 140 (3): 940–946.

173 Palata, O., Andronova, A., Šámal, M. et al. (2022). Synthesis of (di)thiahelicenes and dithiophenohelicenes by [2 + 2 + 2] cycloisomerization of alkynes. *Helv. Chim. Acta* 105 (3): e202100225.

3

Enantioselective Synthesis of Helicenes

Ken Tanaka

Department of Chemical Science and Engineering, Tokyo Institute of Technology, Tokyo, Japan

3.1 Introduction

The classical helicene synthesis is based on the oxidative photocyclization of stilbene-type precursors; however, this method suffers from low efficiency and cannot be applied to the enantioselective synthesis. To obtain sufficient amounts of enantioenriched helicenes, non-photochemical synthetic methods have been developed and applied to the enantioselective helicene synthesis [1]. These methods allow us to introduce electronically and sterically diverse substituents and clarify the structure–chiroptical property relationship.

For the most part, the enantioselective helicene syntheses are based on the transition-metal-catalyzed aromatic ring construction [2]. The most reported method is the transition-metal-catalyzed [2 + 2 + 2] cycloaddition of alkynes, and excellent enantioselectivities have been accomplished [3]. Besides, the transition-metal-catalyzed intramolecular hydroarylation of alkynes has also been applied to the catalytic enantioselective helicene synthesis [4].

This chapter summarizes the methods for the enantioselective helicene synthesis. In particular, transition-metal-catalyzed enantioselective aromatic ring construction reactions will be highlighted. In addition, several other enantioselective aromatic ring construction reactions including organocatalytic reactions [5] will be discussed.

3.2 Metal-Catalyzed [2 + 2 + 2] Cycloaddition

A [2 + 2 + 2] cycloaddition strategy is highly attractive for the helicene synthesis because of its convenient operation and applicability to the catalytic enantioselective synthesis. Rhodium [6] and nickel [7] complex-catalyzed reactions have been most frequently used for the catalytic enantioselective helicene synthesis. Both the rhodium(I) [6] and nickel(0) [7] complexes can catalyze the intramolecular [2 + 2 + 2] cycloaddition of triynes, giving the helicenes. However, only the rhodium(I) complexes can catalyze the intermolecular [2 + 2 + 2] cycloaddition

Helicenes: Synthesis, Properties, and Applications, First Edition.
Edited by Jeanne Crassous, Irena G. Stará, and Ivo Starý.

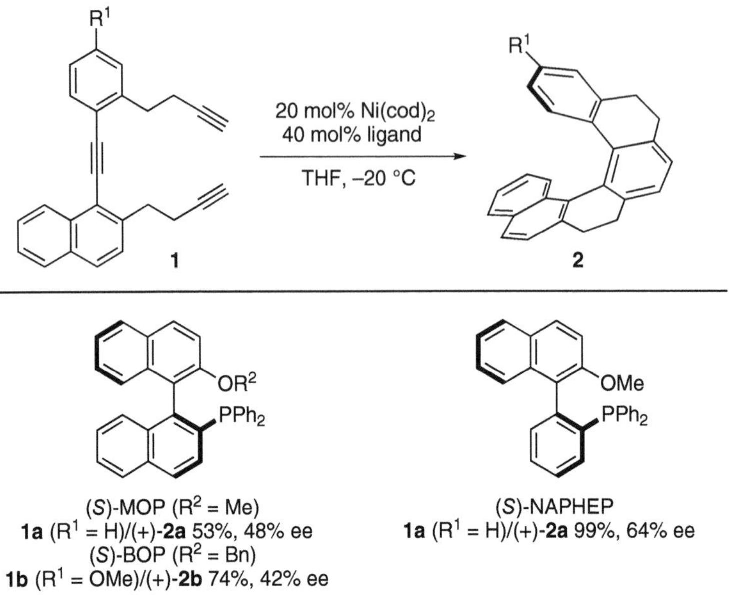

Scheme 3.1 Enantioselective synthesis of [6]helicenes by Ni-catalyzed intramolecular [2 + 2 + 2] cycloaddition.

between tetraynes and 1,3-diynes, giving the helicenes. This superior ability of the rhodium(I) catalyst allows the diversity-oriented helicene synthesis by changing the substituents of each coupling partner. Palladium(0) complexes are suitable catalysts for the intermolecular [2 + 2 + 2] cycloaddition involving arynes. This valuable transformation was applied to the enantioselective helicene synthesis, although the successful examples are few.

3.2.1 Ni- and Co-catalyzed [2 + 2 + 2] Cycloaddition

The application of the transition-metal-catalyzed [2 + 2 + 2] cycloaddition to the catalytic enantioselective helicene synthesis was pioneered by Stará, Starý, and coworkers [8]. The Ni(0)/(*S*)-MOP complex-catalyzed intramolecular [2 + 2 + 2] cycloaddition of triyne **1a** gave the desired tetrahydro[6]helicene **2a** with moderate yield and ee value (Scheme 3.1) [8]. The same catalyst gave a tetrahydro[7]helicene with moderate yield and ee value (41%, 40% ee) [9]. The use of (*S*)-BOP as a ligand improved the yield of tetrahydro[6]helicene **2b** but decreased the ee value [10]. Finally, the use of (*S*)-NAPHEP as a ligand gave **2a** in quantitative yield with an improved ee value of 64% [11], while the use of sulfonamide-substituted BOP ligand also afforded **2a** with 64% ee albeit in low yield [12].

A Ni(0)/(*R*)-QUINAP complex catalyzed the intramolecular [2 + 2 + 2] cycloaddition of arene-linked triynes **3a–c** to give the corresponding dibenzo[6]helicenes **4a–c** with high yields and ee values (Scheme 3.2) [13, 14]. The same catalyst also gave a sterically demanding hexabenzo[7]helicene **4d** with moderate ee value, although the product yield was low [15].

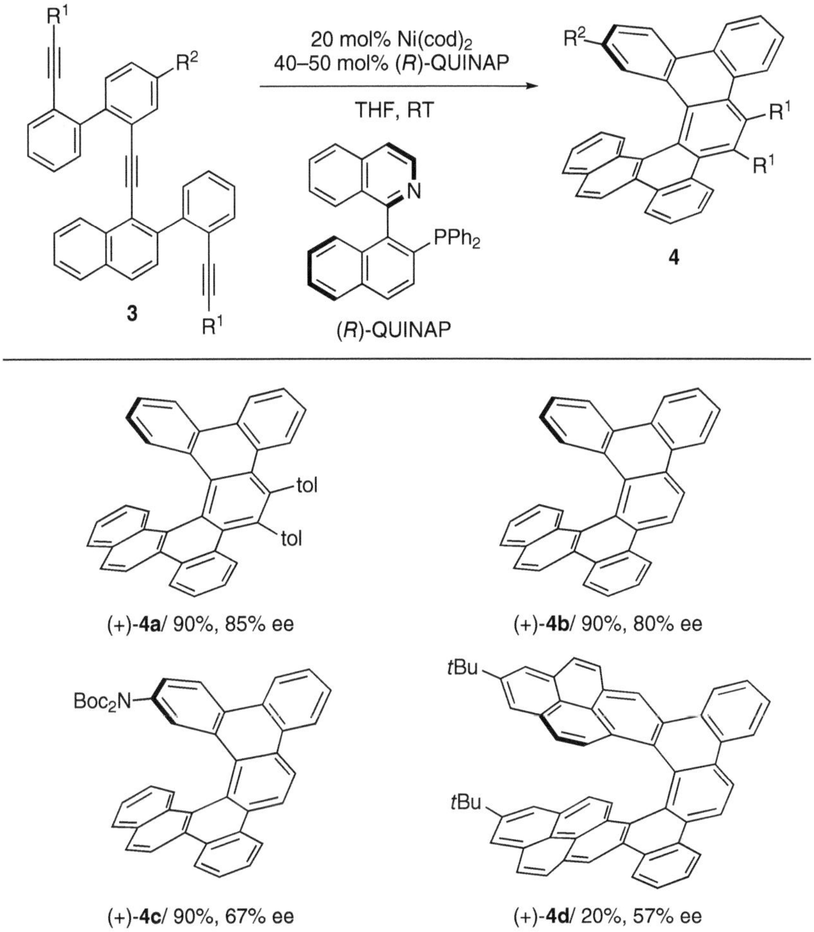

Scheme 3.2 Enantioselective synthesis of [7]helicenes by Ni-catalyzed intramolecular [2 + 2 + 2] cycloaddition.

A Ni(0)/(*M,R,R*)-(*M,R,R*)-NHC complex catalyzed the intramolecular [2 + 2 + 2] cycloaddition of phenylnaphthalene-linked triyne to give the corresponding dibenzo[7]helicene **4e** with high yield and ee value (Scheme 3.3) [16]. Although dibenzo[6]helicene **4b** could also be synthesized by using this Ni(0)/chiral NHC catalyst, the product ee value was decreased to 66% [16].

A chiral neomenthylindene Co(I) complex was also tested in the intramolecular [2 + 2 + 2] cycloaddition of triynes **1**. Although the desired nonracemic tetrahydro[6] and [7]helicenes **2a** and **2c** were obtained in quantitative yields, their ee values were low (Scheme 3.4) [11].

3.2.2 Rh- and Ir-Catalyzed [2 + 2 + 2] Cycloaddition

In the cationic Rh(I) complex-catalyzed cross-[2 + 2 + 2] cycloaddition reactions [17], the use of a biaryl bisphosphine as a ligand is important to gain high reactivity

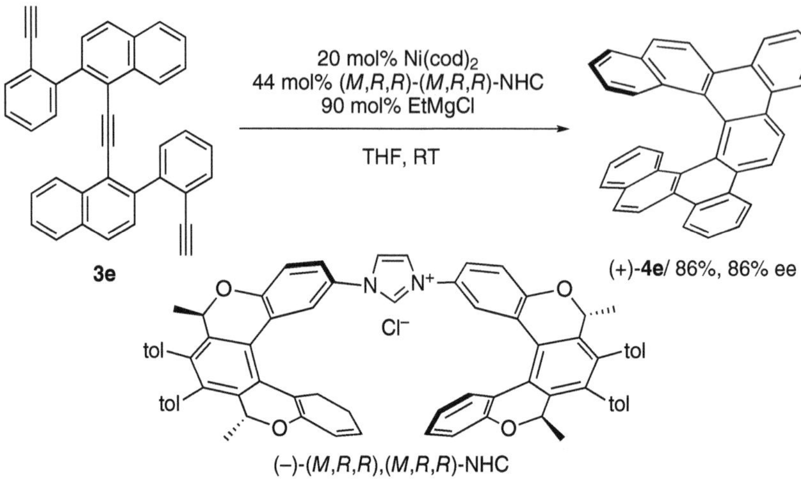

Scheme 3.3 Enantioselective synthesis of [7]helicene by Ni/NHC-catalyzed intramolecular [2 + 2 + 2] cycloaddition.

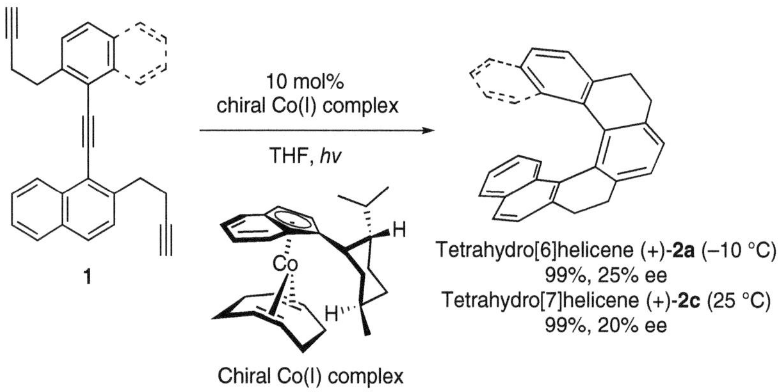

Scheme 3.4 Enantioselective synthesis of [6] and [7]helicenes by Co-catalyzed intramolecular [2 + 2 + 2] cycloaddition.

[18, 19]. However, interestingly, the enantioselective intramolecular [2+2+2] cycloaddition of triyne **5a** proceeded by using not the biaryl bisphosphine but a non-biaryl bisphosphine [(*R*,*R*)-Me-DuPhos] as a ligand to give the corresponding [7]helicene-like molecule **6a** in good yield and ee value (Scheme 3.5) [20]. Although the reactivity of electron-rich triyne **5b** was lower than that of electron-poor triyne **5a**, the corresponding [7]helicene-like molecule **6b** was obtained with improved ee value. When the biaryl bisphosphine ligands were used for **5a**, an unexpected ladder-type molecule **7a** was generated along with **6a** presumably through the unusual [2+1+2+1] cycloaddition. Among the ligands examined, the use of xyl-Segphos gave **7a** in the highest yield.

A possible mechanism of the rhodium-catalyzed [2+2+2] and [2+1+2+1] cycloadditions is shown in Scheme 3.6. [7]Helicene-like molecule **6** is generated

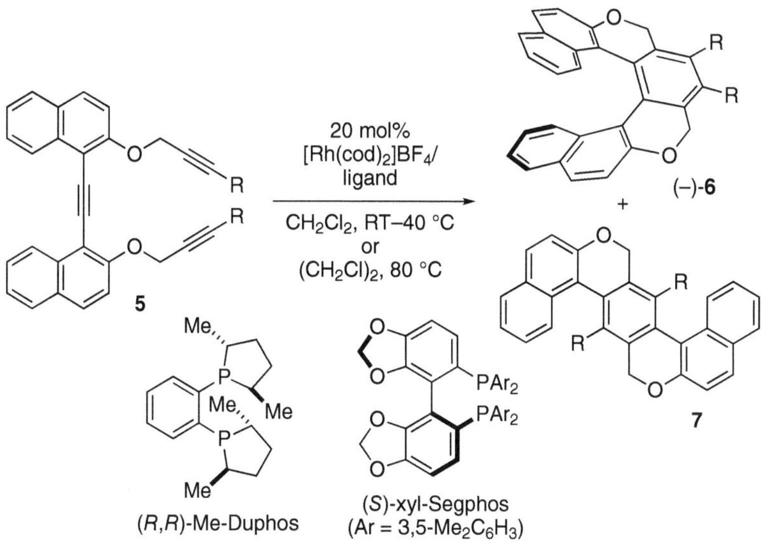

5 (R)	Ligand	Temp	(–)-6 /% yield (% ee)	7 /% yield
5a (CO_2Me)	(*R*,*R*)-Me-Duphos	RT	80 (71)	<1
5b (*n*-Bu)	(*R*,*R*)-Me-Duphos	40 °C	71 (85)	10
5a (CO_2Me)	(*S*)-xyl-Segphos	80 °C	39 (31)	34

Scheme 3.5 Enantioselective synthesis of [7]helicene-like molecules by Rh-catalyzed intramolecular [2 + 2 + 2] cycloaddition. Source: Modified from [20].

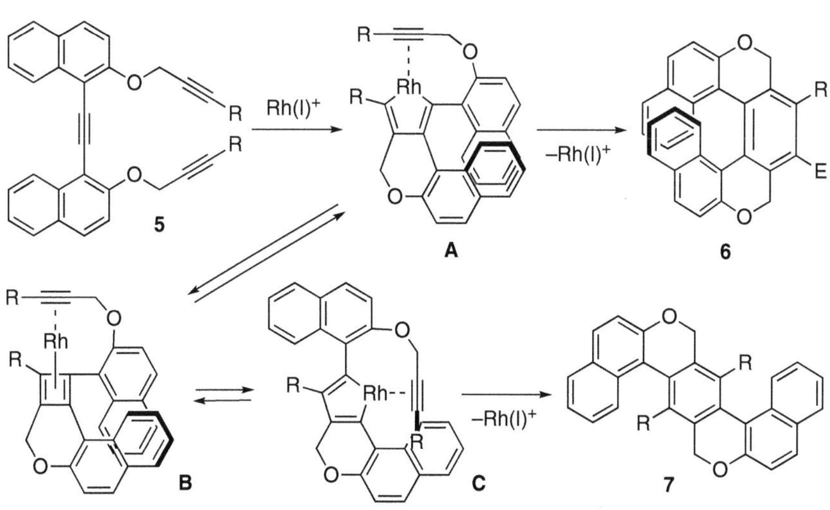

Scheme 3.6 Mechanisms for Rh-catalyzed intramolecular [2 + 2 + 2] and [2 + 1 + 2 + 1] cycloadditions.

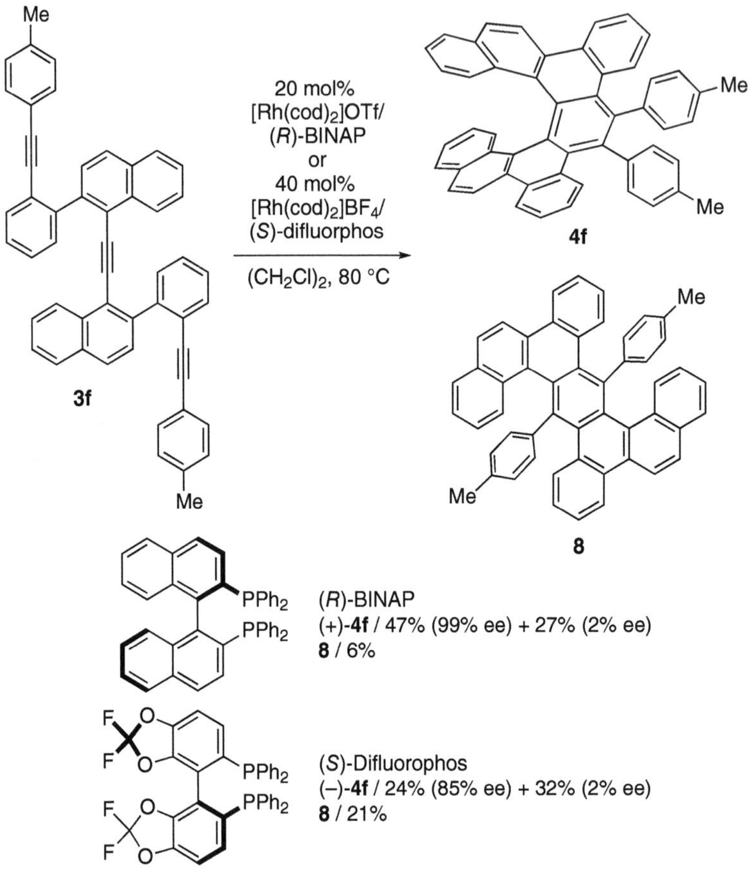

Scheme 3.7 Enantioselective synthesis of [7]helicenes by Rh-catalyzed intramolecular [2 + 2 + 2] cycloaddition.

from sterically demanding rhodacycle **A**. To release the steric hindrance, rhodacycle **A** may be converted to rhodacycle **C** through Rh(I)/cyclobutadiene complex **B**, which furnishes ladder molecule **7** [21, 22].

Carbohelicenes can also be synthesized by the rhodium-catalyzed intramolecular [2 + 2 + 2] cycloaddition. Dibenzo[7]helicene **4f** was synthesized with up to 99% ee along with [2 + 1 + 2 + 1] cycloaddition product **8** by the rhodium(I)/(*R*)-BINAP complex-catalyzed enantioselective intramolecular [2 + 2 + 2] cycloaddition of 2-phenylnaphthalene-linked triyne **3f** (Scheme 3.7) [23]. The use of (*S*)-Difluorphos instead of (*R*)-BINAP decreased the yield of **4f** and increased the yield of **8**.

The enantioselective synthesis of [9]helicene-like molecules **6c** and **6d** was achieved by the cationic Rh(I)//(*R*)-H_8-BINAP or (*R*)-BINAP complex-catalyzed intramolecular [2 + 2 + 2] cycloadditions of 3-phenanthrenol-linked triynes **5c** and **5d** (Scheme 3.8) [24]. Importantly, [2 + 1 + 2 + 1] cycloaddition products were not generated in these reactions.

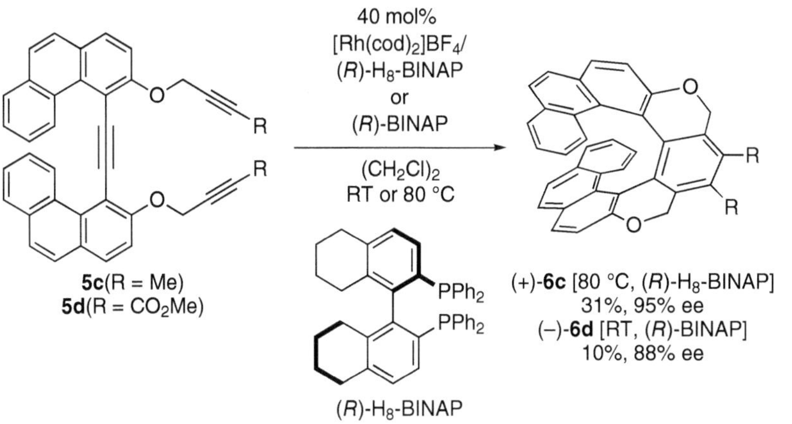

Scheme 3.8 Enantioselective synthesis of [9]helicene-like molecules by Rh-catalyzed intramolecular [2 + 2 + 2] cycloaddition.

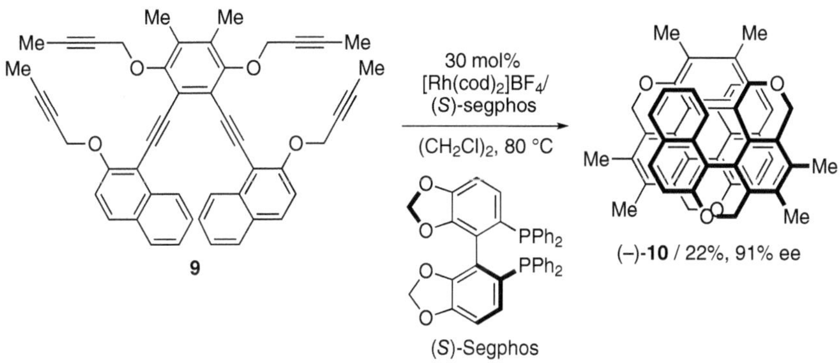

Scheme 3.9 Enantioselective synthesis of [11]helicene-like molecule by Rh-catalyzed intramolecular [2 + 2 + 2] cycloaddition.

Sterically more demanding [11]helicene-like molecule **10** was also synthesized by the rhodium-catalyzed double intramolecular [2 + 2 + 2] cycloadditions of hexayne **9** (Scheme 3.9) [25]. The desired [11]helicene-like molecule **10** was obtained with high ee value by using (*S*)-Segphos as a ligand, although the product yield was low.

The asymmetric synthesis of double helicenes that enhance chiroptical properties is an important target. Chiral S-shaped double helicene-like molecule (−)-**12** was synthesized with high ee value by the cationic Rh(I)/(*R*)-Difluorphos complex-mediated intramolecular double [2 + 2 + 2] cycloadditions of hexayne **11** (Scheme 3.10) [26]. However, this molecule is configurationally unstable, and epimerization proceeded to give more stable *meso*-**12** as a major product.

Almost enantiopure S-shaped double helicene-like molecule **14** (>99% ee), possessing stable helical chirality, was synthesized by the Rh(I)/(*S*)-Difluorphos complex-catalyzed highly diastereo- and enantioselective intramolecular double [2 + 2 + 2] cycloadditions of 2-naphthol- and benzene-linked hexayne **13**

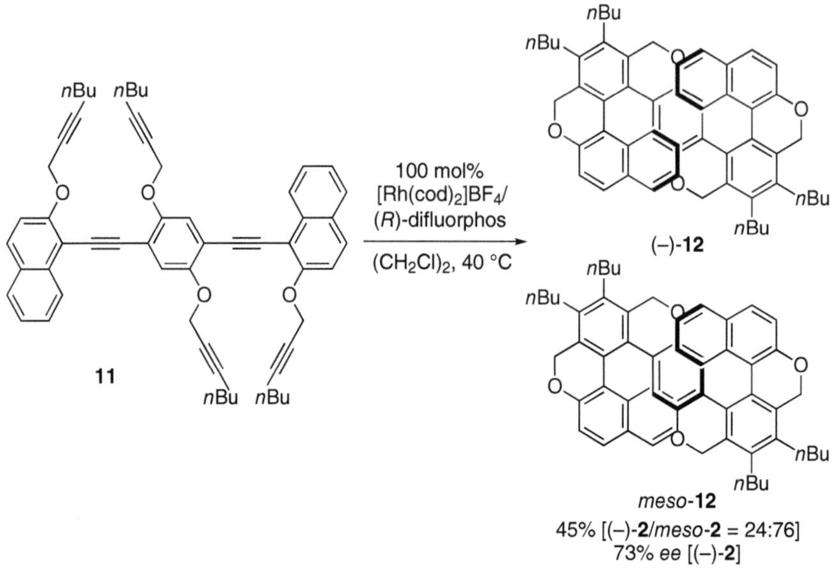

Scheme 3.10 Enantioselective synthesis of S-shaped double helicene-like molecule by Rh-catalyzed intramolecular [2 + 2 + 2] cycloaddition.

(Scheme 3.11) [27]. The introduction of two additional fused benzene rings significantly increases the configurational stability. Thus, epimerization and racemization were not observed even at 100 °C.

In general, intermolecular reactions are more synthetically useful than intramolecular reactions for expanding product diversity. The intermolecular [2 + 2 + 2] cycloaddition giving helicenes and helicene-like molecules were achieved by using the highly active cationic rhodium(I)/axially chiral biaryl bisphosphine complexes as catalysts and tetraynes and 1,3-diynes as coupling partners.

The regio- and enantioselective intermolecular double [2 + 2 + 2] cycloadditions of electron-rich tetraynes **15** with electron-poor dialkynylketones **16** proceeded in the presence of a cationic Rh(I)/(*R*)-Segphos or (*R*)-H_8-BINAP catalyst to give the corresponding [9]helicene-like molecules **17**, although both the product yields and ee values were low to moderate (Scheme 3.12) [28]. Additionally, the reactions of electron-poor ester-linked tetraynes **18** with **16** also proceeded in the presence of a cationic Rh(I)/(*R*)-Segphos, (*R*)-BINAP, or (*R*)-H_8-BINAP catalyst to give the corresponding [9]helicene-like molecules **19** (Scheme 3.12) [28].

Dialkynylphosphine oxides **21** could also react with tetraynes **20** at room temperature in the presence of a cationic Rh(I)/(*R*)-H_8-BINAP or (*R*)-tol-BINAP catalyst to give the corresponding phospha[7]- and [9]helicene-like molecules **22** in varying yields and ee values (Scheme 3.13) [29].

The regio- and enantioselective intermolecular double [2 + 2 + 2] cycloaddition of biphenyl-linked terminal tetrayne **23** with dialkynyl ketones **16** proceeded at room temperature in the presence of the cationic Rh(I)/(*S*)-xyl-Segphos catalyst to furnish the corresponding carbo[7]helicenes **24** in good yield with

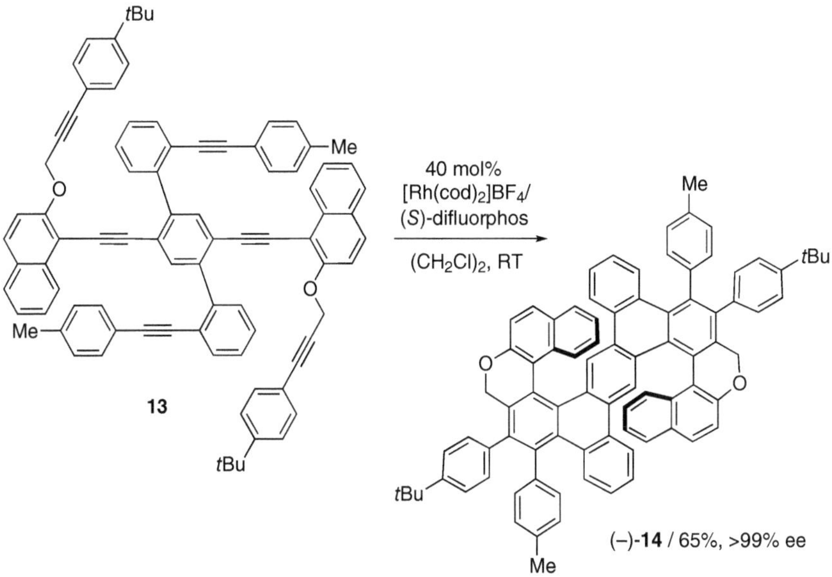

Scheme 3.11 Enantioselective synthesis of configurationally stable S-shaped double helicene-like molecule by Rh-catalyzed intramolecular [2 + 2 + 2] cycloaddition.

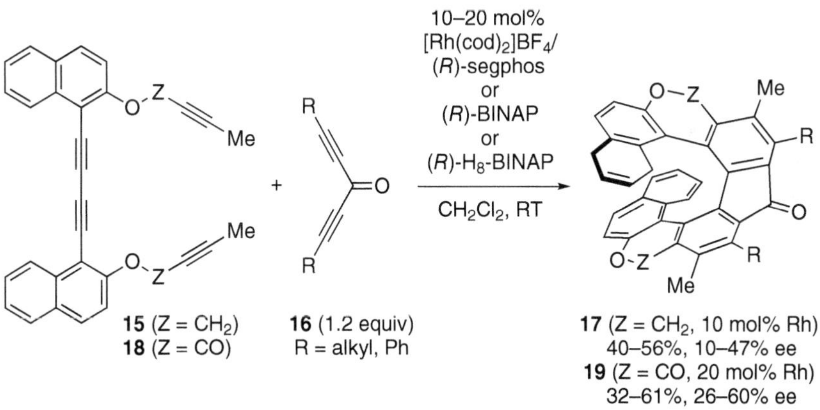

Scheme 3.12 Enantioselective synthesis of [9]helicene-like molecules by Rh-catalyzed intermolecular [2 + 2 + 2] cycloaddition.

high ee values (Scheme 3.14) [30]. Both electron-rich terminal tetrayne **23** and electron-poor internal tetrayne **25** reacted with dialkynyl ketones **16** in the presence of 10–20 mol % of a cationic Rh(I)/(*S*)-Difluorophos complex to give the corresponding carbo[7]helicenes **26** improved yields, although ee values decreased (Scheme 3.14) [30].

Carbo[9]helicene **28** was also synthesized with high ee value from binaphthyl-linked terminal tetrayne **27** and dialkynyl ketone **16** by using the cationic

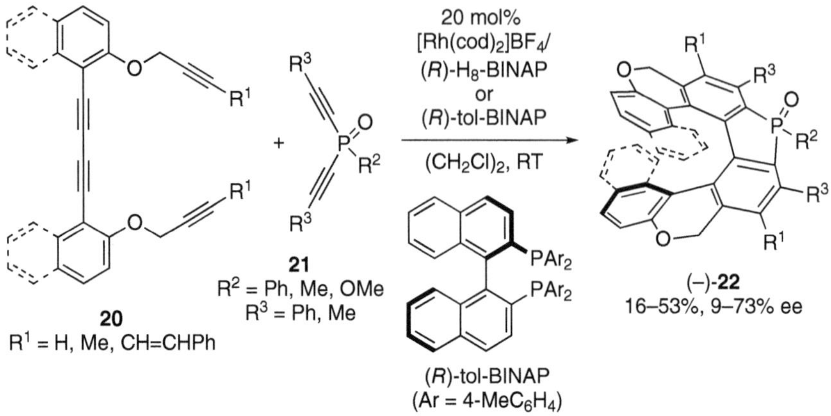

Scheme 3.13 Enantioselective synthesis of phospha[7]- and [9]helicene-like molecules by Rh-catalyzed intermolecular [2 + 2 + 2] cycloaddition.

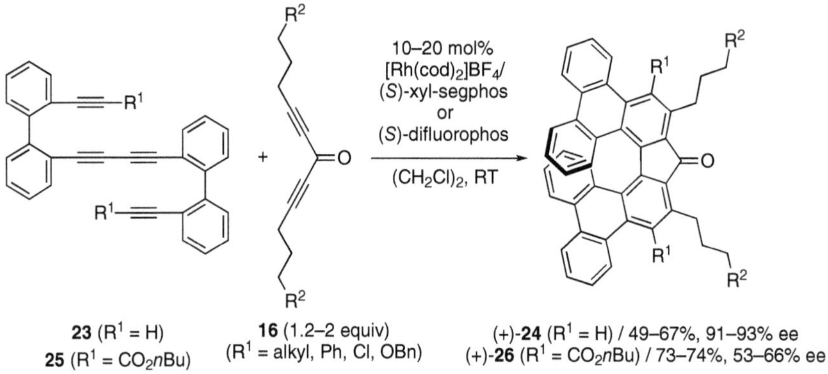

Scheme 3.14 Enantioselective synthesis of [7]helicenes by Rh-catalyzed intermolecular [2 + 2 + 2] cycloaddition.

Rh(I)/(*S*)-H_8-BINAP complex as a catalyst, but the product yield was very low (Scheme 3.15) [31].

The use of dialkynylphosphine oxides **21** instead of dialkynyl **phosphonates 21** afforded phospha[7]helicenes **29** with moderate yields and ee values (Scheme 3.16) [30].

Sila[7]helicene **31** was also synthesized with high ee value by using silicon-linked bis(propargylic alcohol) **30** and the cationic Rh(I)/(*S*)-Segphos catalyst, although the product yield was low (Scheme 3.17) [32].

An Ir(I) catalyst was also applied to the enantioselective helicene synthesis as shown in Scheme 3.18 [33]. The reaction of tetrayne **32** with diyne **33** in the presence of an Ir(I)/(*S,S*)-Et-FerroTANE complex gave [2 + 2 + 2] cycloaddition product (triyne) **34**, but double [2 + 2 + 2] cycloaddition product (helicene) **35** was not generated at all even at 100 °C. The targeted sila[7]helicene **35** was obtained by the treatment of **34** with a stoichiometric amount of the Ni(0)/PPh_3 complex.

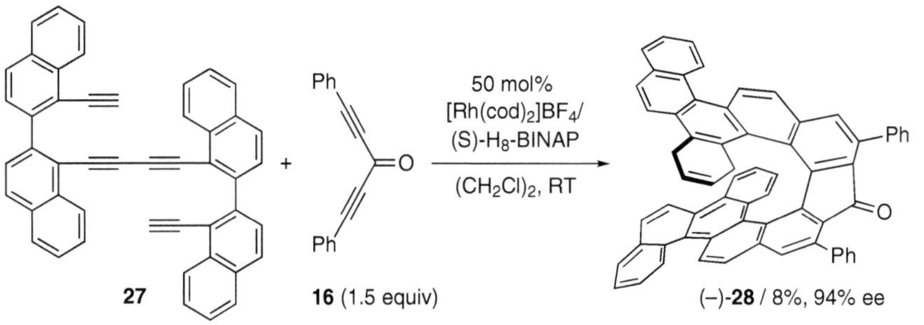

Scheme 3.15 Enantioselective synthesis of [9]helicenes by Rh-catalyzed intermolecular [2 + 2 + 2] cycloaddition.

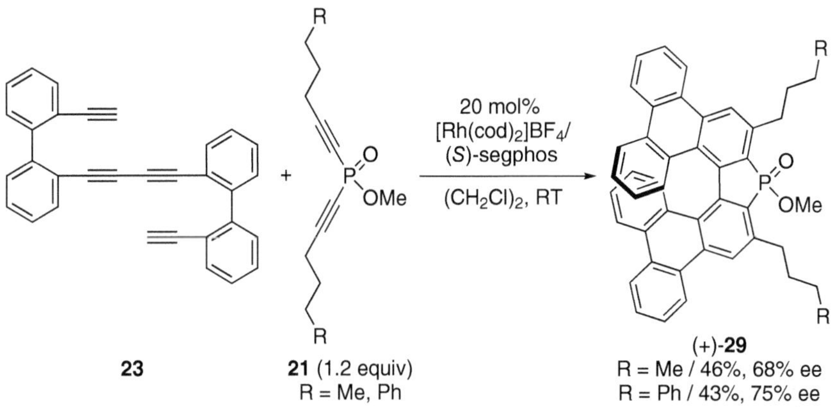

Scheme 3.16 Enantioselective synthesis of phospha[7]helicenes by Rh-catalyzed intermolecular [2 + 2 + 2] cycloaddition.

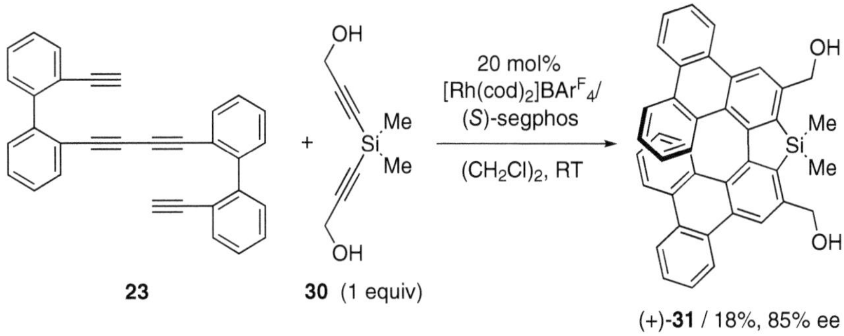

Scheme 3.17 Enantioselective synthesis of sila[7]helicene by Rh-catalyzed intermolecular [2 + 2 + 2] cycloaddition.

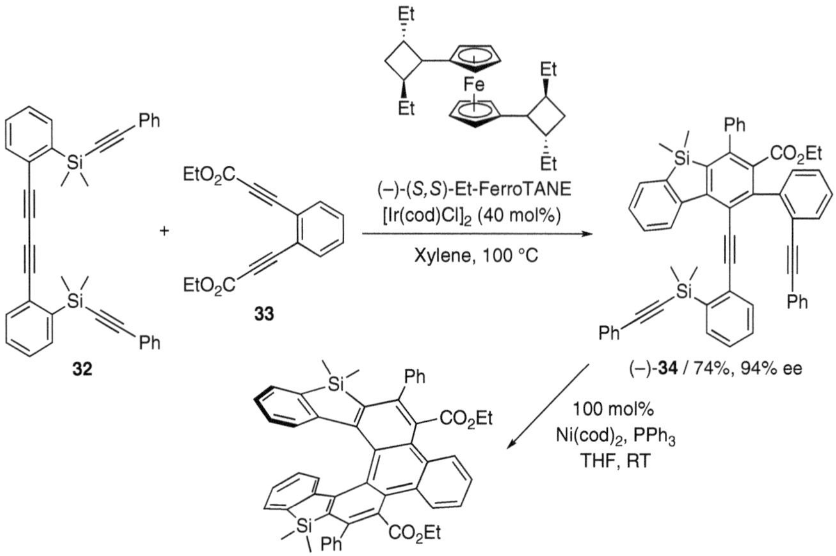

(+)-**35** / 94%, 93% ee

Scheme 3.18 Enantioselective synthesis of disila[7]helicene by Ir-catalyzed and Ni-mediated [2 + 2 + 2] cycloadditions.

3.2.3 Pd-catalyzed [2 + 2 + 2] Cycloaddition

Palladium(0) complexes show high catalytic activity in the intermolecular [2 + 2 + 2] cycloaddition involving arynes. Thus, the enantioselective helicene synthesis by the Pd(0)-catalyzed intermolecular [2 + 2 + 2] cycloaddition of two arynes with alkynes was investigated, although few examples have been reported so far.

The enantioselective intermolecular [2 + 2 + 2] cycloaddition of two arynes generated *in situ* from a fluoride ion-induced elimination of trimethylsilyl and trifluoromethanesulfonate-substituted naphthalene **36** with dimethylacetylene dicarboxylate (**37**) proceeded at room temperature in the presence of a Pd(0)/(*R*)-BINAP catalyst to give the corresponding [5]helicene **38** with moderate ee values, but the product yields were low due to the formation of undesired regioisomers (Scheme 3.19) [34].

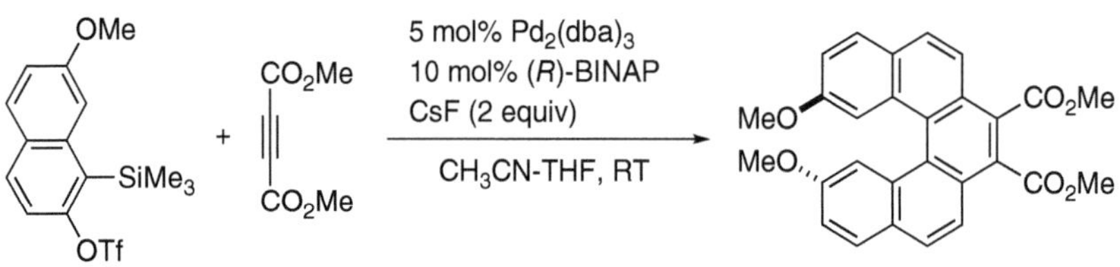

36 **37** (1.4 equiv) (+)-**38** / 35–60% ee

Scheme 3.19 Enantioselective synthesis of [5]helicene by Pd-catalyzed intermolecular [2 + 2 + 2] cycloaddition.

Optically active triple helicenes **40** were synthesized by the Pd(0)/(*R*)-QUINAP complex-catalyzed enantioselective intermolecular [2+2+2] cycloaddition of two helicenyl arynes, which are generated *in situ* from trimethylsilyl and trifluoromethanesulfonate-substituted [5]helicenes **39**, with dialkyl acetylene dicarboxylates **37** (Scheme 3.20) [35]. Ee values were high (84–96%), and the absolute stereochemistry of **40** (R = Me) was revealed to be (*M, P, M*) by a single-crystal X-ray diffraction analysis.

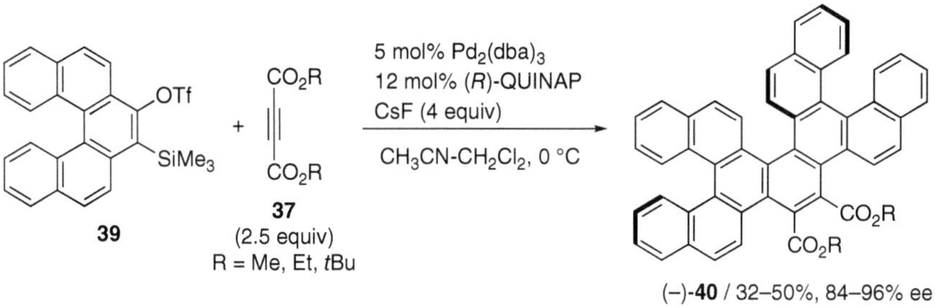

Scheme 3.20 Enantioselective synthesis of triple helicenes by Pd-catalyzed intermolecular [2 + 2 + 2] cycloaddition.

3.3 Metal-catalyzed Hydroarylation

Both the transition-metal-catalyzed [2+2+2] cycloaddition of alkynes and the transition-metal-catalyzed intramolecular hydroarylation of alkynes are useful methods for the aromatic ring construction and have also been applied to the catalytic helicene synthesis [4]. Although this transformation is limited to the intramolecular reaction, it has an advantage that the substrate synthesis becomes simple because it is a C–H bond functionalization.

As a first example, helically chiral aza[6]helicenes **43** were synthesized with moderate ee values by the cationic Au(I)/(*R*)-BINAP complex-catalyzed enantioselective sequential intramolecular hydroarylation of diynes **41** through axially chiral biaryls **42** (Scheme 3.21) [36]. Importantly, the use of an excess amount of less ionic [OTf] anion is crucial to promote the desired sequential hydroarylation giving **43**.

The use of the cationic Au(I)/(*R*)-BINAP complex also catalyzed the enantioselective sequential intramolecular hydroarylation of diynes **44**, leading to helically chiral fully benzenoid carbo[6]helicenes **45** (Scheme 3.22) [37]. As the free hydroxy-helicene was unstable under air, the products were isolated after methylation.

The Au(I)-catalyzed quadruple intramolecular hydroarylation of tetrayne **46** also proceeded at room temperature to give S-shaped double azahelicene (+)-**47**, although a high catalyst loading was required (Scheme 3.23) [36]. Enantiopure S-shaped double azahelicene (+)-**47** and a mixture of *racemic*-**47** and *meso*-**47** were separately isolated through a silica gel preparative thin-layer chromatography (PTLC). Removal of the 4-alkoxybenzyl group on the nitrogen followed by

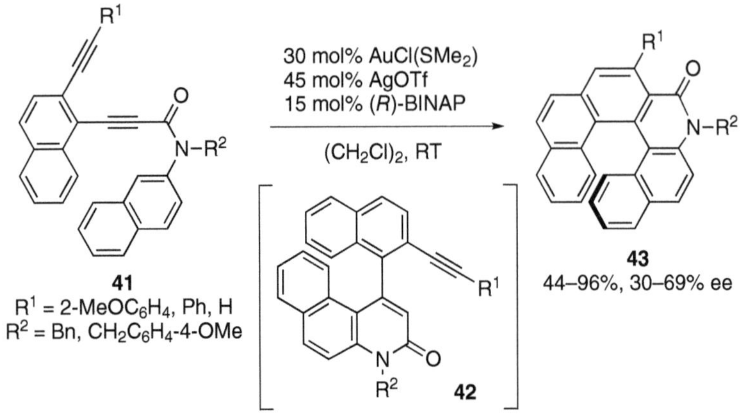

Scheme 3.21 Enantioselective synthesis of aza[6]helicenes by Au-catalyzed intramolecular hydroarylation.

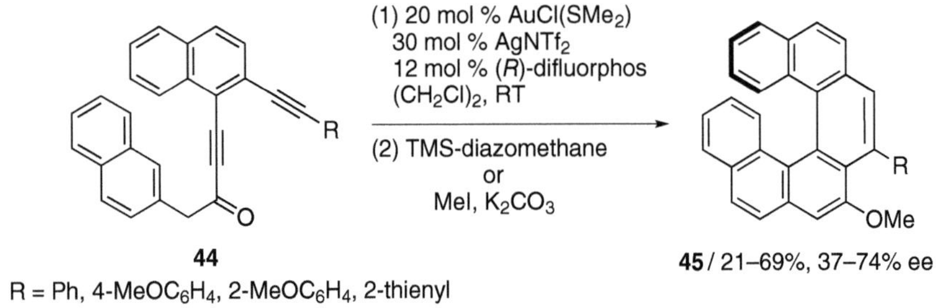

Scheme 3.22 Enantioselective synthesis of carbo[6]helicenes by Au-catalyzed intramolecular hydroarylation.

chlorination gave S-shaped double azahelicene (+)-**48**, possessing two pyridine units (Scheme 3.23) [36].

The diastereo- and enantioselective synthesis of double carbo[6]helicenes **50** was also accomplished by the cationic Au(I)/(*R*)-BINAP complex-catalyzed quadruple hydroarylation followed by methylation of bis(diynones) **49** with moderate yields and ee values without forming meso isomers (Scheme 3.24) [37].

The enantioselective intramolecular quadruple hydroarylation of 2,7-diaminonaphthalene-based tetrayne **51** in place of 1,8-diaminonaphthalene-based tetrayne **46** afforded diaza[10]helicene (−)-**52** with high ee value at room temperature by using a cationic Au(I)/(*R*)-Difluorphos catalyst, although the yield was low (Scheme 3.25) [38]. This diaza[10]helicene (−)-**52** was successfully converted into fully aromatic diaza[10]helicene (−)-**53**.

Both pyridine-fused helicenes and coumarin-fused helicene **55** were enantioselectively synthesized by the cationic Au(I)/(*R*)-MeO-DTB-BIPHEP complex-catalyzed intramolecular hydroarylation of [5]helicenyl propiolate **54** (Scheme 3.26) [39].

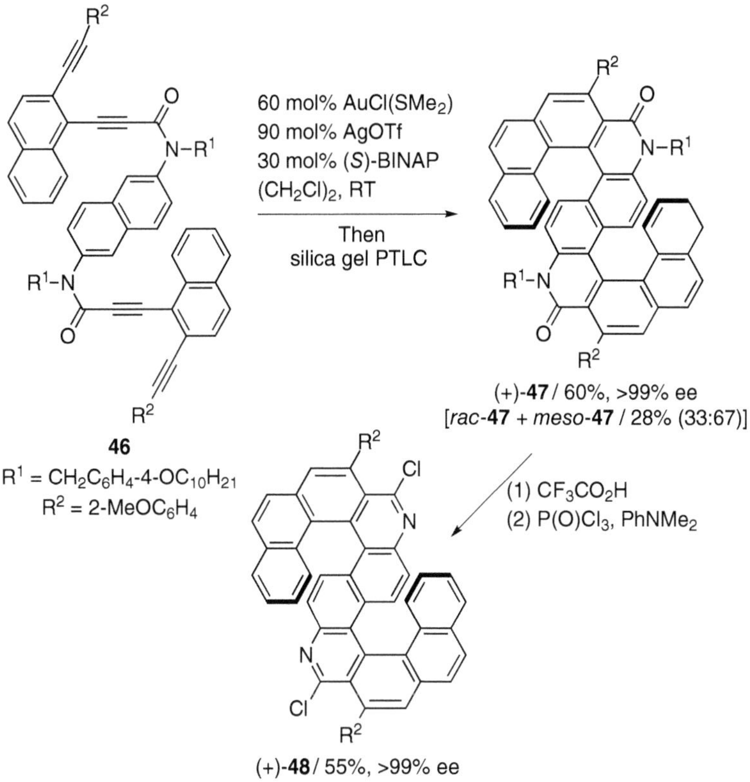

Scheme 3.23 Enantioselective synthesis of double azahelicenes by Au-catalyzed intramolecular hydroarylation.

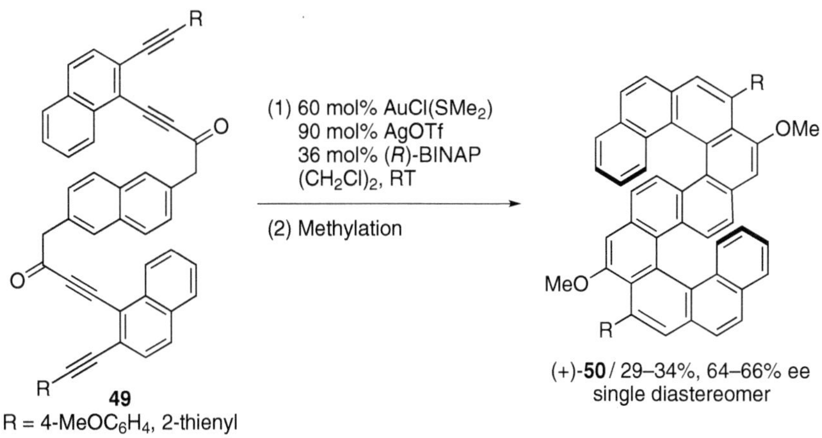

Scheme 3.24 Enantioselective synthesis of double carbohelicenes by Au-catalyzed intramolecular hydroarylation.

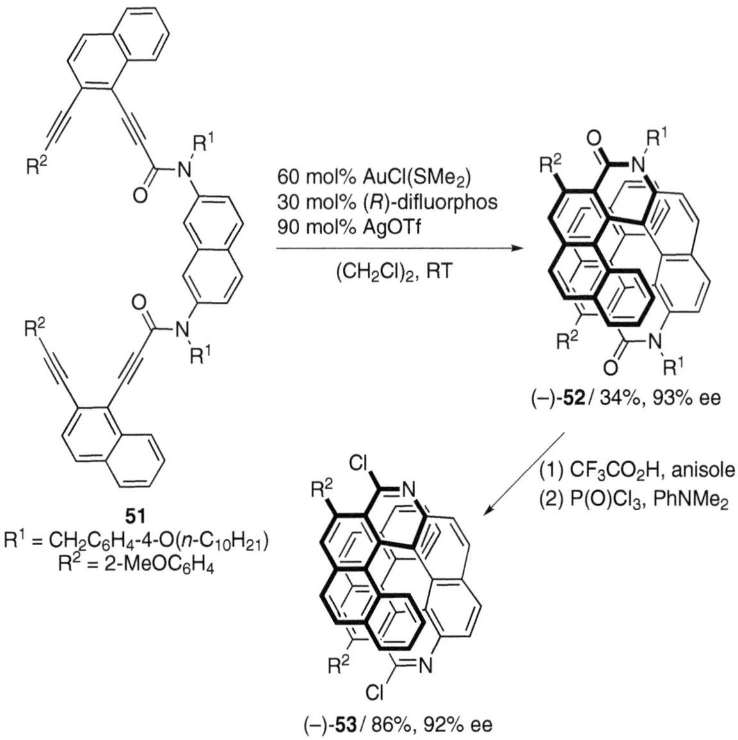

Scheme 3.25 Enantioselective synthesis of aza[10]helicenes by Au-catalyzed intramolecular hydroarylation.

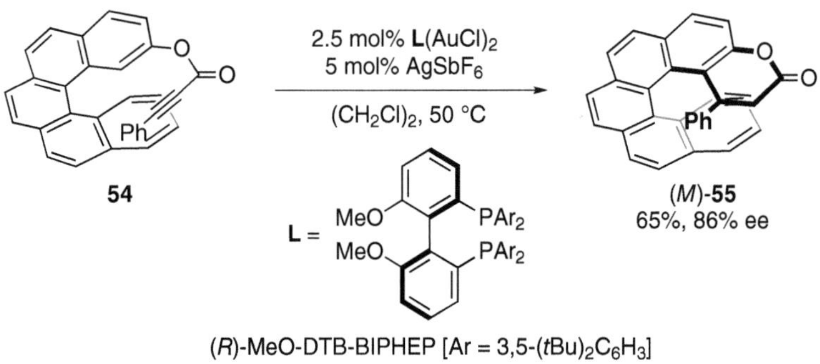

Scheme 3.26 Enantioselective synthesis of coumarin-fused helicene by Au-catalyzed intramolecular hydroarylation.

Gold(I) complexes **56** bearing α-cationic phosphonites, derived from TADDOL, as ancillary ligands catalyzed the regio- and enantioselective intramolecular hydroarylation of diynes **57** at −20 °C to give carbo[6]helicenes **58** with good to high ee values along with their isomers **59** (Scheme 3.27) [40, 41]. Complex **56a** catalyzed the

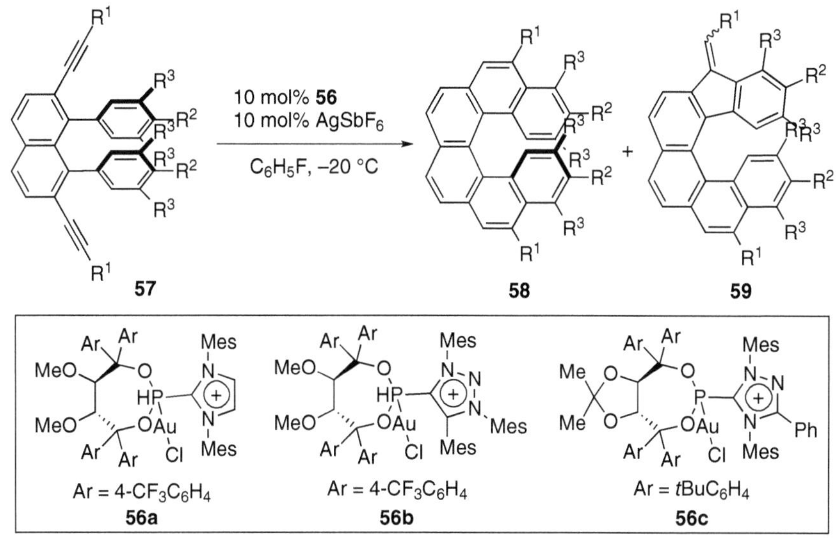

57 (R^1, R^2, R^3)	**56**	**58** + **59** / % yield	**58/59**	**58** / % ee
57a (Ph, H, H)	**56a**	77	95:5	**58a** / 92
57b (*p*-tol, H, H)	**56a**	88	93:7	**58b** / 91
57c (4-$MeOC_6H_4$, H, H)	**56a**	90	96:4	**58c** / 81
57d (4-FC_6F_4, H, H)	**56a**	90	93:7	**58d** / 90
57e (*p*-tol, OMe, H)	**56b**	99	88:12	**58e** / 78
57f (*p*-tol, H, Me)	**56a**	84	88:12	**58f** / 75
57g (Me, H, H)	**56c**	95	100:0	**58g** / 70

Scheme 3.27 Enantioselective synthesis of carbo[6]helicenes by Au/α-cationic phosphonite-catalyzed intramolecular hydroarylation.

hydroarylation of **57a** to give **58a** in good yield with high ee value. Complex **56a** also catalyzed the hydroarylation of **57b–d** having the aryl group at the alkyne terminus to give **58b–d** in good yields with high ee values. The hydroarylation of **57e** having the 4-methoxyphenyl group at the 1- and 8-positions of the naphthalene ring, leading to **58e** and **59e**, could be catalyzed by 1,2,3-triazolium-based gold(I) complex **56b** while regio- and enantioselectivities decreased. Substitution with the *m*-tolyl groups of the naphthalene ring (**57f**) lowered the regio- and enantioselectivities. For the dimethyl-substituted diyne **57g**, the 1,2,4-triazolium-based chiral gold(I) complex **56c** was a suitable catalyst, and the desired hydroarylation proceeded in excellent yield with perfect regioselectivity, but the product ee value was moderate.

The use of 1,2,4-triazolium-based chiral gold(I) complex **56d** as a catalyst enabled the regio- and enantioselective synthesis of 1,12-disubstituted carbo[4]helicenes **61** by the enantioselective sequential intramolecular hydroarylation of diynes **60** at −20° (Scheme 3.28) [42]. These transformations are highly regioselective, and thus regioisomers **62** were generated in low yields.

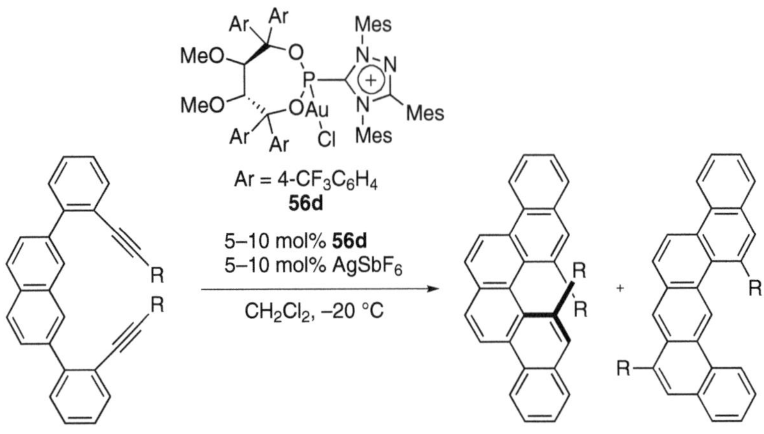

60
R = Ph, tol, 4-MeOC$_6$H$_4$

61 **62**
84–98% (**61**/**62** = 93:7–98:2), 89–99% ee

Scheme 3.28 Enantioselective synthesis of carbo[4]helicenes by Au/α-cationic phosphonite-catalyzed intramolecular hydroarylation.

3.4 Metal-catalyzed Oxidative Annulation

The catalytic enantioselective synthesis of oxa[9]helicenes **64** (up to 94% ee) was achieved by the chiral vanadium complex (**65**)-catalyzed oxidative annulation of benzo[*c*]phenanthren-2-ols **63** (Scheme 3.29) [43]. In this process, the vanadium complex **65** works as both a redox catalyst and a Lewis acid catalyst, allowing the sequential reaction via an oxidative coupling/intramolecular cyclization sequence.

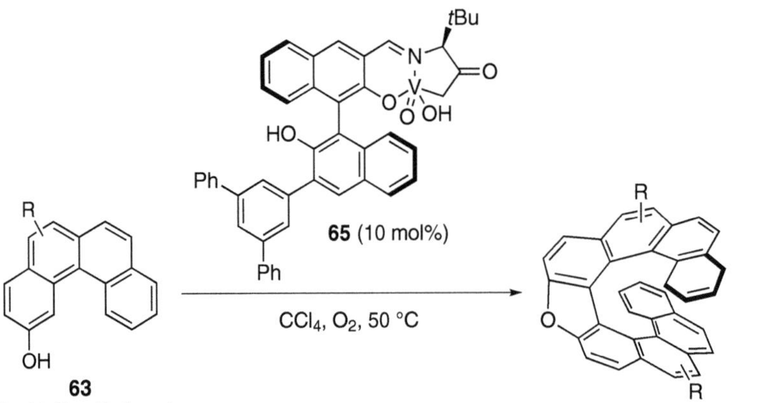

63
R = H, Br, alkyl, aryl

(−)-**64** / 56–84%, 44–94% ee

Scheme 3.29 Enantioselective synthesis of oxa[9]helicenes by vanadium-catalyzed oxidative annulation.

3.5 Organocatalyzed Annulation

Organocatalytic aromatic ring construction reactions are also attractive for the enantioselective helicene synthesis because many chiral organocatalysts are readily available [5]. However, only a few examples have been reported so far regarding the application of organocatalyzed annulation reactions to the enantioselective helicene synthesis.

Although it is not an organocatalyzed example, the enantioselective helicene synthesis by using a stoichiometric amount of a chiral amine [(−)-sparteine] was reported. The enantioselective synthesis of oligothiophene-based [7]helicene **67** was accomplished by the (−)-sparteine-mediated annelation of the axially chiral bis(aryllithium) **66**-Li_2 with electrophilic sulfur equivalent, although the product yields (20–37%) and ee values (19–47% ee) were low (Scheme 3.30) [44]. The kinetic resolution of 7,8-dipropyltetrathia[7]helicene was accomplished by (−)-sparteine-mediated asymmetric silylation [45] and asymmetric formylation with chiral formamides [46], while their selectivities were low to moderate.

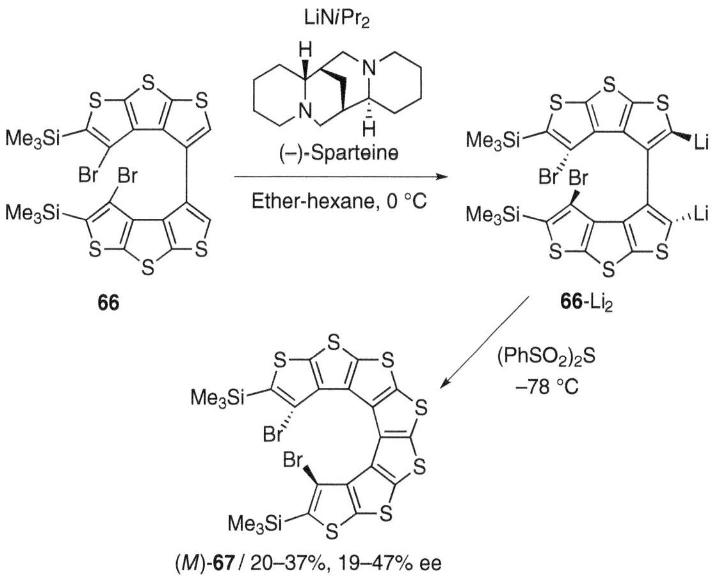

Scheme 3.30 Enantioselective synthesis of oligothiophene-based [7]helicene by sparteine-mediated annelation.

As a first example of the enantioselective helicene synthesis by the organocatalytic aromatic ring construction, the enantioselective synthesis of dihydro-aza[6]helicene **70** was achieved by the Fischer indolization of hydrazines **68** with aromatic ketone **69** catalyzed by SPINOL [1,1′-spirobiindane-7,7′-diol]-derived chiral phosphoric acid (*S*)-**71** (Scheme 3.31) [47]. Dihydro-aza[7]helicenes were also successfully synthesized, although their ee values (75–80% ee) were lower than that of the dihydro-aza[6]helicene **70** [47].

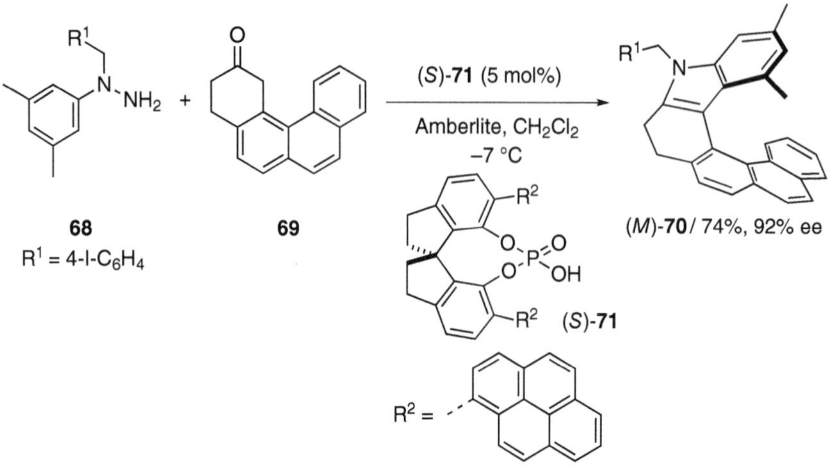

Scheme 3.31 Enantioselective synthesis of dihydro-aza[6]helicene by chiral phosphoric acid-catalyzed annulation.

On the other hand, the double Fischer indolization of aromatic ketone **72** with hydrazine **73** catalyzed by the same chiral phosphoric acid catalyst **71** followed by chloranil oxidation afforded aza[9]helicene **74** with a high ee value of 88% (Scheme 3.32) [47].

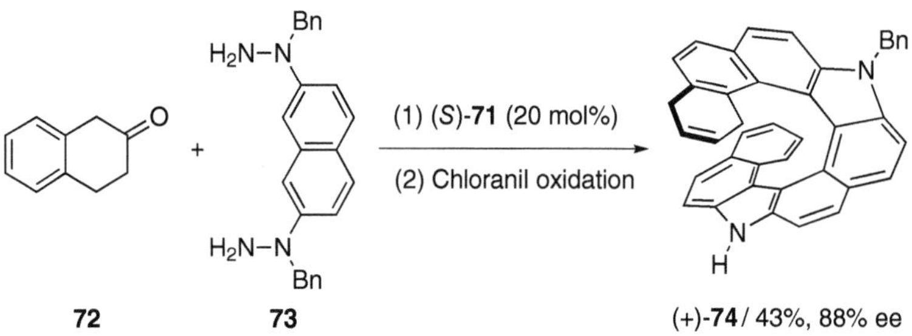

Scheme 3.32 Enantioselective synthesis of diaza[9]helicene by chiral phosphoric acid-catalyzed annulation.

Simultaneous construction of axial and helical chiralities was accomplished by the organocatalytic aromatic ring construction reaction. The highly diastereo- and enantioselective synthesis of [6]helicenes **76** bearing two axial stereogenic elements was accomplished by the sequential annulation of diynes **75** catalyzed by chiral organocatalyst **77a** (Scheme 3.33) [48].

Recently, expedient synthetic access to a new family of configurationally stable dioxa[6]helicenes from simple achiral precursors was reported (Scheme 3.34) [49]. The organocatalyzed domino Michael/*C–O* alkylation of oxa[5]helicenol **78** and nitroalkenes **79** gave dihydro-dioxa[6]helicenes **80** possessing both two stereogenic carbon atoms and a helical shape with high yields and ee values. The subsequent base-promoted elimination of HNO_2 gave dioxa[6]helicenes **81** with almost retaining helical chirality.

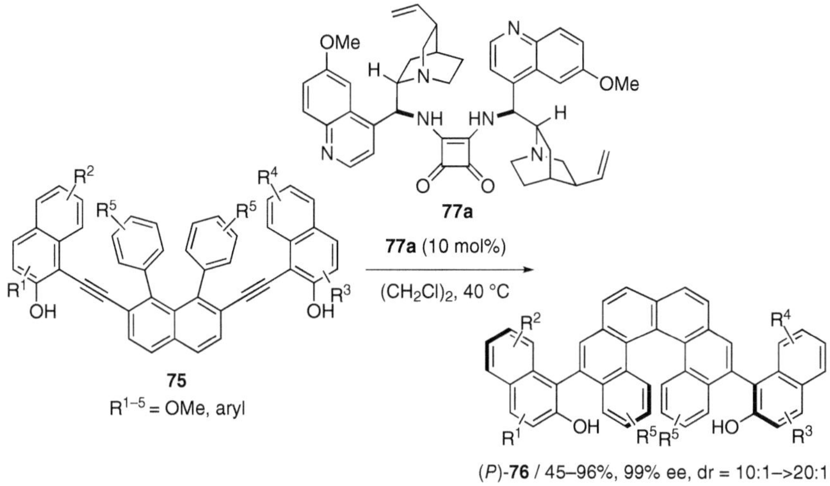

Scheme 3.33 Organocatalytic simultaneous construction of axial and helical chiralities.

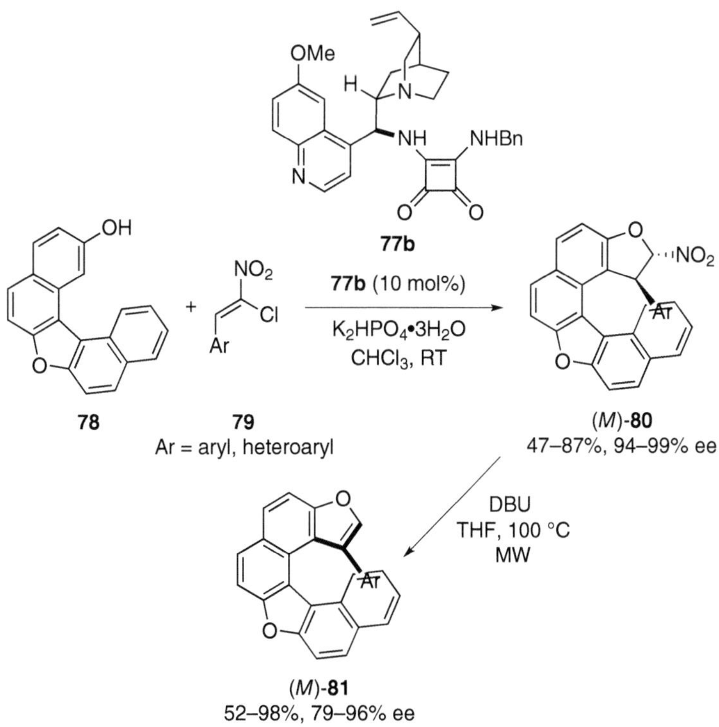

Scheme 3.34 Enantioselective synthesis of dioxa[6]helicenes by organocatalyzed domino Michael/*C–O* alkylation.

3.6 Conclusion

This chapter overviewed the enantioselective synthesis of helicenes and helicene-like molecules. Various aromatic ring construction reactions have been developed and applied to the enantioselective helicene synthesis. Four types of aromatic ring construction reactions (type-1: [2+2+2] cycloaddition, type-2: intramolecular alkyne hydroarylation, type-3: oxidative annulation, and type-4: organocatalyzed annulation) were disclosed. Among these, the transition-metal-catalyzed [2+2+2] cycloaddition might be the most powerful strategy. Two different reaction patterns, intramolecular and intermolecular [2+2+2] cycloadditions, were developed: (i) the intramolecular reactions of triynes constructing three rings and (ii) the intermolecular reactions between tetraynes and diynes constructing five rings. As the transition-metal-catalyzed [2+2+2] cycloaddition allows to construct multiple aromatic rings, this method enabled the enantioselective synthesis of long helicenes and helicene-like molecules. On the other hand, other aromatic ring construction reactions (types-2, 3, and 4) can construct only one or two rings. However, as these new methodologies are still in their infancy, the further development of more efficient, powerful, and selective catalysts and new reaction patterns would broaden the available helicene structures. Future targets for the enantioselective helicene synthesis will be helicenes with challenging large steric hindrances, helicenes with multiple chiralities, helicenes with excellent functionality, and so on.

References

1 Gingras, M., Félix, G., and Peresutti, R. (2013). *Chem. Soc. Rev.* 42: 1007.
2 Tanaka, K. (ed.) (2013). *Transition-Metal-Mediated Aromatic Ring Construction.* Hoboken: Wiley.
3 Tanaka, K. (ed.) (2013). *Transition-Metal-Mediated Aromatic Ring Construction.* Hoboken: Wiley Chapter 10.
4 Hendrich, C., Sekine, K., Koshikawa, T. et al. (2021). *Chem. Rev.* 121: 9113–9163.
5 Shirakawa, S., Liu, S., and Kaneko, S. (2016). *Chem. Asian J.* 11: 330.
6 Tanaka, K., Kimura, Y., and Murayama, K. (2015). *Bull. Chem. Soc. Jpn.* 88: 375.
7 Stará, I.G. and Starý, I. (2020). *Acc. Chem. Res.* 53: 144.
8 Stará, I.G., Starý, I., Kollárovič, A. et al. (1999). *Tetrahedron Lett.* 40: 1993.
9 Alexandrová, Z., Sehnal, P., Stará, I.G. et al. (2006). *Collect. Czech. Chem. Commun.* 71: 1256.
10 Teplý, F., Stará, I.G., Starý, I. et al. (2003). *J. Org. Chem.* 68: 5193.
11 Heller, B., Hapke, M., Fischer, C. et al. (2013). *J. Organomet. Chem.* 723: 98.
12 Stará, I.G., Andronova, A., Kollárovič, A. et al. (2011). *Collect. Czech. Chem. Commun.* 76: 2005.
13 Jančařík, A., Rybáček, J., Cocq, K. et al. (2013). *Angew. Chem. Int. Ed.* 52: 9970.
14 Karras, M., Holec, J., Bednárová, L. et al. (2018). *J. Org. Chem.* 83: 5523.
15 Buchta, M., Rybácek, J., Jančařík, A. et al. (2015). *Chem. A Eur. J.* 21: 8910.
16 Sánchez, I.G., Šámal, M., Nejedlý, J. et al. (2017). *Chem. Commun.* 53: 4370.

17 Shibata, Y. and Tanaka, K. (2012). *Synthesis* 44: 323.

18 Tanaka, K. and Shirasaka, K. (2003). *Org. Lett.* 5: 4697.

19 Takeishi, K., Sugishima, K., Sasaki, K., and Tanaka, K. (2004). *Chem. A Eur. J.* 10: 5681.

20 Tanaka, K., Kamisawa, A., Suda, T. et al. (2007). *J. Am. Chem. Soc.* 129: 12078.

21 Gleiter, R. and Kratz, D. (1990). *Angew. Chem. Int. Ed.* 29: 276.

22 Tran-Van, A.-F., Götz, S., Neuburger, M., and Wegner, H.A. (2014). *Org. Lett.* 16: 2410.

23 Yamano, R., Shibata, Y., and Tanaka, K. (2018). *Chem. A Eur. J.* 24: 6364.

24 Yamano, R., Hara, J., Murayama, K. et al. (2017). *Org. Lett.* 19: 42.

25 Kimura, Y., Fukawa, N., Miyauchi, Y. et al. (2014). *Angew. Chem. Int. Ed.* 53: 8480.

26 Kimura, Y., Shibata, Y., and Tanaka, K. (2019). *Eur. J. Org. Chem.*: 1390.

27 Kinoshita, S., Yamaono, R., Shibata, Y. et al. (2020). *Angew. Chem. Int. Ed.* 59: 11020.

28 Tanaka, K., Fukawa, N., Suda, T., and Noguchi, K. (2009). *Angew. Chem. Int. Ed.* 48: 5470.

29 Fukawa, N., Osaka, T., Noguchi, K., and Tanaka, K. (2010). *Org. Lett.* 12: 1324.

30 Sawada, Y., Furumi, S., Takai, A. et al. (2012). *J. Am. Chem. Soc.* 134: 4080.

31 Murayama, K., Shibata, Y., Sugiyama, H. et al. (2017). *J. Org. Chem.* 82: 1136.

32 Murayama, K., Oike, Y., Furumi, S. et al. (2015). *Eur. J. Org. Chem.*: 1409.

33 Shibata, T., Uchiyama, T., Yoshinami, Y. et al. (2012). *Chem. Commun.* 48: 1311.

34 Caeiro, J., Pena, D., Cobas, A. et al. (2006). *Adv. Synth. Catal.* 348: 2466.

35 Yubuta, A., Hosokawa, T., Gon, M. et al. (2020). *J. Am. Chem. Soc.* 142: 10025.

36 Nakamura, K., Furumi, S., Takeuchi, M. et al. (2014). *J. Am. Chem. Soc.* 136: 5555.

37 Satoh, M., Shibata, Y., and Tanaka, K. (2018). *Chem. A Eur. J.* 24: 5434.

38 Tanaka, M., Shibata, Y., Nakamura, K. et al. (2016). *Chem. A Eur. J.* 22: 9537.

39 Usui, K., Yamamoto, K., Ueno, Y. et al. (2018). *Chem. A Eur. J.* 24: 14617.

40 González-Fernández, E., Nicholls, L.D.M., Schaaf, L.D. et al. (2017). *J. Am. Chem. Soc.* 139: 1428.

41 Nicholls, L.D.M., Marx, M., Hartung, T. et al. (2018). *ACS Catal.* 8: 6079.

42 Hartung, T., Machleid, R., Simon, M. et al. (2020). *Angew. Chem. Int. Ed.* 59: 5660.

43 Sako, M., Takeuchi, Y., Tsujihara, T. et al. (2016). *J. Am. Chem. Soc.* 138: 11481.

44 Rajca, A., Miyasaka, M., Pink, M. et al. (2004). *J. Am. Chem. Soc.* 126: 15211.

45 Doulcet, J. and Stephenson, G.R. (2015). *Chem. A Eur. J.* 21: 18677.

46 Doulcet, J. and Stephenson, G.R. (2015). *Chem. A Eur. J.* 21: 13431.

47 Kötzner, L., Webber, M., Martínez, J.A. et al. (2014). *Angew. Chem. Int. Ed.* 53: 5202.

48 Jia, S., Li, S., Liu, Y. et al. (2019). *Angew. Chem. Int. Ed.*: 18496.

49 Liu, P., Bao, X., Naubron, J.-V. et al. (2020). *J. Am. Chem. Soc.* 142: 16199.

4

Cationic Triarylcarbenium Helicenes: Synthesis, Resolution, and Applications

Johann Bosson[1,2], Niels Bisballe[3], Bo W. Laursen[3], and Jérôme Lacour[1]

[1] *Department of Organic Chemistry, University of Geneva, Geneva, Switzerland*
[2] *Department of Chemistry, Koç University, Sarıyer/Istanbul, Turkey*
[3] *Department of Chemistry, University of Copenhagen, Copenhagen, Denmark*

4.1 Introduction

Cationic triarylmethyl Ar_3C^+ dyes (Figure 4.1, panel a) have been topics of importance since the beginning of academic and industrial dyestuff applications [1]. In classical structures, steric repulsions occur among the aryl groups and a twist of the aromatic rings is enforced that reduce the efficacy of charge delocalization and electronic stabilizations. By the bridging of *ortho*-positions of the aromatic rings, the rich photophysics and redox chemistry of the cationic π systems can be enhanced, and, of interest for this particular review, stable helical triarylcarbenium ions can be prepared with unique properties compared to other [n]helicene series. Herein, a general overview of this cationic [n]helicene domain is presented including most recent references that have appeared since the previous review [2]. This book chapter will cover synthetic approaches to the general structures in Figure 4.1, panel b and extensions thereof, physicochemical properties, and various applications from biology to material sciences.

The first examples of triarylcarbenium with two *ortho*-bridges, namely, dioxa **1** and gem-dimethyl carbo **2** [4]helicenes were reported by Neunhoeffer and Haase (1958) [3] and Hellwinkel (1978) [4], respectively. However, their relatively low chemical stability and lack of configurational stability precluded a general use and subsequent applications. In 2000, the synthesis of chemically and configurationally stable dimethoxyquinacridinium **3** (DMQA, Figure 4.1, panel c) was reported, and it constituted a game changer [5]. Using a simple and flexible one-pot process, the gram scale preparation of a [4]helicene was now possible; the compound displays, in addition, absorption and fluorescence in the red part of the visible spectrum and multiple stable redox states. This cationic DMQA derivative **3** opened up this area of heterohelicene chemistry. As it will be shown in the following paragraphs, the side chains that are attached to the nitrogen atoms (R or Ar groups) do not influence significantly most of the chemical and electronic properties of the diaza [4]helicene. As such,

Helicenes: Synthesis, Properties, and Applications, First Edition.
Edited by Jeanne Crassous, Irena G. Stará, and Ivo Starý.

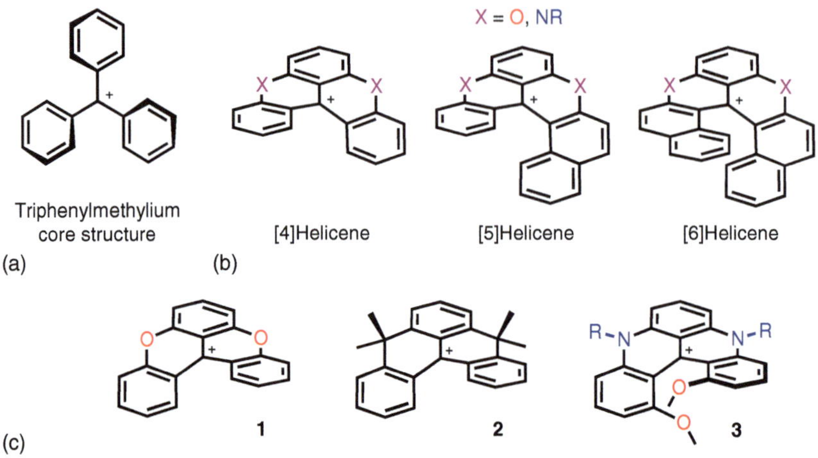

Figure 4.1 (a) Triphenylmethylium cation. (b) Core structure of cationic [4]-, [5]-, and [6]helicenes. (c) First reported cationic helicenes **1** and **2** and configurationally stable cationic [4]helicene **3**, (*M*)-configuration shown.

structure **3** stands for a core DMQA skeleton with any formal side chains. Compounds with specific arms will be identified separately in the course of the chapter. This general comment on the (lack of) influence of the pendant chains holds for all azahelicenes. Also, for practical reasons and, more importantly, because synthesis and properties are usually unaffected by the nature of the negative counterions, the associated anions will not be specified unless necessary for targeted synthesis or applications.

4.2 Synthesis and Resolution of Cationic [4]-, [5]-, and [6]Helicenes

4.2.1 [4]Helicenes

As mentioned, cationic [4]helicene **3** can be obtained in a one-pot reaction from the readily available triarylcarbenium salt **4** [6] by reaction with primary amines (Scheme 4.1) [5]. This aromatic nucleophilic substitution of the *ortho*-methoxy groups leads to formation of nitrogen bridges forming first the acridinium system **5** and subsequently DMQA **3**. As nitrogen bridges are introduced, the (electrophilic) reactivity of the cations decreases, allowing for stepwise and selective introductions of different amines if necessary. At forcing conditions, third additional oxygen or nitrogen bridges can be formed by cleavage of a methoxy group or by addition of an extra amine moiety, yielding the planar triangulenium dyes triazatriangulenium (TATA) **6** and diazaoxatriangulenium (DAOTA) **7**, respectively, which are beyond the scope of this review.

The first nitrogen bridge is usually obtained by reaction of **4** with 1.0–2.5 equivalents of aliphatic amines at room temperature (or anilines at approx.

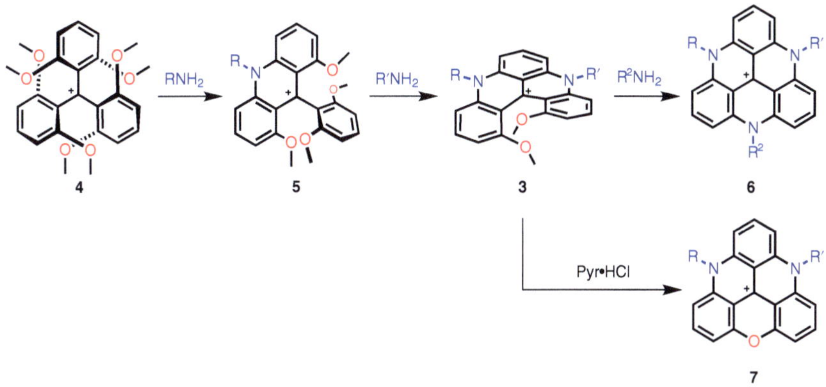

Scheme 4.1 Direct synthesis of **3** starting from triarylcarbenium salt **4** and extension to triangulenes TATA **6** and DAOTA **7**.

50–100 °C) [5b, 7], while the second substitution is most conveniently performed with excess of aliphatic amines (*ca.* 25 equivalents) and elevated temperatures (90–100 °C). If hydrazine NH_2NH_2 is used as nucleophile in one of the two cyclization steps, then helical pH-sensitive quinacridines are obtained instead (see, for instance, Scheme 4.9) [8]. The first chiral resolution of DMQA **3** was achieved by selective precipitation of one diastereomeric ion pair with enantiopure BINPHAT anion [9] (bis(tetrachlorobenzenediolato)mono([1,1′]binaphthalenyl-2,2′-diolato)phosphate(V) anion) [10]. While trivial in the case of simple derivatives bearing *n*-propyl arms, this procedure requires optimization of the precipitation conditions with other side chains (hydrophilic, hydrophobic, functionalized, etc.). In addition, only the precipitated fraction is diastereomerically pure giving rise to a single enantiopure fraction of **3** after chiral anion removal. A more general method for large-scale separation of both enantiopure (*P*)- and (*M*)-**3**, of right- and left-handed configurations, respectively, was later reported [11]. It consisted of (i) the addition of an enantiopure sulfoxide moiety to the cationic center of the [4]helicene followed (ii) by simple chromatographic separation of the two neutral diastereomers **8** (ΔR_f up to 0.3) and (iii) regeneration of the DMQA cation by cleaving of the chiral sulfoxide auxiliary under acidic/electrophilic conditions (Scheme 4.2). Regeneration of the helical cation, however, relies on an uncommon Pummerer fragmentation related to the unusually high cation stability of DMQA (and related triangulenium systems, $pK_{R+} > 14.5$, Table 4.2) [16]. As such, it is not equally efficient for helicenes with one or two oxygen bridges ($pK_{R+} < 14.5$). It also fails for particularly small side chains [17]. Nevertheless, for most derivatives **3**, the process is highly reliable and can afford large quantities of enantiopure (*P*)- and (*M*)-**3** derivatives, even on gram scale. With recent improvements and general availability of chiral stationary phases (CSP) that are amenable to a large range of solvents and additives, researchers can now rely with confidence on analytical and semi-preparative CSP-HPLC separations, which have become the method of choice for resolutions of new cationic helicenes on milligram scale [8a, 18].

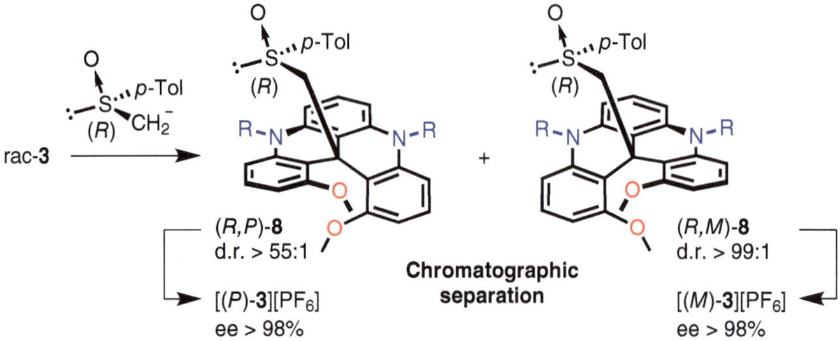

Scheme 4.2 Gram-scale resolution of (*P*)- and (*M*)-**3** by a chiral auxiliary approach.

Several [4]helicene congeners of DMQA with bridging heteroatoms other than nitrogen have been obtained through various synthetic pathways (Scheme 4.3). To make mixed azaoxa and dioxa [4]helicenes, **10** and **11**, respectively, two synthetic routes were proposed in each case. For dimethoxychromenoacridinium **10** (DMCA), one route considered an oxygen ring formation first to derivatives **9** [18b], then final N-ring closure affording **10** along with minor amounts of azadioxatriangulenium (ADOTA) as a by-product [5b]. Otherwise, it was possible to do the contrary and close the N-ring first (acridinium **5**) [13]. Then, partial oxygen ring closure under ether cleavage conditions afforded **10** in moderate to good yields plus small amounts of ADOTA [14]. Dioxa [4]helicene **11** (dimethoxychromenoxanthenium DMCX) was

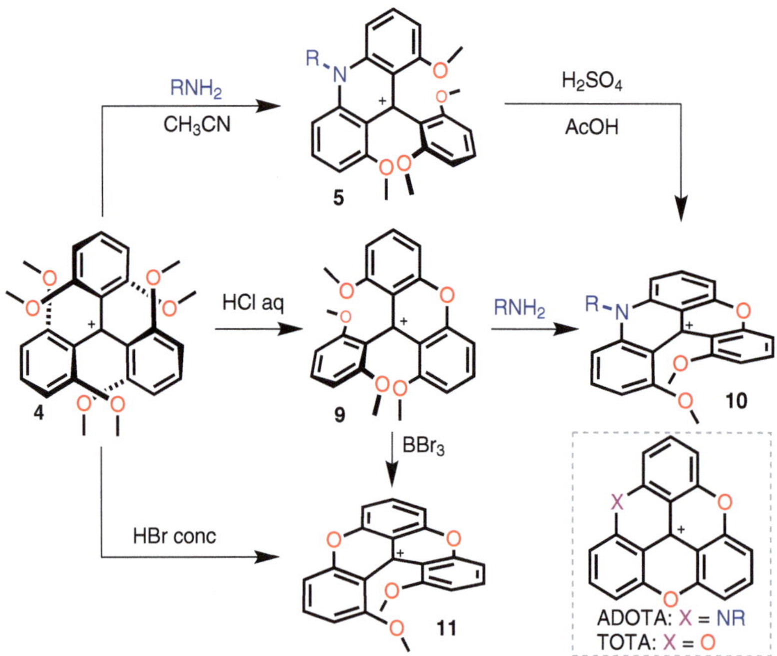

Scheme 4.3 Synthetic approaches to azaoxa and dioxa [4]helicenes **10** and **11**.

prepared in *circa* 15% yield both by a direct synthesis from **4** in HBr [12] and in a five-step procedure [19] applying partial demethylation of **9** with BBr_3. In both cases, low yields are due to the favorable formation of the fully ring closed trioxatriangulenium (TOTA) system.

Sulfur bridged cationic [4]helicene **13** could not be obtained by nucleophilic replacement of the methoxy groups of **4** by sulfur nucleophiles as in the DMQA synthesis (Scheme 4.1). A stepwise process was necessary using a conventional synthesis of thioxanthenium **12** (Scheme 4.4) before the substitution strategy could be used to introduce the nitrogen bridge yielding azathia [4]helicene **13** [20].

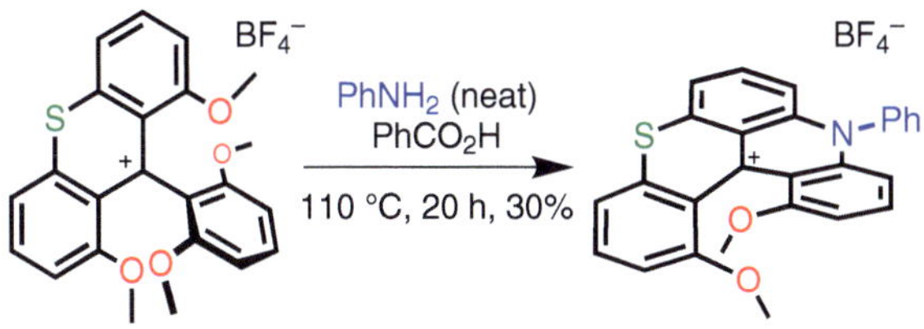

Scheme 4.4 Synthesis of azathia [4]helicene **13**.

During these studies, care was taken to expand, in parallel, the [4]helicene family (**3, 10, 11**) to higher order [5]- and [6]helicene derivatives, by the formal substitution of one or two of the terminal OMe groups by fused benzo moieties. In practice, the extended [6]helicenes were prepared prior to the [5] series, and this review will respect this chronological aspect.

4.2.2 [6]Helicenes

The construction of cationic [6]helicenes required the preparation of an advanced triaryl precursor **19** via xanthone **18** (Scheme 4.5, panel a) [15]. In short, 2-methoxy-1-naphthaldehyde (**14**) was reacted with *in situ* prepared (2,6-dimethoxyphenyl)lithium to afford the resulting alcohol **15**, which was then oxidized to the corresponding ketone **16**. Selective demethylation (BBr_3) to ketone **17** followed by a thermal cyclization (neat) afforded xanthone **18**. For the next step, $CeCl_3$ was necessary to achieve the third arylation of **18** with (2-methoxynaphthalen-1-yl)lithium as nucleophile. Then, after acidic treatment with HBF_4, cationic precursor **19** was provided in 58% overall yield for the five steps (up to 10 g scale of product). Then, with **19** in hand, treatment in molten pyridinium hydrochloride salt (224 °C, 2 minutes) afforded dioxa [6]helicene **20** (95%). Azaoxa derivative **21** (40%) was prepared by addition of propyl amine followed, after solvent removal, by a thermal treatment at 200 °C of the non-isolated intermediate. Diaza derivative **22** was formed directly from **19** at high temperature in the presence of an excess of primary amine (microwave irradiation, 170 °C, 5 minutes) in moderate yields (43–47%, Scheme 4.5, panel b). Similar to **DMQA**, diaza **22** could be resolved into single enantiomers via a diastereoselective precipitation after counterion

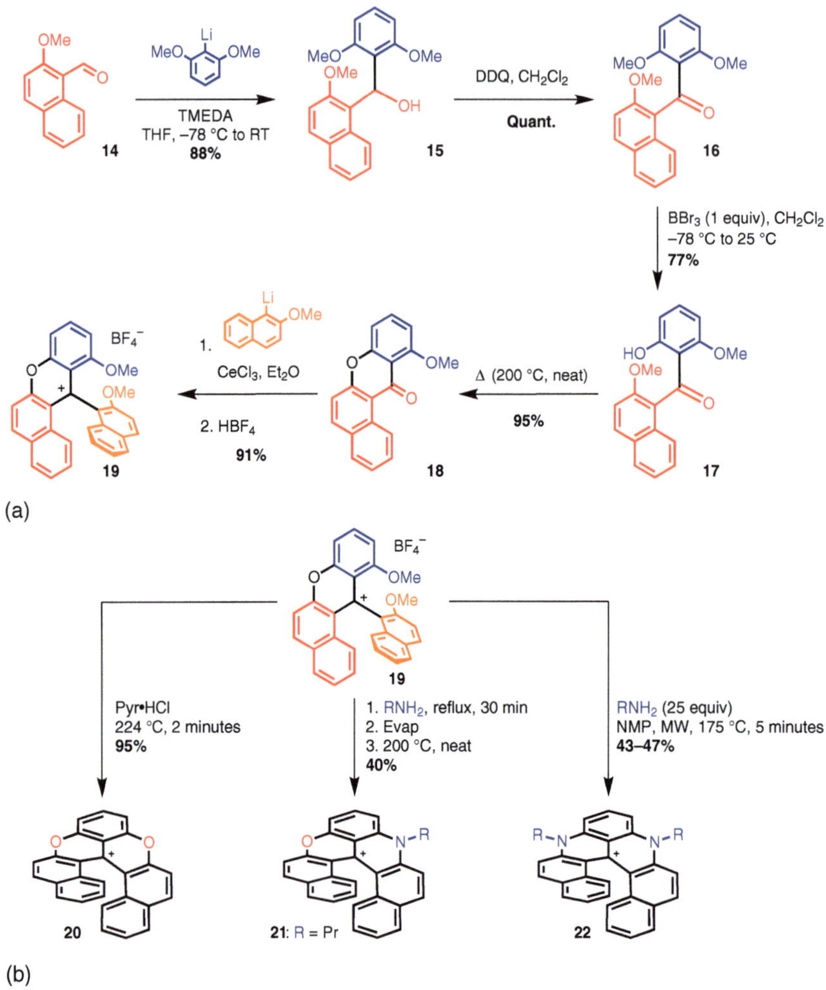

Scheme 4.5 (a) Preparation of common precursor **19**. (b) Formation of cationic dioxa, azaoxa, and diaza [6]helicenes **20**, **21**, and **22**.

metathesis using enantiopure BINPHAT anion. The resolution of **20**, **21**, and **22** could also be performed by CSP-HPLC [18c]. Replacement of the O atoms inside the helices by more electron-donating N atoms induces a notable enhancement of the chemical stability of the cations (see the succeeding texts) [15b] and of the configurational stability as well, with $\Delta G^{\ddagger}$ values of 125, 153, and >155 kJ mol^{-1} at 130, 170, and 180 °C for **20**, **21**, and **22**, respectively (see Table 4.1 in section 4.4.1).

More recently, the direct transformation of dioxa **20** to azaoxa **21** and diaza **22** was disclosed (Scheme 4.6). The use of benzoic acid to promote ring closures and the overall mild reaction conditions (up to 70 °C) permits the introduction of a variety of primary amines, including thermally sensitive ones [24]. Moreover, reaction **21** → **22** generalizes the formation of unsymmetrically substituted diaza [6]helicenes if necessary (R ≠ R′). In the enantiopure series, the conversion of either (*M*)- or (*P*)-dioxa

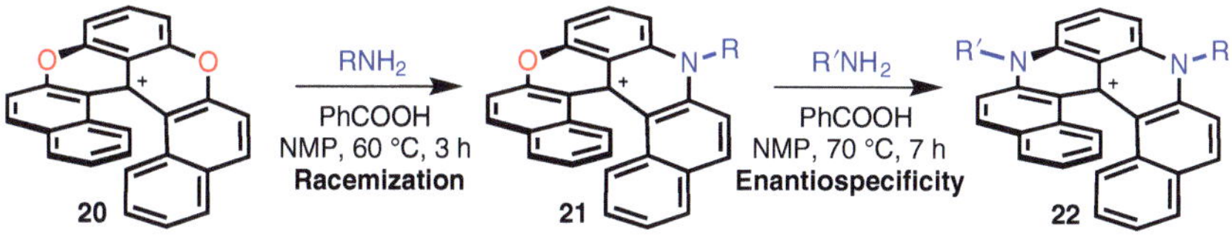

Scheme 4.6 Direct transformation of dioxa **20** to azaoxa **21** and diaza **22** under mild reaction conditions, (*P*)-enantiomers shown.

20 to azaoxa **21** led to racemic mixtures essentially, whereas the transformation from azaoxa **21** to diaza **22** revealed to be enantiospecific with retention of configuration. This was rationalized via a combined experimental and computational investigation of the reaction mechanism, pointing to the formation of propeller triarylcarbenium ion intermediates in the reaction **20** → **21**, those species being not configurationally stable under the reaction conditions.

4.2.3 [5]Helicenes

The first cationic [5]helicenes of this series were synthesized as intermediates in the preparation of extended benzo-fused diazatriangulenium dyes (BDATA **29**, Scheme 4.7) [25]. The xanthone precursor **18**, also used in the [6]helicene synthesis (Scheme 4.5), is now obtained by a simplified three-step method. Additions of 2-methoxyphenyllithium or 2,6-dimethoxyphenyllithium to **18** proceeded without $CeCl_3$ and gave the corresponding xanthene derivatives **23** and **24**, respectively, upon acidification. Under acidic methylether cleavage conditions in molten pyridine hydrochloride, plain dioxa [5]helicene **25** was obtained. Subsequent treatment with methylamine at 90–95 °C for 18 hours gave access to the diaza [5]helicene **26** in modest yields. Under identical conditions, **24** could be converted directly to the diaza [5]helicene **27**, featuring a methoxy substituent in the fjord region. The corresponding dioxa [5]helicene **28** was obtained by treatment of **24** in refluxing in

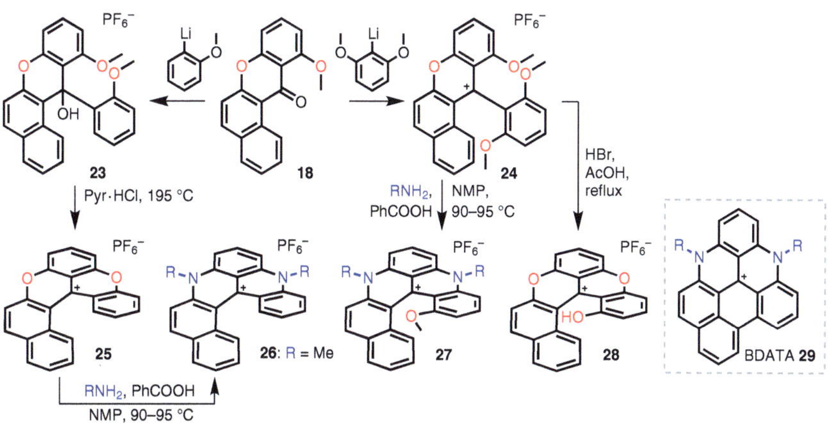

Scheme 4.7 Synthesis of the first reported cationic [5]helicenes.

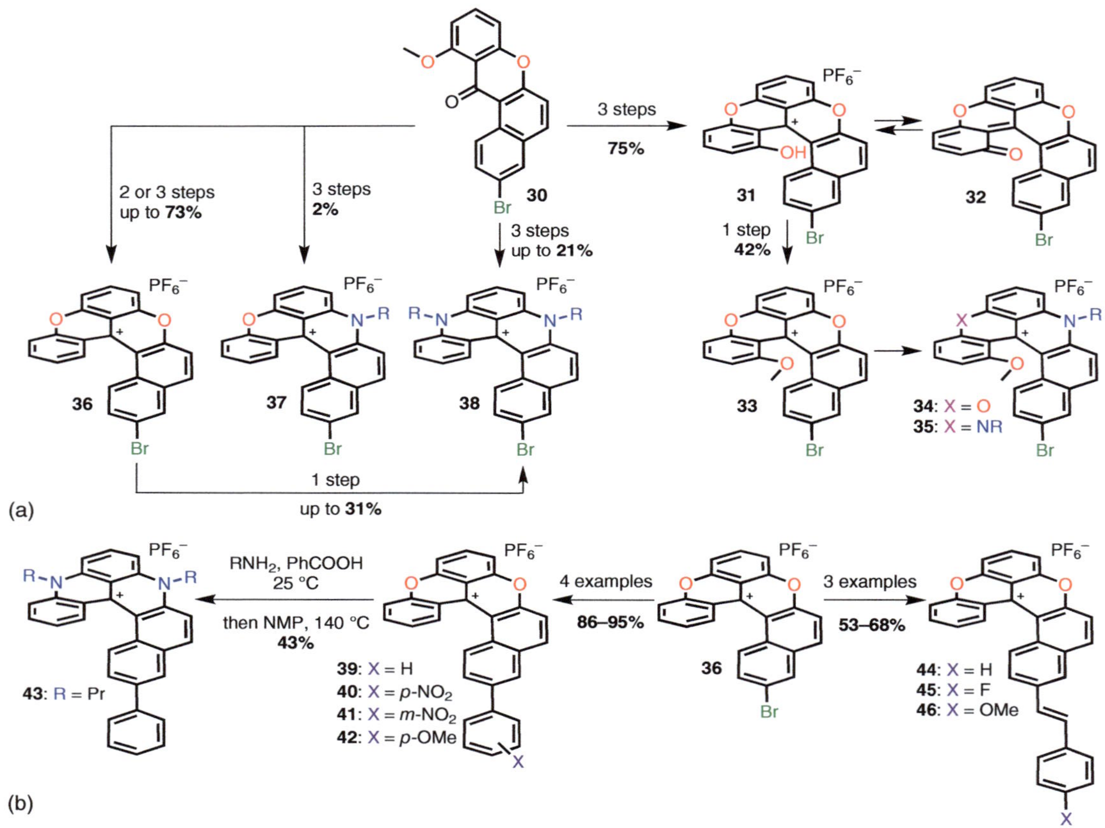

Scheme 4.8 (a) Bromo-functionalized cationic [5]helicenes. (b) Derivatization by cross-coupling reactions.

HBr/acetic acid, the reaction conditions leading to a hydroxyl group in the fjord region. Planarization of the [5]helicenes to yield BDATA **29** was achieved for both **26** and **27** by heating in polyphosphoric acid (PPA) [25]. This process was significantly accelerated by the presence of the methoxy group in **27** [25a]. A similar reactivity was, however, not observed for the dioxa variants **25** and **28**.

The cationic [5]helicenes have been further functionalized by arylation and styrylation (Scheme 4.8, panel a) [22]. In this instance, a brominated benzo[a]xanthone **30** was prepared in four steps and served as key synthetic intermediate to afford the corresponding helicenes with the bromine atom positioned on the extended naphthyl ring. It was possible to obtain both dioxa and diaza [5]helicenes **36** and **38**, respectively, and even the mixed azaoxa derivative **37** was isolated in small quantities of a single regioisomer. Derivatives carrying a methoxy group in the fjord region could be synthesized from **30** as well. Hydroxyl-containing dioxa **31** was obtained in three steps from **30** (75% combined yield), this derivative yielding its neutral counterpart **32** by simple deprotonation. Methylation of the hydroxyl group of **31** afforded dioxa [5]helicene **33** (42%), which was then converted to the corresponding azaoxa and diaza [5]helicenes **34** and **35**, respectively. Starting from the brominated dioxa [5]helicene **36**, Suzuki–Miyaura cross-coupling reactions with a selection of aryl and styrylboronic acids gave the corresponding functionalized dioxa [5]helicenes **39–42** and **44–46** (Scheme 4.8, panel b). Derivatives carrying electron donating and withdrawing groups (electron donating group [EDG]s and electron withdrawing group [EWG]s, respectively), for both the aryl and styryl variants, were obtained. Treating the phenylated dioxa [5]helicene **39** with excess of a primary amine and benzoic acid at 140 °C gave access to the diaza derivative **43** in moderate 43% yield.

4.3 Regioselective Functionalization

4.3.1 Generalities: Nomenclature

In this part, the development of transformations to expand the scope of cationic helicenes, and series of [4]- and [6]helicenes in particular, by regioselective late (and early)-stage functionalization of the helical core will be described. In terms of terminology, in a certain analogy with the language used in natural product synthesis, geographical expressions such as "north," "south," or even "tropical" will be used to indicate the positions of the introduced functional groups (Figure 4.2).

4.3.2 Late-stage Functionalization

4.3.2.1 North Functionalization of Cationic [4]- and [6]Helicenes

Somewhat counterintuitively, cationic diaza [4]- and [6]helicenes are electron-rich compounds. As mentioned briefly, the chemical stability of those derivatives arises from the large delocalization of their positive charge, as depicted by their extended pK_{R+} values (see the succeeding texts). Further description of their remarkable stability will be presented later in the course of this chapter. In fact, the

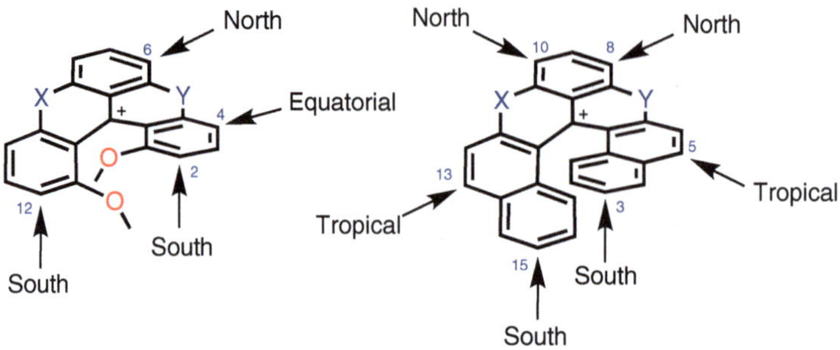

Figure 4.2 Functionalization pattern of cationic [4]- and [6]helicenes.

northern benzene ring in compounds **3** and **22**, flanked by the two intramolecular amino groups, is particularly electron rich in comparison with the other rings. As a consequence, electrophilic aromatic substitutions happen on this north side, i.e. in positions *ortho/para* to the amino groups that are position 6 for diaza [4]helicene **3** and 8 and 10 for diaza [6]helicene **22** (Figure 4.2). In the case of diaza [4]helicene **3** [26], nitration reactions occur readily by treatment under biphasic conditions (HNO_3:CH_2Cl_2), providing mononitro derivative **47** (Figure 4.3). The nitro group can be further derivatized, for instance, via reduction reactions to the amino group and further manipulations. Also, Vilsmeier–Haack reactions afford mono carboxaldehyde **48**. This derivative represents a particularly valuable synthetic intermediate easily transformed into many different [4]helicenes carrying functional groups ranging from alcohols to acids [27], esters, and amides [28], but also into nitriles, alkenes, or alkynes [26]. In fact, the aldehyde can be oxidized to the corresponding carboxylic acid through a Pinnick Kraus oxidation, opening the way to esterification and amidation reactions. Amide and ester derivatives will be particularly useful for the development of specific applications (see the

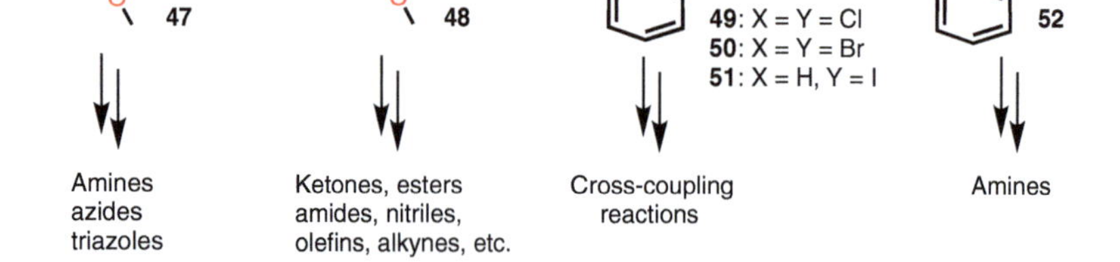

Figure 4.3 Functionalization of position 6 of cationic diaza [4]helicene **3**, leading to monofunctionalized [4]helicenes **47** and **48**, and of positions 8 and/or 10 of **22**, leading to functionalized diaza [6]helicenes **49-52**, and possible further derivatizations.

succeeding texts). Other transformations provided many other auxochromes that are detailed in the original publication [26]. Diaza [4]helicene **53** was also effectively formylated in position 6 providing carboxaldehyde **54** in good yields (Scheme 4.9) [29]. Compounds of type **54** were then converted to chiral BODIPY derivatives **55** in a one-pot two-step procedure (combined yields of 28–82%). Focusing on diaza [6]helicene **22** [15], regioselective halogenation reactions are uniquely possible in positions 8 and/or 10. More specifically, dichloro and dibromo derivatives **49** and **50** can be obtained by simple treatment with *N*-chlorosuccinimide (NCS) and *N*-bromosuccinimide (NBS), respectively, while *N*-iodosuccinimide (NIS) afforded mono iodo derivative **51**. Dibromo **50** was further engaged into palladium-catalyzed Sonogashira and Suzuki coupling reactions, allowing a wide scope of derivatizations at positions 8 and 10. Finally, nitration reactions could also be promoted yielding dinitro derivative **52**.

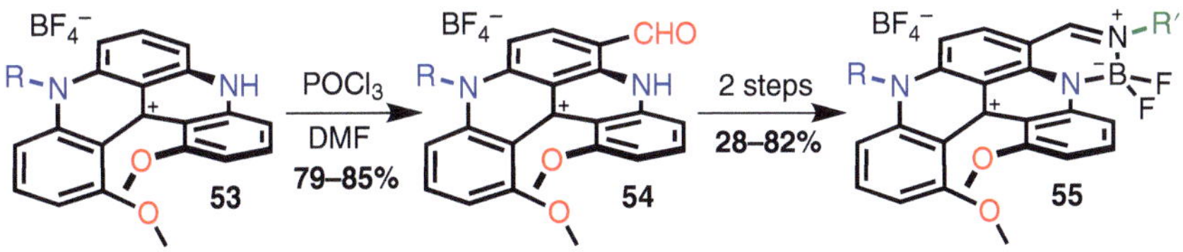

Scheme 4.9 Synthesis of chiral BODIPY derivatives **55**.

4.3.2.2 South Functionalization of Cationic [4]- and [6]helicenes

Interestingly, and somewhat surprisingly, the regioselectivity of the electrophilic substitution reactions can be drastically altered if they are performed under strongly acidic conditions. In fact, such conditions are known to promote the reversibility of substitution reactions [30]. Then, considering the steric encumbrance created by the lateral chains on the N-atoms, substitutions at positions 6, 8 and 10 of cationic diaza [4]- and [6]helicenes **3** and **22** are not thermodynamically favored, and, conversely, substitutions occurring at positions 2 and 12 of **3** and 3 and 15 of **22** are preferred (Figure 4.4). More specifically, Friedel–Crafts alkylation, acylation, and sulfonylation reactions performed in PPA or in Eaton's reagent (7.5% P_2O_5 in CH_3SO_3H) [31] with **3** as substrate lead over time to derivatives functionalized at position 2 and/or 12 exclusively, to give **56–58** [32]. Under similar reaction conditions, alkylation, acylation, and sulfonation reactions of **22** happened at position 3 exclusively, to give **59–61**.

Furthermore, acylation of **3** with salicylic or thiosalicylic acid derivatives allowed for the construction of extended cationic [5]helicenes (Scheme 4.10). From the acetylated diaza [4]helicene **62**, upon basification of the reaction medium, nucleophilic aromatic substitutions of one of the methoxy groups by the alcohol or thiol functional group occurred and generated the corresponding (thio)xanthone **63**. Furthermore, this *in situ* acylation/S_NAr sequence is enantiospecific. Then, under even more strongly acidic conditions (7.5% P_2O_5 in CF_3SO_3H) [33] and using α-halogenated benzoic acids, further acylations were permitted and could lead,

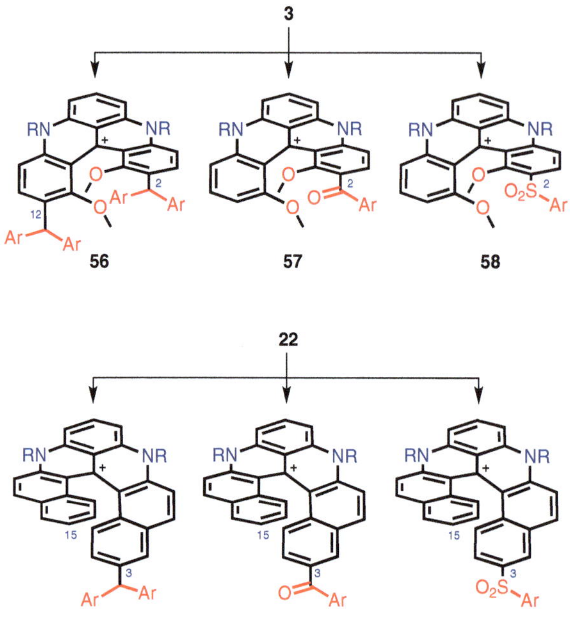

Figure 4.4 Functionalization of positions 2 and/or 12 of cationic diaza [4]helicenes and of positions 3 of cationic diaza [6]helicenes promoted by strongly acidic conditions.

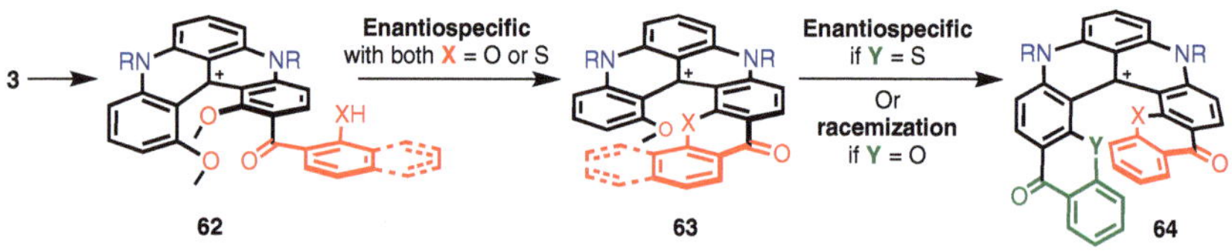

Scheme 4.10 Enantiospecific access to extended [5]- and [6]helicenes.

after heteroatom exchanges, to the formation of extended [6]helicenes **64**. In this particular situation, moving from [5]helicenes **63** to [6]helicenes **64**, the enantiospecificity strongly depends upon the nature of the heteroatoms X and Y that are introduced in the first [4] → [5] and second [5] → [6] ring closures. It leads to full enantiospecificity or complete racemization; the reasons are detailed in the original publication [32].

Such a strategy was used for the construction of helical tetracenes and pentacenes, obtained in three-step sequences from core structures **3** or **22** [18d]. Selective mono- or bisacylation reactions using 2-bromobenzoic acid as reagent afforded mono- (**65** and **67**) and bisketones (**66** and **68**). Those derivatives were then engaged in Buchwald–Hartwig cross-coupling reactions with aniline; final oxidative Cu-catalyzed ring closures afford the targeted helicene-acene hybrids (Figure 4.5). It is important to note that in pentacenes **69** and **70**, the aniline groups (in red) are positioned on the helicene rim, whereas in tetracenes **71** and **72** they are pointing

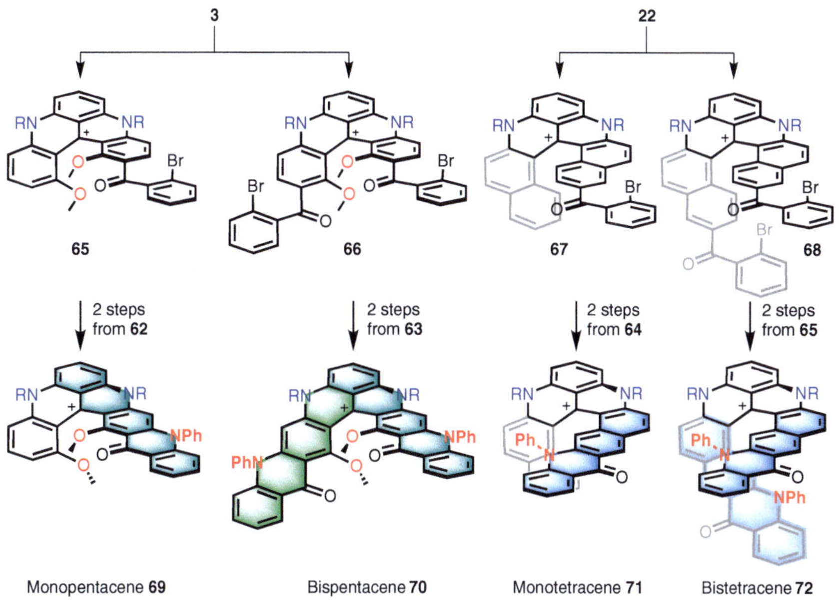

Figure 4.5 Synthesis of mono- and bispentacenes **69** and **70** and of mono- and bistetracenes **71** and **72**.

toward the fjord region. The position of these substituents drastically impacts the helical pitch and the chiroptical properties of the molecules. These two aspects will be discussed in detail in the following paragraphs.

4.3.2.3 Tropical Functionalization Cationic [6]Helicenes

In diaza [4]- and [6]helicenes **3** and **22**, the "north" and "south" functionalization patterns rely on the electron-rich and nucleophilic nature of these moieties. A totally different reactivity and regioselectivity can be afforded starting from electron-poor helicenes instead, and dioxa [6]helicene **20** in particular [18e]. In fact, this derivative can be successfully engaged in oxidative nucleophilic aromatic substitutions. Under such reaction mechanism, reactions happen at *para* position(s) of the formal positive charge of the triarylcarbenium core [34]. Experimentally, treatment of dioxa **20** with strongly nucleophilic 1,3,3-trimethyl-2-methyleneindoline yields mono- or bisaddition products **73** and **74** functionalized at position(s) 5 (and 13), the selectivity being controlled by the stoichiometry of the reagent and the reaction conditions (Scheme 4.11, panel a) [18e]. In a similar fashion, treatment of **20** with di*iso*propylethylamine (DIPEA, Hünig's base) affords enamine derivative **76** (Scheme 4.11, panel b). In this case, the alkyl amine is oxidized *in situ* to the corresponding enamine by **20**, which is reduced to the neutral adduct **75**. In other words, **20** acts both as oxidant and electrophile in this transformation. Derivatives **73**, **74**, and **76** are characterized by strong bathochromic and hyperchromic shifts of their optical properties compared with parent **20**, which relies on the extension of the conjugation in these derivatives. This feature will be discussed later.

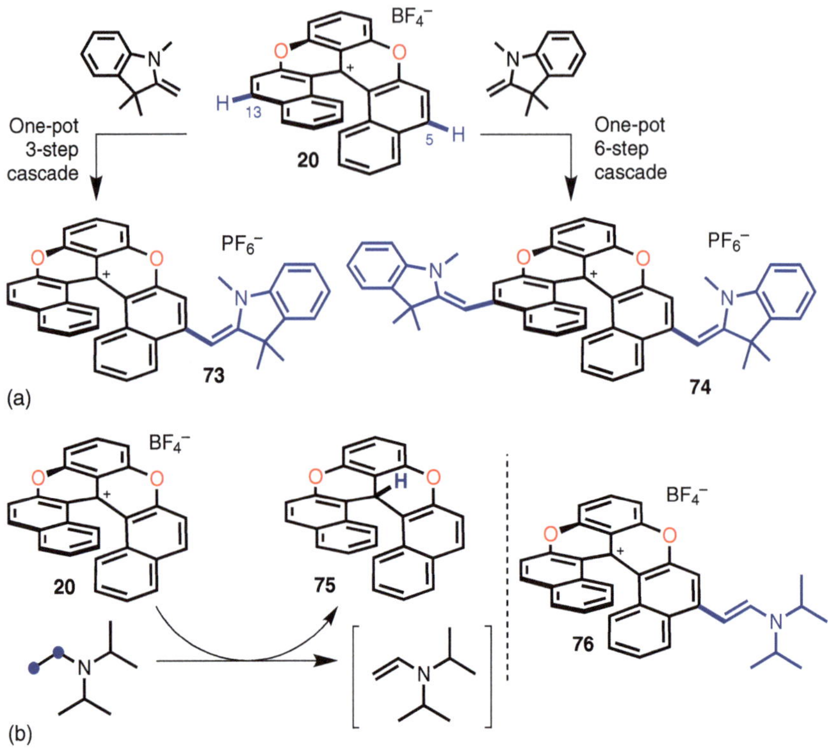

Scheme 4.11 (a) Functionalization of dioxa [6]helicene **20** at positions 5 and 13 with indoline-derived enamines. (b) Oxidation of Hünig's base by **20**, ultimately leading to the formation of **76** by reaction between the enamine and another molecule of **20**.

4.3.2.4 Central Nucleophilic Additions on [4]-, [5]-, and [6]Helicenes

In previous paragraphs, it was stated that cationic diaza [4]- and [6]helicenes, **3** and **22**, are electron-rich species able to react with electrophiles. However, in the presence of strong nucleophiles such as hydrides or organolithium reagents (i.e. H^-, R^-), the cationic derivatives can still undergo nucleophilic additions to their central carbon. Such a reactivity leads to the formation of neutral species, often referred as *leuco* adducts due to their lack of absorption in the visible domain. Prototypical examples are sulfoxides **8** (Scheme 4.2) [11], mentioned in the context of the chiral resolution of **3**, or by-product **75** formed during the oxidation of DIPEA by **20** (Scheme 4.11) [18e]. In various other studies, it was shown that **3**, **10**, **20**–**22**, and **33**–**38** can react with $NaBH_4$ or milder $NaBH_3CN/NaBH(OMe)_3$ to form the corresponding *leuco*-H derivatives; these neutral products are quite easier to purify by chromatography on silica gel or to handle in CSP-HPLC resolutions [14, 18b, c, 22]. Also, with *leuco*-H **77** (obtained from **3**), an umpolung to the corresponding carbanion was possible by deprotonation with *n*-BuLi; further trapping with (soft) electrophiles allows the formation of derivatives **78** (Figure 4.6, panel a) [35]. In other studies, the carbanion $NC\text{-}CH_2^-$, generated by deprotonation of CH_3CN with NaH, was shown to react with **3** to generate the corresponding *leuco*-CH_2CN **79** [10]. Further elaboration

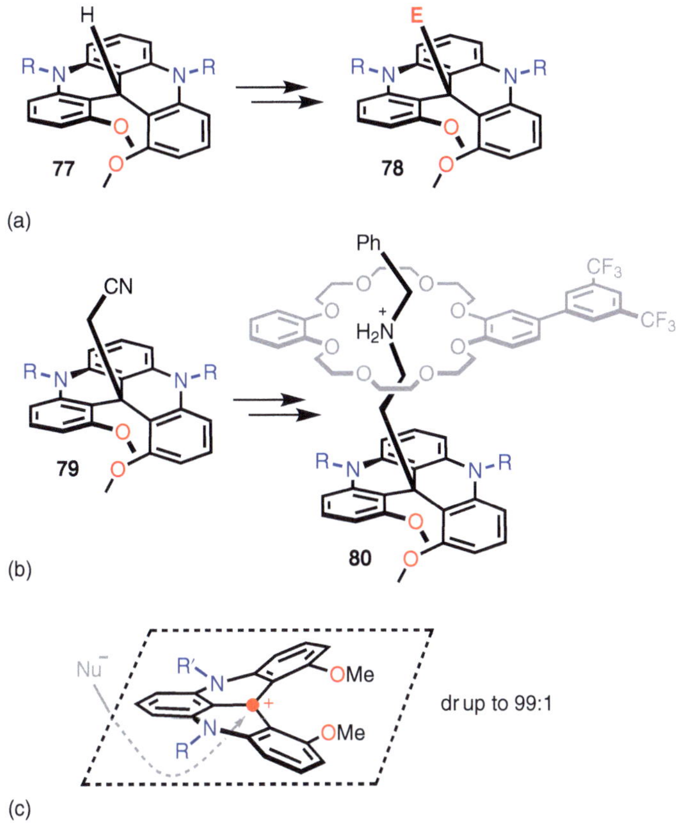

Figure 4.6 (a) Functionalization of *leuco* **77** with electrophiles E^+ via an umpolung strategy. (b) Structure of *pseudo*rotaxanes **80**. (c) Diastereoselectivity during the formation of *leuco* adducts with R ≠ R′.

of **79** permitted the formation chiral *pseudo*rotaxanes **80** (Figure 4.6, panel b), low diastereoselectivity being, however, observed upon the threading of the axis with a nonsymmetrical crown-ether ring [36]. Finally, if non C_2-symmetrical **3** are used (with two different nitrogen side chains R and R′), then the two faces of the [4]helicenes are diastereotopic. Consequently, nucleophilic additions can occur on one preferential *si* or *re* face, and excellent levels of diastereoselectivity were achieved (up to 99:1, Figure 4.6, panel c) if side chains R and R′ are different enough in nature (small methyl *vs.* larger propyl, or small propyl *vs.* larger mesityl). The origin of the selectivity lies probably with the fact that one of the nitrogen atom prefers to remain planar (sp^2 configuration) while the other can bend (pyramidal configuration) to accommodate the newly created stereogenic center [7c, 13].

4.3.3 Early-stage Equatorial Functionalization of [4]Helicene

In recent work, the previously not accessed 4 positions ("equatorial," Figure 4.2) on the diaza [4]helicene **3** were functionalized through selective early-stage chlorination of the acridinium precursor **5** (Scheme 4.12) [37]. This strategy offers an interesting supplement to the well-developed selective electrophilic "north" and "south" functionalization of electron-rich diaza [4]- and [6]helicenes described previously. Electrophilic chlorination of **5** with NCS in CH_3CN gave selectively both mono- and disubstitutions, **81** and **82**, on the dimethoxyphenyl group that is oriented perpendicularly and hence only weakly conjugated to the electron deficient acridinium ring [5b, 38]. Interestingly, formation of the diaza [4]helicene system from **81** by reaction with propylamine under standard conditions gave exclusively the 4-chloro isomer **84**, due to activation of the neighboring methoxy-substituted position in the ring closure reaction. The reaction intermediate **83** for the ring-closing reaction was identified to have the amine first substituted on the acridinium part, and the selectivity arises from the intramolecular S_NAr being significantly faster for the methoxy group neighboring the chlorine.

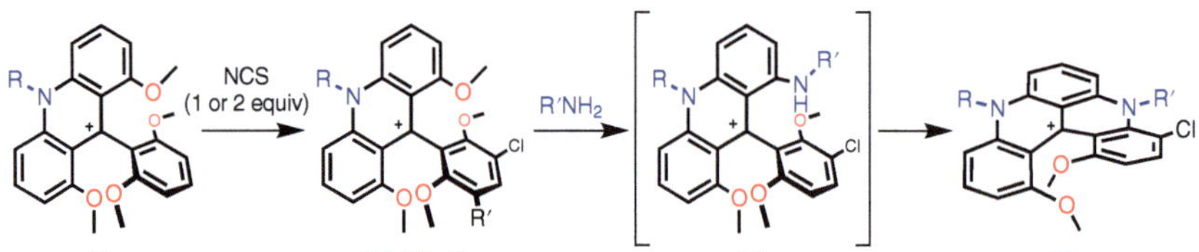

Scheme 4.12 Early-stage functionalization by electrophilic chlorination of acridinium **5**.

4.4 Structural Properties

4.4.1 Helical Pitch and Configurational Stability

The helical structure and configurational stability of the [4]helicene DMQA **3** is ensured by the two methoxy groups at the inner rim of the helix. X-ray crystal structures confirm a helical pitch of 2.7 Å between the two oxygen atoms in the system [10, 18d]. For [5]helicenes (Scheme 4.8), the pitch increases to 2.9 Å without even substituents inside the inner rim (compounds **36** and **38**) and then to 3.1 Å for **33** and the corresponding azaoxa [5]helicene **34** with one methoxy group in the fjord region [22]. In the [6]helicene series helical pitch of 3.1–3.3 Å have been reported [15a, 18d]. These values are quite comparable to those found for simple [6]carbohelicenes (3.2–3.3 Å) [39]. The configurational stability of cationic helicenes has been reported for a range of [4]-, [5]-, and [6]helicenes in terms of racemization barriers; the compilation of these data in Table 4.1 shows several interesting trends. The dimethoxy [4]helicene DMQA **3** is remarkably stable ($\Delta G^{\ddagger} > 173$ kJ mol^{-1}) with a racemization barrier surpassing that of regular [6]carbohelicene [10]. The barrier determined for the dioxa [4]helicene **11** was, however, found to be significantly lower than that of **3**, emphasizing a large impact of the heteroatom bridges on the stability [40] and an increased flexibility brought by oxygen bridges compared with nitrogen [19]. This was later confirmed by theoretical studies of the racemization mechanism of various

82 Saleh, N., Moore, B., Srebro, M. et al. (2015). *Chem. A Eur. J.* 21: 1673–1681.
83 Shen, C., Anger, E., Srebro, M. et al. (2013). *Chem. A Eur. J.* 19: 16722–16728.
84 Anger, E., Rudolph, M., Shen, C. et al. (2011). *J. Am. Chem. Soc.* 133: 3800–3803.
85 Bourissou, D., Guerret, O., Gabbai, F.P., and Bertrand, G. (2000). *Chem. Rev.* 100: 39–92.
86 Hopkinson, M.N., Richter, C., Schedler, M., and Glorius, F. (2014). *Nature* 510: 485–496.
87 Smith, C.A., Narouz, M.R., Lummis, P.A. et al. (2019). *Chem. Rev.* 119: 4986–5056.
88 Hellou, N., Jahier-Diallo, C., Basle, O. et al. (2016). *Chem. Commun.* 52: 9243–9246.
89 Hellou, N., Srebro-Hooper, M., Favereau, L. et al. (2017). *Angew. Chem. Int. Ed.* 56: 8236–8239.
90 Lee, J., Chen, H.-F., Batagoda, T. et al. (2016). *Nat. Mater.* 15: 92–98.
91 Macé, A., Hellou, N., Hammoud, J. et al. (2019). *Helv. Chim. Acta* 102: e1900044.
92 Hafedh, N., Favereau, L., Caytan, E. et al. (2019). *Chirality* 31: 1005–1013.
93 Gauthier, E.S., Abella, L., Hellou, N. et al. (2020). *Angew. Chem. Int. Ed.* 59: 8394–8400.
94 Morgante, P., Captain, B., Chouinard, C.D. et al. (2020). *Tetrahedron Lett.* 61: 152143.
95 Karras, M., D'browski, M., Pohl, R. et al. (2018). *Chem. A Eur. J.* 24: 10994–10998.
96 Gomez-Sandoval, Z., Pena, E., Fonseca Guerra, C. et al. (2009). *Inorg. Chem.* 48: 2714–2716.
97 Roy, M., Berezhnaia, V., Villa, M. et al. (2020). *Angew. Chem. Int. Ed.* 59: 3264–3271.
98 Zhou, Z., Fu, L., Hu, Y. et al. (2020). *Angew. Chem. Int. Ed.* 59: 15923–15927.
99 Katz, T.J. and Slusarek, W. (1979). *J. Am. Chem. Soc.* 101: 4259–4267.
100 Pammer, F., Sun, Y., Sieger, M. et al. (2010). *Organometallics* 29: 6165–6168.
101 Pammer, F., Sun, Y., and Thiel, W.R. (2011). *Inorg. Chim. Acta* 374: 205–210.
102 Pammer, F., Sun, Y., Pagels, M. et al. (2008). *Angew. Chem. Int. Ed.* 47: 3271–3274.
103 Latorre, A., Urbano, A., and Carreno, M.C. (2011). *Chem. Commun.* 47: 8103–8105.
104 Urbano, A., del Hoyo, A.M., Martínez-Carrión, A., and Carreño, M.C. (2019). *Org. Lett.* 21: 4623–4627.
105 Sudhakar, A. and Katz, T.J. (1986). *J. Am. Chem. Soc.* 108: 179–181.
106 Akiyama, M. and Nozaki, K. (2017). *Angew. Chem. Int. Ed.* 56: 2040–2044.
107 Gilbert, A.M., Katz, T.J., Geiger, W.E. et al. (1993). *J. Am. Chem. Soc.* 115: 3199–3211.
108 Sudhakar, A., Katz, T.J., and Yang, B.-W. (1986). *J. Am. Chem. Soc.* 108: 2790–2791.

109 Katz, T.J., Sudhakar, A., Teasley, M.F. et al. (1993). *J. Am. Chem. Soc.* 115: 3182–3198.

110 Rose-Munch, F., Li, M., Rose, E. et al. (2012). *Organometallics* 31: 92–104.

111 Alvarez, C.M., Barbero, H., Garcia-Escudero, L.A. et al. (2012). *Inorg. Chem.* 51: 8103–8111.

112 Fuchter, M.J., Schaefer, J., Judge, D.K. et al. (2012). *Dalton Trans.* 41: 8238–8241.

113 Ito, H., Sakai, H., Okayasu, Y. et al. (2018). *Chem. A Eur. J.* 24: 16889–16894.

114 Mori, T. (2021). *Chem. Rev.* 121: 2373–2412.

6

Tetrathiahelicenes: An Infinite Source of Inspiration

Silvia Cauteruccio[1], Andreas Dreuw[2], Emanuela Licandro[1], and Patrizia Romana Mussini[1]

[1] *Department of Chemistry, University of Milan, Milan, Italy*
[2] *Interdisciplinary Center for Scientific Computing, Heidelberg University, Heidelberg, Germany*

6.1 Introduction

An appropriate structural description for tetrathiahelicenes is the following: "members of a class of poliheteroaromatic chiral molecules containing alternated thiophene and benzene rings assuming a very elegant and intriguing helical shape, which confers them several important properties which allow for interesting applications." However, if from the previous sentence, we remove the word "thiophene," the same definition can be applied for several other classes of helicenes containing different heteroaromatic rings. So, the question could be: why write a chapter specifically on tetrathiahelicenes? Which features does the combination of thiophene and arene rings in helical systems define?

To understand the peculiarity of tetrathiahelicenes, it is useful to consider analogies and differences between the two thiophene and benzene molecules. They are the perfect couple: they share some similarities but, at the same time, they have complementary or different behavior. Thiophene and benzene have the same appearance, both are colorless liquids with almost the same pleasant odor, have similar boiling point but different reactivity that allows selectivity in reactions in which both rings are present. And this is indeed the interesting and peculiar feature of this class of heterohelicenes. The contemporary presence of benzene and thiophene in the helical scaffold confers specific chemical, geometric, and electronic properties and at the same time allows a high regioselective functionalization of the helix with consequent obtainment of tetrathia[7]helicene (7-TH) variously substituted with groups of different nature and properties. This synthetic flexibility allows the modulation of specific properties, making it easier to exploit their potential applications.

This is a reason why we can speak of "a class of tetrathiahelicenes," and this is the reason why this class of compounds has attracted and still attract a lot of interest within the scientific community.

Helicenes: Synthesis, Properties, and Applications, First Edition.
Edited by Jeanne Crassous, Irena G. Stará, and Ivo Starý.

6.2 Synthesis of 7-TH Scaffold

The synthesis of tetrathiahelicene (7-TH) scaffold is the challenging aspect of all the chemistry of 7-TH. In fact, the investigation of properties and behavior of this sulfur-containing helicene requires the availability of efficient, reproducible, and versatile synthetic methodologies to have enough material for studying potential applications. So far, two general synthetic strategies have been reported in the literature for their preparation, namely:

(i) the oxidative cyclization of bis(benzodithienyl)ethenes.
(ii) annulation of 3,3′-bis(benzo[1,2-*b*:4,3-*b*′]dithiophene) derivatives (Scheme 6.1).

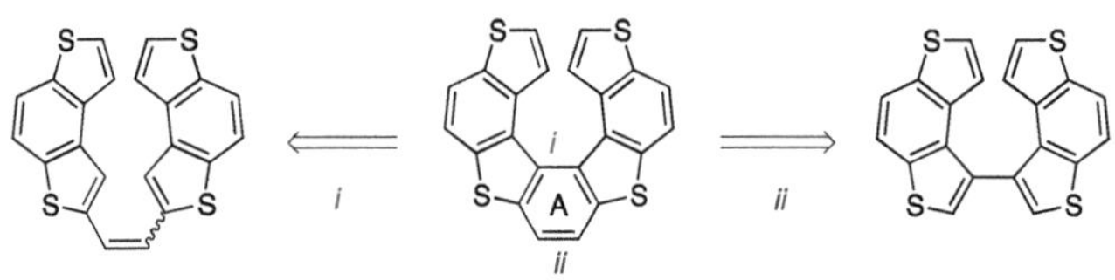

Scheme 6.1 General synthetic strategies for the synthesis of 7-TH scaffold.

The great interest for the synthetic aspects is also related to the fact that these molecules are chiral and for several applications, the enantiopure isomer is required. So far, no enantioselective synthesis has been proposed and we think that this is the real challenge for the next future that will open new horizons in the utilization of these eclectic and promising helical systems.

Here the two synthetic methodologies mentioned earlier will be discussed in detail.

6.2.1 Oxidative Cyclization of Bis(benzodithienyl)ethenes

This methodology is based on the construction of the central arene ring of 7-TH starting from bis(benzodithienyl)ethenes. To obtain 7-TH, two types of oxidative cyclization have been developed:

(a) Photochemical Oxidative Cyclization
(b) Non-photochemical Oxidative Cyclization

Both methods start from the same type of precursor, that is 1,2-bis(benzodithienyl) ethene or its derivatives, that already contain the four thiophene rings of the final 7-TH, on which new arene ring(s) are built. These two methodologies are quite versatile and reproducible and have been widely used to obtain even functionalized 7-TH. The only limitation is the presence of oxidants (O_2 and I_2 in method *(a)* and chemical oxidants in method *(b)*) that render the methodology not suitable when functional groups sensitive to oxidation are present. A further limitation in the case of method *(b)* is the fact that thiophene rings, with a free α-position, are susceptible to oxidation leading to polymerization products.

6.2.1.1 Photochemical Oxidative Cyclization

The first example of a 7-TH prepared through a photochemical oxidative cyclization dates back to 1971 [1] when Wynberg applied the well-established Mallory reaction to the preparation of the title compounds. Standard conditions for the photocyclization are the use of a medium pressure Hg lamp as the source of irradiation or the use of an RPR-100 Rayonet photochemical reactor irradiating in the visible range between 400 and 800 nm [2].

Since then, almost all papers dealing with 7-TH reported the same methods to obtain even functionalized systems [2–13]. The starting compounds in these reactions are 1,2-bis(benzodithienyl)ethenes or analogous alkenes that can be used as a mixture of (*E*)- and (*Z*)-isomers. Of course, only the (*Z*)-isomer is the appropriate diastereoisomer for the photochemical cyclization, but one can take advantage of the photochemical interconversion of the two isomers and the mixture of the two can be used to obtain a good yield of the helicene. However, it has been observed that the use of stereodefined and more soluble (*Z*)-isomers greatly facilitates the photochemical ring closure that, in this case, can be performed in solution instead of in a slurry, using a medium-pressure Hg lamp (125 W) in benzene and in the presence of a catalytic amount of iodine with a significative reduction of the reaction time. This procedure is described later and reported in Scheme 6.4.

The only solvents reported for these reactions are benzene or toluene, which proved to give the best results in terms of yield with a minor amount of by-products. In addition, the use of a catalytic amount of I_2 in the presence of air as an oxidant assures the completion of the reaction. In these conditions, the oxidation of thiophene rings was never observed. A general consideration of this reaction is that, although normally quite efficient, it requires appropriate photochemical apparatus as well as diluted solutions and the scale-up to obtain a large amount of product is therefore not so trivial and feasible. An alternative is offered by the chemical oxidation as described in Paragraph 1.1.2.

In turn, the synthesis of 1,2-bis(benzodithienyl)ethenes or analogous alkenes precursors of 7-TH has been realized mainly with three methods:

- Wittig reaction
- McMurry oxidative coupling
- Cross-coupling reactions

In the earlier-cited seminal paper of Wynberg of 1971 [1], the first tetrathiahelicene **1** was obtained starting from the olefin **2**, in turn, prepared through a double Wittig condensation of the dialdehyde **3** and phosphonium salt **4** (Scheme 6.2).

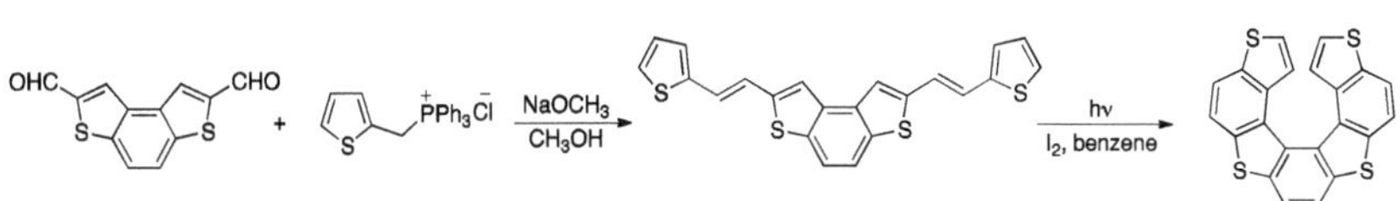

Scheme 6.2 First synthesis of 7-TH **1**.

Since then, the synthesis of olefins as direct precursors of 7-TH was realized using this method that is certainly efficient and reproducible [2–6]. However, this methodology requires several steps to prepare both aldehyde and phosphonium salt.

Only late in 2003 [7], the 1,2-bis-(2-benzo[1,2-*b*;4,3-*b*′]dithiophenyl) ethene **5** was introduced (Scheme 6.3).

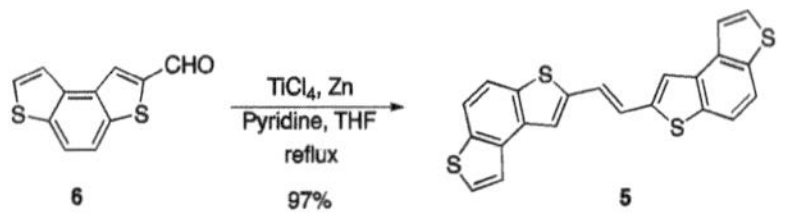

Scheme 6.3 Synthesis of ethene **5**.

The rationale is that even if the Wittig reaction is efficient and reproducible, several steps are required to prepare both aldehyde and phosphonium salt, and sometimes unsatisfactory yields are observed. The phosphonium salt is prepared from the aldehyde itself, which is preliminarily reduced to alcohol, transformed into the corresponding halide, and, in turn, reacted with the triphenylphosphine. Owing to the symmetric substitution on the double bond of olefin **5**, the McMurry coupling, which just utilizes the aldehyde as the starting material, is advantageous because of the reduced number of steps with respect to Wittig condensation. With this methodology, a more convenient procedure for the whole synthesis of 7-TH was introduced. The McMurry reaction is highly versatile being applicable to variously substituted benzodithiophene aldehydes [10, 11].

Both methods described earlier, Wittig condensation and McMurry reductive coupling developed for the preparation of olefins precursors of 7-TH, are certainly valuable and feasible ways, even if, in both cases, alkenes are mainly obtained as *trans*-isomer or as a mixture of *cis* and *trans*-isomers, depending on the nature of the substituent on the carbonyl group.

The insolubility of *trans*-alkene **5** requires that the photochemical isomerization to the *cis*-isomer and the subsequent cyclization to helicene **1** take place in a suspension, and this involves the use of a very large amount of solvent and a very long reaction time.

Looking for a solution to these problems, a simple basic idea was proposed, which is the stereoselective synthesis of *cis*-alkenes in order to verify whether they could be photocyclized to helicenes more efficiently.

The synthesis of *cis*-alkenes was easily realized through a double stereospecific palladium-mediated Suzuki coupling between stereodefined *Z*-diboronic esters and the benzodithiophene (Scheme 6.4) [9].

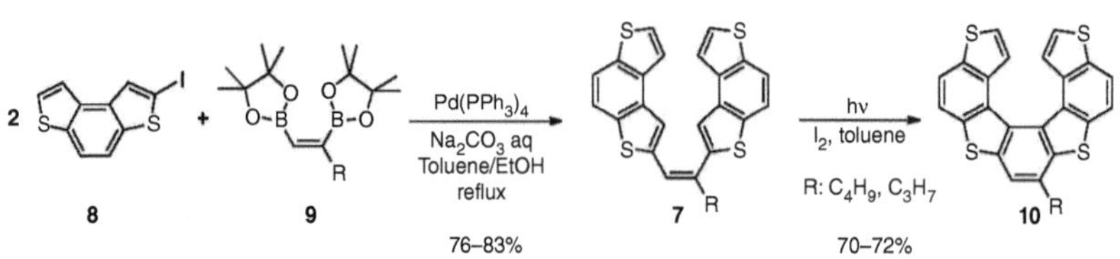

Scheme 6.4 Synthesis of 7-THs **10**.

Despite the fact that this Pd-catalyzed coupling method is generally highly efficient and stereospecific, and is, therefore, a potent tool widely used to assemble aromatic and heteroaromatic systems with unsaturated molecules to obtain highly conjugated π-systems in one step, it was never used to build up alkenes such as benzodithiophene **7**.

The reaction is run with two equivalents of the 2-iodo-benzodithiophene **8** and one equivalent of commercially available (*Z*)-diboronic acid esters **9** in toluene/ethanol, in the presence of a Na_2CO_3 aqueous solution and $Pd(PPh_3)_4$ (3%) as the catalyst (Scheme 6.4). The corresponding alkenes **7** are obtained in good yield as pure *cis* isomers and result very soluble in a number of organic solvents thanks to the nonplanar shape and the presence of alkyl chains. This not only greatly facilitates their spectroscopic characterization but also, more important, considerably facilitates their oxidative photochemical cyclization to the corresponding substituted tetrathia[7]helicenes **10**.

The same concept of forcing in a *cis* configuration double bonds of olefin precursors of 7-TH led to design alkene **11** (Scheme 6.5) containing the maleimide functionality, which imparts the right *cis* stereochemistry for efficient photocyclization and higher solubility [12, 13].

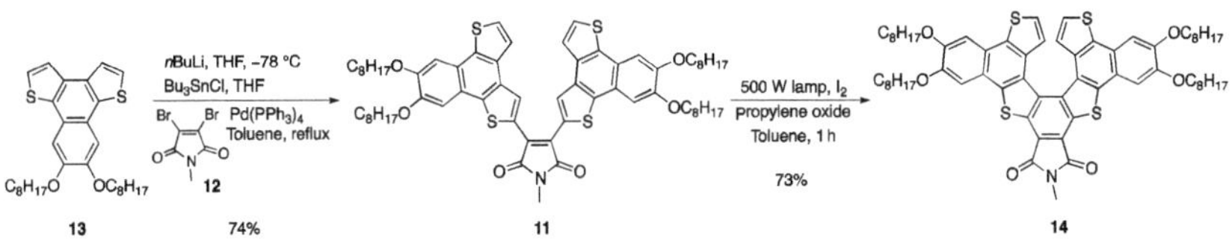

Scheme 6.5 Synthesis of functionalized 7-TH **14**.

Olefin **11** was prepared through a Stille coupling from dibromomaleimide **12** and benzodithiophene **13**.

1.1.2 Non-photochemical Oxidative Cyclization

Although the photocyclization method described earlier is frequently used for the synthesis of 7-TH, some inherent problems preclude a full exploitation of this reaction especially for the scale-up of the synthesis of 7-TH exhibiting interesting functional properties. Problems are the dilute reaction conditions required, the incompatibility with acid-sensitive groups, and nitro and amino groups that quench the singlet photo-induced state formed in the ring closure. These drawbacks can be overcome using an alternative and more simple synthetic methodology for ring closure that uses chemical oxidants. This strategy successfully leads to the construction of aromatic rings even in elaborated polycyclic systems. The application of this methodology to the synthesis of thiahelicenes [14], including the 7-TH derivative **15** [15], has been reported using the cheap $FeCl_3$ (Scheme 6.6).

Unfortunately, this approach has not been tested for large-scale preparation even if the use of $FeCl_3$ allows a simple workup procedure. The only limitation of this method is the necessity to protect the α and α′ positions of the terminal thiophene rings. In fact, if these positions have no substituents, the thiophene radical cation generated at the α-position leads to the formation of polymers [16, 17].

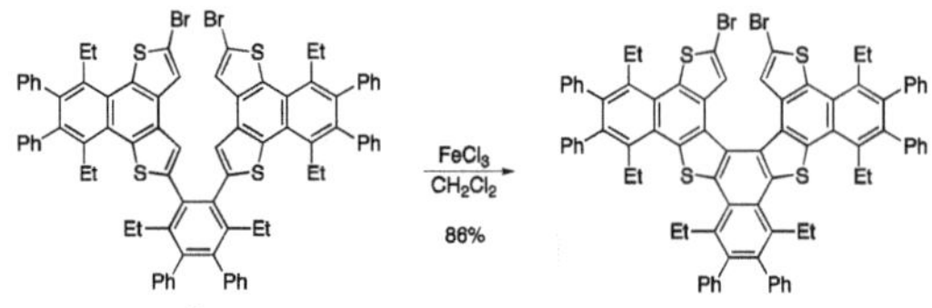

Scheme 6.6 Synthesis of 7-TH **15** through $FeCl_3$-mediated oxidative cyclization.

Another oxidant that can be used for the intramolecular oxidative cyclization to 7-TH is the dichlorodicyanoquinone (DDQ)/acid system [18], described by Scholl many years ago for realizing the intramolecular oxidative carbon–carbon bond formation between two benzenoid rings to give biaryl systems and extensively utilized for the synthesis of planar polycyclic aromatic compounds.

An example for the synthesis of a tetrathiahelicene, such as the compound **17**, is reported in Scheme 6.7, where three new carbon–carbon bonds are formed in a one-pot procedure starting from the intermediate **18**.

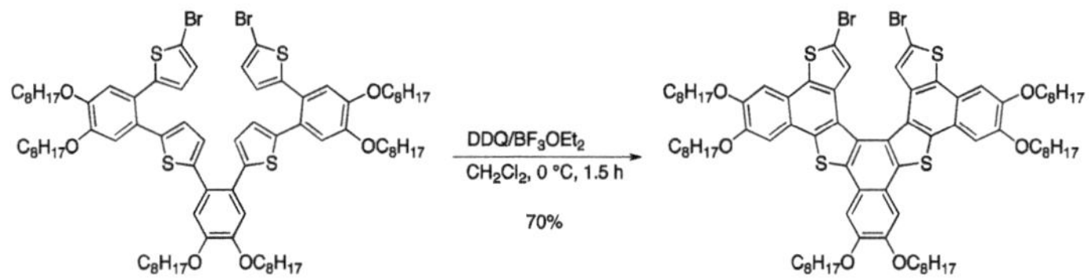

Scheme 6.7 Synthesis of 7-TH **17** through Scholl-type cyclization.

Even in this case, to suppress the polymerization, the blocking of the α-positions of thiophene rings is required, and in these examples, two bromine atoms have been inserted for this purpose.

6.2.2 Annulation of 3,3′-Bis(benzo[1,2-*b*:4,3-*b*′]dithiophene) Derivatives

All methods reported earlier for the synthesis of the 7-TH scaffold are based on the formation of a carbon–carbon bond between the two *β*-positions of thiophene rings, which generates an arene ring. These synthetic pathways are quite efficient and reproducible. However, there are some limitations: (i) no new functional groups can be introduced in these reactions, (ii) no diastereo-/enantioselective version of the reaction can be envisaged, and (iii) quite harsh oxidative conditions are needed. For these reasons, complementary methods that proceed under mild conditions and extend the range of application are very useful. In this category, there are methods that involve the construction of an arene ring starting from biheteroaryl precursors, such as 3,3′-bis(benzodithiophene) species on which the two carbon atoms' fragment necessary to build the aryl ring is introduced either through (i) intramolecular McMurry coupling on dialdehyde or through (ii) cross-coupling reactions.

These two approaches represent the most suitable and promising way to achieve highly functionalized 7-TH systems, especially in enantioenriched form taking advantage of the potential axial chirality of bis(benzodithiophene) species.

Two examples of cyclization of 3,3′-bis(benzo[1,2-*b*:4,3-*b*′]dithiophene)dialdehyde through an intramolecular McMurry coupling have been reported for the synthesis of the enantiopure 2,13-dimethyltetrathia[7]helicene (Scheme 6.8) [19] and of a pentathia[7]helicene [20].

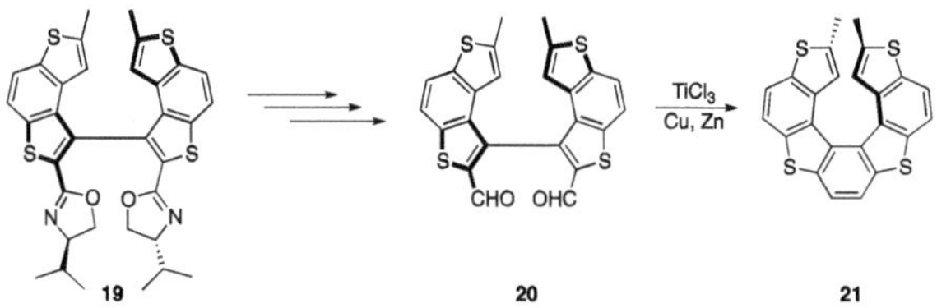

Scheme 6.8 Synthesis of 7-TH **21** by the intramolecular McMurry coupling of dialdehyde **20**.

More recently, efficient and versatile synthetic routes of functionalized tetrathia[7]helicenes through method (ii), that is metal-mediated cross-coupling reactions, have been described [21]. The key intermediates of these methodologies are 2-bromo-3,3′-bibenzo[1,2-*b*:4,3-*b*′]dithiophenes (**22**) synthesized through a palladium-catalyzed homocoupling reaction between two benzo[1,2-*b*:4,3-*b*′] dithiophene (BDT) units followed by a regioselective α-bromination. Direct palladium-catalyzed annulation of bromides **22** with internal alkynes provides a set of 7,8-disubstituted 7-THs **23** in moderate to good yields (46–80%). Otherwise, monosubstituted 7-THs **24** can be easily prepared through Sonogashira coupling of **22** with terminal alkynes, followed by platinum- or indium-promoted cycloisomerization of alkynyl intermediates **25**. Finally, the versatility of bromides **22** is also demonstrated by using them for the preparation of benzo (hetero) fused 7-TH derivatives **26** via Suzuki coupling with (hetero)arylboronic acids and the photocyclization of the obtained intermediates **27** (Scheme 6.9).

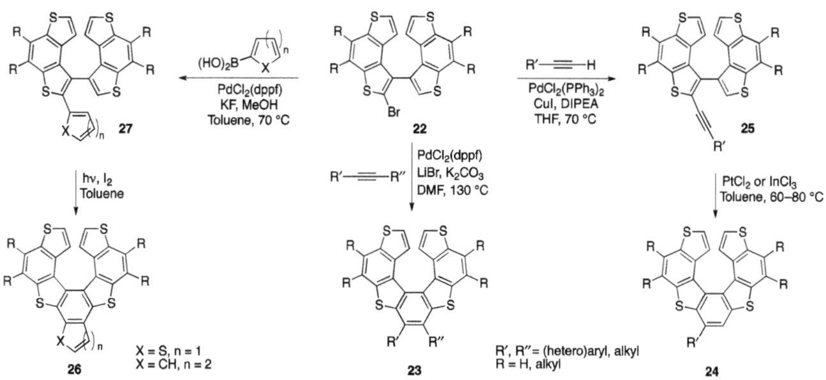

Scheme 6.9 Syntheses of functionalized 7-THs by metal-catalyzed cross coupling reactions.

6.3 Functionalization

One of the most interesting aspects of tetrathiahelicene is the possibility to selectively functionalize them in the two 2 and 13 positions. This is an important point because the study of functional properties of these systems requires methods to have access to different typologies of tetrathiahelicenes containing substituents able to modulate electronic and steric properties. The all-conjugated π system of 7-TH can be perturbed introducing both electron-withdrawing and/or electron-donating groups, thus giving access to thiahelicenes suitable for different applications. In this way, it is in principle possible to obtain tetrathiahelicenes with a push-pull structure by inserting an electron-donor moiety on one terminal thiophene ring and an electron-acceptor one on the other side. Again, it is possible to transform the 7-TH in a suitable phosphorus-based ligand for catalysis by inserting phosphorus substituents on one or both 2,13 positions, or insert groups able to coordinate metals.

All this chemistry is possible thanks to the peculiar reactivity of 7-TH with respect for example to carbohelicenes and taking advantage of the acidity of protons α to the sulfur atom in the two terminal thiophene rings. Therefore, as widely demonstrated by the numerous examples described later, the most convenient way to introduce substituents is the deprotonation of 2 and 13 positions with a strong base and reaction of the anion(s) with electrophiles. One of the first examples of this chemistry is the mono-formylation of 7-TH **1** performed by reacting the anion of the helicene with dimethylformamide or *N*-methylformanilide in $POCl_3$ (Scheme 6.10) [7].

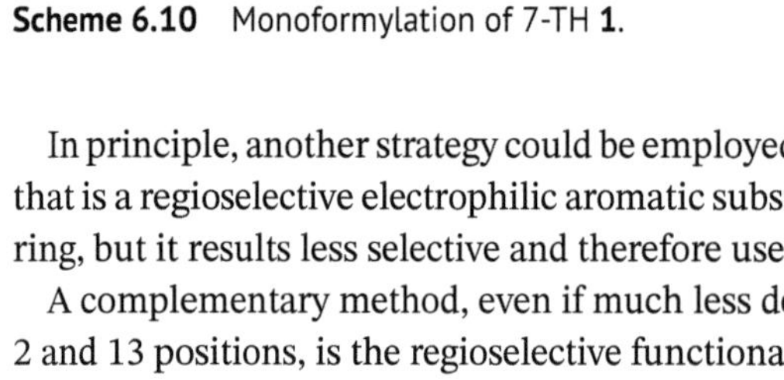

Scheme 6.10 Monoformylation of 7-TH **1**.

In principle, another strategy could be employed to obtain the same compound **28**, that is a regioselective electrophilic aromatic substitution on the terminal thiophene ring, but it results less selective and therefore useless for this purpose.

A complementary method, even if much less developed, to insert substituents on 2 and 13 positions, is the regioselective functionalization of the terminal thiophene rings of the alkene precursor of helicene. An example is the reaction of the anion of alkene **29** with the Weinreb amide **30** as the electrophile [10]. In this particular case using one equivalent of *n*-BuLi, a mixture of the mono- and disubstituted alkenes **31a** and **31b** in 47 and 33% yield, respectively, was isolated as pure *cis*-isomers. These alkenes can be separated by column chromatography. Interestingly, after six months at −20 °C, the presence of 10% of the corresponding *trans*-isomers was detected in the two alkenes, thus indicating the tendency of the *cis*-isomers to isomerize in the solid state to the *trans*-isomers. The subsequent photocyclization of alkenes **31a,b** gives the functionalized helicenes **32a,b** (Scheme 6.11).

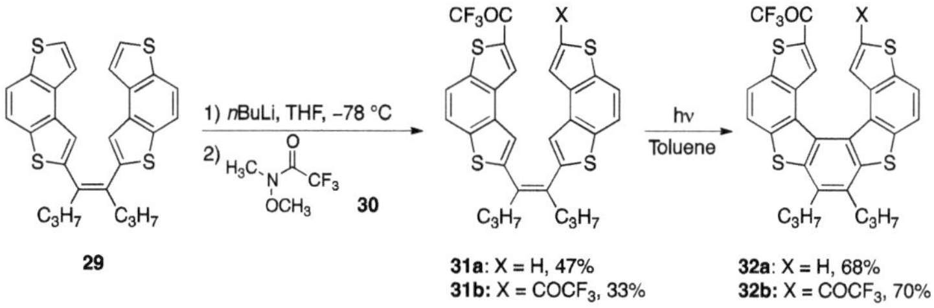

Scheme 6.11 Synthesis of functionalized 7-TH **32**.

The mono- and diformylated 7-TH **28** and **33** are useful starting compounds to obtain more elaborate systems. An example is the preparation of mono- and dinitrile derivatives **34** and **35** through the elaboration of the formyl group into the cyano one (Scheme 6.12) [22].

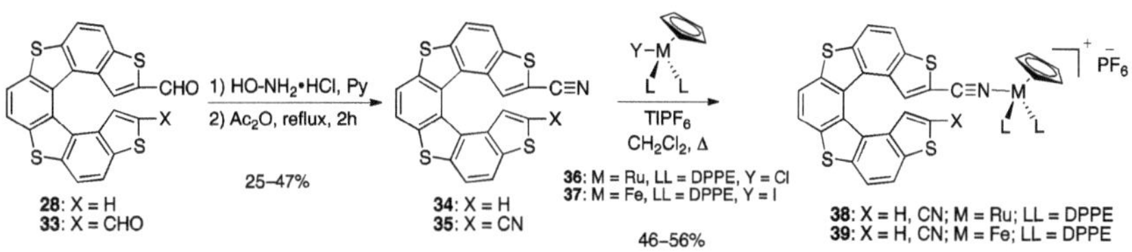

Scheme 6.12 Synthesis of ruthenium and iron complexes **36** and **37**.

The mono- and disubstituted 7-TH nitriles **34** and **35** are suitable ligands for the coordination with ruthenium and iron complexes **36** and **37** through a halide abstraction reaction (Scheme 6.12), obtaining 7-TH organometallic complexes **38** and **39** as potential organometallic materials with nonlinear optics (NLO) properties.

A quite systematic study on the reactivity of anion(s) of 7-TH with different electrophilic reagents allowed the obtainment of several substituted 7-TH. In particular, in addition to the already mentioned introduction of formyl group(s) [10], a variety of tetrathia[7]helicenes bearing electron-donor and electron-acceptor groups in positions 2, 13 are easily accessible using the same strategy [23]. In particular, the substituted 7-TH **40–42** reported in Figure 6.1 can be prepared using the appropriate electrophilic reagents.

Figure 6.1 Structure of 7-TH derivatives **40–42**.

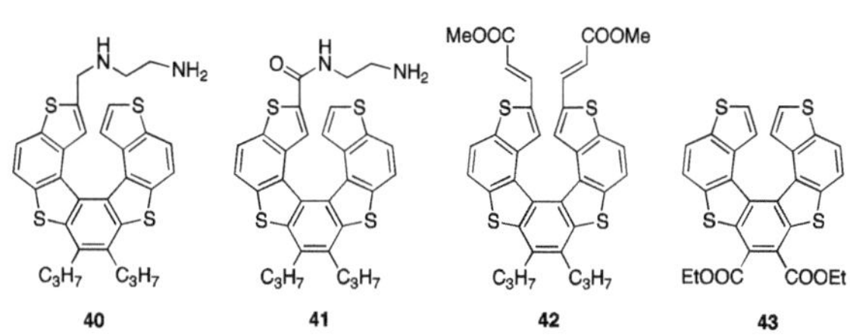

Even positions 7 and 8 of the helical scaffold can be endowed with substituents (**43**), but in this case, these groups must be already present on the olefin precursor of tetrathiahelicene [23].

A very interesting study that takes advantage of the selective functionalization of 7-TH in 2 and 13 positions describes the synthesis, structures, and spectroscopic properties of bridged helicenes with a variety of helical pitches [24]. Racemic and optically active [7]thiaheterohelicenes with different helical pitches **44** were prepared by reaction of 2,13-bis(chloromethyl)[7]thiaheterohelicene **45** with alkanedithiolates of different lengths (Scheme 6.13) in order to obtain elongation and contraction of the helicene molecule and to study the flexibility of the molecular spring in solution.

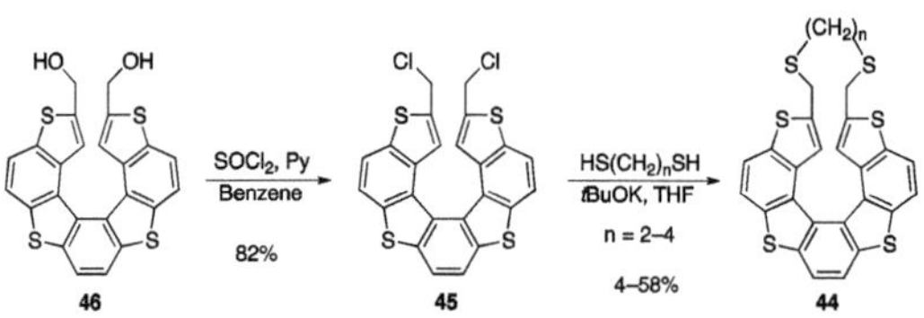

Scheme 6.13 Synthesis of bridged 7-TH derivative **44**.

The length of the linker between the two terminal thiophene rings determines a very different dihedral angle between terminal thiophene rings. It has been found that the angle between two terminal thiophene rings of the helicenes can range from 21.9° to 59.2°; therefore, the helical pitch of the molecular spring can be fine-tuned by proper choice of a spacer group.

Another very useful functionalization of the 7-TH scaffold, such as **1** and **47**, concerns the introduction of halogen atoms, such as the iodine (Scheme 6.14); the corresponding 7-TH mono- and diiodides **48** and **49** can be used as convenient intermediates for the introduction of π-conjugated moieties (e.g. alkenyl or alkynyl pendants) through palladium-catalyzed cross-coupling reactions (Heck reaction [23] and Sonogashira reaction [25]).

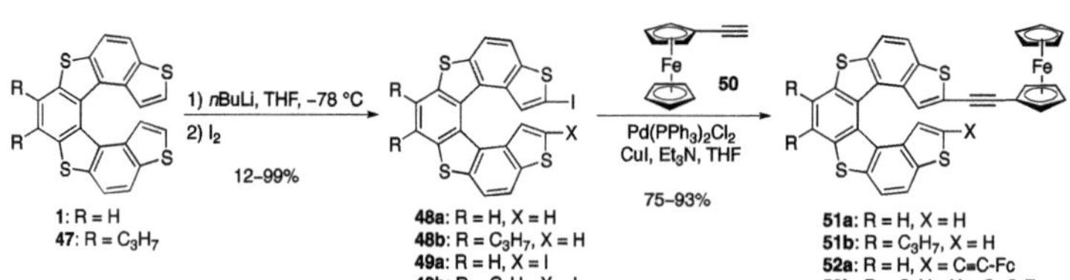

Scheme 6.14 Synthesis of Fc-based organometallic complexes **51** and **52**.

Noteworthy, the Sonogashira couplings between iodides **48** and **49** with the ferrocenyl (Fc)-based terminal alkyne **50** allow the synthesis of the corresponding organometallic complexes **51** and **52**, which exhibit an interesting electrochemical behavior,[25] as discussed in Paragraph 4.

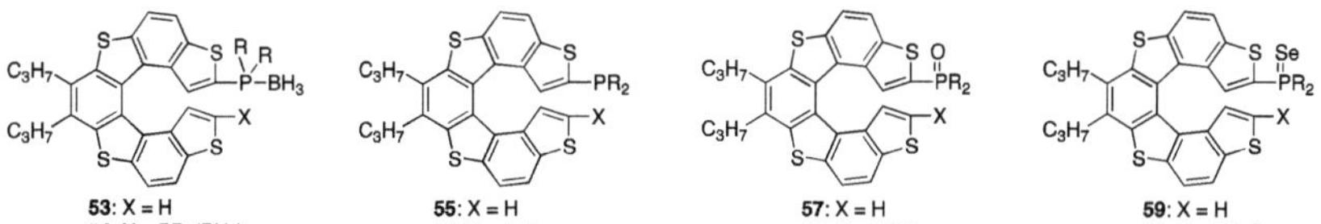

Figure 6.2 Structure of phosphorus-based 7-TH derivatives.

Interesting is the introduction of different phosphorus-based substituents that allows the obtention of a set of tetrathia[7]helicenes containing phosphine-borane adducts **53,54** [26, 27], phosphanes **55,56** [26, 28], phosphine oxides **57,58** [28, 29], and phosphine selenides **59,60** [28]. All these 7-TH derivatives can be synthesized according to the general procedure that involves the introduction of the phosphorus group on the 7-TH scaffold through the selective deprotonation of the 2,13 positions, followed by the *in situ* reaction of the corresponding mono- and dianion with the appropriately chlorinated phosphorus electrophiles (Figure 6.2).

Most of these derivatives have been tested as chiral ligands in organic [28, 29] and organometallic catalysis [26, 30], as discussed in Paragraph 7, while a 7-TH phosphine oxide has been used as a hard ditopic O=P donor in rhenium-based supramolecular coordination complexes for potential applications in sensors, light harvesters, and nanomaterials [31].

6.4 Resolution of Enantiomers

The obtainment of optically pure 7-TH derivatives is an essential objective in order to use these systems as chiral elements in the manifold fields in which they are usually involved, taking into account their configurational stability even at a high temperature [32]. Since very few examples of enantio- and diastereoselective procedures have been so far reported for the synthesis of the 7-TH scaffold, the optical resolution of the racemate is still today a fundamental step to obtain enantiopure 7-TH. Thus, in the last decades, much attention has been devoted to developing and optimizing resolution methods not only especially based on chromatographic techniques but also through enzymatic and kinetic resolution procedures.

6.4.1 Chiral HPLC

The first example of the resolution of the parent 7-TH **1** by high-performance liquid chromatography (HPLC) has been reported in 1981, using a stationary phase consisting of silica gel covalently bound to the chiral charge transfer complexing agent (tetranitrofluorenylidene)amino]oxy]propionic acid (TAPA) [33]. Indeed, TAPA represents one of the best selective chiral complexing agents for the chromatographic separations of racemic carbohelicenes and heterohelicenes [34], due to the presence of the electron-accepting tetranitrofluorenylidene group able to form a charge-transfer complex with the electron-donating helicene skeleton, along with the ability of the oxypropionic acid moiety to recognize the chiral helical framework [35].

Nowadays, chiral stationary phases with good performances can be successfully used for the resolution of differently substituted 7-TH derivatives, especially based on polysaccharides coated or immobilized on silica gel, such as the Chiralcel® OD column [11, 36] and the Chiralpack® IA column, respectively [26, 27, 29]. Among different resolution methods so far developed, chiral HPLC still represents one of the most practical and general methods to obtain enantiopure 7-TH derivatives.

6.4.2 Enzymatic Resolution

The enzymatic resolution of a functionalized 7-TH, the 2,13-(hydroxymethyl) tetrathia[7]helicene (**46**), can be accomplished through the lipase-catalyzed enantioselective transesterification with vinyl acetate [37, 38]. In this study, the first example of enantiomeric enhancement of a 7-TH derivative was observed when a sample of an enantiomeric-enriched compound was submitted to chromatographic purification on an achiral phase.

Pseudomonas cepacia-mediated transesterification of (±)-**46** affords the (*P*)-**46** with high optical purity (98% enantiomeric excess [*ee*]) along with a mixture of the corresponding (*M*)-monoacetate and (*M*)-diacetate, while the same reaction with *Candida antarctica* provides the enantiomer (*M*)-**46** in 92% *ee* besides (*P*)-monoacetate and (*P*)-diacetate. Although this method allows the obtainment of both enantiomers in high optical purity through an operationally simple methodology, further in-depth studies are needed to achieve the enzymatic resolution of 7-TH on a large scale.

6.4.3 Formation of Diastereoisomers

The 2,13-(hydroxymethyl)tetrathia[7]helicene (**46**) is also effectively separated into its antipodes by means of the formation of the corresponding diastereoisomers **61** in turn obtained by the photocyclization of alkene **62** bearing a chiral camphor-based pendant (Scheme 6.15) [4].

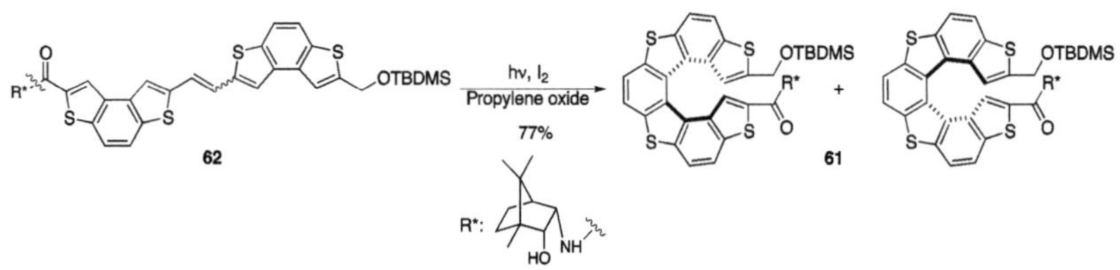

Scheme 6.15 Photocyclization of chiral alkene **62**.

The removal of the chiral auxiliary allows the isolation of both enantiomers (*P*)-**46** and (*M*)-**46** in excellent *ee* (c. 96%).

6.4.4 Kinetic Resolution

Single and double kinetic resolution is an interesting and alternative approach to obtain functionalized 7-TH in high optical purity [39, 40]. The enantioselective mono- and diformylation of (±)-**47** can be realized through the deprotonation with *n*BuLi following the treatment with chiral formamides as formyl donors [38]. By this way, the corresponding mono- and diformylated 7-TH are isolated in moderate to good *ee* (up to 73%) by single and double kinetic resolution, respectively. Again, the use of (−)-sparteine to promote the enantioselective lithiation of (±)-**47** affords samples of mono- and disubstituted silylated 7-TH derivatives in excellent *ee* (up to 98%).[40]

6.5 Electrochemical Properties and Applications of Tetrathiahelicenes

Electrochemistry can provide important clues on the tetrathiahelicene electronic properties, to be compared with computational and spectroscopic data. Moreover, tetrathiahelicenes are attractive tools from the electrochemical perspective, too, as racemate electroactive organic semiconductors and, above all, as enantiopure inherently chiral selectors for application in chiral electroanalysis.

6.5.1 Investigation of Tetrathiahelicene Electroactivity on a Systematic Compound Series

A systematic voltammetric study on ten tetrathiahelicenes **1,10a,b**, **47**, **63–67** plus parent **BDT** and a hexathiahelicene **68** for comparison (Figure 6.3) [41] provided the first rationalization of redox features and electrooligomerization ability in the tetrathiahelicene family, discussing them with the linear α-oligothiophene series α-T_n [42, 43] as a benchmark, and comparing them with UV-vis spectroscopy features.

While, as reported by Rajca et al., thiahelicenes entirely made of condensed thiophene rings with the S atoms pointing outside the helix [44, 45] exhibit properties close to those of single thiophenes, pointing to negligible conjugation efficiency, tetrathiahelicenes, alternating condensed thiophene and benzene rings with homotopic thiophene terminals, feature much higher conjugation efficiency and are much more electrochemically active.

After normalization vs. the intersolvental $Fc^{+}|Fc$ reference couple, the first oxidation and first reduction peak potentials of unsubstituted tetrathiahelicene **1** in CH_2Cl_2+ 0.1 M tetrabutylammonium hexafluorophosphate (TBAP) look close to

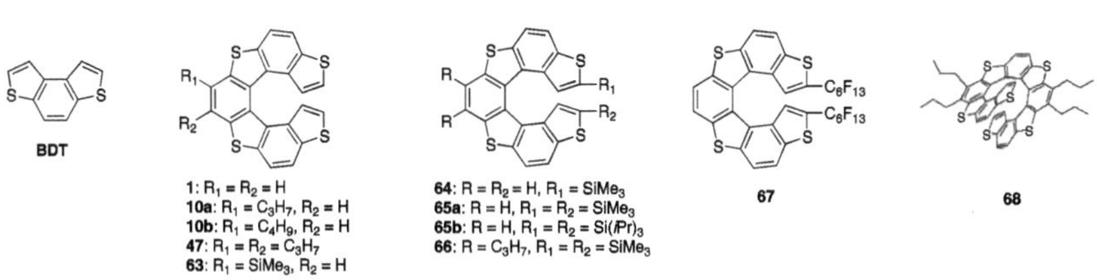

Figure 6.3 The collection of tetrathiahelicenes electrochemically investigated in Ref. [41] (plus **BDT** and a hexathiahelicene, for the sake of comparison). Source: Adapted from [41].

the 2,2′-bithiophene (α-T_2) and 2,2′:5′,2″-terthiophene (α-T_3) ones, respectively, and the resulting electrochemical highest occupied and lowest unoccupied molecular orbitals (HOMO-LUMO) gap looks intermediate between the α-T_2 and α-T_3 ones. This appears consistent with a 7-TH system with partially impaired conjugation, on account of the torsional angle and/or of the ring connectivity. It also points to possible concurrent radical cation formation on both molecule moieties [41]. Notably, the spectroscopic HOMO-LUMO gap of 7-TH is intermediate between the α-T_3 and 2,2′:5′,2″:5‴,2‴-quaterthiophene (α-T_4) ones instead; this could be justified in terms of the orbitals involved in the intramolecular electron transition not coinciding with those involved in electron transfer to/from the electrode with the formation of net charged species in the solvent + electrolyte medium.

Within the same 7-TH series, alkyl or trialkylsilyl substituents result in small potential shifts, the rationalization of which apparently requires to consider not only inductive effects but possibly also other ones, such as sterical hindrance and modulation of the torsional angle [41]. Instead, a remarkable electron-withdrawing inductive effect is observed with fluoroalkyl substituents, with a significant shift of both first oxidation and reduction to more positive potentials [41].

The first oxidation peak features are consistent with a chemical follow-up, making the process chemically irreversible. In particular, when at least one thiophene terminal with free α position is available, coupling processes (dimerization or oligomerization *via* radical cation coupling) [43] can take place, with the formation of an electroactive layer, more oxidable than the starting monomer (as pointed out by its less positive onset potentials), of increasing thickness with increasing cycle number (as accounted for by the increasing current in subsequent cyclic voltammetry [CV] patterns). This is particularly evident for unsubstituted 7-TH, while alkyl or trialkylsilyl substituents can significantly hinder the process, as a consequence of their enhancing solubility and/or resulting in increasing sterical hindrance.

Such ability to form regularly growing electroactive films is a great advantage for the preparation of chiral functional surfaces with respect to C_5, C_6, and C_7 carbohelicenes [46–48]. In fact, the latter ones, although more active than the earlier mentioned all-thiophene helicenes by Rajca et al. [44, 45], are much more difficult to electrooligomerize, to the point that in 2014 Hrbac et al. [46] proposed to electrooxidize 3-([7]-helicen-9-yl)-thiophene instead of [7]-helicene, with the helicene practically acting as a chiral pendant of the single thiophene, the moiety undergoing oligomerization (although a significant oxidation potential shift to less positive potentials with respect to unsubstituted thiophene points to significant conjugation between the thiophene and the helicene building blocks). Later on, the same authors showed that even carbohelicenes as such can be electrooligomerized after all, but the first oxidation peak potential is unfortunately close to the background, and thus the potential range for cycling must be carefully managed, to avoid overoxidation and subsequent loss of electrochemical activity (as highlighted by gradually decreasing currents in subsequent redox cycles) [47]. If such a condition is achieved, an electroactive film is formed (with currents gradually increasing), but its growth appears slow and quite limited with respect to the tetrathiahelicene case [48].

With both thiophene terminals hindered, no coupling can, of course, take place, and the first oxidation peak, either a unique one, or a system of two nearly merging ones (depending on possible 3D interactions between radical cations on the helicene terminals) becomes chemically reversible [41]. A limiting benchmark case is provided by the hexathiahelicene compound, the terminal thiophenes of which are completely overlapping with π stacking interactions, resulting in a couple of chemically and electrochemically reversible twin oxidation peaks with a large potential difference.

6.5.2 Investigation of Electronic Properties of Electroactive 7-TH Adducts With Advanced Functional Properties

Further electrochemical studies took into consideration 7-TH as building blocks in electroactive molecules with advanced functional properties, including 7-TH ferrocene derivatives **51a,b** and **52a,b** (Scheme 6.14), 7-TH phosphorus derivatives **53–60** (Figure 6.2), and push-pull 7-TH adducts **69** and **70** for NLO (Figure 6.4).

In 2012, Rose-Munch et al. [25] discussed 7-TH derivatives **51a,b** and **52a,b** with one or both terminals linked by a triple bond to a ferrocenyl group, providing a much higher HOMO and thus a much smaller HOMO-LUMO gap. Accordingly, their CV patterns feature a first chemically and electrochemically reversible oxidation, obviously corresponding to the Fc moiety/moieties, followed by a second irreversible oxidation peak at much more positive potentials, obviously corresponding to the 7-TH moiety. Somehow unexpectedly, reasoning in terms of increased overall conjugation only, the Fc peak is located at significantly more positive potentials than with free Fc; at the same time, the 7-TH peak is at significantly less positive potential than in free 7-TH. Moreover, the reduction peak, also corresponding to the 7-TH moiety, is located at a more positive potential than with free 7-TH. Such features indicate effective communication between Fc and 7-TH via the triple bond linker with partial electron density delocalization from the Fc site to the 7-TH one. Attractively, the adduct with a single Fc group, having one free 7-TH terminal, undergoes fast electrodeposition of an electrochemically active film. Of course, this requires to cycle the potential around the second, irreversible, oxidation peak (corresponding to 7-TH oxidation to radical cation); however, intriguingly, the oligomerization CV pattern appears as a continuum with no solution of continuity between the reversible Fc process and the faraway 7-TH irreversible one [25].

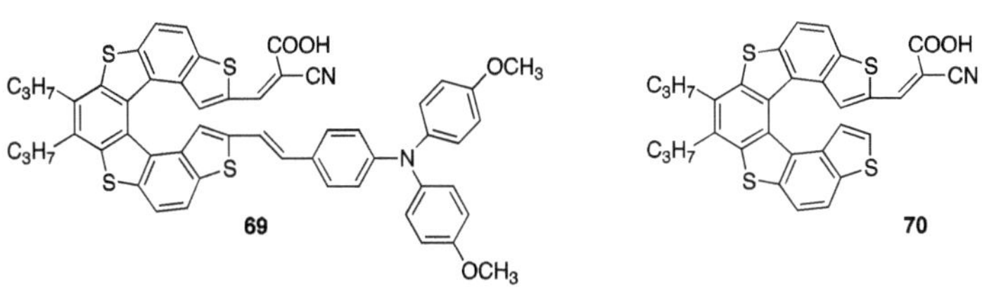

69 **70**

Figure 6.4 Electroactive molecules with 7-TH building blocks, electrochemically investigated in Ref. [25] (7-TH-ferrocene adducts), Ref. [28] (7-TH-phosphanes), and Ref.[49, 50] (push-pull 7-TH adducts for NLO).

In 2016, Cauteruccio et al. discussed in view of chiral organocatalysis applications a large systematic family of 7-TH derivatives with one or both terminals directly conjugated to alkyl phosphanes, phosphane oxides, phosphane boranes, and phosphane selenides [28].

In all phosphane cases, the first, chemically irreversible oxidation peak could safely be assigned to the phosphane terminals (with their substituents modulating the peak potential), while the following one and the first reduction peak should correspond to oxidation and reduction of the helicene building block to radical cation and radical anion, respectively [28].

Protecting the phosphine groups with borane, the earlier irreversible first oxidation peak disappears in favor of a reversible one, corresponding to the formation of a stable radical cation on the 7-TH moiety. Concurrently, consistent with the borane electron-withdrawing nature, reduction peaks shift to less negative potentials and, in some cases, a second one can also be observed in the accessible potential window [28]. Analogously, converting phosphines into oxides, a reversible first oxidation peak related to 7-TH takes the place of the irreversible one related to phosphine; concurrently, a positive shift is observed for both oxidation and reduction peaks, even more remarkable than in the former case. Interestingly, some *in situ* regeneration of the phosphine can also be observed performing convenient potential cycles [28].

Phosphine selenides show a behavior similar to phosphane oxides concerning reduction, consistent with an electron-withdrawing substituent. Instead, first oxidation takes place remarkably before the 7-TH one, pointing to the selenide group itself acting as the first oxidation site [28].

Finally, in 2018/2019, Cauteruccio et al. discussed two push-pull systems for solar cells with a dipropyl-7-TH building block conjugated through a double bond to either a cyanoacrylic acid or a triarylamine moiety. In the first case, the cyanoacrylic moiety provides a more favorable reduction site, while the first oxidation is still attributed to the 7-TH moiety (the second one could correspond to the cyanoacryl one). Vice versa, in the second case, the first oxidation peak, chemically and electrochemically reversible, is attributed to the triarylamine moiety [49, 50].

6.5.3 Application of 7-TH-based Inherently Chiral Films for Enantioselective Electroanalysis

Although tetrathiahelicenes have interesting electrochemical features even as racemates, their by far utmost attractiveness consists in the possibility of exploiting their enantiopure antipodes as chiral, and specifically "inherently chiral," selectors, to achieve effective enantiodiscrimination in electroanalysis and, more generally, in electrochemistry.

Enantiomers of electroactive probes have the same physicochemical properties and therefore identical CV patterns, unless they undergo the electron transfer at an electrode|solution interphase endowed with chirality, thus achieving diastereomeric and thus energetically different situations, enabling observation of different potentials for the two probe enantiomers. Many approaches have been so far proposed for this aim, mostly based on either electrode surface modification with a chiral selector or adoption of a medium including a chiral selector [51–53]. However, most approaches do not look resolutive, suffering from significant drawbacks, and, in many cases, depending on selectors with weak chirality manifestations [51, 52]. This problem typically concerns selectors in which chirality originates from one or more localized stereocenter(s) external to the main molecule backbone responsible for key functional properties. Instead, powerful chirality manifestations, in both chiral electroanalysis [51–56] and chiroptics, [57, 58] are exhibited by molecules endowed with "inherent chirality." This concept implies chirality and key functional properties to originate from the stereogenic element, which can coincide with the main molecular backbone on account of a tailored torsion with an associated energy barrier too high to be overcome at room temperature. This can be typically achieved with either atropisomeric molecules of C_2 axial stereogenicity, or, even more ideally, with molecules of helical stereogenicity. Moreover, in both cases, regioregular design can enable to transfer and amplify the selector properties at macro- or supramolecular level.

Actually, very large potential differences for the enantiomers of various electroactive probes have been observed working with many inherently chiral atropisomeric selectors, either as monomers to be electrooligomerized, thus modifying an achiral electrode surface (for example, in Ref. [54, 56]), or as cations of molecular salts to be employed as inherently chiral ionic liquid media (if liquid at room temperature) and/or as additives in achiral ionic liquids (for example, in Ref [55]), the ionic liquid media having been chosen on account of their intrinsically very high structural order at the charged interphase, locally even mimicking bulk features of liquid crystals.

In that context, it was overdue to test chiral selectors of *helical* stereogenicity, providing as earlier underlined the most ideal model of inherent chirality. A first successful investigation appeared in 2019 [59], performed with the enantiopure monomers of 7-TH, exploiting the latter inherent chirality jointly with its earlier-discussed high ability to produce by electrooxidation inherently chiral electroactive films on achiral electrode films.

Neat potential differences, of some hundreds of mV, were observed in CV experiments on such modified electrode surfaces, for the enantiomers of three different chiral probes, a model ferrocenyl one, plus dopa and tyrosine.

Moreover, working on indium tin oxide (ITO) or screen-printed single-walled-nanotube electrodes modified with a thin layer electrodeposited from 7-TH monomer solutions, neat potential differences were observed in CV experiments with the achiral ferrocyanide/ferricyanide redox couple in aqueous solution under application of north pole/south pole (N/S) vs. S/N magnetic fields [59]. Such phenomenon, just before observed with atropisomeric selectors too [60], is similar to the magnetic circular dichroism (CD) obtained for achiral samples under magnetic field [61], and could be looked at as the manifestation of a strong, truly chiral asymmetric magnetic field corresponding to the electrodeposited film.

The potentialities of inherently chiral helical selectors were also very recently confirmed by Fontana et al. with successful enantiodiscrimination CV tests performed in achiral ionic liquids with inherently chiral aza and diazahelicenium salts as inherently chiral additives [62, 63].

6.6 Computational Studies on Tetrathiahelicenes

6.6.1 General Remarks

The 7-TH scaffold is a prototypical organic molecule, and hence, it possesses a closed-shell electronically stable electronic ground state. In general, a plethora of different and suitable theoretical methods are thus available for computational studies. For geometry optimization, frequency calculations and other ground-state properties adapted semiempirical methods are useful, like, for example, the recently developed GFN2 method [64] or standard ground-state density functional theory (DFT) [65] in combination with established exchange-correlation (xc) functionals and medium-sized basis sets. But also single-reference wavefunction-based methods like Møller-Plesset perturbation theory or (approximate) coupled-cluster approaches deliver reliable results for ground-state properties [66].

When however excited states or absorption spectra or electron-detached or attached states of 7-TH or its derivatives are to be studied, special care has to be taken with respect to the choice of the theoretical methodology. Due to its molecular size, linear-response time-dependent DFT is the method of choice for excited electronic states of 7-TH since it is computationally efficient and can be sufficiently accurate for qualitative interpretations in combination with a suitable xc-functional [65]. In general, whenever time-dependent density functional theory (TDDFT) is to be applied a thorough testing and benchmarking of xc-functionals is required by comparison either to results of higher-level quantum chemical benchmark methods or to experimental data. Due to the molecular size of 7-TH, however, theoretical benchmark values are generally not accessible, and therefore at least consistency checks of theoretical results can be performed. For example, when conceptually different xc-functionals yield similar results for the investigated 7-TH in combination with TDDFT, it is most likely suitable for its investigation. The best xc-functional can then be chosen by comparison with experimental data. In our studies so far, long-range corrected xc-functionals have turned out to be a good choice for the computation of excited, electron-attached and -detached states (see later) [27, 28, 30, 49].

In contrast to TDDFT, the method like the approximate coupled cluster scheme of second order (CC2) or the algebraic diagrammatic construction scheme (ADC(2)) has been shown to achieve a quantitative agreement when optical spectra and excited states of 7-TH are investigated (see later). Typically, these methods possess an established accuracy for excited electronic states of organic molecules of about 0.2 eV and do not depend on the choice of an xc-functional [67]. In addition, these families of ab initio methods can be successively improved by increasing the employed basis set or the level of electron correlation treatment.

6.6.2 Electron-Detached and Attached States of 7-TH Phosphorus Derivatives

Experimental electrochemical investigations provided relationships between the electronic and structural properties of 7-TH phosphorus derivatives, which have

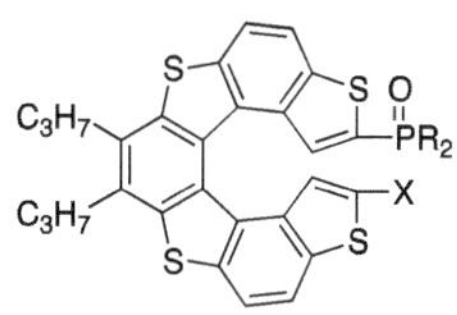

57a: R = Ph, X = H
58a: R = Ph, X = $POPh_2$
58b: R = *t*Bu; X = $PO(tBu)_2$

Figure 6.5 Structure of investigated 7-TH phosphine oxides.

been supported by quantum chemical investigations of the oxidation and reduction processes [28]. Therefore, the first ionization potential (IP), accounting for oxidation, and the first electron affinity (EA), accounting for a reduction process, have been computed for selected aryl phosphine oxides **57a**, **58a**, and **58b** (Figure 6.5). DFT in combination with the xc-functional wB97XD and the basis set cc-pVDZ was chosen, as it is known that pure DFT functionals are usually insufficient for calculating IEs or electronic excitations with large charge-transfer character and an increasing amount of Hartree-Fock exchange improves the results significantly. To capture the effect of solvation effects, a polarizable continuum model (PCM) using integral equation formalism variant (IEFPCM) was applied to consider dichloromethane (DCM) and water as solvents.

The vertical first IP and first EA of **57a**, **58a**, and **58b** have been calculated at the optimized ground-state structures as the difference between the total electronic energies of the neutral and cationic or anionic systems, respectively. The EA of 7-TH phosphine oxides has been found to be in the order of 2 eV in solution and the IP has a value of roughly 6 eV in solution (Table 6.1). It becomes apparent that the IEs are decreased and the EAs are increased, as expected, when a solvation model is included in the calculation owing to stabilization of the charged molecule in the environment. The stabilization is slightly more pronounced in the polar solvent water, in which the EAs are about 0.2 eV higher than in DCM.

To understand in which molecular region of the 7-TH derivatives the oxidation or reduction process takes place, one usually inspects the HOMO and LUMO. This is justified since according to Koopman's theorem, the negative orbital energy of the HOMO corresponds to the IP and of the LUMO to the EA. In most organic molecules, this is a reasonable approximation and appropriate. In fact, this holds very well for **57a** and **58a**, and their squared HOMO and LUMO do represent the density of the

Table 6.1 Ionization potentials and electron affinities of **57a**, **58a**, and **58b** calculated in the gas phase, in dichloromethane (DCM) and aqueous solution using DFT/ωB97XD/cc-pVDZ and an appropriated polarizable continuum model for solvation.

	57a	**58a**	**58b**
Ionization potential			
Gas phase	6.18	6.00	7.10
DCM	6.02	6.09	6.13
Water	5.90	5.97	6.06
Electron affinity			
Gas phase	1.47	1.83	0.53
DCM	1.86	1.94	1.83
Water	2.05	2.14	2.06

detached and attached electron very well [28]. The densities of the detached and attached electrons in an oxidation or reduction process can be easily calculated as the difference of the electron densities of neutral and the cationic or anionic species. Thereby, a largely negative (detachment) density is obtained for the cationic species and a largely positive (attachment) density for the anionic species. It is instructive to note that positive (negative) contributions in detachment (attachment) densities reveal large orbital relaxation effects during the detachment (attachment) process.

The frontier orbitals of **57a** and **58a** are practically indistinguishable from the ones of **58b**, which are displayed in Figure 6.6, independent of the inclusion of a PCM or the solvent. All of them are located on the 7-TH skeleton and are hardly influenced by the substituents. This is comparable to other 7-TH derivatives. None of them has significant contributions on the phosphine oxide or the alkyl side chains.

In the case of 7-TH **58b**, however, the HOMO does not represent the detachment density of the ionization process (Figure 6.6). Instead, the squared HOMO-1 corresponds to the ionization process and reflects the detachment density most closely. In other words, the first ionization of **58b** corresponds to the removal of an electron from the HOMO-1 and not the HOMO, which is an illustrative example for the breakdown of the simple orbital interpretation.

Using the same theoretical methodology, also higher ionization potentials and electron affinities have been calculated for these 7-TH derivatives, and thereby the

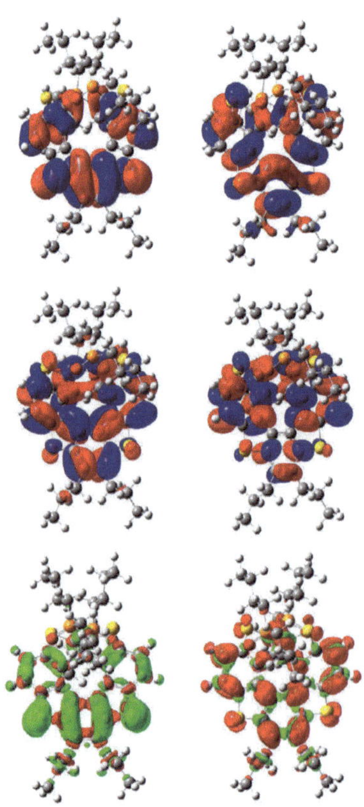

Figure 6.6 HOMO-1 (top left), HOMO (top right), LUMO (middle left), LUMO+1 (middle right), and the difference densities of the cationic and anionic forms of 7-TH derivative **58b** (bottom left and bottom right). Source: Adapted from Ref. [28].

higher oxidation and reduction processes could be understood. The agreement between the calculated trends in the IPs and EAs and the measured oxidation and reduction potentials has been found to be very good. It needs to be pointed out here that for a comparison to CV measurements, geometries of the electron-detached and attached species need to be separately optimized at the corresponding level of theory. Our theoretical investigations unambiguously revealed a clear dominance of the 7-TH scaffold on the electronic properties of all investigated molecules. P=O or P=Se groups have been found to influence the reduction or oxidation processes of 7-TH derivatives only marginally.

6.6.3 Optical Properties of Chiral Tetrathiahelicene-Based Alkyl Phosphine–Borane Complexes

The lack of in-depth studies of the chiroptical properties of 7-TH derivatives prompted us to investigate the electronic CD spectrum of a prototypical 7-TH-based alkyl phosphine–borane complex **54a** (see inset of Figure 6.7, R = *n*Bu) [27]. For the calculation of the optical spectra, the ground-state equilibrium structures of helical (*P*)- and (*M*)-enantiomers have been separately optimized at DFT/ωB97XD/cc-pVDZ level of theory, and the excitation energies and corresponding oscillator and rotatory strengths have been computed at the RI-CC2/cc-pVDZ level.

For evaluation of the quality of the calculated excitation energies, their values are shifted to lower energies by −0.325 eV, accounting for systematic errors, and then compared to the measured absorption spectrum (Figure 6.7 left). The energy shift has been chosen such that the first excitation energy matches the first absorption band. Thereby, an overall good agreement of the computed excitation energies at the chosen resolution of identity coupled clusters approximated single and double (RICC2) level of theory and the experimental absorption spectrum is obtained.

Based on these calculations, the individual excited states have been analyzed in detail. It turned out that the first electronically excited state (S1) is mainly

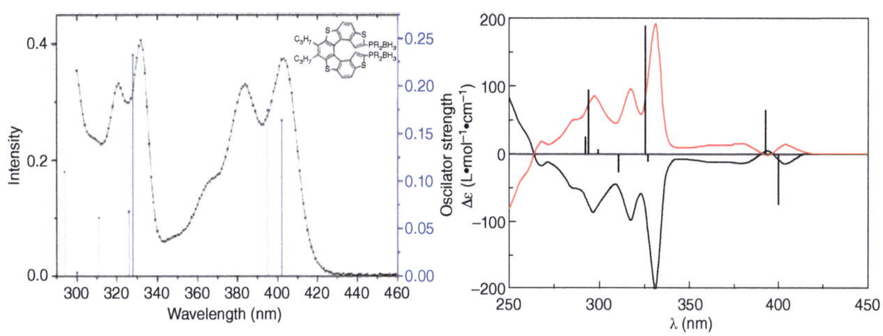

Figure 6.7 Comparison of the computed absorption spectrum (left) and electronic circular dichroism spectrum (right) of the *P*-enantiomer of the investigated 7-TH alkyl phosphine–borane complex **54a** (see inset, R = *n*Bu) at the theoretical level of RI-CC2/cc-pVDZ (vertical lines, shifted by −0.325 eV) with the experimental spectra (solid lines) measured in $CHCl_3$ (10^{-4} M). Source: Adapted from Ref. [27].

characterized by a HOMO-1 to LUMO transition (80.3%) while the second excited state (S2) is characterized as HOMO to LUMO transition (89.6%). Again, it contradicts the intuition gained from the simple orbital picture that the HOMO to LUMO transition is not the S_1 state as the ionization of the 7-TH derivative **58b** did not occur from the HOMO as well. The involved molecular orbitals are practically identical to those depicted in Figure 6.6 and thus also the lowest excited states are strictly located at the central 7-TH helicene part. The observed small variations in the absorption spectra of different 7-TH derivatives are then due to minor electronic and steric effects of side groups or heavy atoms. Since the S_1 and S_2 states can in principle both contribute to the first experimental absorption band at 400 nm wavelengths (Figure 6.7), vibrationally resolved absorption spectra of the $S_0 \rightarrow S_1$ and the $S_0 \rightarrow S_2$ transitions were calculated at TDDFT/ωB97XD/cc-pVDZ level of theory. It turned out that practically only the S_1 state contributes to the peak in the experimental absorption spectrum, while the S_2 state is very broad and shallow [27].

In addition to the linear absorption spectrum, also the CD spectra of the 7-TH derivative have been simulated for the (*P*)- and (*M*)-enantiomers (Figure 6.7). Therefore, the rotatory strength of the corresponding electronic transitions has been computed at RICC2 level and compared to the measured CD spectrum. The comparison of the experimental CD spectra of the (+) and (−)-enantiomers (red and black lines in Figure 6.7 bottom) with the computed CD spectrum for the *P*-helical enantiomer allowed for the first time for an unambiguous assignment of *P* and *M* helicity to the (+) and (−)-enantiomers of the investigated 7-TH derivative.

6.6.4 Computational Study on Vibrational Raman Optical Activity of Helicenes

Vibrational Raman optical activity (VROA) spectroscopy, based on the difference of Raman scattering for incident right and left circularly polarized lights, is a powerful tool to unravel the configurations as well as the conformations of molecules. VROA spectra of four helicenes, including hexahelicene, the parent 7-TH **1** and its pyrrole and furan analogs, have been simulated and interpreted to unravel signatures of their helicity combined with π-electron conjugation [68].

All helicenes show intense VROA peaks attributed to their π-conjugated structure and associated with collective vibrational modes, and for the three heterohelicenes, upon changing the heteroatoms, similar modes presenting comparable atomic contribution patterns have been observed, though the vibrational and electronic properties are modified. Some fingerprints could therefore be associated with the helicity of the system, and, in particular, most of the VROA bands are positive for left-handed helicenes.

6.7 Materials

6.7.1 NLO

The design and development of NLO materials based on organic molecules is an important field of research that has attracted a lot of interest thanks to important applications in optoelectronic technology. Aromatic π-systems, in particular, are highly attractive because of their large and fast NLO responses joined with the flexibility of organic synthesis that allows the design of a huge number of chromophores. With respect to other push-pull π-systems used in NLO, in helicenes the p-conjugated segment is the helix itself and this opens up the way to enhance NLO responses. In addition, the absence of centrosymmetry at the supramolecular level of thiahelicenes is a condition for second-order phenomena.

As widely described in Paragraph 2, 7-TH can be functionalized with donor and acceptor groups on the two terminal thiophene rings leading to chiral helical push-pull systems. In this field, both theoretical and experimental studies have been done that are described here.

A seminal work in this field reports the first experimental hyperpolarizability γ value of unsubstituted 7-TH which resulted to be larger than that of carbon disulfide and therefore very promising. Of course, the introduction of donor and acceptor substituents on the helicene scaffold should improve the performance as NLO chromophore [69].

Semiempirical and *ab initio* calculations of the polarizability and hyperpolarizabilities of enantiomerically pure substituted (*M*)-tetrathia-[7]-helicenes show that the best NLO properties are obtained with the presence of nitro (NO_2) and the amino (NH_2) pair substituents, which result in the best electron-withdrawing and donating groups, respectively, in the investigated series. This type of calculation requires a relatively low computational effort and therefore can be useful to perform a first screening of optical properties of functionalized helicenes to help the synthesis of new efficient NLO materials [70].

A systematic experimental and theoretical investigation of the electric field-induced second harmonic (EFISH) response using the EFISH generation technique of tetrathiahelicene substituted with NO_2, CH=CHCN, and $COCF_3$ units as accepting groups shows how these functional groups can tune the first and second hyperpolarizabilities.

Substituents on 7-TH are located on both the terminal thiophene positions and the central benzene ring of the helicene, as reported in structures **32, 71–73** (Figure 6.8).

In acceptor-monosubstituted helicenes, an increase of the second-order NLO response $\mu\beta_{||}$ (i.e. μ is the dipole moment and $\beta_{||}$ is the projection of the vector part of β on the dipole moment vector) is found as a function of increasing acceptor strength, while the Λ-shape (systems in which two D/A pairs intersect at the donor or acceptor group forming donor-acceptor-donor (DAD)- or acceptor-donor-acceptor (ADA)-like structures) structural feature yields a larger $\mu\beta_{||}$ response than its analogue derivative, monosubstituted at one terminal thiophene position.

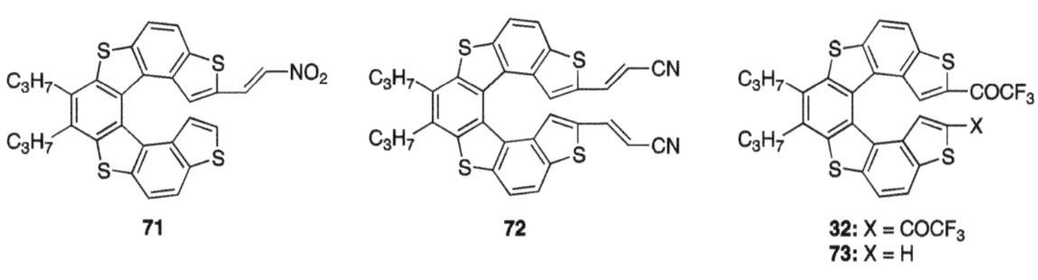

Figure 6.8 Structure of substituted 7-TH **32, 71-73**.

These results are substantially in line with those observed in donor- and acceptor-group-substituted benzene and stilbene derivatives [71] and their relative effectiveness in inducing asymmetric polarizabilities using EFISH and third-order harmonic generation (THG) measurements is established [72, 73].

A computational study using the time-dependent Hartree–Fock approach and the Austin Model 1 semiempirical Hamiltonian reports the oxidation effects on the hyper-Rayleigh scattering (HRS) second-order NLO responses of homo- and heterohelicenes. Upon oxidation, in the case of 7-TH, a substantial increase of the HRS response associated with a reduction of the first dipole-allowed excitation energy is predicted. Moreover, in the specific case of helicenes bearing thiophene units, the increase of HRS response upon oxidation is associated with the presence of a substantial axial contribution to the first hyperpolarizability, whereas the radial component is otherwise dominant [74].

In another theoretical work, the importance of the contribution of the magnetic interactions to the second-order NLO responses of chiral helicenes is reported. In particular, the macroscopic second-order NLO properties of their oriented films or isotropic solutions are demonstrated by orientationally averaging microscopic molecular properties evaluated *ab initio* at the random phase approximation level.

When the helicene has no or weak D/A substituents, the mixed electric-magnetic responses are of the same order of magnitude as the pure electric-dipole counterpart, i.e. when the pure electric-dipole response is small. When strong D/A substituents are present, the pure electric-dipole second harmonic generation (SHG) response increases substantially and more than its mixed electric-magnetic counterpart. Consequently, the ratio between the mixed electric-magnetic and pure electric responses decreases. It is therefore confirmed the possibility of tuning the mixed electric-magnetic response by employing appropriate substituted chiral thiahelicenes [75].

6.7.2 Self-Assembly

Studies of properties of thiahelicenes on nanoscale show that as in the case of thiahelicene polymers, they can self-assemble, and this behavior can change in a positive way the electro-optical properties of the single molecule of thiahelicene.

Films of tetrathia[7]helicene **1** and 7,8 dipropyl tetrathia[7]helicene **47**, when deposited at 25 °C under vapor-phase conditions on a bare silica surface (SiO_2), and on a silanized one with hexamethyldisilazane (HMDS) show, as inferred from atomic force microscopy (AFM) images, an interesting tubular morphology. These preliminary results can be a starting point for further investigations on the applications of thiahelicene chemistry in nanosciences [76].

Self-assembly of helicenes on metal surfaces under high-vacuum conditions is another interesting phenomenon, whose investigation by means of scanning tunneling microscopy (STM) is helpful to understand the principles underlying processes involving the spontaneous resolution and chiral recognition of helicenes [77].

The first STM images of a material obtained by exposing the Au(111) surface to the vapor of the parent 7-TH **1** in a closed vessel at 70 °C dates back to 1991 [78], and reveals a quasi-ordered and almost hexagonal arrangement of the helical system with the molecular planes parallel to the metal surface. The interaction between the 7-TH and the gold surface is rather weak. The enantiopure (*M*)-7-TH diol **46** molecules also adsorb on the Au(111) surface yielding sporadic isolated monomers, dimers and trimers along with self-assembled twin rows, whose formation seems strictly related to the temperature [79]. The molecular adsorption model proposed for these twin rows does not involve the formation of hydrogen bond between the hydroxyl groups of neighboring molecules or a strong π–π stacking interaction. Noteworthy, (*M*)-**46** molecules take on two different orientations within the self-assembled layer, as shown by high-resolution STM images. Again, STM studies demonstrate that the adsorption of a racemic mixture of 7-TH dialdehyde **33** is strongly influenced by the different nature of the metal surfaces [80]. Indeed, while the adsorption of **33** molecules on Au(111) provides self-assembled twin rows or single rows depending on the molecular coverage, neither the formation of self-assembled twin rows nor the molecular arrangement into different domains can be obtained on Cu(001) and NiAl(110) surfaces, in which only molecular clusters are observed. An STM study reported in 2020 [81] shows that two classes of dithia[7]helicenes, namely helicenes bearing thiophene end rings with either *exo* or *endo* topologies, can be deposited on gold (111) surfaces at room temperature, and the effect of the sulfur atom is to strengthen their binding to the electrodes. The position of the sulfur atom in the helical framework and its binding interaction with the gold surface also play a crucial role. An ordered layer of clusters over the gold is observed for the *exo*-dithia[7]helicene, presumably due to a tighter binding with the surface, while for the *endo*-dithia[7]helicene there is a higher number of molecules per cluster, as a result of enlarged reticular energy.

6.7.3 Molecular Spring

7-TH **1** can be considered a spring due to the stretching properties of the helix. In fact, despite the rigidity arising from the fused benzene rings, the presence of substituents on the terminal thiophene rings can vary the dihedral angle of the helical scaffold. The stretching property of a 7-TH derivative substituted with a disulfide moiety and a carboxy group at the end of the two thiophene rings has been studied by AFM-based single-molecule force spectroscopy (SMFS) and electrostatic state calculations.

Also, electrostatic state calculations showed that the 7-TH framework is more flexible than structures like, for example, biphenyls [82].

6.7.4 Photoswitches

Photochromism is studied because of its potential applications in devices, such as switches and optical memories. Photochromic [7]-thiahelicenes based on the 1,2-dithienylcyclopentene backbones satisfy the requirements of successful chiroptical photoswitches [83–86]. They are thermally stable in both ring-closed form and ring-open form, display high stereoselectivity in their photocyclization reactions [84–86], and exhibit large changes in their CD and optical rotatory dispersion (ORD) spectral properties.

6.8 Catalysis

As discussed in Paragraph 2, the versatility of the optimized procedures for the selective functionalization of the 2 and 13 positions of 7-TH scaffold allows the introduction of different phosphorus-containing groups, including phosphine-borane adducts, phosphanes, phosphine oxides, phosphine-selenides, and a diphosphonate. Thanks to the easy access to this wide range of derivatives, the catalytic behavior of some of them has been investigated in both organometallic and organic catalyses [87].

6.8.1 Organometallic Catalysis

The use of tetrathia[7]helicenes with appended phosphorus functions as chiral ligands in homogenous transition metal catalysis has been investigated in the rhodium-catalyzed asymmetric hydrogenation of prochiral alkenes as a benchmark reaction [26]. The cationic rhodium (I) complexes (±)-**74a** and (±)-**74b**, containing the 7-TH diphosphane (±)-**56a**, are prepared in excellent yields by a ligand-exchange reaction using an equimolar amount of the two Rh(I) species $[Rh(COD)_2]BF_4$ and $[Rh(COD)_2]BARF$, respectively (Scheme 6.16): COD = 1,5-cyclooctadiene.

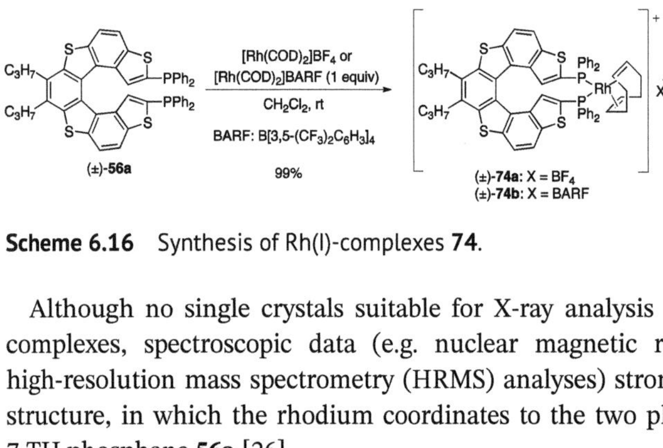

Scheme 6.16 Synthesis of Rh(I)-complexes **74**.

Although no single crystals suitable for X-ray analysis are available for these complexes, spectroscopic data (e.g. nuclear magnetic resonance (NMR) and high-resolution mass spectrometry (HRMS) analyses) strongly suggest a chelated structure, in which the rhodium coordinates to the two phosphorus pendants of 7-TH phosphane **56a** [26].

First catalytic studies of these complexes, aimed to verify their behavior in standard hydrogenation reactions, show that the enantiopure Rh(I) complex (*P*)-(+)-**74a** is able to promote the reduction of the itaconic acid and methyl 2-acetamidoacrylate in high conversion (67–99%), and the reduced products are obtained in moderate *ee* (32–40%) [26]. Asymmetric hydrogenation of similar substrates promoted by a rhodium complex containing carbo[6]helicene-based diphosphane provides comparable results (39% *ee*) [88].

Afterward, 7-TH mono and diphosphane (±)-**55a** and (±)-**56a** are successfully used as phosphorus-based ligands to prepare the mono and dinuclear gold(I) complexes (±)-**75** and (±)-**76**, respectively, which are obtained in quantitative yield using Au(tht)Cl as gold(I) source (Scheme 6.17) [30].

The structure of both complexes is confirmed by a complete analytical and spectroscopic characterization, and a weak intramolecular aurophilic interaction

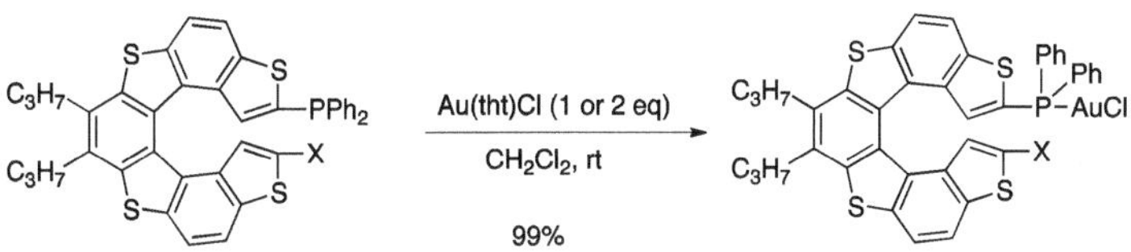

Scheme 6.17 Synthesis of Au(I)-complexes **75** and **76**.

Figure 6.9 Structure of **HelPhos-S** derivatives.

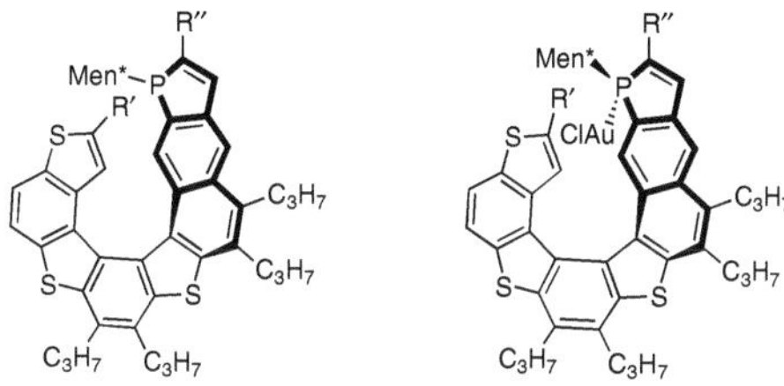

between the two gold atoms (3.18 Å) is observed for dinuclear complex **76** by X-ray analysis [30].

A preliminary study on the catalytic behavior of both complexes is performed in common gold(I)-promoted cycloisomerization reactions, in which both (±)-**75** and (±)-**76** are efficient pre-catalysts, for example, in the hydroarylation of allenes and the hydroxycarboxylation of allene-carboxylates, providing the final products in good to excellent yield (65–99%) [30]. Although these catalysts display great efficacy, the corresponding enantiopure systems provide unsatisfactory results in terms of enantioselectivity. High levels of enantioinduction in gold(I)-catalyzed cycloisomerizations can be achieved by means of structural modifications of the 7-TH skeleton, in which a terminal phosphole unit is incorporated into thiahelicene-based scaffolds (**HelPhos-S**, Figure 6.9) [89, 90]. In this case, the presence of the phosphorus atom in the internal edge of the helix represents one of the key factors for the high enantioselectivity of the corresponding gold(I) complexes **77** (Figure 6.9).

Different enyne cycloisomerizations are performed using enantiopure pre-catalysts **77**, which afford the cyclized products in good yield (up to 91%) and enantioselectivity (ee up to 96%) [89, 90].

6.8.2 Organocatalysis

The 7-TH diphosphine oxides [29] **58a,c,d** and 7-TH monophosphanes [28] **55a,c** have been tested as chiral Lewis bases in stereoselective organocatalytic reactions (Figure 6.10).

Enantiopure 7-TH tertiary diphosphine oxides (+)-(*P*)-**58a,c,d** are used as organocatalysts in polyhalosilane-mediated reactions, such as $SiCl_4$-mediated aldol-type reactions and $HSiCl_3$-mediated stereoselective reduction of imine [29]. Although the final products of these reactions are obtained with low enantioselectivity (up to 23%), aryl 7-TH diphosphine oxides **58a** and **58d** show to be very

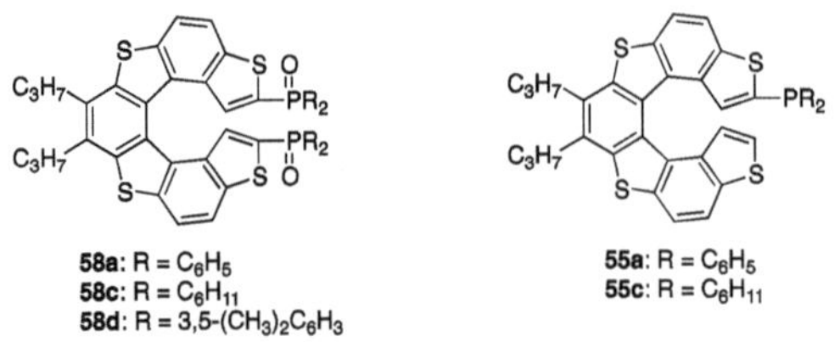

Figure 6.10 Structure of 7-TH phosphine oxides **58a,c,d** and 7-TH phosphanes **55a,c**.

active organocatalysts, affording the final products in good yield and excellent diastereoselectivity [29].

On the contrary, alkyl diphosphine oxides **58c** are not able to promote this reaction, presumably due to the highly steric hindrance of the four cyclohexyl moieties that hamper the formation of the catalytically active hypervalent silicon intermediate [29].

The catalytic behavior of 7-TH monophosphanes (±)-**55a** and (±)-**55c** as Lewis base organocatalysts have been investigated in some [3+2]-annulation reactions [28]. Alkyl 7-TH phosphane **55c** is able to promote these cycloaddition reactions yielding the final products in high conversion (72–82%), otherwise aryl 7-TH phosphane **55a** does not provide any products, presumably due to its lower nucleophilic character. Overall, as expected, the catalytic activity as well as the stereoselectivity of 7-TH phosphanes and phosphine oxides are strongly affected by their electronic and steric properties. So, proper modifications on phosphorus groups and/or 7-TH scaffold could lead to enhanced catalytic performances. Again, among the plethora of chiral phosphorus ligands so far developed in homogeneous catalysis, these first studies confirm the potential of these helical phosphanes and phosphine oxides as ligands in transition metal catalysis and organocatalysis.

6.9 Interaction with Biomolecules

The ability of the 7-TH scaffold to interact with naturally occurring compounds is one of the peculiar features of this class of heterohelicenes, and this behavior renders these helical molecules promising drugs for therapeutic application in the biomedical field. In comparison to other fields of applications, this area is still in its infancy, and a limited number of studies have been so far reported, mainly involving the chiral recognition of 7-TH with biomolecules [91–93] (e.g. carbohydrates, deoxyribonucleic acid (DNA), and more complex structures), and, more recently, the development of delivery nanosystems for model 7-TH [94, 95] as proof-of-concept for potential applications as therapeutic agents.

6.9.1 Chiral Recognition of Biomolecules

The first study on the chiral recognition of small biomolecules with 7-TH systems was reported in 2001, using alkyl β-D-pyranoside micelles, such as octyl β-D-glucopyranoside (C8G) and dodecyl β-D-maltopyranoside (C12M) micelles, that are able to solubilize lipids and hydrophobic compounds [91].

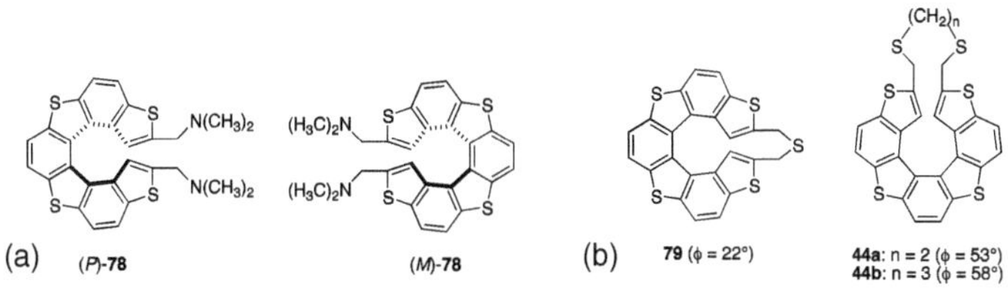

Figure 6.11 Structure of biologically active 7-TH derivatives.

The behavior of the parent 7-TH **1** in these micelles was investigated by NMR and CD spectroscopy, and in the micellar solution of C12M, it was found that ^{1}H-NMR spectra of both enantiomers of **1** displayed different chemical shifts, due to the presence of diastereomeric interactions. Noteworthy, the chemical shift variations of the (*M*)-**1** protons were larger than those of the (*P*)-**1** protons, suggesting a stronger interaction of C12M micelles with the *M* enantiomer than the *P* enantiomer.

CD experiments done on the (*P*)-2,13-bis(dimethylaminomethyl)-tetrathia[7]helicene (**78**) (Figure 6.11a) demonstrate that it is an interesting enantioselective ligand able to selectively bind to the left-handed conformer Z-DNA over the right-handed Watson–Crick B-form [92], inducing the conversion of the B-DNA into the Z-DNA. On the contrary, the enantiomer (*M*)-**78** is a nonselective ligand in binding to Z-DNA since its binding constant for Z-DNA is significantly lower than that of (*P*)-**78**, and very similar to that for B-DNA [92]. Noteworthy, when the amino groups are replaced by hydroxyl groups, no selectivity is detected, confirming the crucial rule of the protonated amino moieties to bind Z-DNA. Despite the peculiar recognition of Z-DNA by (*P*)-**78** is not yet clarified, this work is an interesting starting point to design 7-TH-based inhibitors of biological functions related to Z-DNA, whose biological roles are still debated but represent an active research field.

Again, thiahelicenophanes **79** and **44** (Figure 6.11b) are able to display enantioselective recognition toward G-quadruplex structures, which involve chiral-rich spaces and Z-DNA and are known to inhibit telomerase activity [93]. The (*M*)-thiahelicenophane **79**, with a shorter linker and the lowest dihedral angle ($\phi = 22°$), shows enantioselective binding to telomeric G-complexes and the inhibition of the enzymatic activity of telomerase. From a structure-activity relationship point of view, the dihedral angle of the helical scaffold seems to be a key parameter for the development of novel helical-based G-quadruplex ligands, though no models of binding of (*M*)-**79** to G-quadruplexes have been identified yet, and further studies are needed. However, these results certainly demonstrate the potential of 7-TH systems as small-molecule inhibitors of telomerase, whose inhibition represents a promising approach to cancer therapy.

6.9.2 Delivery Systems for 7-TH

The potential use of tetrathia[7]helicenes as therapeutics in the biomedical field is closely related to their bioavailability and their ability to penetrate the cell membrane, which are essential prerequisites for the interaction of the drug with the cell components and to fulfill its biological effect. In this context, the study of proper drug

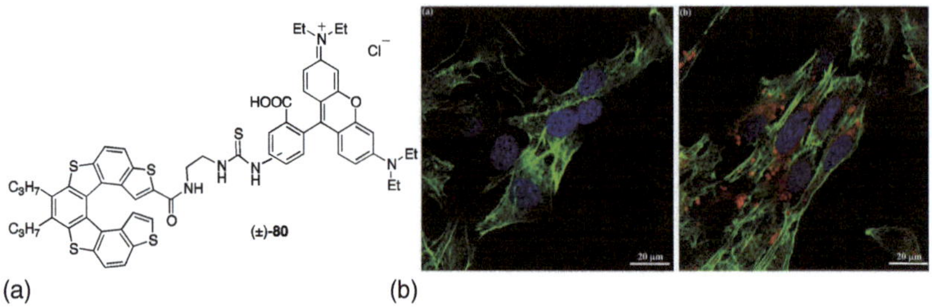

Figure 6.12 (a) Structure of luminescent 7-TH; (b) confocal laser scanning microscopy imagines of Balb/3T3 clone A31 cells: (a) untreated; (b) treated with **80**-PLGA nanosystem.

delivery systems that ensure an efficient uptake and a controlled release in cells of hydrophobic 7-TH derivatives is of great interest.

A nanoconstruct based on organic nanoparticles of poly(lactic-co-glycolic acid) (PLGA), loaded with a luminescent 7-TH derivative (±)-**80** (Figure 6.12a), selected as a model therapeutic 7-TH, displays a very narrow particle size distribution, a high electric charge on the surface that prevents the aggregation of the nanoparticles, and it allows a controlled release of (±)-**80** from the nanoparticles, without an initial burst step [94]. Noteworthy, *in vitro* experiments on Balb/3T3 clone A31 fibroblasts show that the cytotoxicity of this nanoconstruct is significantly lower than that of the free 7-TH **80**, and an efficient and rapid internalization into the cells is observed within 1 hour (Figure 6.12b), presumably via a micropinocytosis mechanism, which prevents lysosomal degradation of the nanosystem and therefore protects the biological activity of the loaded 7-TH [94]. All these features clearly indicate the suitability of PLGA as a biocompatible and biodegradable nanovector for the delivery of 7-TH derivatives.

More recently, halloysite nanotubes (HNTs), which are naturally aluminosilicate clay extracted in the form of nanotubes, turned out as an alternative drug delivery nanosystem for the intracellular and controlled delivery of 7-TH [95].

A novel nanoconstruct **HNT-7TH**, in which the 2-formyltetrathia[7]helicene (±)-**81** (Figure 6.13) is covalently anchored to properly functionalized HNTs (**HNT-APTES**) through an imine bond shows an efficient and controlled release of the 7-TH under mild acidic pH conditions, which are usually required to promote the hydrolysis of the imine bond [95].

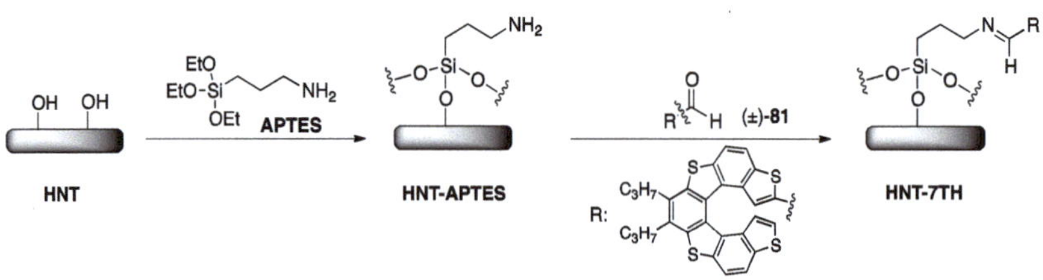

Figure 6.13 Functionalization of HNTs with 7-TH aldehyde **81**.

This behavior is also confirmed by in vitro experiments on two tumoral cell lines with different extracellular pH, confirming the potential of this novel nanoconstruct for the triggered release of 7-TH derivatives for potential therapeutic applications.

6.10 Conclusion

As stated in the title of this chapter, tetrathiahelicenes have been, are, and surely will be an infinite source of inspiration for chemists, and active in the field of synthetic methodologies, catalysis, chiral sensors, nanoscience, photophysics, biology, computational study, materials, and perhaps other disciplines. In fact, as can be seen from what is described in this chapter, the peculiar and unique combination of their functional, behavioral, and structural features renders this class of polycyclic heteroaromatic chiral systems unique and very versatile both for basic research as well as for potential applications in a different field. This is the reason why a huge amount of chemistry has been developed since 1971 till today for this class of helical systems, and it can be predicted this trend will continue and, hopefully, will lead to new applications. Recent achievements in novel synthetic strategies for their preparation let predict the possibility to have access to enantiomerically pure/enriched systems, and this will speed up a lot of those applications based on their chirality. Authors believe that the "thiahelicenes era" is far to be concluded, and they hope that other book chapters will be published in the future describing novel chemistry and applications, perhaps not even imaginable today.

References

1 Wynberg, H., Groen, M.B., and Schadenberg, H. (1971). Synthesis and resolution of some heterohelicenes. *J. Org. Chem.* 36: 2797–2809. https://doi.org/10.1021/jo00818a016.
2 Caronna, T., Catellani, M., Luzzati, S. et al. (2001). Molecular crystal architecture and optical properties of a thiohelicenes series containing 5, 7, 9, and 11 rings prepared via photochemical synthesis. *Chem. Mater.* 13: 3906–3914. https://doi.org/10.1021/cm010093z.
3 Lehman, P.G. and Wynberg, H. (1971). Simplified synthetic routes to symmetrical heterohelicenes. *Recueil* 90: 1113–1118. https://doi.org/10.1002/recl.19710901009.
4 Tanaka, K., Osuga, H., Shogase, Y., and Suzuki, H. (1995). Convergent synthesis of optically active bifunctionalized [7]thiaheterohelicene. *Tetrahedron Lett.* 36: 915–918. https://doi.org/10.1016/0040-4039(94)02371-H.
5 Caronna, T., Sinisi, R., Catellani, M. et al. (2000). Photochemical synthesis and structural properties of high membered thiohelicenes. *Chem. Commun.* 1139–1140. https://doi.org/10.1039/B002581J.

6 Caronna, T., Sinisi, R., Catellani, M. et al. (2001). Photochemical synthesis and optical properties of high membered thiohelicenes. *Synth. Met.* 119: 79–80. https://doi.org/10.1016/S0379-6779(00)00749-9.

7 Maiorana, S., Papagni, A., Licandro, E. et al. (2003). A convenient procedure for the synthesis of tetrathia-[7]-helicene and the selective α-functionalization of terminal thiophene ring. *Tetrahedron* 59: 6481–6488. https://doi.org/10.1016/S0040-4020(03)01056-1.

8 Osuga, H., Nagai, S., and Tanaka, K. (2005). Synthesis and properties of polyalkylated thiaheterohelicenes. *Phosphorus, Sulfur Silicon Relat. Elem.* 180: 1499–1500. https://doi.org/10.1080/10426500590913375.

9 Baldoli, C., Bossi, A., Giannini, C. et al. (2005). A novel and efficient approach to (Z)-1,2-bis(benzodithienyl)ethene -precursors of tetrathia[7]helicenes. *Synlett* 1137–1141. https://doi.org/10.1055/s-2005-865211.

10 Licandro, E., Rigamonti, C., Ticozzelli, M.T. et al. (2006). Synthesis and functionalization of novel tetrathia[7]helicenes as new push-pull systems. *Synthesis* 3670–3678. https://doi.org/10.1055/s-2006-950222.

11 Bossi, A., Maiorana, S., Graiff, C. et al. (2007). Silyl-substituted tetrathia[7]helicenes: synthesis, X-ray characterization and reactivity. *Eur. J. Org. Chem.* 2007: 4499–4509. https://doi.org/10.1002/ejoc.200700494.

12 Waghray, D., Nulens, W., and Dehaen, W. (2011). Efficient synthesis of benzo fused tetrathia[7]helicenes. *Org. Lett.* 13: 5516–5519. https://doi.org/10.1021/ol202236r.

13 Waghray, D. and Dehaen, W. (2013). A fragment based approach toward thia[n]helicenes. *Org. Lett.* 15: 2910–2913. https://doi.org/10.1021/ol400825w.

14 Larsen, J. and Bechgaard, K. (1996). Direct oxidative cyclization of 1,2-bis(benzothiophene-2-yl)ethylenes as a replacement of photocyclization in the synthesis of thiaheterohelicenes. *J. Org. Chem.* 61: 1151–1152. https://doi.org/10.1021/JO951277Y.

15 Pei, J., Zhang, W.-Y., Mao, J., and Zhou, X.-H. (2006). Helical polycyclic aromatics containing thiophenes: synthesis and properties. *Tetrahedron Lett.* 47: 1551–1554. https://doi.org/10.1016/j.tetlet.2006.01.006.

16 Tovar, J.D. and Swager, T.M. (2001). Poly(naphthodithiophene)s: robust, conductive electrochromics via Tandem cyclization–polymerizations. *Adv. Mater.* 13: 1775–1780. https://doi.org/10.1002/1521-4095(200112)13:23<1775::AID-ADMA1775>3.0.CO;2-N.

17 Tovar, J.D., Rose, A., and Swager, T.M. (2002). Functionalizable polycyclic aromatics through oxidative cyclization of pendant thiophenes. *J. Am. Chem. Soc.* 124: 7762–7769. https://doi.org/10.1021/ja0262636.

18 Waghray, D., de Vet, C., Karypidou, K., and Dehaen, W. (2013). Oxidative transformation to naphthodithiophene and thia[7]helicenes by intramolecular Scholl reaction of substituted 1,2-bis(2-thienyl)benzene precursors. *J. Org. Chem.* 78: 11147–11154.

19 Tanaka, K., Suzuki, H., and Osuga, H. (1997). Non-photochemical route to chiral disubstituted [7]thiaheterohelicenes via biaryl- and carbonyl-coupling reactions. *J. Org. Chem.* 62: 4465–4470. https://doi.org/10.1021/jo970272t.

9

Helical Nanographenes: Synthetic and Chiroptical Achievements

Sandra Míguez-Lago[1], Juan P. Mora-Fuentes[1], Carlos M. Cruz[2], and Araceli G. Campaña[1]

[1] *Departamento de Química Orgánica, Unidad de Excelencia de Química aplicada a Biomedicina y Medioambiente (UEQ), Facultad de Ciencias, Universidad de Granada, Granada, Spain*
[2] *Department of Chemistry, Faculty of Science, University of Zurich, Zurich, Switzerland*

9.1 Introduction

Centuries ago, pencils were created to write history, and still nowadays the carbon-based hexagonal layers that compose their nuclei continue to revolutionize them. Graphene, as this two-dimensional (2D) honeycomb structure is known, constitutes the first carbon allotrope whose isolation was awarded with the Nobel Prize in Physics [1]. This historical event was celebrated as close in time as in 2010, and the winners of this award, physicists Andre Geim and Konstantin Novoselov, were the engineers of the Scotch tape-based micromechanical exfoliation method going from graphite, the mineral composed by a multilayered array of graphene sheets held together by noncovalent interactions, to graphene, constituted by one individual layer of hexagonally distributed sp^2-hybridized carbon atoms [2, 3]. Despite the rush in the number of publications regarding graphene, mushrooming from 157 in 2004, when everything started, to 2500 in 2010, skeptics claim that the promise of its glory has not turned into facts to date. But the truth is that, in comparison with silicon, one of the most famous materials on Earth, whose application into chips was made to wait 157 years from its discovery, investment in graphene-based technologies seems close to be paid off in the near future. Example of the efficient academic research–industrial application bridging is shown in the myriad industries such as the pharmaceutical, automotive, aerospace, coatings, energy, chemical, and electronics among others, which are regular customers of the global graphene manufacturers.

Interest in graphene naturally burgeoned by virtue of its astonishing physical properties, from its conductivity (way better than copper) to its flexibility and resistance (much higher than steel) and last but not least its transparency, which allows the implementation as large sheets into windows and panels for optoelectronic applications. All these features mirror its structural configuration, affected by not only size and curvature but also chemical substitution at the rim, heteroatom

Helicenes: Synthesis, Properties, and Applications, First Edition.
Edited by Jeanne Crassous, Irena G. Stará, and Ivo Starý.

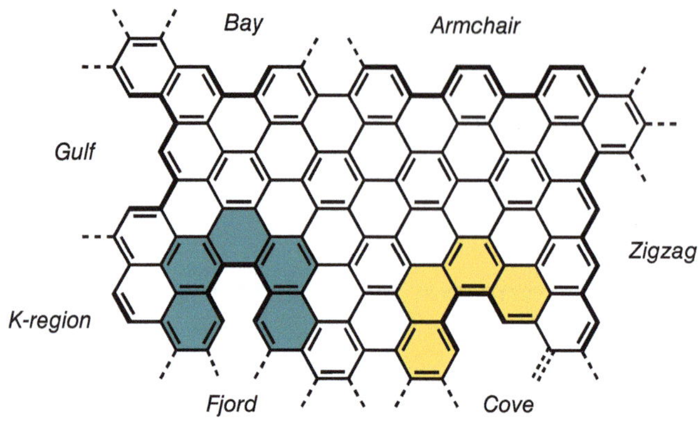

Figure 9.1 Edge topologies of graphene (*zigzag*, *armchair*, *bay*, *cove*, *fjord*, and *gulf*). Cove and fjord highlighted as masked [4, 5]helicenes. Source: Based on [9].

doping, and edge topology. Besides, when considering multilayer graphite flakes, thickness plays a crucial role in the final material properties, responding the latter to the theory of bands of graphite, where graphene was taken as a model [4]. Concerning the size, its influence in the bandgap has been demonstrated, being inversely proportional to the width in the case of graphene nanoribbons (GNRs) [5–7]. Thus, regarding the shape of the rim, there are two basic edge motifs known as *zigzag* and *armchair* [8], which give rise to four main topologies named *bay*, *fjord* (both from *armchair* rim), cove, and gulf (from *zigzag* rims) (Figure 9.1) [9]. The presence of these topologies within the structure of graphene dramatically alters not only its optoelectronic properties [10] but also its stability [11–14].

The classical ways of obtaining graphene, mostly based on *top-down* approaches, begin with macroscopic materials. Therefore, chemical vapor deposition (CVD) [15], whose main characteristic is getting films of homogeneous thickness, succeeded in obtaining millimeter-sized graphenes, achieving carrier mobilities greater than 7100 $cm^2V^{-1}s^{-1}$ [16]. A second *top-down* option was the Scotch tape method for the mechanical exfoliation of graphite, which, despite being awarded with the Nobel Prize in Physics, is yet quite laborious to be implemented at large scale [1, 17]. A third classical *top-down* approach is lithography, based on cutting larger GNRs by applying high voltages through, for instance, scanning tunneling microscope (STM) [18]. In general, despite encompassing the cheapest and large-scale implementable methodologies, *top-down* approaches suffer from low product definition and lack of reproducibility of the procedures, highly dependent on the preparation route, not to mention that such approaches often originate irregular grain boundaries, which act to the detriment of charge carrier mobilities, among others [19]. Along with those challenges, the size of the graphene flakes obtained is usually too large to be solution processed, making the product difficult to be implemented into assembly line manufactured real devices.

In order to overcome those issues, scientists envisioned the design of models of graphene of nanometric size, coined nanographenes (NGs), whose synthesis was mainly accomplished following *bottom-up* approaches, and founded on the chemical assembly of smaller building blocks. Below NGs, within the subnanometric scale, are common polycyclic aromatic hydrocarbons (PAHs), and in between, considered the smallest member of the family of NGs, is hexa-*peri*-hexabenzocoronene (HBC) [20]. HBC was firstly synthesized by Clar and coworkers in 1958 by a final heating at 481 °C of tetrabenzoperopyrene [21, 22], and, later on, Halleux [23] and Schmidt [24] reported alternative methods although also using quite harsh conditions. Alternative synthesis reported by Müllen *et al.* based on cyclotrimerization of diphenylacetylenes, combined with the oxidative cyclodehydrogenation (Scholl reaction) as key steps, was presented years after [25]. These last strategies, together with Diels–Alder reaction followed by Scholl reaction, constitute the two main methodologies through which most of the large NGs and "superacenes" reported have been synthesized [26]. More recent strategies toward NGs, based on benzannulation [27], chemistry of short-lived aryne intermediates [28], annulative π-extension (APEX) over the K-region of nonactivated PAHs [29], dehydrative π-extension (DPEX) reactions [9], and either Ullman coupling or Pt/Pd cycloaromatization, both followed by final oxidative cyclodehydrogenation [30–32], on-surface synthesis [33, 34], or photocyclization reactions [35], have been also developed for this purpose (Figure 9.2).

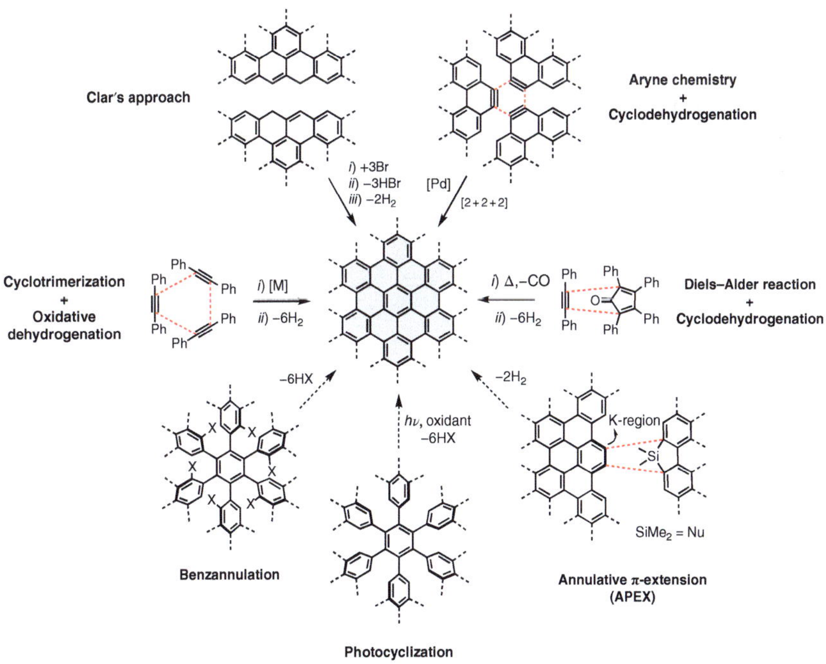

Figure 9.2 Selected synthetic *bottom-up* strategies toward generic nanographenes.

Thus, atomic precision achieved by means of *bottom-up* approaches allows not only for the construction of defined structures but also for the introduction of *à la carte* defects within graphene monolayers. In fact, the presence of defects based on the advent of non-hexagonal rings on the graphene network was previously visualized in highly oriented pyrolytic graphite (HOPG) by means of transmission electron microscopy (TEM) [36, 37] and in epitaxial graphene through STM [38], among others, and its appearance was both theoretically [39] and experimentally [40] proven to cause a modification in the electronic properties of graphene, more specifically in the electronic bandgap. Defects object of these reports, whose origin typically resides in either the formation of a vacancy, the inclusion of an extra atom, or atoms' rearrangements such as the so-called Stone–Wales rearrangement within graphene network, are classified as endoskeletic defects (Figure 9.3). Thus, vacant defects have been theoretically examined with regard to their relative stability [41], and the effect of the introduction of extra carbon atoms within the hexagonal network has been experimentally reported [42, 43]. Completing the set, one last type of defects, denominated exoskeletic, which is caused by steric hindrance, originates a deviation of graphene from its original planarity (Figure 9.4). These induced twists give rise to helical structures appropriately called helicenes.

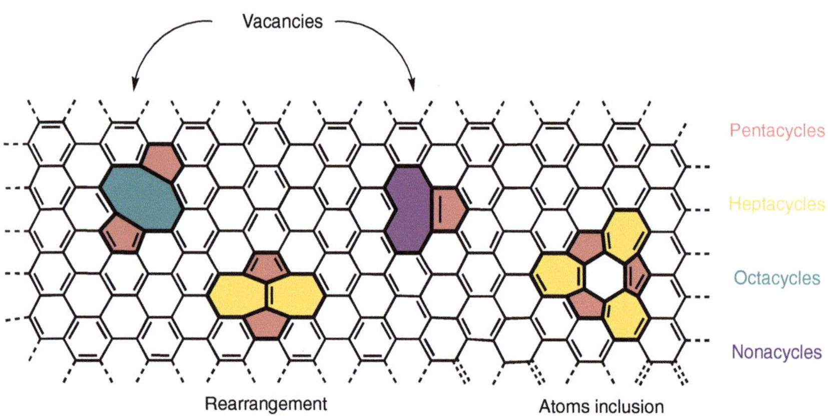

Figure 9.3 Examples of endoskeletic defects.

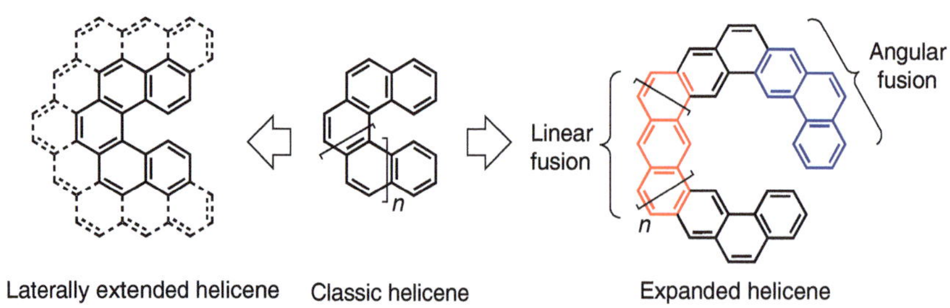

Figure 9.4 Types of helicenes according to length, extension, and expansion.

Actually, the concept of helicene was previously introduced within this section when discussing the *cove*- and *fjord*-like topologies of graphene's edges. Thus, helicenes arise from the *ortho*-fusion of benzene rings in an angular array, consequence of the steric repulsion of their terminal rings. Depending on their length, helicenes can be classified with a numerical descriptor resulting in [n]helicenes, where n is the number of *ortho*-fused rings forming the helix (Figure 9.4). These helical molecules can be also labeled as "expanded helicenes" in case they alternately combine both linear and *ortho*-fusion of the benzenic rings, as originally suggested by Tilley and coworkers [44, 45]. By virtue of the extension of the radial π-system, helicenes can be as well categorized as "laterally extended helicenes." Hence, the combination of the two aforementioned features gives rise to the so-called "laterally extended expanded helicenes."

Synthetically speaking, many approaches have been reported for the realization of both purely carbonated [46, 47] and main-group atoms-containing helicenes [48], without extension and expansion within their structures. However, they are out of the scope of this chapter, and, therefore, we will focus on helical NGs, highlighting those ones with high degree of lateral expansion and/or extension. Thus, along with the multifarious synthetic methodologies directed toward the synthesis of NGs briefly introduced in former paragraphs, a detailed discussion regarding particular cases and the properties they exhibit will be given in the corresponding sections of this chapter.

Among the diverse contributions of the incorporation of helicenes into NGs, the first noteworthy feature would be the increase in solubility, traditionally enhanced for insoluble NGs by the introduction of solubilizing groups such as lateral alkyl chains. The mechanism of action follows nevertheless the same principle, weakening the cooperative π–π interactions between flat sheets of pristine NGs. On top of that, helicity is also responsible for altering the optoelectronic properties of NGs, even if it is a slight disruption of the electronic conjugation along the system [49]. Furthermore, inspired by the groundbreaking discovery of Jarillo-Herrero and collaborators, where a NG bilayer was endowed with either insulating or unconventional superconducting properties by simply rotating one layer over the other at a magical angle (θ) (Figure 9.5a) [50–52] one might imagine that introducing helical moieties into NGs may allow the construction of NG bilayers in which the helicene constitutes the hinge around which a certain twist angle can be defined, tuning this way the material behavior (Figure 9.5b). Those helical NGs might also be envisioned as simple models of graphene nanocoils with properties as electromagnets mimicking a Riemann surface [53].

Beyond that, helicenes provide NGs with helicity, a type of chirality that could be present in organic molecules (Figure 9.6). Chirality, also known as dissymmetry, is ubiquitous in nature, and its main characteristic is the impossibility of superimposing a molecule with its corresponding mirror image, both necessarily lacking an improper axis of rotation (S_n). In the particular case of helicity, chiral descriptors P (plus) and M (minus) denote right- and left-handed helices, respectively [54]. Thus, chirality is present in most of the biomolecules, as, for instance, amino acids, constituting every single protein in nature in L configuration, and carbohydrates, mostly

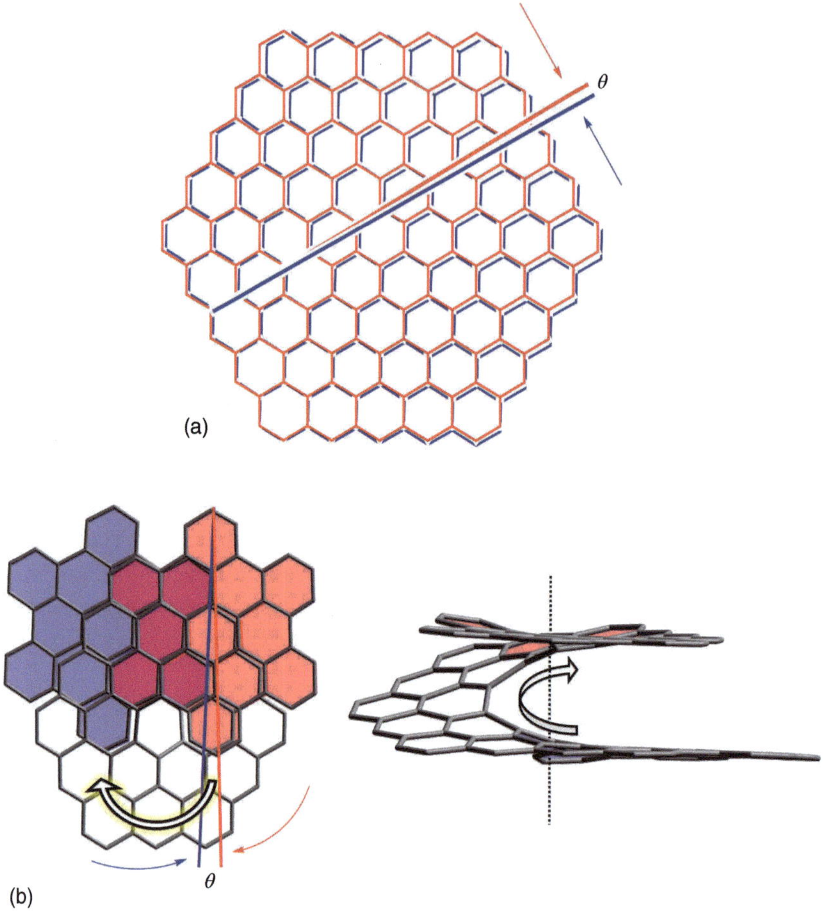

Figure 9.5 (a) Image of the twisted nanographene bilayer of Jarillo-Herrero and coworkers and (b) analogous setup with helicene in between as hinge.

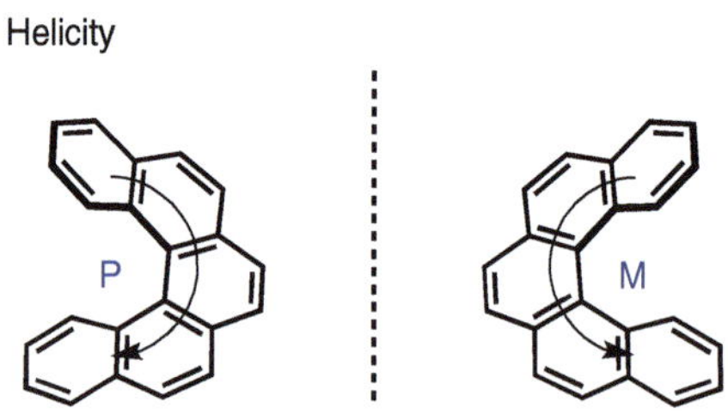

Figure 9.6 Helicity present in helicenes. *M* and *P* chiral descriptors.

found in D configuration. This peculiarity makes the blend of helicenes and NGs an appealing tool envisioning applications including novel chiroptical properties, in order to avoid the attachment of extra chiral subunits, as happened in the case of other zero-dimensional (0D) to 2D carbon allotropes [55–57].

Among the chiroptical properties habitually arising from the presence of helicenes as chiral moieties in organic molecules, optical rotation (OR) and spectroscopies such as electronic circular dichroism (ECD) and circularly polarized luminescence (CPL) can be found. ECD is defined as the difference between the absorption of left and right circularly polarized light (CPLight) for a chiral sample, where the interaction of this CPLight with any chiral compound results in one diastereomeric type, being the degree of diastereomeric discrimination quantified through the g-factor (g_{abs}), defined as the ratio between the ECD ($\Delta\varepsilon$) and the absorbance (ε) [58]. On the other hand, CPL is the emission analog of ECD, unraveling additional features of the excited electronic state, and with a corresponding luminescence dissymmetry factor (g_{lum}) described as the ratio of such emitted left- and right-handed intensity difference ($\Delta I = I_L - I_R$) divided by the average emitted intensity ($I_L + I_R/2$) [59]. Besides these linear optical properties, helicenes have their place in nonlinear optics (NLO) [60], which has been demonstrated to be of great interest in applications [61] such as three-dimensional (3D) microfabrication, achieving better resolved lithography processes, optical logic, autocorrelation, pulse reshaping, and optical power limiting. Moreover, NLO is postulated to be relevant for imaging techniques employing half energetic photons in the far infrared with high penetration power through human tissues, reaching thereby the corresponding fluorescent targets [62]. Two-photon absorption (TPA), which was the first nonlinear optical process to be observed along with its two-photon upconverted emission (TPE) counterpart, has been examined for purely organic systems [63]. Among them, noteworthy are some of the first publications on helicenes experimentally reporting TPA and two-photon circular dichroism (TPCD) [64, 65] and the first helicene-containing NG combining both TPA and CPL [66]. TPCD, theoretically enunciated by Tinocco back in the 1970s [67], is grounded on the simultaneous absorption of two circularly polarized photons of less energy than the needed for a certain electronic transition to occur. In the particular case of two isoenergetic photons, the TPCD is defined as the difference between cross sections for left and right circular polarization [68]. Thus, TPCD presents advantages such as deeper penetration power, better resolution and signal-to-noise ratio, and reduced photodamage [69]. On the other hand, there are scarce examples of the emission analog, the TPA upconverted CPL (TP-UCPL), which has only been achieved in chiral perovskite nanocrystals [70], and, hence, helical NGs represent a promising platform for further developments.

Concerning helicenes, high configurational stability is required in order to resolve and isolate stereoisomers. In order to increase configurational stability, appropriate substitution at the bay region in case of [4]helicenes and of the fjord region in the case of [5]helicenes is advised. Moreover, shape persistency is also desired to increase the degree of discrimination between the left- and right-handed CPLight, related to rotational strength (R) defined as the scalar product of both electric (μ_e) and magnetic (μ_m) transition dipole moments. Recently, many efforts have been directed in the group of Mori and coworkers toward the unraveling of the structure–property relationships of chiral chromophores [71].

In light of all the aforementioned, in this chapter our intention is to put the spotlight on π-extended helical NGs with (by definition) [20] core size equal to or larger than 1 nm, emphasizing the publications dealing with the synthesis and chiroptical properties of configurationally stable enantioenriched and resolved chiral helical NGs, along with a selection of their tangible and potential applications.

9.2 All Carbon Helical Nanographenes

The preparation of well-defined NGs [20] embedding helicene moieties is an emerging field, given the solubility improvement with respect to their planar counterparts, among other advantages. Therefore, the high efficiency and specificity of the commonly used methodologies has led to well-defined curved structures and the establishment of unequivocal structure–property relationships. Thus, the design of specific NG geometries plays a fundamental role in the development of potential electronic applications, including organic field effect transistors (OFET) or organic photovoltaics (OPV).

Helicene-based NGs are also a platform to blend twisted helicoid structures with appealing electronic and chiroptical properties. However, an essential feature of helicenes to be taken into account is their racemization barrier [72], which mainly depends on the steric hindrance between the terminal *ortho*-fused rings of the [*n*]helicene. Unsubstituted carbohelicenes with $n < 6$ exhibit, in general, low configurational stability at room temperature according to their racemization barrier. Thus, generally speaking, carbo[4]helicene derivatives cannot be isolated as enantioenriched mixtures at room temperature, while carbo[5]helicenes could be resolved into enantiomers, which tend to racemize over several days under ambient conditions. For this reason, the introduction of substituents at the *cove* or *fjord* positions of the [*n*]helicenes with $n < 6$ increases their steric hindrance and, therefore, their racemization barriers, providing an efficient way to improve the configurational stability. However, [*n*]helicenes with $n > 6$ generally show higher configurational stabilities without further functionalization.

On the other hand, an alternative strategy to increase the configurational stability in helicene-based NGs is the introduction of multiple helicenes in the structure (Figure 9.7) [73]. As described by Nuckolls [74], it is possible to reach isomerization barriers up to 18.5 kcal mol^{-1} for multiple carbo[4]helicenes, much higher than simple [4]helicene (4 kcal mol^{-1}) [75]. Similarly, the isomerization barriers of double carbo[5]helicene derivatives, ranging from 27.4 [76] to 31.8 [77] kcal mol^{-1}, are found to be higher compared with the 24.1 kcal mol^{-1} reported for carbo[5]helicene [72]. This trend is also observed on extended double helicenes, such as double [6]helicene and double [7]helicene. Their isomerization barriers were calculated as 43.5 [78] kcal mol^{-1} and 46.0 [79] kcal mol^{-1}, respectively, showing a remarkable increase compared with their corresponding monohelicenes, with calculated isomerization barriers of 37.3 [78] kcal mol^{-1} and 41.7 [78, 80] kcal mol^{-1} for carbo[6]helicene and carbo[7]helicene, respectively.

As a first step on the exploration of compounds that resemble to helically curved graphene, numerous fully carbonated helical NGs have been synthesized and reported in literature. Thus, this section will focus firstly on achiral or racemic helical graphene molecules, followed by chiral and enantioenriched

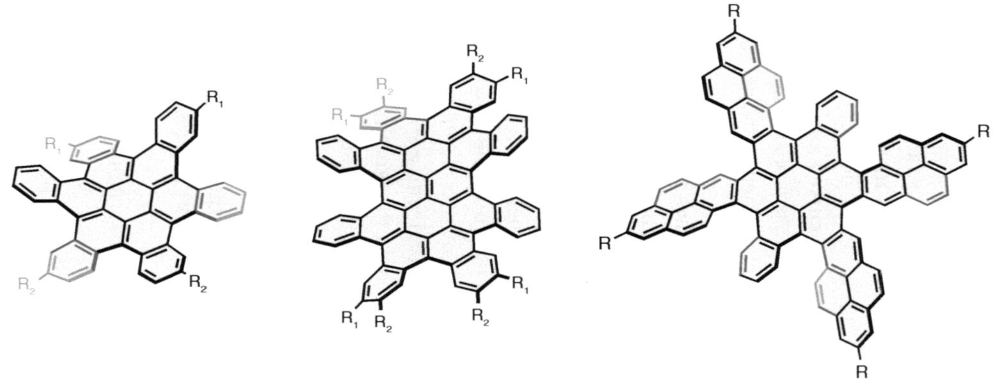

c-HBC: $R_1 = R_2 = H$
1a: $R_1 = R_2 = OC_{12}H_{25}$
1b: $R_1 = H$, $R_2 = OC_{12}H_{25}$

2a: $R_1 = R_2 = H$
2b: $R_1 = OC_{12}H_{25}$; $R_2 = H$
2c: $R_1 = R_2 = OC_{12}H_{25}$

3a: $R = H$
3b: $R = {}^tBu$
3c: $R = OC_{12}H_{25}$

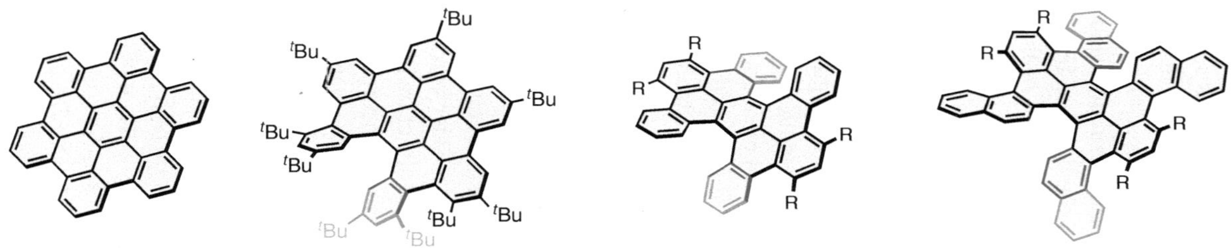

HBC

4

5a: $R = OCH_3$
5b: $R = OC_6H_{13}$

6a: $R = OC_3H_7$
6b: $R = OC_6H_{13}$

Figure 9.8 HBC, *c*-HBC, and their single and multiple helicene-containing derivatives.

and *anti-* (*P,M*) conformers of **5b** were separated. The two diastereoisomers of **5b** exhibited different optoelectronic and electrochemical properties, being the HOMO–LUMO optical gap of the *twisted* conformer 0.39 eV smaller than the *anti-* one, which exhibited a value of 2.80 eV [76]. In the case of **6b**, both *twisted* and *anti-* conformers were observed among the 24 possible diastereoisomers, being investigated through ^{1}H NMR spectroscopy, DFT calculations, and X-ray diffraction analysis. The *twisted* conformer interconverts to the *anti-* conformation at elevated temperature with an experimental isomerization barrier of 27.7 kcal mol^{-1}, in good agreement with the theoretical value of 29.0 kcal mol^{-1} [87].

Together with the progress in helical NGs, the new synthetic strategies toward the controlled introduction of five- and seven-membered rings into NG systems allow the introduction of positively and negatively curved topologies, respectively. In combination with the presence of helical moieties, novel curved aromatics are created, leading to bowl-helix and saddle–helix hybrid NGs. As a result, the large perturbation of their molecular symmetry has great impact on their optoelectronic properties.

In this context, in 2011, Sygula and coworkers reported a bowl-helix hybrid PAH containing three corannulene subunits fused by a central benzene ring and thus generating three units of carbo[5]helicene (Figure 9.9, **7**) [88]. The synthetic strategy is based on a Pd-catalyzed cyclotrimerization reaction from corannulene-based aryne precursors. X-ray diffraction and computational calculations agreed on the C_1 symmetric conformer as the most stable, with theoretical isomerization barriers in the range of 8.4–17.3 kcal mol^{-1} between stereoisomers, matching the experimental value of 8.5–8.6 kcal mol^{-1}. More recently, Martín and coworkers synthesized a novel bowl-helix hybrid NG consisting of a helical HBC fused with a corannulene unit (Figure 9.9, **8**) [89]. The synthetic procedure is based on the bromination and subsequent Sonogashira coupling over pristine corannulene. The resulting alkyne is subjected to Diels–Alder reaction, originating the HBC backbone and creating the fully conjugated structure by a final Scholl reaction using $FeCl_3$ at low temperatures. In addition, the incorporation of a corannulene unit at the backbone of an NG results in both easier oxidation and reduction processes and thus smaller optical HOMO–LUMO gaps (2.61 eV) in comparison with HBC (2.65 eV) as a result of the extension in the effective conjugation.

In addition, combination of pentagonal and heptagonal rings along with carbohelicenes has been successfully implemented in the same NG. More specifically, Campaña and coworkers reported the preparation of a series of functionalized heptagon-containing HBC analogs [90] that were achieved by a sequence of Co-catalyzed alkyne cyclotrimerization and cyclodehydrogenation reactions. In the case of compound **9**, an edge-shared pentagon–heptagon combination is obtained together with the carbo[5]helicene moiety, as revealed by X-ray crystallography (Figure 9.9, **9**). By the end of 2019, Liu, Feng, and coworkers synthesized two novel azulene-containing helical NGs belonging to the family of "superhelicenes" (term coined after Jux and coworkers as helicenes based on HBC units) [91], comprising two HBC units fused by a pentagon–heptagon combination in a helical fashion (Figure 9.9, **10**) [92]. The synthetic strategy is based on an ICl-promoted

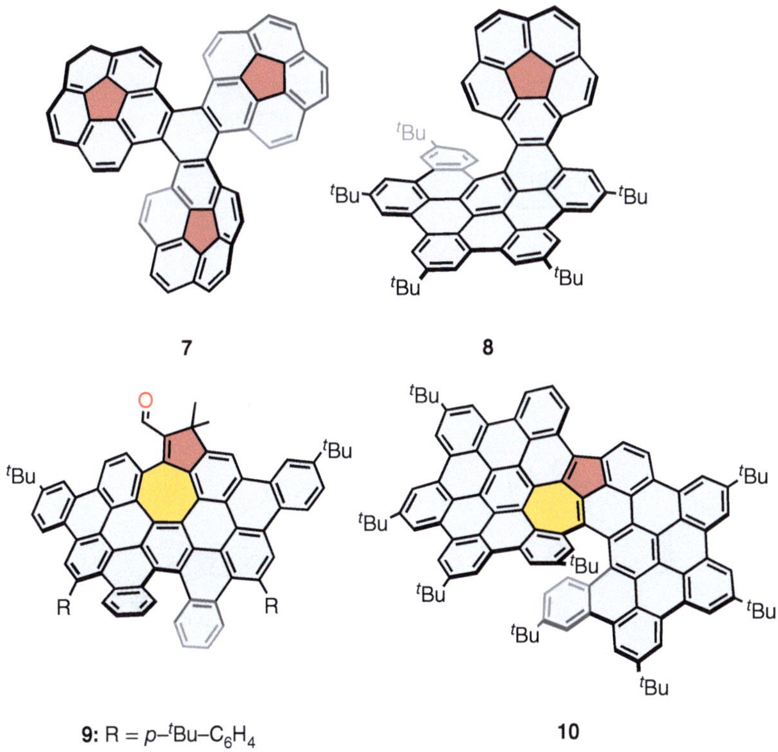

Figure 9.9 Pentagon- and heptagon–pentagon-containing helical NGs.

cycloaromatization of *ortho*-substituted diynes. Interestingly, the azulene moiety was formed during the final cyclodehydrogenation reaction, as a consequence of the steric hindrance presented at the oligophenylene precursors, generating a central heptagon-containing carbo[6]helicene moiety. The optoelectronic and electrochemical properties were investigated. Therefore, it was demonstrated how the extension of the π-system affects their optoelectronic properties, with an increase of the HOMO energy level (−4.99 eV) and, therefore, a decrease of its HOMO–LUMO energy optical gap (1.88 eV) with respect to less conjugated analogs comprising an azulene unit. However, no racemic resolution was reported.

Moving to saddle–helix hybrid NGs, Durola and coworkers serendipitously obtained a highly distorted and, thereby, soluble substituted NG bearing a carbo[5]helicene surrounding a seven-membered carbocycle (Figure 9.10, **11**) [93]. The unexpected heptagonal ring was obtained by two consecutive rearrangements during an intramolecular Scholl reaction from a hindered precursor.

Within the saddle–helix hybrid NG family, several examples including both heptagonal rings and carbo[4]helicene moieties have been reported. Miao and coworkers contributed to this family with the preparation of the tetrabenzo[7]circulene **12** and with **13a** and **13b**, which contain two seven-membered

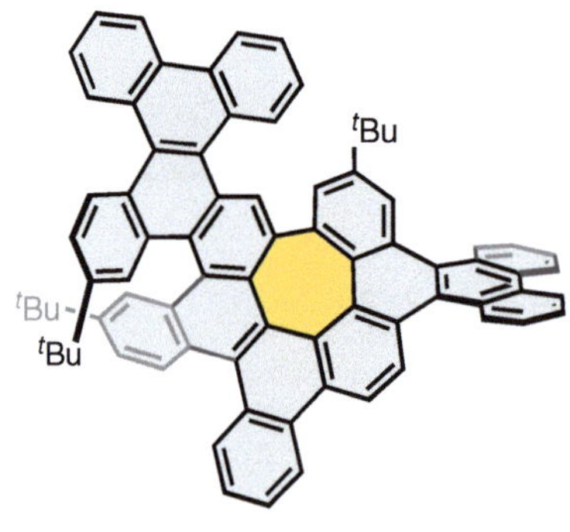

11

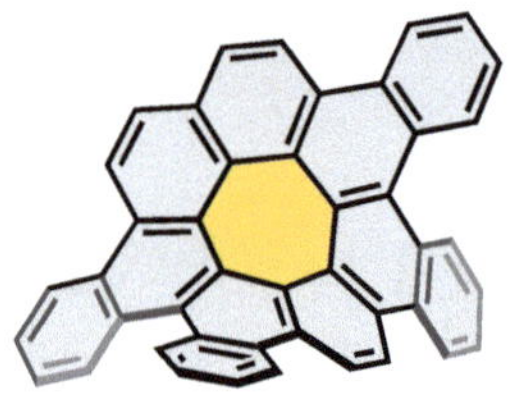

12

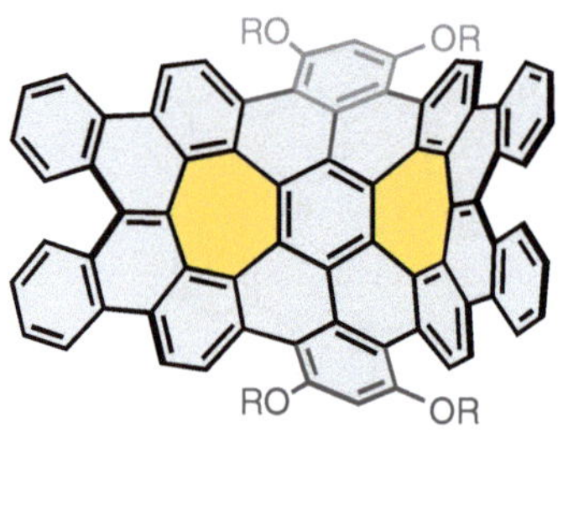

13a: R = C_8H_{17}
13b: R = C_6H_{13}

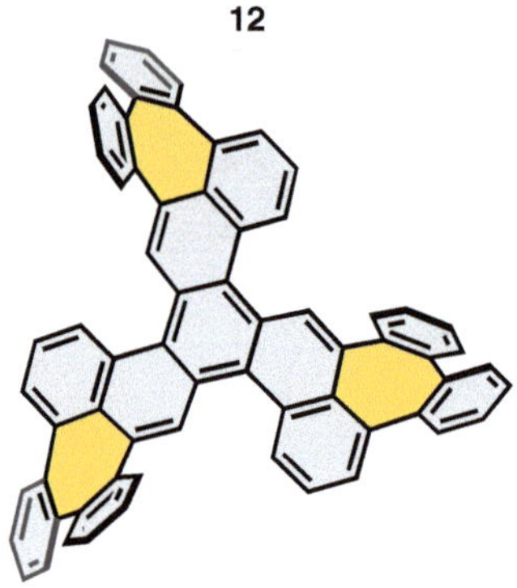

14

Figure 9.10 Heptagon-containing distorted nanographenes.

rings and two carbo[4]helicenes (Figure 9.10) [94, 95]. In those cases, the heptacycles are generated by Friedel–Crafts acylation, and the π-system is expanded through carbonyl derivatization by Ramirez–Corey–Fuchs reaction and final either Pd-catalyzed cyclization (**12**) or oxidative cyclodehydrogenation (**13a** and **13b**). Itami and coworkers also presented the synthesis of a C_3-symmetric propeller-shaped NG comprising three seven-membered rings and three carbo[4]helicenes (Figure 9.10, **14**) [96]. Compound **14** was synthesized by intramolecular C–H arylation, leading to the closure of the non-hexagonal rings instead of the hexagonal ones. Among the 24 possible diastereoisomers, only two relatively stable conformations were observed by means of ^{1}H NMR.

Campaña and coworkers also reported the influence of heptagonal carbocycles in the nonlinear optical properties of a family of well-defined ribbon-shaped helical NGs [97]. This series of NGs are formed by three fused HBC units bearing four carbo[5]helicenes and differ on the presence or absence of a heptagonal cycle into their structure (Figure 9.11, **15a**, **15b**, and **16**). Saddle–helix hybrid NGs (**15a** and **15b**) exhibited lower optical bandgaps and higher TPA in comparison with the purely hexagonal analog **16**, pointing out at the influence of the curvature created by the heptagonal ring in the final properties of NGs.

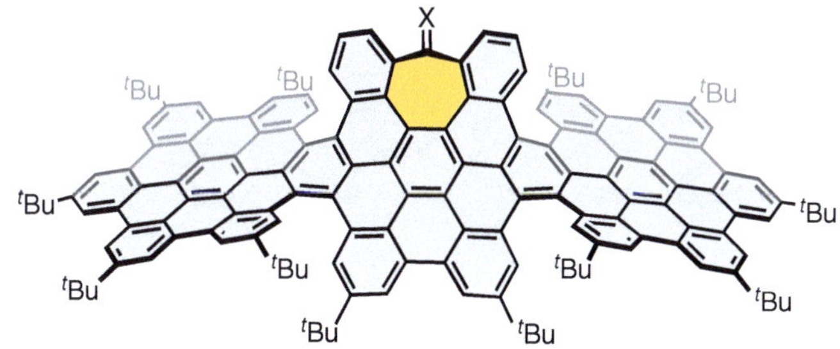

15a: X = O
15b: X = CH_2

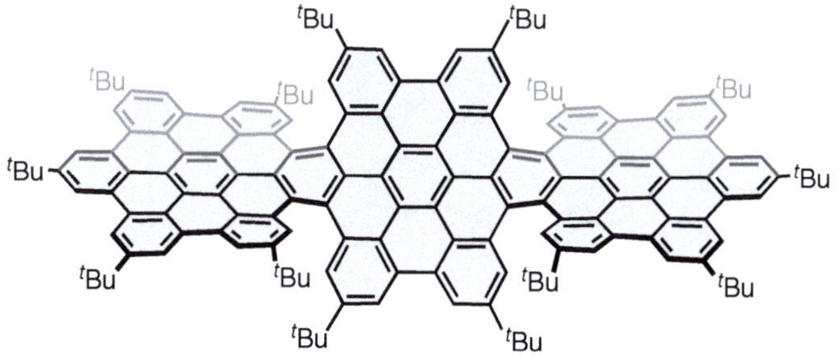

16

Figure 9.11 Heptagon-containing and fully hexagonal analog of ribbon-shaped NGs containing three HBC units.

9.2.2 Chiral and Enantioenriched Helical Nanographenes

The simplest chiral enantioenriched example of a helical NG was reported by Matsuda and coworkers in 2018, when the authors synthesized a hexa-*peri*-hexabenzo[7] helicene as a helical analog to HBC (Figure 9.12, **17**) [98]. The chemical structure of **17**, bearing three fused perylene units, was confirmed by X-ray diffraction crystallography. Compound **17** showed a weak absorption band from 550 to 800 nm as a consequence of π–π interactions between the terminal rings, which resulted in a small HOMO–LUMO optical bandgap ($\Delta E_{\text{H-L}}$ = 0.86 eV). Furthermore, the chiroptical properties of its enantiomers were measured by means of ECD spectroscopy ($|\Delta\varepsilon| = 74\ \text{M}^{-1}\ \text{cm}^{-1}$ at 680 nm) yielding a high g_{abs} value, higher than 10^{-2} in the near-infrared region, which was one order of magnitude larger than simple [7]helicene ($g_{abs} = 5.0 \times 10^{-3}$ at 268 nm) [99].

Recently, Asako, Müllen, Narita, and coworkers obtained a novel helical NG of similar size containing also a single carbohelicene, this time a carbo[5]helicene moiety additionally distorted by the presence of two seven-membered rings (Figure 9.12, **18**) [100]. The authors serendipitously obtained this unusual structure during their investigations on the preparation of a helical HBC derivative. Remarkably, during the last Scholl reaction (promoted by the combination of DDQ and TfOH), the authors observed an aryl group rearrangement creating the two heptagonal rings.

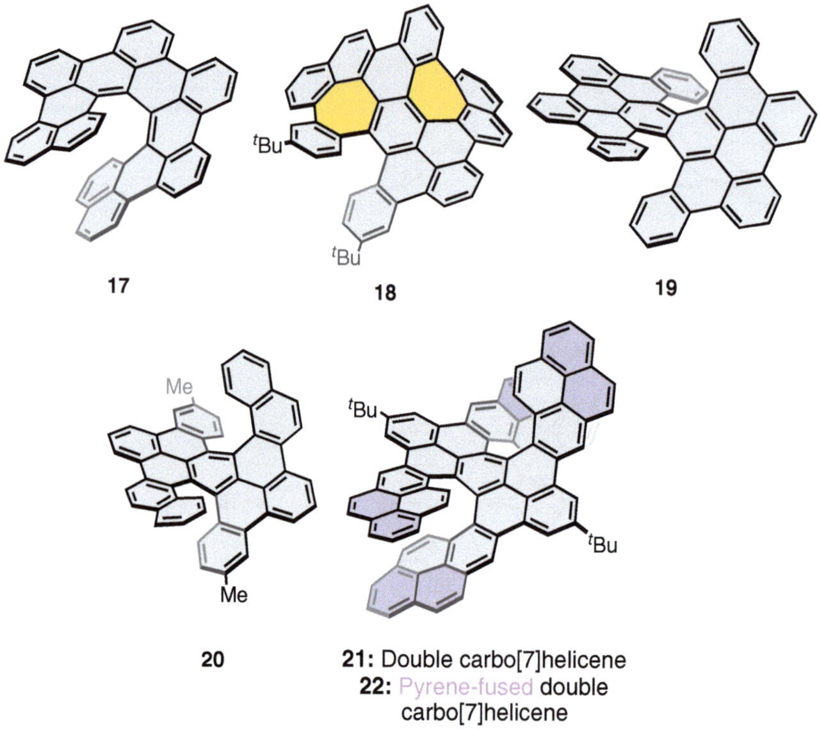

Figure 9.12 Chiral and enantioenriched examples of NGs containing one or two carbo[*n*]helicenes.

The reaction mechanism of this unexpected rearrangement was investigated by DFT and confirmed to occur either through a radical cation mechanism or through an arenium cation mechanism. The extended carbo[5]helicene containing a heptagonal ring displayed a racemization barrier of 25.4 kcal mol^{-1}, slightly higher than that reported for carbo[5]helicene (24.1 kcal mol^{-1}) [80, 101]. The structure was resolved by X-ray crystallography and its emission spectra measured both in solution and crystalline state. The enantiomers of **18** were resolved by chiral stationary phase high-performance liquid chromatography (CSP-HPLC) and their ECD spectra measured, showing $|\Delta\varepsilon|$ values of 148 M^{-1} cm at 350 nm.

Moving to double helicenes in NGs of similar size, in 2015, Itami and coworkers reported the synthesis of π-extended double carbo[6]helicene **19** (Figure 9.12) [78] by fourfold oxidative C–H biphenylation of naphthalene followed by Scholl reaction. Two isomers, namely, *twisted* chiral (*P*,*P*)/(*M*,*M*) and *meso* diastereoisomers were isolated. The structures of chiral enantiomers were also confirmed by X-ray diffraction crystallography and resolved by CSP-HPLC. Furthermore, DFT calculations estimate an isomerization barrier of 43.5 kcal mol^{-1} from the thermodynamically more stable *twisted* isomer to the *meso* one, significantly higher than the one estimated for simple [6]helicene (37.3 kcal mol^{-1}). The electronic absorption spectrum of the *twisted* isomer (λ_{max} at 401 nm) appears bathochromically shifted in comparison with that of *meso* isomer (λ_{max} at 382 nm), which was also observed in the photoluminescence spectra (λ_{em} at 525 nm *vs* 496 nm for *twisted* and *meso* isomers, respectively).

Within this family of highly stable double helicenes, Tanaka and coworkers described the isolation of a double carbo[6]helicene by enantioselective Rh-catalyzed [2+1+2+1] cycloaddition reaction of a triyne followed by a final Scholl reaction (Figure 9.12, **20**) [102]. Optoelectronic and chiroptical properties of **20** were measured, showing high fluorescence quantum yield ($\Phi_F = 0.75$) and absorption and luminescence dissymmetry factors of 9.3×10^{-4} at 490 nm (g_{abs}) and 7.5×10^{-4} at 529 nm (g_{lum}), respectively. In addition, the OR was also measured with a value of $[a]_D^{20} = +590°$, which resulted smaller than that of simple (*P*)-[6]helicene ($[a]_D^{25} = +3707°$) [103].

Later on, Müllen, Narita, and coworkers reported the synthesis and characterization of two structurally related double carbo[7]helicenes, **21** [79], and its π-extended version **22** [104] including terminal pyrene-fused units (Figure 9.12). Both compounds show higher theoretical interconversion barriers (46.0 and 46.4 kcal mol^{-1}, respectively) with respect to the corresponding mono[7]helicenes (41.7 kcal mol^{-1}) [80], also suggesting that the laterally π-extension may increase the interconversion barrier of multiple helicenes. The electronic absorption spectrum of **22** appears bathochromically shifted in comparison with that of **21** as a consequence of the lateral extension of the π-system reaching an NIR emission at 700 nm, which results in a decrease of its optical bandgap, with a value of 1.92 eV *vs* 2.25 eV. In addition, these same authors together with Petrukhina and coworkers reported the isolation of three different reduced species of **21** ($\mathbf{21}^{2-}$, $\mathbf{21}^{3-}$, $\mathbf{21}^{4-}$) [105], which were completely characterized by X-ray diffraction crystallography analysis and NMR spectroscopy and supported by theoretical calculations. Crystals were formed by sublimation in vacuum using alkali metals (*K* and *Rb*) at elevated temperatures, allowing the first chemical reduction of a double [7]helicene. This increasing negative charge in **21** resulted in structural changes, and, as a consequence, the peripheral dihedral angle ranged from 16.5° (**21**) [79] to 20.7° ($\mathbf{21}^{4-}$) [105].

Following the trend of introducing helicity in HBC, several helical HBC analogs including multiple helicene moieties have been reported. The first of them, including three carbo[5]helicenes into a single PAH, was reported in 2000 by Peña and coworkers where each helicene shares three benzene units with the other two (Figure 9.13, **23**) [106, 107]. The authors reported the synthesis of hexabenzotriphenylene (HBT, **23**), based on the palladium-catalyzed cyclotrimerization of 9,10-didehydrophenanthrene. A conformational study afforded a detailed energy profile analysis of the different diastereoisomers, finding a free energy barrier of 26.2 kcal mol^{-1} from the C_2 to the D_3 symmetric diastereomer.

A propeller-shaped triple carbo[5]helicene in which the three helicenes share two benzene rings with his neighbor was reported by Watanabe and coworkers (Figure 9.13, **24**) [108]. Trinaphthotryphenylene **24** (TNT) was isolated as a mixture of diastereomers, being the one with all the helicenes in the same configuration the major product ((*P,P,P*)/(*M,M,M*)-**24**) and the most stable one. In this case, the experimental value of the isomerization barrier (from (*P,P,P*) to (*P,P,M*)) was

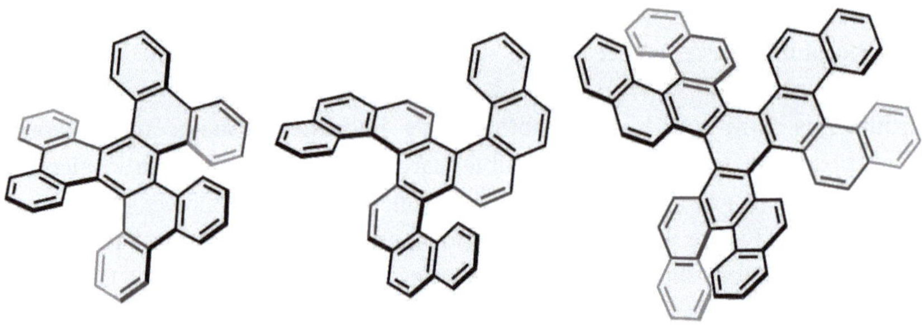

23 24 25

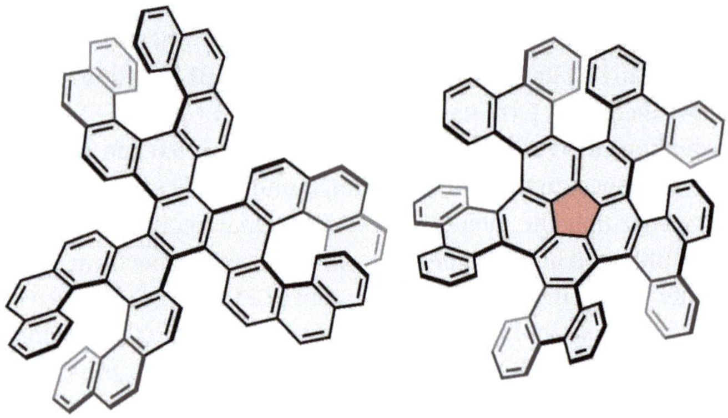

26 27

Figure 9.13 Enantioenriched examples of propeller-shaped NGs.

established as 27.4 kcal mol^{-1}, which is comparable with the value of **23**, previously reported by Peña and coworkers [107]. More recently, Mori and coworkers presented a study on the optical and chiroptical properties of a series of triple carbo[5]helicene NGs, including **23** and **24** [109]. Interestingly, chiroptical properties measurements supported by DFT calculations revealed that **23** showed larger g_{abs} (1.8×10^{-3}) and g_{lum} values (1.3×10^{-3}) than any other configurationally stable single [*n*]helicene ($n \geq 6$) or other triple [5]helicene with different symmetry. These experimental values were rationalized, comparing the magnitude and angle formed by the electric and magnetic transition moments.

Next, expanding the structure of **23**, Kamikawa [110] and Gingras [111] independently and simultaneously reported the first hexapole[5]helicene, bearing six carbo[5]helicene moieties, three inner (as in **23**), and three external helicenes (Figure 9.13, **25**). Two different synthetic approaches, both starting from 7,8-dibromo[5]helicene, were developed. While Kamikawa used a Pd-catalyzed [2+2+2] cycloaddition of an aryne intermediate precursor to generate triphenylene derivatives, Gingras and coworkers reported a synthetic procedure based on a Ni-catalyzed Yamamoto-type cyclotrimerization reaction. The most stable conformers are those with the same configuration on the inner helicenes and the opposite on the outer ones, showing a D_3 symmetry and being the most stable isomer among the 10 possible ones as shown by DFT calculations. CSP-HPLC allowed the racemic resolution of the most stable isomer with alternant helicene configuration ((−)-**25** bears *M* configuration at inner helicenes and *P* configuration at outer ones). The subsequent study by vibrational circular dichroism (VCD) and ECD, reaching a dissymmetry factor g_{abs} of 4.8×10^{-3} at 324 nm, shows a similar value as that reported for single [5]helicene (4.2×10^{-3}) [99]. One year later, Peña and coworkers also reported the synthesis of **25** by aryne chemistry, although in this case it was used as substrate for the on-surface cyclodehydrogenation of the six carbohelicenes to prepare an extended NG [112].

A further extension of **25** led to the extended analog **26**, bearing three external carbo[7]helicenes instead of carbo[5]helicenes, which was reported by Ravat [113] and by Coquerel and Gingras [114]. The enantiospecific synthesis of the propeller-shaped NG reported by Ravat and coworkers is based on an Ni-catalyzed Yamamoto-type [2+2+2] cyclotrimerization of the configurationally stable (*M*)-9,10-dibromo[7]helicene. Thus, the *M* configuration is maintained in the outer [7]helicenes, and the opposite *P* configuration is obtained in the generated inner [5]helicenes, leading to the D_3-symmetric enantiomer named (−)-(*M*,*M*,*M*,*P*,*P*,*P*)-D_3-**26**. Similar result was presented by Coquerel and Gingras who also reported the synthesis of opposite enantiomer (+)-(*P*,*P*,*P*,*M*,*M*,*M*)-D_3-**26** starting from the (*P*)-9,10-dibromo[7]helicene with opposite configuration together with the diastereoselective synthesis of racemic C_2-**26** from the racemic 9,10-dibromo[7]helicene. Ravat and coworkers reported that the absorption and emission maxima of **26** appear bathochromically shifted in comparison with **25** [110, 111] and, therefore, resulted in a slightly decreased optical gap (2.33 eV *vs* 2.38 eV). In addition, the TPA spectrum was recorded rendering maximum TPA cross-section values of 436 and 517 GM in toluene and THF, respectively. The chiroptical properties were also spotlighted by means of ECD spectroscopy ($|\Delta\varepsilon| = 72\ \mathrm{M^{-1}cm^{-1}}$ at 460 nm) with a dissymmetry factor g_{abs} value of 3.3×10^{-3}. Remarkably, both studies reported that the OR value for (−)-(*M*,*M*,*M*,*P*,*P*,*P*)-D_3-**26** resulted much smaller ($[a]_D^{23} = -222°$ [90], $[a]_D^{25} = -300°$ [91]) than that of single (*M*)-[7]helicene ($[a]_D^{20} = -6200°$) or single (*P*)-[5]helicene ($[a]_D^{25} = +1670°$), which can be rationalized in terms of the outbalancing between the inner (*P*)-[5]helicenes and outer (*M*)-[7]helicenes. Furthermore, Coquerel and Gingras [114] also include a detailed computational study on the aromaticity of **26** using several methods revealing the limitation of the models of aromaticity when applied to nonplanar distorted PAHs. Nucleus-independent chemical shift (NICS) and anisotropy of the current induced density (ACID) methodologies resulted in a nonaromatic character of the triphenylene core, which could be related to the great distortion of the rings. Finally, the capability of **26** to bind to silver(I) ions in the bay regions and establish silver(I)-NG hybrids was also presented.

Also as a propeller-shaped multihelical NG, recently, Itami and coworkers reported the first example of a quintuple carbo[6]helicene fused by a corannulene core (Figure 9.13, **27**) [115]. In this case its X-ray crystal structure exhibited a C_5-symmetric propeller shape. The synthesis of **27** was carried out through a Pd-catalyzed intramolecular direct arylation of pentakis(2'-chlorobiphenyl-2-yl)corannulene that was previously obtained from prepared pentaborylcorannulene by fivefold Suzuki coupling with 2-bromo-2'-chlorobiphenyl. The propeller structure of the racemic **27** was confirmed by X-ray diffraction analysis. Thanks to experimental and computational studies, the different isomerization energy barriers of the eight possible stereoisomers were investigated. Among them, (*P*,*P*,*P*,*P*,*P*)/(*M*,*M*,*M*,*M*,*M*)-**27** was found to be the most stable diastereoisomer with an experimental isomerization barrier of 34.2 kcal mol^{-1}. Eventually, both enantiomers were successfully separated by CSP-HPLC, and the absolute configuration was identified by ECD spectroscopy combined with TD-DFT calculations. The ECD spectrum of (*P*,*P*,*P*,*P*,*P*)-**27** exhibited positive Cotton effects from 264 to 336 nm and from 353 to 535 nm and a negative Cotton effect between 336 and 353 nm.

Moving to the family of superhelicenes (helicenes based on HBC units), in 2018, Martín and coworkers communicated a novel helical bilayer NG in which two HBC units are linked by a carbo[8]helicene (Figure 9.14, **28**) [116]. Its synthesis was achieved in three steps from an enantiopure (*M*)-11,14-bis((trimethylsilyl)ethynyl) hexahelicene. After the deprotection of both alkynes, a twofold Sonogashira coupling with *p*-tert-butyliodobenzene followed by a double Diels–Alder reaction with *tetrakis*-(*p*-*tert*-butylphenyl)-cyclopentadienone led to the oligophenylene precursor. The final Scholl cyclodehydrogenation yielded (*M*)-**28** with an enantiomeric excess of 93%, determined by CSP-HPLC. X-ray diffraction crystallography revealed a structure where both HBCs are arranged almost in a parallel fashion. The ECD spectrum of a dichloromethane solution displays a positive Cotton effect at 380 nm followed by a negative one around 450 nm. The shape of the ECD spectrum is concentration dependent, evolving to a single negative band at 520 nm when the concentration is multiplied by 255, which is consistent with the sign usually observed for (*M*)-carbohelicenes [117].

Furthermore, between 2018 and 2020, Wang and coworkers carried out an outstanding work in this field reporting a series of multihelicene-containing NGs based on HBCs. The first reported example was a hexapole[7]helicene, which is based on a central HBC unit fused with six surrounding carbo[7]helicene units (Figure 9.14, **29**) [118] and a subsequent extended version, a hexapole[9]helicene with six peripheral carbo[9]helicenes [119]. The synthetic methodology is based in both cases on an alkyne cyclotrimerization reaction of diphenylacetylene derivatives to create a polyphenylene precursor based on hexaphenylbenzene followed by final Scholl oxidation. X-ray diffraction revealed the propeller structure of those twisted chiral NGs. UV–vis spectra of both hexapole [7]/[9]helicenes reach the near infrared (IR) region with a major band centered at 618 nm/605 nm and up to 800 nm, leading to remarkably low optical bandgap values (1.51/1.55 eV). Chiral resolution of **29** was also achieved by CSP-HPLC, with no signal of racemization up to 270 °C. The chiroptical properties of both (*P*,*P*,*P*,*P*,*P*,*P*)/ (*M*,*M*,*M*,*M*,*M*,*M*)-**29** were evaluated, giving rise to a mirror ECD spectra from 300 up to 800 nm. Both compounds show great ability to stabilize oxidized species, as cyclic voltammetry (CV) showed up to six reversible single-electron oxidation waves between 0.28 and 1.05 V for hexapole[7]helicene **29** and between 0.44 and 1.40 V (*vs* Fc/Fc^{+}) for hexapole[9]helicene.

Finally, following a synthetic route based on that reported by Müllen for the preparation of a graphite disk containing 222 carbon atoms [120], Wang and coworkers

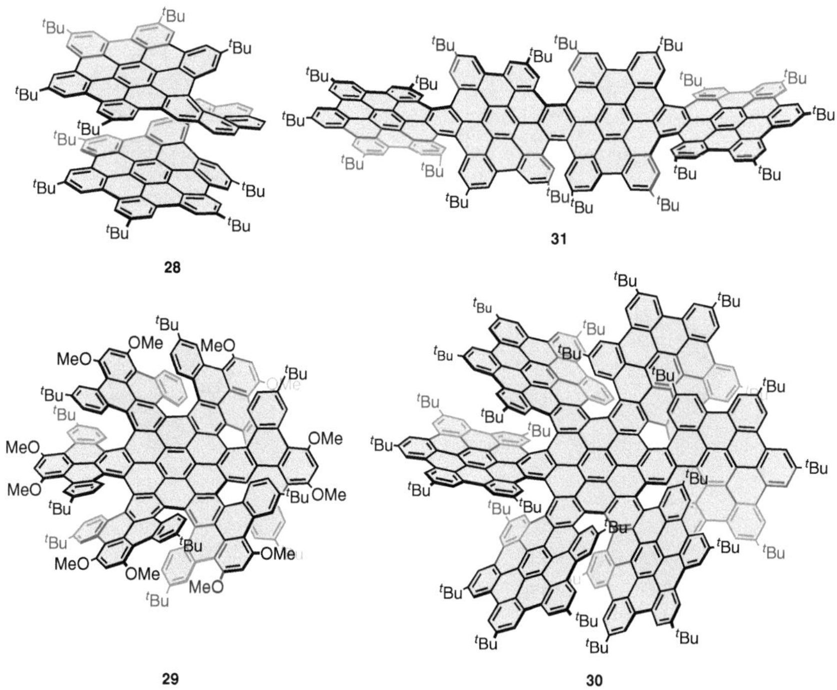

Figure 9.14 Examples of "superhelicenes" arranged in different geometries.

reported the synthesis of a flagship NG propeller containing 258 conjugated carbon atoms, which represents the largest 3D conjugated PAH ever prepared using scalable solution chemistry (Figure 9.14, **30**) [121]. The NG molecule consists on a central HBC fused with six surrounding HBC units, where the crowdedness led to the generation of six peripheral carbo[7]helicenes. The structure of the reported compound was confirmed by means of NMR, MS, and X-ray diffraction despite its size (3.50 nm width and 1.25 height). The compound showed an absorption spectrum ranging from 300 to 800 nm with extinction coefficients of the order of 10^5 $\text{M}^{-1}\text{cm}^{-1}$. Although the fluorescence quantum yield was low (0.016), the compound exhibits an emission spectrum ranging from 750 to *ca.* 1200 nm. Remarkably, the ECD spectra of both enantiomers (*P,P,P,P,P,P,P*)/(*M,M,M,M,M,M,M*) showed a record-high Cotton effect in the visible spectrum with $|\Delta\varepsilon|$ values of 1182 and 1090 M^{-1} cm^{-1} at 374 and 405 nm, respectively.

The same research group recently broadened the family of the HBC-based multihelicenes by reporting a new member of "supertwistacene" (Figure 9.14, **31**) [122], based in four linearly fused HBC units sharing one aromatic ring two by two and reaching 4.3 nm in length. To this point, the authors resolved the obtained racemic mixture (*M,M,M,M,M,M*)/(*P,P,P,P,P,P*) into their enantiomers displaying a UV-vis spectrum in dichloromethane with absorption bands up to 625 nm and an ECD spectrum with $|\Delta\varepsilon|$ values of 760 M^{-1} cm^{-1} at 450 nm.

In the same vein, saddle–helix hybrid superhelicenes can be also found in literature. In 2018, Campaña and coworkers reported the synthesis and chiroptical

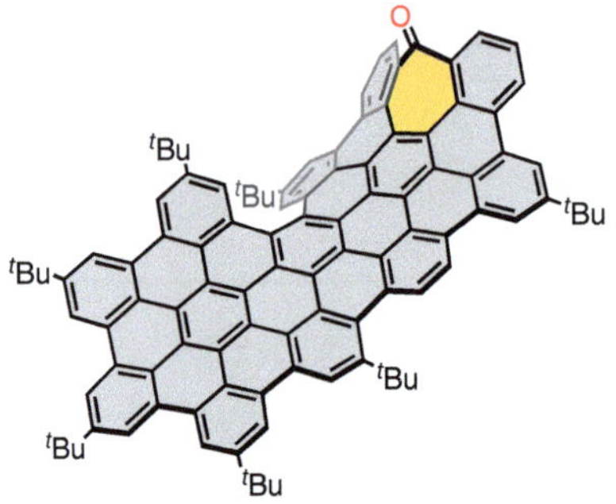

32

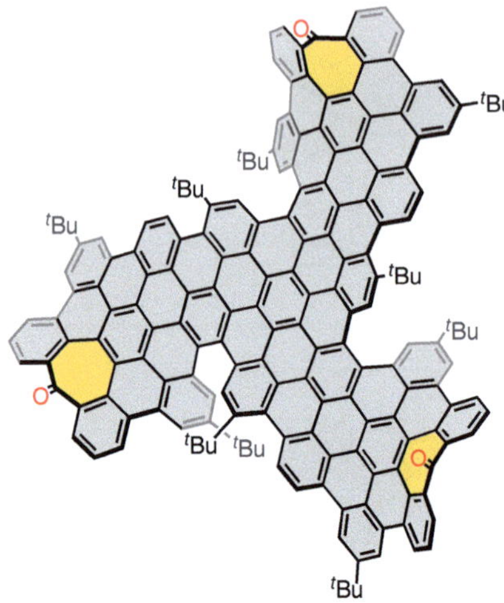

34

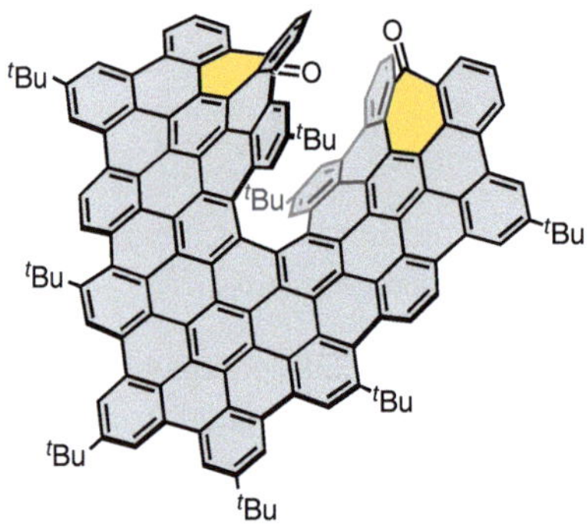

33

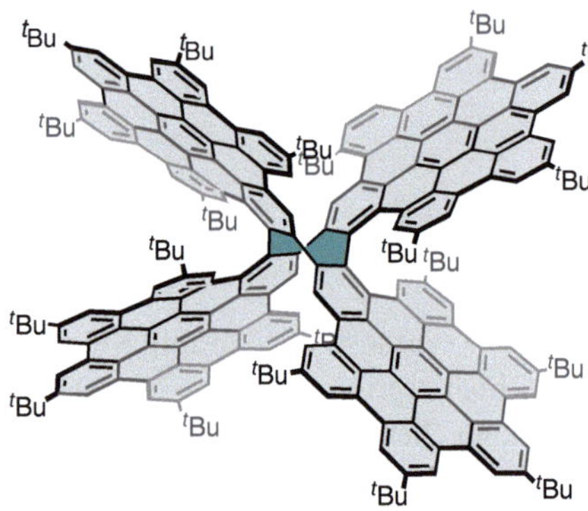

35

Figure 9.15 Examples of saddle–helix hybrid NGs.

properties of an enantiopure distorted ribbon-shaped NG constituted by two HBC-like units (Figure 9.15, **32**) [66]. X-ray crystal structure showed a distorted conformation comprising a central π-extended carbo[5]helicene and a saddle-curved end due to the introduction of a heptacycle. The racemic resolution was achieved by CSP-HPLC, followed by characterization of both enantiomers by means of TPA, ECD, and CPL, observing a phenomenon of upconverted emission, and being this the first enantiopure NG presenting CPL response up to that point. Remarkably, the average fluorescence lifetime reached 21.5 ns, longer than many perylene bisimides [123]. Besides, moved by the interest in shedding light into the resulting properties of blending NGs, helicity, and saddle curvature, this research group also presented the first fully π-extended enantiopure carbo[7]helicene (Figure 9.15, **33**) [124]. The skeleton of **33** is formed by three HBC-like units containing two saddle-shaped heptagonal rings at the edges and generating a central undecabenzo[7]helicene unit. Thus, the resulting ribbon NG is arranged in a fully helical shape. Both enantiomers were successfully resolved by CSP-HPLC, and their chiroptical properties were studied by ECD and CPL spectroscopies, giving rise to dissymmetry factors of *ca.* 2×10^{-3} both in absorption and emission.

Comprising three octabenzo[5]helicenes, Campaña and coworkers also described the preparation of another saddle–helix hybrid superhelicene constituted by four fused HBC units arranged in a triskelion shape (Figure 9.15, **34**) [125]. The arrangement of its three ^{t}Bu-substituted carbo[5]helicenes gave rise to two pairs of diastereoisomers, (*P*,*P*,*P*)/(*M*,*M*,*M*)-**34** with C_3 symmetry and (*P*,*P*,*M*)/(*M*,*M*,*P*)-**34** with C_1 symmetry. The absorption spectra showed optical energy gaps of 2.01 and 2.06 eV, respectively, while the emission exhibited a remarkable quantum yield of 27.6% at 645 nm. Each diastereoisomer was successfully resolved by CSP-HPLC, and their chiroptical properties were studied by ECD and CPL spectroscopies, displaying higher dissymmetry factor for the enantiomeric forms of the C_3-symmetric diastereoisomer ($|g_{abs}| = 8.2 \times 10^{-3}$ at 446 nm) than for the C_1-symmetric one ($|g_{abs}| = 8.8 \times 10^{-4}$ at 425 nm) and reaching luminescence dissymmetry factors of 3×10^{-4} at 643 nm.

Within this family of distorted NGs, a recent example presented by Molina-Ontoria, Guldi, Martín, and coworkers must be considered [126]. In this case, a purely carbonated, electron-rich, and stable NG, embedding a quadruple [7]helicene linked through a central cyclooctatetraene moiety and four HBC units, was presented (Figure 9.15, **35**). The HPLC-mediated chiral resolution as well as ECD characterization of both non-assigned enantiomers was achieved, showing a spectrum in dichloromethane that extends up to *ca.* 450 nm, whose main feature is a shouldered Cotton effect at 368 nm. Besides, the cyclic voltammogram of **35** shows two oxidations at + 0.62 and + 0.85 V corresponding to a four-electron loss process [126].

9.3 Heteroatom-Containing Helical Nanographenes

The field of fully carbonated helical NGs is growing fast nowadays as it has been demonstrated in the previous section. However, it is also important to remark that, along with the helical graphene molecules, an important number of heteroatom-containing helical NGs are also being prepared and studied. Considering the wide range of heteroatom-containing single and multiple helicene systems reported [48], this section will be also focused on those compounds with, at least, similar size to HBC, that is, going from 1 nm size as they could be considered as NGs. In this kind of extended aromatic compounds, the properties are usually directly related to the conjugated structure. Thus, the introduction of heteroatoms into the sp^2 carbon lattice produces a modification on the electronic structure that has an effect on their properties. Moreover, these particular properties can be achieved by the introduction of different kinds of heteroatoms or combinations of them [127–129]. In this way, a wide variety of helical NGs containing heteroatoms of groups 15 and 16 have been recently reported and their properties studied.

9.3.1 Helical Nanographenes Containing Heteroatoms of Group 15 (N, P)

Helical NGs containing heteroatoms of group 15 are the most common ones since their synthetic approaches are based on the chemistry of aromatic nitrogen compounds, which has been an extensively developed field in organic chemistry. Nevertheless, novel synthetic strategies toward extended nitrogen-containing PAHs are still being developed nowadays [130–133]. The nitrogen–carbon exchange into an NG hexagonal lattice introduces doping sites that have an effect on the structure of the resulting NG as well as on its electronic and electrochemical properties. Typical exchanges are not only reduced to introduce nitrogen or phosphorous atoms instead of carbon (i.e. introduction of pyridine or pyrazine rings), but, sometimes, the doping is also associated with a modification of the hexagonal lattice, introducing non-hexagonal rings (i.e. pyrrole rings). Thus, the different hybridization shown by the heteroatoms along with the possibility of a ring size modification would produce a variety of structural changes in the NGs. Moreover, the presence of these heteroatoms would have an impact on the electronic and electrochemical properties, such as modifying their spectroscopic properties or introducing easily accessible oxidized states.

The deliberated introduction of heteroatoms of group 15 in helical NGs has its first expression on the work reported by Nakamura, Hatakeyama, and coworkers in 2014 [134]. The authors reported the preparation of an NG containing both two helicenes and two phosphorous atoms in the aromatic structure (Figure 9.16, **36a–c**). The synthesis is described starting from hexabromobenzene and achieving the target molecule in four steps. The structure of the phosphine sulfide **36a** was confirmed by X-ray crystallography, where the same configuration is found for both helicene moieties as well as a *cis* arrangement of the two sulfur atoms. Remarkably, the combination of two helicene moieties and the tetrahedral geometry of the phosphorous atoms causes a tremendous distortion into the central benzene ring, which exhibits a boat-like distorted conformation with a bending angle of approximately 23°. This high distortion is also observed in the decreased aromaticity of this central benzene, which shows a less negative value of NICS (GIAO-B3LYP/6-311+G) than the rest of benzenoid rings. Additionally, the authors succeeded in either removing or exchanging the sulfur atoms by oxygen, obtaining their phosphine (**36b**) and phosphine oxide (**36c**) derivatives, respectively. With this work, the authors introduce a novel approach to distort NGs by the inclusion of heteroatoms in a double helicene scaffold.

Related to compound **36b**, a double [5]helicene-containing NG doped with nitrogen atoms as part of a pyrazine moiety was later reported (Figure 9.16, **37**) [135]. Compound **37** was prepared after the coupling of two units of dibenzo-9a-azaphenalene and a subsequent oxidation with DDQ. Structurally, the least energetic configuration was that where both helicenes show opposite configuration, probably due to a better accommodation of the nitrogen geometries. Regarding aromaticity, NICS calculations were performed, confirming the antiaromaticity of the central ring. This antiaromaticity could be the cause of

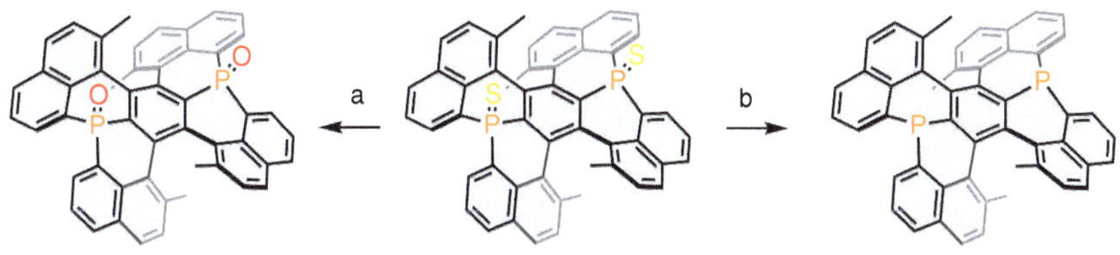

36c **36a** **36b**

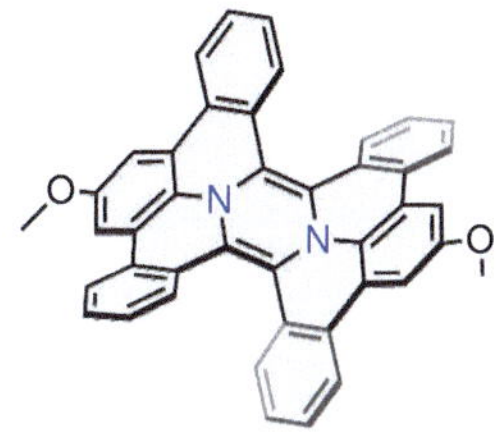

37

Figure 9.16 Reported P- and N-containing fully hexagonal NGs. Reagents and conditions: (a) MCPBA (3.0 eq.), CH_2Cl_2, RT, 24 hours, 75% and (b) Et_3P (20 eq.), *o*-dichlorobenzene, 60 °C, 12 hours, 89%. Source: Based on [135].

the high instability of **37**, which requires preparation under strict deaerated conditions.

However, when the pyrazine unit is substituted by a pyrrolopyrrole moiety, the resulting structures gain on stability and, therefore, allow a full characterization of their optical and electrochemical properties. Based on their previous works on the preparation of π-expanded pyrrole[3,2-b]pyrroles [136], Gryko and coworkers prepared a series of butterfly-shaped heteroatom-doped NGs (Figure 9.17, **38**, **39**, and **40**) [137, 138]. The synthetic strategy not only allowed them to prepare nitrogen-doped helical NGs but also to introduce thiophene units, generating different nitrogen–sulfur-doped NGs (**39** and **40**) or seven-membered rings (**41**) [137, 138]. The structures of these helical NGs were confirmed by means of NMR and X-ray crystallography. In the case of **38**, low temperatures ^{1}H NMR analysis showed that the most stable isomer is that where their helicenes possess opposite configuration, contrary to what was observed for pyrazine **37**. Regarding the optical properties, a similar absorption profile was found for the three members of the family (**38**, **39**, and **40**). The fluorescence quantum yield (Φ_F) range from 0.12 to 0.32 depending on the solvent and the compound, with an intense solvatochromic effect in case of compound **38** (shifting the emission maximum to the red up to 54 nm). More recently, theoretical studies gave insights on how the presence of nitrogen atoms and seven-membered rings has an effect on the local and global aromaticity in comparison with their fully hexagonal or non-helical counterparts [139].

Pyrrole–benzene exchange is an effective tool for the introduction of nitrogen atoms into the structure of NGs, placing the nitrogen atom either as part of the central structure or in the periphery. When taking part of the core of the aromatic structure, nitrogens can only act as doping points, while if the pyrrole ring is placed in a way that nitrogen is located at the edge of the structure, a new reactive site

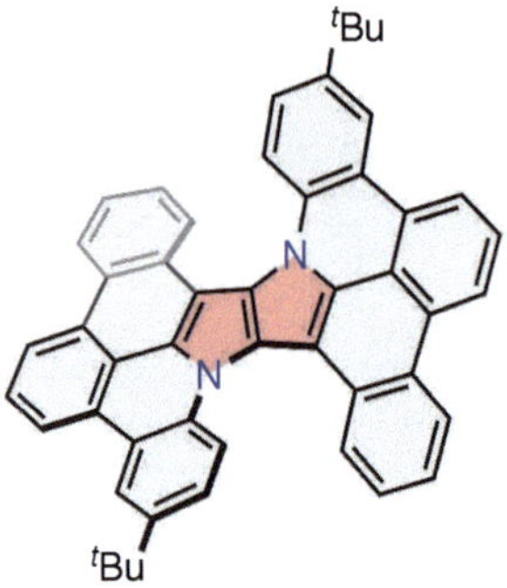

38

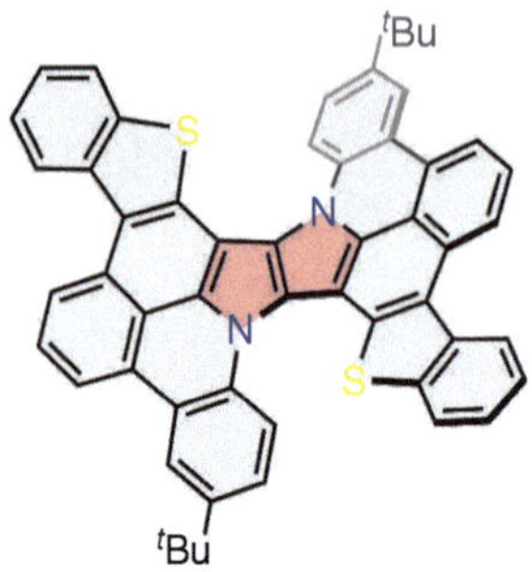

39

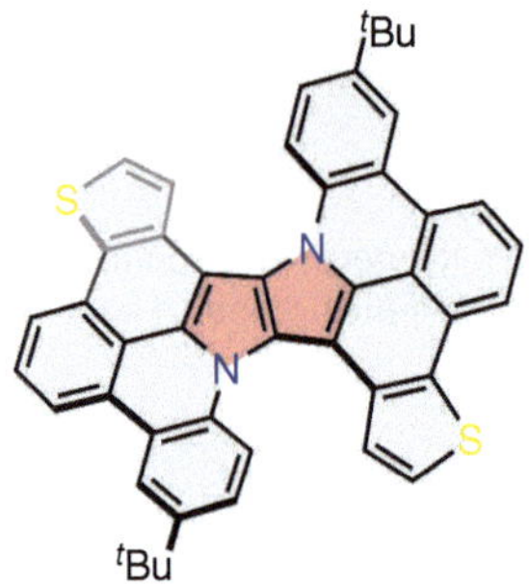

40

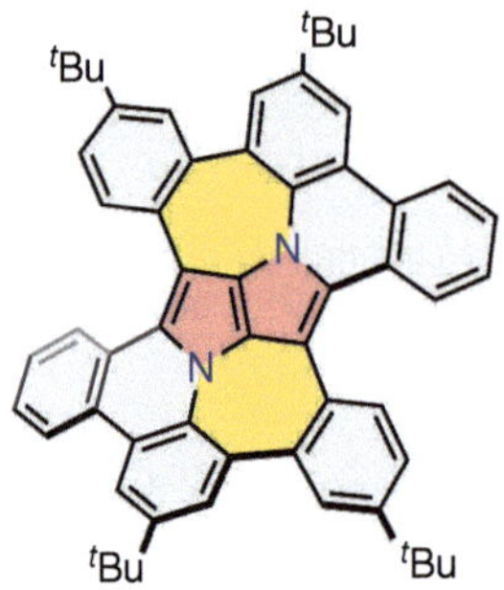

41

Figure 9.17 Butterfly-shaped heteroatom-doped NGs.

appears, and further functionalization is possible. Moreover, the location of such pentagonal rings plays a key role in the topology of the products, being possible to induce the formation of helical structures after Scholl reaction, for instance. An illustrative example was given by Jux and coworkers [140], where the preparation of pyrrole-containing hexaarylbenzene (HAB) derivatives is reported either by [2+2+2] alkyne cyclotrimerization or Diels–Alder reaction. Scholl reaction over the HAB containing one pyrrole unit formed the partial cyclodehydrogenation products where one C–C bond is missing, connecting the pyrrole exclusively through the α position (Figure 9.18, **42a–b**). The obtained N-doped helical HBC analogs showed a slightly lower inner pitch (3.1 Å) than those from pristine carbohelicenes (3.2 Å) as observed in their X-ray diffraction structures. Recently, helical HBC-naphthalene diimide (NDI) hybrids have been reported by the group of Hirsch [141]. The HBC and the NDI units are linked through benzimidazole ring, generating a diaza[7]helicene moiety. The chiroptical properties of the isolated enantiomers were studied, finding a $|g_{abs}|$ value of 1.5×10^{-3} at 650 nm.

The introduction of six pyrrole units in the periphery of a coronene molecule leads to an N-doped analog of HBC, namely, hexapyrrolohexaazacoronene (Figure 9.18, HPHAC). Its first isolation was reported in 2007 by Müllen and coworkers [142], combining a sixfold nucleophilic aromatic substitution of hexafluorobenzene with pyrrolyl sodium salts and Scholl reaction. Furthermore, the subsequent oxidation

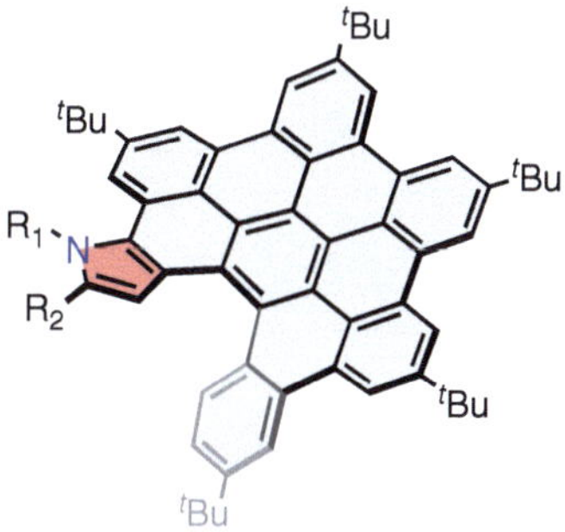

42a: $R_1 = SO_2Ph$; $R_2 = H$
42b: $R_1 = H$; $R_2 = CO_2Et$

HBC

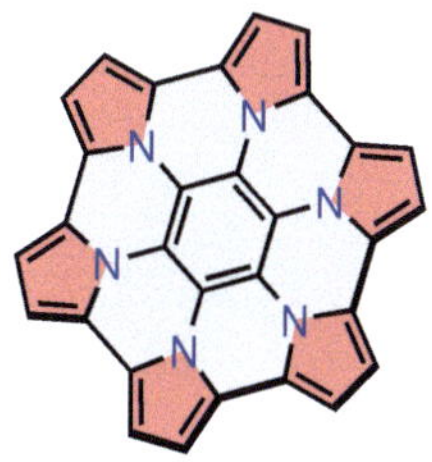

HPHAC

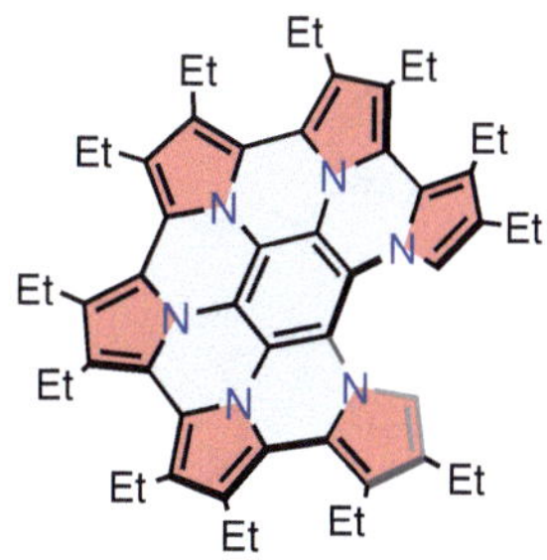

43

Figure 9.18 Structures of HBC, hexapyrrolohexaazacoronene (HPHAC), and their partially fused analog as an N-doped helical NGs. Source: Based on [144].

step was supported by a comprehensive study previously made by Jouini and coworkers [143]. In the HPHAC, the presence of nitrogen atoms at the core of the aromatic surface was found to stabilize higher oxidation states. Thus, it was possible to characterize both its corresponding radical cation and dication [142].

Besides the possibility of introducing a single helicene moiety in the structure of HPHAC by controlling the amount of oxidant during the Scholl reaction (Figure 9.18, **43**) [144], the HPHAC structure has been successfully used for the preparation of propeller-shaped helical NGs. In this sense, the group of Stępień has developed a comprehensive study on a radially extended system with an HPHAC core unit (Figure 9.19, **44**) [145]. Compound **44** was prepared according to previously reported methodologies, linking six naphthalene monoimide (NMI) units to a central HPHAC. Remarkably, the resulting structure mixes the electron-rich character of the HPHAC core and the electron-poor behavior of the NMI groups. The resulting compound can be also considered as a hexapole[7]helicene, where the helicenes are created after the Scholl reaction. As confirmed by NMR spectroscopy, the nonfunctionalized helicenes showed a low interconversion barrier and underwent rapid configurational switching in solution. However, the X-ray diffraction structure confirmed a D_{3d} symmetric saddle-shaped geometry with an alternated configuration of the helicenes. This radially extended HPHAC structure

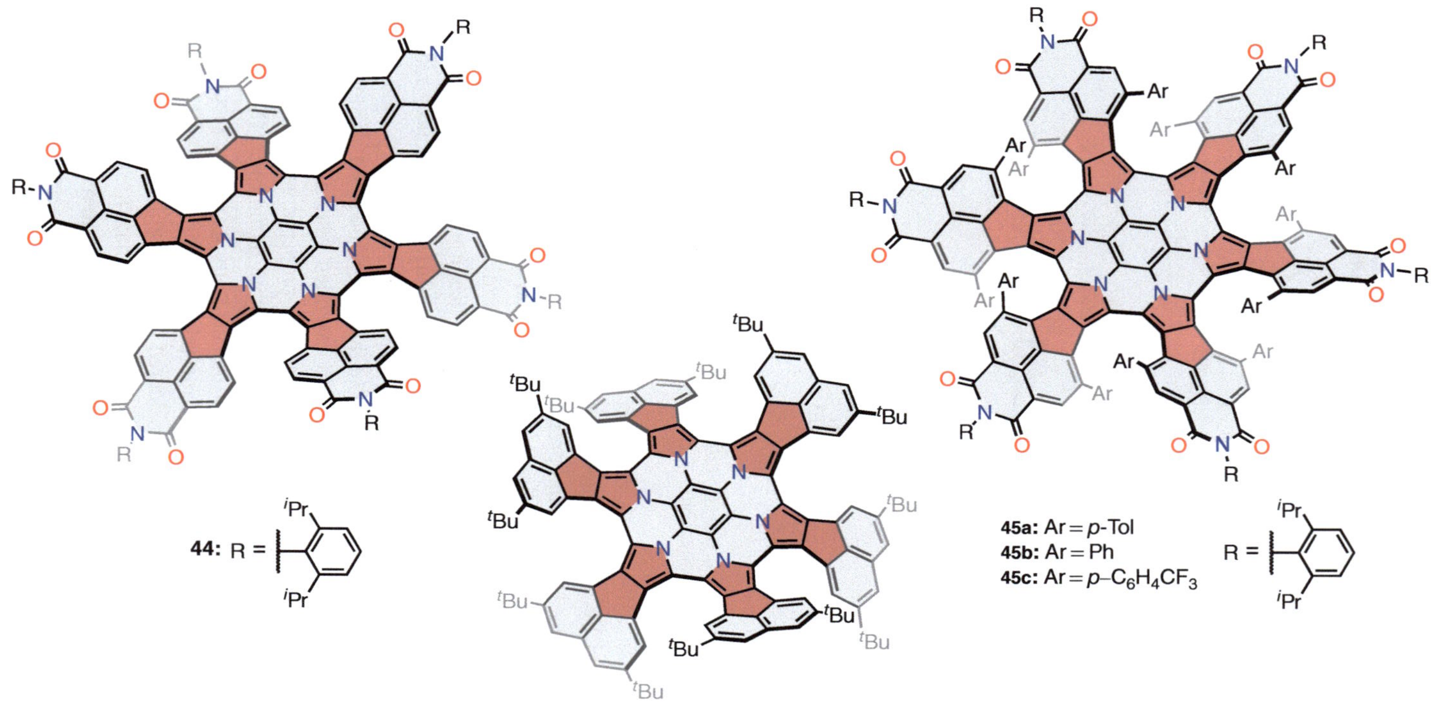

Figure 9.19 Structures of the reported HPHAC-based extended NGs. Source: Based on [148].

showed an extraordinary redox behavior, being possible to identify up to two oxidized and ten reduced states, yielding anions with low bandgaps. When the NMI units are functionalized with bulky groups, i.e. aryl groups, the interconversion is blocked, and a propeller-like structure can be obtained (Figure 9.19, **45a–c**) [146]. Remarkably, for the preparation of compounds **45a–c**, the reagents involved in the Scholl reaction are replaced by Br_2 or NBS, giving rise to different stereoselectivity. Thus, for the reaction with NBS, the D_6 symmetric propeller-shaped with all the helicenes showing equal configuration was the obtained isomer, while when using Br_2, one of the helicene moieties exhibits opposite configuration obtaining the C_2 symmetric isomer. Remarkably, the $|\Delta\varepsilon|$ values are higher in the D_6 symmetric isomers compared with the C_2 symmetric ones, suggesting an additive effect of the helicene configuration to the ECD intensity. Last but not least, the electrochemical bandgap was evaluated as *ca.* 1.5 eV for the three propeller analogs studied, showing also an extraordinary electron acceptor behavior as reported for its non-configurationally stable analog (**44**). More recently, the group of Stępień has also reported a nanosandwich-type structure created after sigma dimerization of two units of heptagon-containing N-doped NG radical, analog to **44** [147]. This dimerization process was found to be photochemically reversible but remarkably thermally stable. Recently, similar to the examples reported by Stępień, an N-doped NG bearing naphthalene units instead of NMI has been reported by Takase, Kobayashi, Uno, and coworkers (Figure 9.19, **46**) [148]. Similar to **44**, compound **46** shows stable oxidized species (cation and dication) that were confirmed by X-ray crystallography. Remarkably, the authors also reported its association with C_{60} in a 1:1 ratio, showing higher binding affinity with the oxidized species.

The electrochemical properties exhibited by this family of propeller compounds are mainly governed by the NMI moiety, from which the extraordinary electron-acceptor behavior arises. This characteristic is also exhibited by their extended analogs, the perylene diimides (PDI) (or perylene bisimides, PBI). PDI have been known for a long time and have been used successfully in many applications [149], ranging from fluorescent dyes for imaging [150] to organic transistors and photovoltaics [151]. Among the recent literature, it is possible to find the synthesis and characterization of helical [152–159] and propeller [160–163] PDI-based compounds with remarkable chiroptical properties [159] together with their applications in different devices.

Propeller-shaped N-doped NGs are not only limited to extended HPHAC cores; analogs based on extended HBCs have been also studied [164]. Within the remarkable family of propeller-shaped distorted NGs reported by the group of Wang (see above, Section 9.2.2) [118, 119], an N-doped analog is included (Figure 9.20, **47**) [164]. The synthetic methodology is similar to that reported for the hexapole-[7]helicene [118], using dibenzoullazine units instead of dibenzopyrene. Therefore, the doped sites in **47** were located at the "blades," and not at the core structure, as in the HPHAC derivatives, being its structure confirmed by means of HRMS, NMR, and X-ray crystallography, confirming a D_6 symmetry as in the case of its fully carbonated analog. Its geometry closely resembles the one

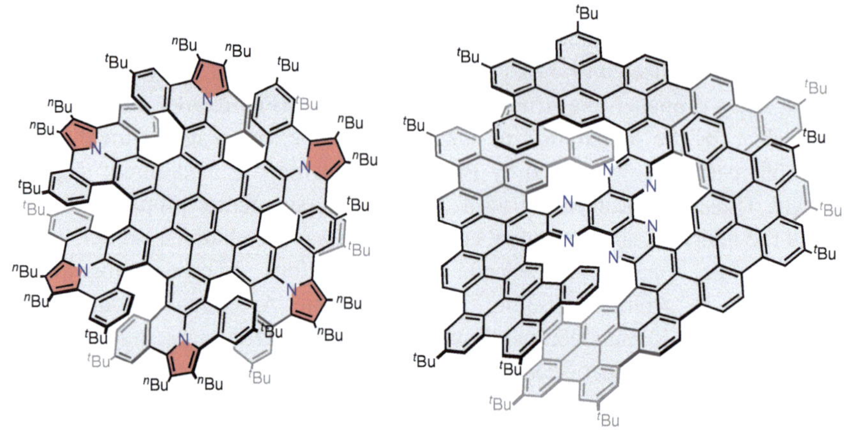

47 48

Figure 9.20 Examples of N-containing helical NGs reported by Wang's or Nierengarten's groups.

of carbonated analog **29**, with similar torsion angle (θ) values at the helicenes. Hence, the differences observed in their properties would be directly related to the replacement of benzene rings for pyrrole rings. The first evidence is reported in the absorption spectrum, where a more intense and red-shifted absorption band is reported for the N-doped analog (λ_{max} = 683 nm, ε = 1.3×10^5 M^{-1} cm^{-1}) with respect to the fully carbon analog (λ_{max} = 593 nm, ε = 8.6×10^4 M^{-1} cm^{-1}). This fact is also observed in the emission spectra, exhibiting emission maxima at 894 nm and 777 nm for **47** and **29**, respectively. Remarkably, no significant changes are observed in the ECD spectra, finding a similar pattern and intensities of the bands for both compounds. According to DFT calculations, the N-doped analog **47** would show a narrower energy gap (1.82 eV) compared with **29** (2.05 eV). This difference in energy is confirmed by electrochemical measurements of CV and differential pulse voltammetry (DPV), where the first oxidation wave is shifted 0.25 V to lower potentials in the case of **47**. Six oxidation waves can be found in the two analogs; however, they were produced by chemical oxidation and studied by spectroscopic methods in the case of the N-doped propeller.

Further N-doped propeller-shaped helical NGs with extended blades can be found in literature, such as the example reported by Nierengarten and coworkers (Figure 9.20, **48**) [165]. In this case, a central hexaazatriphenylene (HAT) unit is extended via sixfold Diels–Alder reaction and further Scholl reaction. The resulting compound is a three-bladed N-doped NG that was obtained as a mixture of diastereoisomers, a minor propeller-shaped D_3 symmetric and a major C_2 symmetric one. The UV-vis spectrum of both diastereoisomers extends up to 800nm with a similar profile. Remarkably, only the higher soluble C_2 diastereoisomer was resolved into its enantiomers. The ECD spectra obtained for D_3-**48** exhibited intense Cotton effects with $|\Delta\varepsilon|$ values of 762 and 768 M^{-1} cm^{-1} at 392 nm for each enantiomer.

9.3.2 Helical Nanographenes Containing Heteroatoms of Group 16 (O, S)

Helicenes containing heteroatoms of group 16, especially those containing oxygen and sulfur, are a well-known family of compounds among chemists. Numerous oxa- and thiahelicene derivatives have been extensively studied with a wide range of applicability [166]. While a broad spectrum of oxa[*n*]helicenes of different shapes and lengths (with the outstanding oxa[19]helicene as the record-breaking one) [167] has been recently reported, their extended analogs are scarce. One of the reasons of this gap could be the presence of these heteroatoms itself. When introducing *O* or *S*, the aromatic surface cannot be extended through them due to their reduced valence compared with *C* or *N*. Therefore, atoms from group 16 are confined to the edge of the aromatic structures. Nevertheless, extended helical NGs can be found in literature, such as the first member of the recently coined family of *superhelicenes*, reported by Jux and coworkers in 2018 (Figure 9.21, **49**) [91]. In this work the authors prepared a helical NG consisting in two HBC units linked by a furan moiety. The followed synthetic pathway was based on a twofold Diels–Alder reaction and a final Scholl reaction. Photophysical analysis by means of UV-vis and fluorescence spectra showed a red-shifted absorbance in comparison with related planar HBC derivatives, which is in concordance with a delocalization of the electron density through the HBC units. Moreover, the fluorescence quantum yield ranged an extraordinary 80–85%. On the other hand, the electrochemical bandgap of **49** was evaluated as 2.4 eV in concordance with that derived from the absorption and fluorescence spectra and lower than that of HBC derivatives. Recently, the authors, in collaboration with Fuchter and coworkers, reported the racemic resolution of **49** and the study of the chiroptical properties both in solution and thin film and embedded within achiral polymers [168]. Remarkably, the $|g_{lum}|$ value of **49** is increased 500 times (from 3×10^{-4} in solution to 1.5×10^{-1} when embedded in poly(9,9-di-*n*-octylfluorenyl-2,7-diyl) (PFO). This amplification of the chiroptical response in the emission was attributed to a Förster resonance energy transfer (FRET) process from the excited state of the achiral polymers to **49**.

Notwithstanding the lack of diversity of helical NGs containing oxygen atoms, the examples of sulfur-containing derivatives are much more common. The addition of sulfur atoms is mainly carried out by introduction of thiophene rings, which can be present in the different building blocks or formed in a further step. A relevant methodology for the formation of the thiophene ring was given by Itami and coworkers in 2016 [169]. In this work, the authors reported the thienannulation of different PAHs with elemental sulfur being possible to prepare propeller-shaped NGs such as compound **50** (Figure 9.21). Analogous to hexabenzotriphenylene (**24**), **50** shows three [5]helicene units where the isomer with the same configuration on each helicene is the most stable one with a racemization barrier of 24.7 kcal mol^{-1}. A similar structure was prepared by the same authors, changing the central benzene ring by a naphthalene unit and the relative position of the sulfur atoms (Figure 9.21, **51**) [170]. This modification in the structure creates two dithia[6]helicene and two carbo[5]helicene moieties. Remarkably the kinetic product after the final

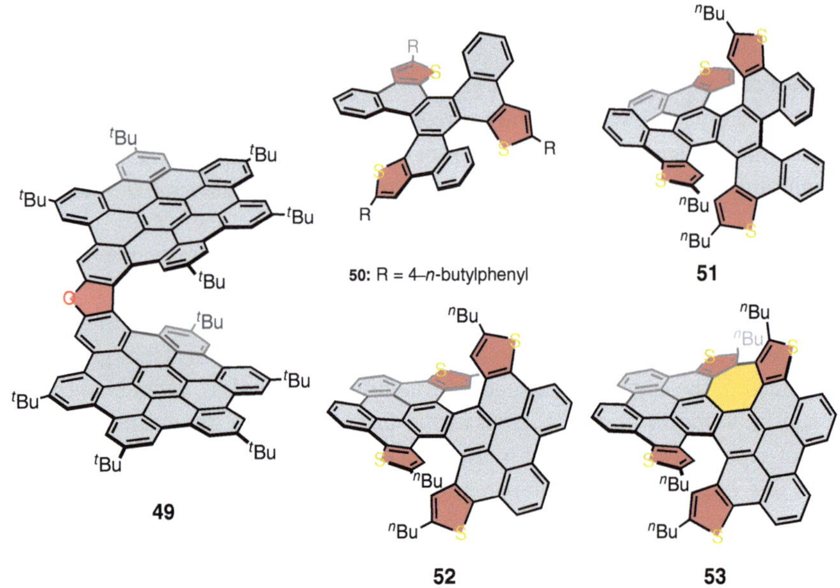

Figure 9.21 O-containing superhelicene (**49**) and S-containing NGs (**50–53**).

Scholl reaction (promoted by $MoCl_5$) was the D_2-symmetric diastereoisomer (*P*,*P*,*P*,*P*)/(*M*,*M*,*M*,*M*)-**51**, which can be converted into the thermodynamically stable (*P*,*P*,*M*,*M*)/(*M*,*M*,*P*,*P*)-**51** by heating the former at 80 °C for two days (isomerization barrier evaluated as 27.8 kcal mol^{-1}). The structures of both diastereoisomers were confirmed by X-ray crystallography and their corresponding enantiomers isolated, enabling the study of their chiroptical properties such as specific rotation and ECD. Furthermore, the racemization barrier for (*P*,*P*,*M*,*M*)/(*M*,*M*,*P*,*P*)-**51** was evaluated as 37.1 kcal mol^{-1} at 298 K. In a further step, when the amount of $MoCl_5$ in the last Scholl reaction is doubled, additional C–C bonds can be formed, resulting in an NG showing an extended double dithia[6]helicene (Figure 9.21, **52**), although chlorination of terminal benzenoid rings is also observed [171]. After palladium-catalyzed dechlorination, the obtained diastereoisomer was the pair (*P*,*P*)/(*M*,*M*)-**52**, not detecting any traces of the *meso* isomer (*P*,*M*)-**52**. Remarkably, the *twisted-to-meso* interconversion barrier was estimated as 50.8 kcal mol^{-1}, higher than that reported for its full-carbon analog **19** (43.5 kcal mol^{-1}). Regarding its electrochemical properties, **52** showed two reversible oxidation waves with half-wave potentials of 0.43 and 0.80 V. Furthermore, the good reversibility of the oxidation waves along with its crystal packing allowed the implementation of **52** in organic FETs, finding its applicability as p-type transistor with hole mobility values of 3.3×10^{-2} $cm^2V^{-1}s^{-1}$. Looking back again to the final Scholl reaction, when the reaction is performed in oxygen-saturated dichloromethane, an additional C–C bond is formed between two thiophene rings, thus creating a seven-membered ring (Figure 9.21, **53**) [172]. Compound **53**, presenting saddle geometry and an embedded helicene into its structure, was the first member of the coined saddle–helix

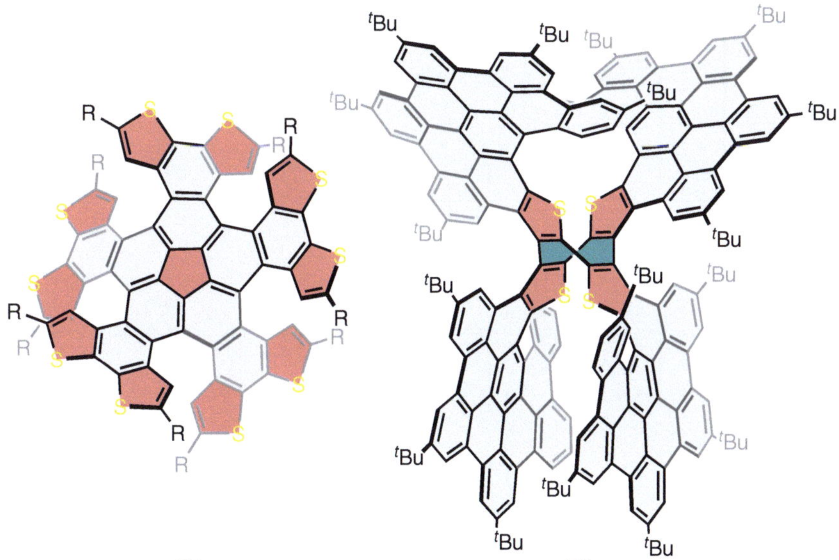

Figure 9.22 Corannulene- and cyclooctatetraene-based S-containing NGs.

hybrid family. The presence of the seven-membered ring seems to have an effect on the isomerization barrier of the dithia[6]helicene, which was estimated as 49.7 kcal mol^{-1}. Compound **53** was resolved into their enantiomers and their ECD spectra evaluated, showing $|\Delta\varepsilon|$ values of 82 $M^{-1}cm^{-1}$ at 501 nm and assigning their corresponding configuration by comparison with the TD-DFT simulated ECD spectra.

A larger thiophene-containing propeller-shaped NG was recently prepared as an analog to the quintuple helicene NG propeller **27** reported in Section 9.2.2 [115]. This decathia-quintuple helicene NG (Figure 9.22, **54**) was prepared from pentaborilated corannulene by Suzuki coupling, achieving the formation of the last C–C bond photochemically [173]. Even so, the obtained diastereoisomer during the synthesis was the C_2-symmetric (*P*,*P*,*M*,*P*,*M*)/(*M*,*M*,*P*,*M*,*P*)-**54**, which was converted into the D_5-symmetric diastereoisomer (*P*,*P*,*P*,*P*,*P*)/(*M*,*M*,*M*,*M*)-**54** upon heating a 1,1,2,2-tetrachloroethane solution at 70 °C for three days. The calculated interconversion barrier was estimated as 33.6 kcal mol^{-1}, high enough to be configurationally stable at room temperature. Unfortunately, the racemic resolution was not achieved, and, therefore, their chiroptical properties have not been reported yet. Nevertheless, the photophysical properties of **54** were studied showing a low quantum yield ($\Phi_F = 0.02$), similar to that of **27**.

The most recent example of a helical NG showing sulfur atoms into its structure has been addressed by Molina-Ontoria, Guldi, Martín, and coworkers [126]. In this work, the authors reported the preparation of a sulfur-containing NG consisting on four HBC-like units linked by a central eight-membered ring (Figure 9.22, **55**). The preparation of compound **55** was achieved starting from tetrabromo

Table 9.1 Summary of the main structural and photophysical properties reported for the compounds gathered on this chapter.

Compound	$\Delta G^{\ddagger}$ (T) (kcal mol^{-1})[a]	ε (M^{-1} cm^{-1}) (λ_{abs}, nm)	$E_{H\text{-}L}^{opt}$ (eV)	λ_{em} (nm)	HOMO (eV)	LUMO (eV)	Φ_F (%)	τ (ns)	$\lvert g_{abs}\rvert$ (λ_{abs},nm)	$\lvert g_{lum}\rvert$ (λ_{abs},nm)	$[\alpha]_D$
c-HBC [82, 83]	—	—	2.5	—	−5.22	−2.72	—	—	—	—	—
2b [84]	—	~155 000 (493)	2.4[h]	549	−5.4[h]	−3.0[h]	—	—	—	—	—
3a [85]	—	59 000 (447) 22 200 (570)	1.92 1.80[h]	584	−5.23[h]	−3.43[h]	—	—	—	—	—
3b [85]		128 700 (452) 53 000 (574)	1.90 1.80[h]	584	−5.17[h]	−3.37[h]	—	—	—	—	—
3c [85]		98 900 (456) 25 400 (596)	1.89 1.88[h]	571	−5.33[h]	−3.45[h]	—	—	—	—	—
4 [86]	—	91 700 (367)	—	475; 506	—	—	—	—	—	—	—
5a [76, 86]	28.68	—	2.73[b, i] 3.16[c, i]	—	−4.78[b, i] −5.05[b, h] −4.92[c, i]	−2.05[b, i] −2.76[b, h] −1.76[c, i]	—	—	—	—	—
5b [76]	—	—	2.41[b] 2.80[c]	554[b] 495[c]	−5.02[b, h] −5.14[c, h]	−2.76[b, h]	—	—	—	—	—
6b [87]	27.72	—	2.00[b, h] 2.35[c, h]	583[b] 530[c]	−5.22[b, h] −5.48[c, h]	−3.22[b, h] −3.13[c, h]	—	—	—	—	—
8 [89]	—	73 000 (346) 22 000 (456)	2.61 2.54[h]	494	−5.17[i] −4.98[h]	−2.16[i] −2.44[h]	50 ±0.05	—	—	—	—
9 [90]	—	—	2.16 2.27[h]	495 590	—	—	7.5	12.9 3.7	—	—	—

10 [92]	—	—	1.88	—	−4.99[h] −4.71[i]	−3.13[h] −2.20[i]	—	—	—	—	—
12 [95]	12.2	—	3.18[i]	520	−5.69[h] −5.30[i]	−2.12[i]	—	—	—	—	—
13a [94]	—	—	2.33	—	−5.37[h]		—	—	—	—	—
14 [96]	19.3[i]	—	3.43[i]	—	−5.14[i]	−1.70[i]	10	—	—	—	—
15a [97]	—	81 000 (577)	2.27[h]	600	—	—	71	—	—	—	—
15b [97]	—	56,000 (575)	2.31[h]	598	—	—	76	—	—	—	—
16 [97]	—	64 000 (571)	2.56[h]	590	—	—	69	—	—	—	—
17 [98]	44.8[i] 50.1[i]	4,700 (675)	2.04[i]	—	−4.61[i]	−2.57[i]	—	0.001–0.002	5.0×10^{-3} (459) 1.6×10^{-2} (680)	—	—
18 [100]	25.4 26.1[i]	—	2.30 3.02[i]	611	−5.17[h] −5.08[i]	−2.06[i]	11	—	—	—	—
19 [78]	43.5[i]	—	2.79[i], [b] 2.82[i], [j]	525[b] 496[j]	−4.88[i], [b] −4.88[i], [j]	−2.09[b], [i] −2.06[i], [j]	5.2[b] 42[j]	1.6[b] 6.5[j]	—	—	−81[g]
20 [102]	—	~35 000 (320)	6.01[i]	529	−6.50[i]	−0.49[i]	75		9.3×10^{-4} (490)	7.5×10^{-4} (529)	+590
21 [79]	46.0[i]	220 000 (269)	2.25[b] 2.65[i], [b] 2.43[j] 3.03[i], [j]	565[b] 538[j]	−5.00[i], [b] −5.16[h], [b] −5.15[i], [j] −5.19[h], [j]	−2.35[b], [i] −2.12[i], [j]	34[b] 11[j]	—	—	—	—
22 [104]	46.3[i]	220 000 (269)	1.92 2.11[h] 2.43[i]	697	−5.02[h] −4.91[i]	−2.91[h] −2.48[i]	3.5	—	—	—	—

Continued

Table 9.1 (Continued)

Compound	ΔG‡ (T) (kcal mol^{-1})[a)]	ε (M^{-1} cm^{-1}) (λ_{abs}, nm)	$E_{H\text{-}L}^{opt}$ (eV)	λ_{em} (nm)	HOMO (eV)	LUMO (eV)	Φ_F (%)	τ (ns)	$\|g_{abs}\|$ (λ_{abs},nm)	$\|g_{lum}\|$ (λ_{abs},nm)	$[\alpha]_D$
23 [106, 107, 109]	26.2 11.7	—	—	483	—	—	1.8	3.6	1.8×10^{-3}	1.3×10^{-3}	—
24 [108, 109]	27.4 29.5[i)]	83 000 (352)	—	469	—	—	2.6	5.2	1.1×10^{-3}	—	—
25 [110, 111]	35.4 [110] 36.3[i)] [110] 35.13 [111] 39.15[i)] [111]	57 800 (393)[d)] [110] 48 200 (393)[e)] [110] 59 000 (392)[e)] [111]	2.38[d), e)] [110] 3.14[d), i)] [110] 3.30[e), i)] [110]	517[d)] [110] 513[e)] [110] 513[e)] [111]	−5.81[d), h)] [110] −5.10[d), i)] [110] −5.82[e), h)] [110] −5.17[e), i)] [110]	−1.96[d), i)] [110] −1.87[e), i)] [110]	4.1[d)] [110] 3.9[e)] [110] 6.0[e)] [111]	5.0[d)] [110] 5.5[e)] [110] 5.6[e)] [111]	2.3×10^{-3} (433)[e)] [110] 4.8×10^{-3} (324)[e)] [110]	—	*−340*[e)] [110] *−496*[e)] [111] *+494*[e)] [111]
26 [113, 114]	38.72 38.38[i)] [114]	77 276 (418)[e)] [113] 90 000 (417)[e)] [114]	2.33[e)] [113] 3.14[e), i)] [113]	538[e)] [114]	−5.06[e), i)] [113]	−1.93[e), i)] [113]	7.3[e)] [113] 5.7[e)] [114]	5.2[e)] [113]	3.3×10^{-3} (460)[e)] [113]	—	−222[e)] [113] +705[d)] [114] −300[e)] [114]
27 [115]	34.2 34.5[i)]	—	3.11[i)]	527	−5.28[i)]	−2.17[i)]	3	—	—	—	—
28 [116]	—	—	—	550	−4.95[i)]	—	—	—	—	—	—
29 [118]	52.1[i)]	92 000 (618) 102 000 (618) [121]	1.51 1.58[h)] 2.00[i)]	—	−4.85[h)] −4.35[i)]	−3.27[h)] −2.34[i)]	—	—	7.0×10^{-4} (618) [121]	—	+500
30 [121]	—	397 000 (367) 179 000 (659)	1.86[i)]	836 934	−4.32[i)]	−2.46[i)]	1.6	6 (836) 4 700 (934)	1.7×10^{-3} (618)	—	—

31 [122]	—	354 000 (448) 205 000 (586)	2.43[i]	601	−4.56[i]	−2.13[i]	72	—	—	—	—
32 [66]	33.0	66 000 (444)	2.22 2.66[h]	560	—	—	13	21.5	2.7×10^{-3} (370) 5.3×10^{-4} (555)	2.3×10^{-4}	−186
33 [124]	—	160 000 (472)	2.20 2.33[h] 2.60[i]	610	—	—	9.8	18	2×10^{-3} (400) 2.5×10^{-3} (580)	2×10^{-3}	−5695
34 [125]	—	297 000[f] (517) 412 000[g] (522)	2.01[f] 2.06[g]	643[f] 650[g]	—	—	27.6	5–6	8.2×10^{-3} (446)[f] 8.8×10^{-4} (425)[g]	3×10^{-4} [f] 2×10^{-4} [g]	+3843[g]
35 [126]	—	403 000 (373)	—	494	−5.72[h]	—	10	4.7, 19.8	—	—	—
36a [134]	—		2.46		−5.23[i]	−2.27[i]	—	—	—	—	—
36c [134]	—	—	2.39	—	—	—	—	—	—	—	—
37 [135]	—	—	1.97	—	—	—	—	—	—	—	—
38[137]	13.38[o]	25 000 (428)	2.75	498	−4.84[h] −4.71[b), i] −4.74[i), j]	−1.30[b), i] −1.29[i), j]	18	—	—	—	—
39 [137]	6.67[o]	26 000 (406)	2.78	461	−4.93[h] −4.82[b), i] −4.84[i), j]	−1.44[b), i] −1.45[i), j]	16	—	—	—	—
40 [137]	9.13[o]	24 000 (429)	2.77	486	−4.88[h] −4.78[b), i] −4.81[i), j]	−1.39[b), i] −1.39[i), j]	13	—	—	—	—

(Continued)

Table 9.1 (Continued)

Compound	ΔG‡ (T) (kcal mol^{-1})[a]	ε (M^{-1} cm^{-1}) (λ_{abs}, nm)	$E_{H\text{-}L}{}^{opt}$ (eV)	λ_{em} (nm)	HOMO (eV)	LUMO (eV)	Φ_F (%)	τ (ns)	$\|g_{abs}\|$ (λ_{abs},nm)	$\|g_{lum}\|$ (λ_{abs},nm)	$[\alpha]_D$
42a [140]	—	29 700 (358)	—	474	—	—	—	—	—	—	—
44 [145]	—	111 000 (335)	2.24[i]	725	−5.74[i]	−3.50[i]	27	5.2	—	—	—
45a [146]	—	64 900 (742)	1.51[h] 1.88[i]	790	−5.16[i]	−3.28[i]	—	—	—	—	—
45b [146]	—	82 500 (730)	1.51[h] 1.58[d), h] 1.93[i]	775	−5.26[i]	−3.33[i]	—	—	—	—	—
45c [146]	—	78 900 (710)	—	—	−5.63[i]	−3.67[i]	—	—	7.0×10^{-3} (350)	—	—
47 [164]	—	128 600 (683)	1.82[i]	894	−4.09[i]	−2.27[i]	2.2	3.7	—	—	—
48 [165]	—	ca. 40 000 (710)[e]	—	—	—	—	—	—	—	—	—
49 [91, 168]	—	64 000 (515)	2.4[h] 2.8[i]	525	−4.80[i]	−2.00[i]	84.3	—	2.0×10^{-3} (509)	3×10^{-4}	—
50 [169]	24.7 25.8[i]	—	—	525	—	—	1	—	—	—	
51 [170]	37.1	59 000 (405)[m] 77 000 (381)[n]	3.07[i, m] 3.2[i, n]	536[m] 514[n]	−4.94[i, m] −5.09[i, n]	−1.87[i, m] −1.89[i, n]	1.7[m] 11[n]	—	—	—	+496[m] +426[n]

52 [171]	50.8	*ca.* 115 000 (405)	2.75[i)]	524 558	−4.83[i)]	−2.08[i)]	20	—	—	—	—
53 [172]	49.7	21 000 (502) 70 000 (394)	2.8[i)]	514 548	−4.85[i)]	−2.05[i)]	23	—	—	—	—
54 [173]	33.6	—	3.12[i)]	502	−4.97[i)]	−1.85[i)]	2	—	—	—	—
55 [126]	—	ca. 17 500 (393)	2.89[i)]	ca. 550	−5.58[h)] −5.19[i)]	−2.30[i)]	9	0.29	—	—	—

a) Isomerization barrier.
b) *Twisted* conformer.
c) *Anti* conformer.
d) C_2-symmetric diastereoisomer.
e) D_3-symmetric diastereoisomer.
f) C_3-symmetric diastereoisomer.
g) C_1-symmetric diastereoisomer.
h) From electrochemistry.
i) Theoretical.
j) *Meso* isomer.
k) (*P*,*M*) diastereoisomer.
l) (*M*,*M*)/(*P*,*P*) diastereoisomer.
m) Propeller-shaped isomer.
n) Saddle-shaped isomer.
o) Unsubstituted analog.
p) Symmetric folded.
q) Asymmetric *twisted*.

cyclooctatetrathiophene, whose aromatic structure was extended through a combination of Sonogashira, Diels–Alder, and Scholl reactions. The structure of **55** was confirmed by means of X-ray crystallography, where two different crystal forms were found. Remarkably, the geometries of the molecule are slightly different from one crystal form to the other. Electrochemical studies over **55** indicated the presence of four oxidation waves at 0.48, 0.66, 0.79, and 1.20 V, which were positively shifted by 0.14 V compared with its fully carbon analog **35**. This fact remarks the stronger electron-donating behavior induced by the thiophene rings in the structure of **55**. The thiophene rings also have an effect on its optical properties, being possible to observe a red shifting both in the absorption and emission spectra of **55** compared with those of **35**. Moreover, transient absorption pump–probe experiments were carried out in both analogs, finding a faster singlet excited state deactivation in **55** due to the combination of the sulfur presence and a higher structural flexibility.

9.4 Summary Table

See Table 9.1.

Acknowledgments

Our contributions to this area of research have been supported by the European Research Council (ERC) under the European Union's Horizon 2020 research and innovation program (ERC-2015-STG-677023), the Ministerio de Ciencia, Innovación y Universidades (MICIU/FEDER/AIE, Spain, PGC2018-101181-B-I00), the Ministerio de Economía y Competitividad (Spain, CTQ2015-70283-P), and the University of Granada (UGR).

References

1 Novoselov, K.S., Geim, A.K., Morozov, S.V. et al. (2004). *Science* 306: 666–669.
2 Geim, A.K. (2011). *Angew. Chem. Int. Ed.* 50: 6966–6985.
3 Novoselov, K.S. (2011). *Angew. Chem. Int. Ed.* 50: 6986–7002.
4 Wallace, P.R. (1947). *Phys. Rev.* 71: 622–634.
5 Son, Y.-W., Cohen, M.L., and Louie, S.G. (2006). *Phys. Rev. Lett.* 97: 216803.
6 Barone, V., Hod, O., and Scuseria, G.E. (2006). *Nano Lett.* 6: 2748–2754.
7 Li, X., Wang, X., Zhang, L. et al. (2008). *Science* 319: 1229–1232.
8 Girit, Ç.Ö., Meyer, J.C., Erni, R. et al. (2009). *Science* 323: 1705–1708.
9 Lungerich, D., Papaianina, O., Feofanov, M. et al. (2018). *Nat. Commun.* 9: 4756.
10 Goto, H., Uesugi, E., Eguchi, R. et al. (2013). *Nano Lett.* 13: 1126–1130.
11 Jiang, D.-e., Sumpter, B.G., and Dai, S. (2007). *J. Chem. Phys.* 126: 134701.

12 Gunlycke, D., Li, J., Mintmire, J.W., and White, C.T. (2007). *Appl. Phys. Lett.* 91: 112108.
13 Boukhvalov, D.W. and Katsnelson, M.I. (2008). *Nano Lett.* 8: 4373–4379.
14 Kim, J., Yamada, Y., Fujita, R., and Sato, S. (2015). *J. Mater. Sci.* 50: 5183–5190.
15 Plutnar, J., Pumera, M., and Sofer, Z. (2018). *J. Mater. Chem. C* 6: 6082–6101.
16 Gao, L., Ren, W., Xu, H. et al. (2012). *Nat. Commun.* 3: 699.
17 Sutter, P.W., Flege, J.-I., and Sutter, E.A. (2008). *Nat. Mater.* 7: 406–411.
18 Tapasztó, L., Dobrik, G., Lambin, P., and Biró, L.P. (2008). *Nat. Nanotechnol.* 3: 397–401.
19 Yazyev, O.V. and Chen, Y.P. (2014). *Nat. Nanotechnol.* 9: 755–767.
20 Chen, L., Hernandez, Y., Feng, X., and Müllen, K. (2012). *Angew. Chem. Int. Ed.* 51: 7640–7654.
21 Clar, E. and Ironside, C.T. (1958). *Proc. Chem. Soc.*: 125–156.
22 Clar, E., Ironside, C.T., and Zander, M. (1959). *J. Chem. Soc.*: 142–147.
23 Halleux, A., Martin, R.H., and King, G.S.D. (1958). *Helv. Chim. Acta* 41: 1177–1183.
24 Hendel, W., Khan, Z.H., and Schmidt, W. (1986). *Tetrahedron* 42: 1127–1134.
25 Müller, M., Kübel, C., and Müllen, K. (1998). *Chem. Eur. J.* 4: 2099–2109.
26 Iyer, V.S., Wehmeier, M., Brand, J.D. et al. (1997). *Angew. Chem. Int. Ed.* 36: 1604–1607.
27 Arslan, H., Uribe-Romo, F.J., Smith, B.J., and Dichtel, W.R. (2013). *Chem. Sci.* 4: 3973–3978.
28 Pozo, I., Guitián, E., Pérez, D., and Peña, D. (2019). *Acc. Chem. Res.* 52: 2472–2481.
29 Ito, H., Segawa, Y., Murakami, K., and Itami, K. (2019). *J. Am. Chem. Soc.* 141: 3–10.
30 Chen, T.-A. and Liu, R.-S. (2011). *Org. Lett.* 13: 4644–4647.
31 Chen, T.-A. and Liu, R.-S. (2011). *Chem. Eur. J.* 17: 8023–8027.
32 Liu, J., Li, B.-W., Tan, Y.-Z. et al. (2015). *J. Am. Chem. Soc.* 137: 6097–6103.
33 Zuzak, R., Pozo, I., Engelund, M. et al. (2019). *Chem. Sci.* 10: 10143–10148.
34 Xu, K., Urgel, J.I., Eimre, K. et al. (2019). *J. Am. Chem. Soc.* 141: 7726–7730.
35 Peurifoy, S.R., Xu, Q., May, R. et al. (2020). *Chem. Sci.* 11: 9978–9982.
36 Hashimoto, A., Suenaga, K., Gloter, A. et al. (2004). *Nature* 430: 870–873.
37 Meyer, J.C., Kisielowski, C., Erni, R. et al. (2008). *Nano Lett.* 8: 3582–3586.
38 Rutter, G.M., Crain, J.N., Guisinger, N.P. et al. (2007). *Science* 317: 219–222.
39 Amelinckx, S. and Delavignette, P. (1960). *Phys. Rev. Lett.* 5: 50–51.
40 Červenka, J. and Flipse, C.F.J. (2009). *Phys. Rev. B* 79: 195429.
41 Lee, G.-D., Wang, C.Z., Yoon, E. et al. (2005). *Phys. Rev. Lett.* 95: 205501.
42 Robertson, A.W., He, K., Kirkland, A.I., and Warner, J.H. (2014). *Nano Lett.* 14: 908–914.
43 Lehtinen, O., Vats, N., Algara-Siller, G. et al. (2015). *Nano Lett.* 15: 235–241.
44 Kiel, G.R., Patel, S.C., Smith, P.W. et al. (2017). *J. Am. Chem. Soc.* 139: 18456–18459.
45 Fujise, K., Tsurumaki, E., Wakamatsu, K., and Toyota, S. (2021). *Chem. Eur. J.* 27: 4548–4552.

46 Gingras, M. (2013). *Chem. Soc. Rev.* 42: 968–1006.
47 Gingras, M., Félix, G., and Peresutti, R. (2013). *Chem. Soc. Rev.* 42: 1007–1050.
48 Dhbaibi, K., Favereau, L., and Crassous, J. (2019). *Chem. Rev.* 119: 8846–8953.
49 Portella, G., Poater, J., Bofill, J.M. et al. (2005). *J. Org. Chem.* 70: 2509–2521.
50 Cao, Y., Fatemi, V., Fang, S. et al. (2018). *Nature* 556: 43–50.
51 Cao, Y., Fatemi, V., Demir, A. et al. (2018). *Nature* 556: 80–84.
52 Gibney, E. (2019). *Nature* 565: 15–18.
53 Xu, F., Yu, H., Sadrzadeh, A., and Yakobson, B.I. (2016). *Nano Lett.* 16: 34–39.
54 Moss, G.P. (1996). *Pure Appl. Chem.* 68: 2193.
55 Maroto, E.E., Izquierdo, M., Reboredo, S. et al. (2014). *Acc. Chem. Res.* 47: 2660–2670.
56 Suzuki, N., Wang, Y., Elvati, P. et al. (2016). *ACS Nano* 10: 1744–1755.
57 Đorđević, L., Arcudi, F., D'Urso, A. et al. (2018). *Nat. Commun.* 9: 3442.
58 Berova, N., Bari, L.D., and Pescitelli, G. (2007). *Chem. Soc. Rev.* 36: 914–931.
59 Longhi, G., Castiglioni, E., Koshoubu, J. et al. (2016). *Chirality* 28: 696–707.
60 Boyd, R.W. (2020). *Nonlinear Optics (Fourth Edition)*, 4ee. Academic Press.
61 Hayat, A., Nevet, A., Ginzburg, P., and Orenstein, M. (2011). *Semicond. Sci. Technol.* 26: 083001.
62 Racknor, C., Singh, M.R., Zhang, Y. et al. (2014). *Methods Appl. Fluoresc.* 2: 015002.
63 Wergifosse, M.d., Houk, A.L., Krylov, A.I., and Elles, C.G. (2017). *J. Chem. Phys.* 146: 144305.
64 Hu, Y., Danilov, E.O., Wex, B., and Neckers, D.C. (2008). *J. Nonlinear Opt. Phys. Mater.* 17: 275–283.
65 Díaz, C., Vesga, Y., Echevarria, L. et al. (2015). *RSC Adv.* 5: 17429–17437.
66 Cruz, C.M., Márquez, I.R., Mariz, I.F.A. et al. (2018). *Chem. Sci.* 9: 3917–3924.
67 Tinoco, I. Jr., (1975). *J. Chem. Phys.* 62: 1006–1009.
68 Toro, C., De Boni, L., Lin, N. et al. (2010). *Chem. Eur. J.* 16: 3504–3509.
69 Vesga, Y. and Hernandez, F.E. (2017). *Chem. Phys. Lett.* 684: 67–71.
70 Chen, W., Zhang, S., Zhou, M. et al. (2019). *J. Phys. Chem. Lett.* 10: 3290–3295.
71 Tanaka, H., Ikenosako, M., Kato, Y. et al. (2018). *Commun. Chem.* 1: 38.
72 Ravat, P., Hinkelmann, R., Steinebrunner, D. et al. (2017). *Org. Lett.* 19: 3707–3710.
73 Li, C., Yang, Y., and Miao, Q. (2018). *Chem. Asian J.* 13: 884–894.
74 Whalley, A.C., Plunkett, K.N., Gorodetsky, A.A. et al. (2011). *Chem. Sci.* 2: 132–135.
75 Grimme, S. and Peyerimhoff, S.D. (1996). *Chem. Phys.* 204: 411–417.
76 Luo, J., Xu, X., Mao, R., and Miao, Q. (2012). *J. Am. Chem. Soc.* 134: 13796–13803.
77 Kashihara, H., Asada, T., and Kamikawa, K. (2015). *Chem. Eur. J.* 21: 6523–6527.
78 Fujikawa, T., Segawa, Y., and Itami, K. (2015). *J. Am. Chem. Soc.* 137: 7763–7768.
79 Hu, Y., Wang, X.-Y., Peng, P.-X. et al. (2017). *Angew. Chem. Int. Ed.* 56: 3374–3378.

80 Janke, R.H., Haufe, G., Würthwein, E.-U., and Borkent, J.H. (1996). *J. Am. Chem. Soc.* 118: 6031–6035.
81 Clar, E. and Stephen, J.F. (1965). *Tetrahedron* 21: 467–470.
82 Xiao, S., Myers, M., Miao, Q. et al. (2005). *Angew. Chem. Int. Ed.* 44: 7390–7394.
83 Kumar, S., Pola, S., Huang, C.-W. et al. (2019). *J. Org. Chem.* 84: 8562–8570.
84 Xiao, S., Kang, S.J., Wu, Y. et al. (2013). *Chem. Sci.* 4: 2018–2023.
85 Chen, Y., Marszalek, T., Fritz, T. et al. (2017). *Chem. Commun.* 53: 8474–8477.
86 Martin, M.M., Hampel, F., and Jux, N. (2020). *Chem. Eur. J.* 26: 10210–10212.
87 Yang, Y., Yuan, L., Shan, B. et al. (2016). *Chem. Eur. J.* 22: 18620–18627.
88 Yanney, M., Fronczek, F.R., Henry, W.P. et al. (2011). *Eur. J. Org. Chem.* 2011: 6636–6639.
89 Fernández-García, J.M., Evans, P.J., Medina Rivero, S. et al. (2018). *J. Am. Chem. Soc.* 140: 17188–17196.
90 Márquez, I.R., Fuentes, N., Cruz, C.M. et al. (2017). *Chem. Sci.* 8: 1068–1074.
91 Reger, D., Haines, P., Heinemann, F.W. et al. (2018). *Angew. Chem. Int. Ed.* 57: 5938–5942.
92 Ma, J., Fu, Y., Dmitrieva, E. et al. (2020). *Angew. Chem. Int. Ed.* 59: 5637–5642.
93 Pradhan, A., Dechambenoit, P., Bock, H., and Durola, F. (2013). *J. Org. Chem.* 78: 2266–2274.
94 Cheung, K.Y., Xu, X., and Miao, Q. (2015). *J. Am. Chem. Soc.* 137: 3910–3914.
95 Gu, X., Li, H., Shan, B. et al. (2017). *Org. Lett.* 19: 2246–2249.
96 Kawai, K., Kato, K., Peng, L. et al. (2018). *Org. Lett.* 20: 1932–1935.
97 Castro-Fernández, S., Cruz, C.M., Mariz, I.F.A. et al. (2020). *Angew. Chem. Int. Ed.* 59: 7139–7145.
98 Nakakuki, Y., Hirose, T., Sotome, H. et al. (2018). *J. Am. Chem. Soc.* 140: 4317–4326.
99 Nakai, Y., Mori, T., and Inoue, Y. (2012). *J. Phys. Chem. A* 116: 7372–7385.
100 Qiu, Z., Asako, S., Hu, Y. et al. (2020). *J. Am. Chem. Soc.* 142: 14814–14819.
101 Barroso, J., Cabellos, J.L., Pan, S. et al. (2018). *Chem. Commun.* 54: 188–191.
102 Yamano, R., Shibata, Y., and Tanaka, K. (2018). *Chem. Eur. J.* 24: 6364–6370.
103 Newman, M.S. and Lednicer, D. (1956). *J. Am. Chem. Soc.* 78: 4765–4770.
104 Hu, Y., Paternò, G.M., Wang, X.-Y. et al. (2019). *J. Am. Chem. Soc.* 141: 12797–12803.
105 Zhou, Z., Fu, L., Hu, Y. et al. (2020). *Angew. Chem. Int. Ed.* 59: 15923–15927.
106 Peña, D., Pérez, D., Guitián, E., and Castedo, L. (1999). *Org. Lett.* 1: 1555–1557.
107 Peña, D., Cobas, A., Pérez, D. et al. (2000). *Org. Lett.* 2: 1629–1632.
108 Saito, H., Uchida, A., and Watanabe, S. (2017). *J. Org. Chem.* 82: 5663–5668.
109 Tanaka, H., Kato, Y., Fujiki, M. et al. (2018). *J. Phys. Chem. A* 122: 7378–7384.
110 Hosokawa, T., Takahashi, Y., Matsushima, T. et al. (2017). *J. Am. Chem. Soc.* 139: 18512–18521.
111 Berezhnaia, V., Roy, M., Vanthuyne, N. et al. (2017). *J. Am. Chem. Soc.* 139: 18508–18511.
112 Zuzak, R., Castro-Esteban, J., Brandimarte, P. et al. (2018). *Chem. Commun.* 54: 10256–10259.

113 Zhang, F., Michail, E., Saal, F. et al. (2019). *Chem. Eur. J.* 25: 16241–16245.

114 Roy, M., Berezhnaia, V., Villa, M. et al. (2020). *Angew. Chem. Int. Ed.* 59: 3264–3271.

115 Kato, K., Segawa, Y., Scott, L.T., and Itami, K. (2018). *Angew. Chem. Int. Ed.* 57: 1337–1341.

116 Evans, P.J., Ouyang, J., Favereau, L. et al. (2018). *Angew. Chem. Int. Ed.* 57: 6774–6779.

117 Furche, F., Ahlrichs, R., Wachsmann, C. et al. (2000). *J. Am. Chem. Soc.* 122: 1717–1724.

118 Zhu, Y., Xia, Z., Cai, Z. et al. (2018). *J. Am. Chem. Soc.* 140: 4222–4226.

119 Wang, Y., Yin, Z., Zhu, Y. et al. (2019). *Angew. Chem. Int. Ed.* 58: 587–591.

120 Simpson, C.D., Brand, J.D., Berresheim, A.J. et al. (2002). *Chem. Eur. J.* 8: 1424–1429.

121 Zhu, Y., Guo, X., Li, Y., and Wang, J. (2019). *J. Am. Chem. Soc.* 141: 5511–5517.

122 Ma, S., Gu, J., Lin, C. et al. (2020). *J. Am. Chem. Soc.* 142: 16887–16893.

123 Würthner, F. (2004). *Chem. Commun.*: 1564–1579.

124 Cruz, C.M., Castro-Fernández, S., Maçôas, E. et al. (2018). *Angew. Chem. Int. Ed.* 57: 14782–14786.

125 Cruz, C.M., Márquez, I.R., Castro-Fernández, S. et al. (2019). *Angew. Chem. Int. Ed.* 58: 8068–8072.

126 Urieta-Mora, J., Krug, M., Alex, W. et al. (2020). *J. Am. Chem. Soc.* 142: 4162–4172.

127 Köster, R., Hattori, S., and Morita, Y. (1965). *Angew. Chem. Int. Ed.* 4: 695–695.

128 Biswas, S., Müller, M., Tönshoff, C. et al. (2012). *Eur. J. Org. Chem.* 2012: 4634–4639.

129 Müller, M., Maichle-Mössmer, C., Sirsch, P., and Bettinger, H.F. (2013). *ChemPlusChem* 78: 988–994.

130 Mateo-Alonso, A. (2014). *Chem. Soc. Rev.* 43: 6311–6324.

131 Cortizo-Lacalle, D., Mora-Fuentes, J.P., Strutyński, K. et al. (2018). *Angew. Chem. Int. Ed.* 57: 703–708.

132 Mora-Fuentes, J.P., Riaño, A., Cortizo-Lacalle, D. et al. (2019). *Angew. Chem. Int. Ed.* 58: 552–556.

133 Wang, X.-Y., Yao, X., Narita, A., and Müllen, K. (2019). *Acc. Chem. Res.* 52: 2491–2505.

134 Hashimoto, S., Nakatsuka, S., Nakamura, M., and Hatakeyama, T. (2014). *Angew. Chem. Int. Ed.* 53: 14074–14076.

135 Wang, X.-Y., Richter, M., He, Y. et al. (2017). *Nat. Commun.* 8: 1948.

136 Krzeszewski, M. and Gryko, D.T. (2015). *J. Org. Chem.* 80: 2893–2899.

137 Krzeszewski, M., Kodama, T., Espinoza, E.M. et al. (2016). *Chem. Eur. J.* 22: 16478–16488.

138 Mishra, S., Krzeszewski, M., Pignedoli, C.A. et al. (2018). *Nat. Commun.* 9: 1714.

139 Benkyi, I., Staszewska-Krajewska, O., Gryko, D.T. et al. (2020). *J. Phys. Chem. A* 124: 695–703.

140 Ammon, F., Sauer, S.T., Lippert, R. et al. (2017). *Org. Chem. Front.* 4: 861–870.

141 Dusold, C., Sharapa, D.I., Hampel, F., and Hirsch, A. (2021). *Chem. Eur. J.* 27: 2332–2341.
142 Takase, M., Enkelmann, V., Sebastiani, D. et al. (2007). *Angew. Chem. Int. Ed.* 46: 5524–5527.
143 Lazerges, M., Jouini, M., Hapiot, P. et al. (2003). *J. Phys. Chem. A* 107: 5042–5048.
144 Oki, K., Takase, M., Mori, S., and Uno, H. (2019). *J. Am. Chem. Soc.* 141: 16255–16259.
145 Żyła-Karwowska, M., Zhylitskaya, H., Cybińska, J. et al. (2016). *Angew. Chem. Int. Ed.* 55: 14658–14662.
146 Navakouski, M., Zhylitskaya, H., Chmielewski, P.J. et al. (2019). *Angew. Chem. Int. Ed.* 58: 4929–4933.
147 Moshniaha, L., Żyła-Karwowska, M., Chmielewski, P.J. et al. (2020). *J. Am. Chem. Soc.* 142: 3626–3635.
148 Sasaki, Y., Takase, M., Kobayashi, N. et al. (2021). *J. Org. Chem.* 86, 5, 4290–4295.
149 Nowak-Król, A. and Würthner, F. (2019). *Org. Chem. Front.* 6: 1272–1318.
150 Sun, M., Müllen, K., and Yin, M. (2016). *Chem. Soc. Rev.* 45: 1513–1528.
151 Nowak-Król, A., Shoyama, K., Stolte, M., and Würthner, F. (2018). *Chem. Commun.* 54: 13763–13772.
152 Yuan, Z., Xiao, Y., and Qian, X. (2010). *Chem. Commun.* 46: 2772–2774.
153 Hartnett, P.E., Matte, H.S.S.R., Eastham, N.D. et al. (2016). *Chem. Sci.* 7: 3543–3555.
154 Zhong, H., Wu, C.-H., Li, C.-Z. et al. (2016). *Adv. Mater.* 28: 951–958.
155 Schuster, N.J., Paley, D.W., Jockusch, S. et al. (2016). *Angew. Chem. Int. Ed.* 55: 13519–13523.
156 Schuster, N.J., Hernández Sánchez, R., Bukharina, D. et al. (2018). *J. Am. Chem. Soc.* 140: 6235–6239.
157 Khokhlov, K., Schuster, N.J., Ng, F., and Nuckolls, C. (2018). *Org. Lett.* 20: 1991–1994.
158 Milton, M., Schuster, N.J., Paley, D.W. et al. (2019). *Chem. Sci.* 10: 1029–1034.
159 Schuster, N.J., Joyce, L.A., Paley, D.W. et al. (2020). *J. Am. Chem. Soc.* 142: 7066–7074.
160 Fu, H., Meng, D., Meng, X. et al. (2017). *J. Mater. Chem. A* 5: 3475–3482.
161 Wang, B., Liu, W., Li, H. et al. (2017). *J. Mater. Chem. A* 5: 9396–9401.
162 Liu, G., Koch, T., Li, Y. et al. (2019). *Angew. Chem. Int. Ed.* 58: 178–183.
163 Meng, D., Liu, G., Xiao, C. et al. (2019). *J. Am. Chem. Soc.* 141: 5402–5408.
164 Guo, X., Yuan, Z., Zhu, Y. et al. (2019). *Angew. Chem. Int. Ed.* 58: 16966–16972.
165 Hahn, U., Maisonhaute, E., and Nierengarten, J.-F. (2018). *Angew. Chem. Int. Ed.* 57: 10635–10639.
166 Shen, Y. and Chen, C.-F. (2012). *Chem. Rev.* 112: 1463–1535.
167 Nejedlý, J., Šámal, M., Rybáček, J. et al. (2017). *Angew. Chem. Int. Ed.* 56: 5839–5843.
168 Wade, J., Brandt, J., Reger, D. et al. (2020). *Angew. Chem. Int. Ed.* 60, 1, 222–227.

169 Meng, L., Fujikawa, T., Kuwayama, M. et al. (2016). *J. Am. Chem. Soc.* 138: 10351–10355.

170 Fujikawa, T., Segawa, Y., and Itami, K. (2016). *J. Am. Chem. Soc.* 138: 3587–3595.

171 Fujikawa, T., Mitoma, N., Wakamiya, A. et al. (2017). *Org. Biomol. Chem.* 15: 4697–4703.

172 Fujikawa, T., Segawa, Y., and Itami, K. (2017). *J. Org. Chem.* 82: 7745–7749.

173 Lin, H.-A., Kato, K., Segawa, Y. et al. (2019). *Chem. Sci.* 10: 2326–2330.

10

Helicene-based Electron Acceptors

Xiao Xiao and Colin Nuckolls

Department of Chemistry, Columbia University, New York, NY, USA

10.1 Introduction

Helicenes are a class of contorted polycyclic aromatic molecules consisting of *ortho*-annulated aromatic rings [1]. These conjugated, chiral molecules have attracted a great deal of attention due to their promising chemical, biological, and chiroptical properties for next-generation materials and technologies. The unique combination of π-conjugation and axial chirality provides the opportunity for numerous applications such as chiroptical nonlinear optics [2], switches/sensors [3], asymmetric catalysis [4], and spin filters [5]. Recently, the development of new synthetic methodologies has further enabled helicenes as promising building blocks for chiral electronic and optoelectronic materials [6, 7]. Helicenes can now incorporate π-surfaces with various sizes [8] and different chemical functionalities [9], which results in distinct and/or complementary functionalities.

The prototypical helicenes – carbo[n]helicenes (where n denotes the number of *ortho*-fused benzene rings, e.g. [6]helicene **1a**, Figure 10.1a) – as well as many other reported chiral polyaromatic hydrocarbons (PAHs) in the literature consist only of benzene rings. Therefore, helicenes are typically regarded as electron donors, as evidenced by their redox activity in cyclic voltammetry [10]. Direct incorporation of heteroatoms like O, S, and N (e.g. **1b**) into helicenes has a dramatic influence on their electronic properties while maintaining the helical backbone with minimal structural changes. These heteroatoms may further enhance or diminish the electron-donating nature of helicenes, depending on how they are incorporated. On the other hand, strategic installment of electron-withdrawing chemical functionalities allows helicenes to accept electrons. These helicenic electron acceptors have unique properties and functionalities compared with their electron-donating counterparts. Notably, these electron-accepting helicenes are not as prevalent as electron-rich helicenes because many traditional synthetic approaches to helicene synthesis (e.g. Friedel–Crafts acylation, Mallory photocyclization, and alkyne trimerization) favor electron-rich precursors. This chapter will highlight the key features of some most recent representative examples of these unusual helicene-based electron acceptors.

Helicenes: Synthesis, Properties, and Applications, First Edition.
Edited by Jeanne Crassous, Irena G. Stará, and Ivo Starý.

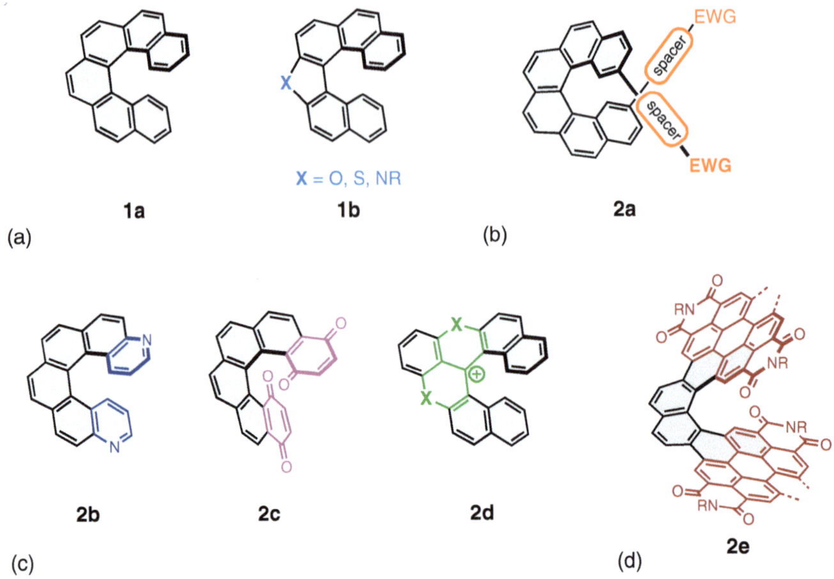

Figure 10.1 (a) The prototypical [6]helicene and hetero[6]helicenes. Three different strategies that render helicenes as electron acceptors: (b) the connection with electron-withdrawing groups via conjugated spacers; (c) judicious incorporation of heteroatoms, also including cationic helicenes; (d) placing helicenes into extended electron-accepting scaffolds. EWG = electron-withdrawing groups.

To provide the readers with a more organized discussion, helicenic electron acceptors are categorized into three types based on how their electron-accepting features are incorporated (Figure 10.1b–d): (i) electron-deficient substituents are remotely connected, yet conjugated, to the main helicene core (e.g. **2a**); (ii) heteroatoms (such as O, B, and N; see **2b** and **2c**) are strategically placed on the helicene backbone, and this includes cationic helicenes, which often contain heteroatoms to stabilize the positive charge (e.g. **2d**); (iii) the helicene is embedded inside of a π-extended framework with electron-accepting nature (e.g. **2e**, the maroon color represents electron-accepting components). Importantly, both the choice of the strategy for rendering helicene electronically deficient and the atomic position of the essential acceptor moieties have a tremendous impact on the properties of helicenes. By assembling some of the key examples here, certain design principles for different applications are discussed.

10.2 Helicenes with Remote Electron-withdrawing Functionalities

The incorporation of remote substituents has the advantage of minimizing structural changes to the helicene core while altering the electronic properties of the system. This is particularly useful for systematic studies of the effect of various substituents on the ultimate properties of the helicenes, which can guide the design

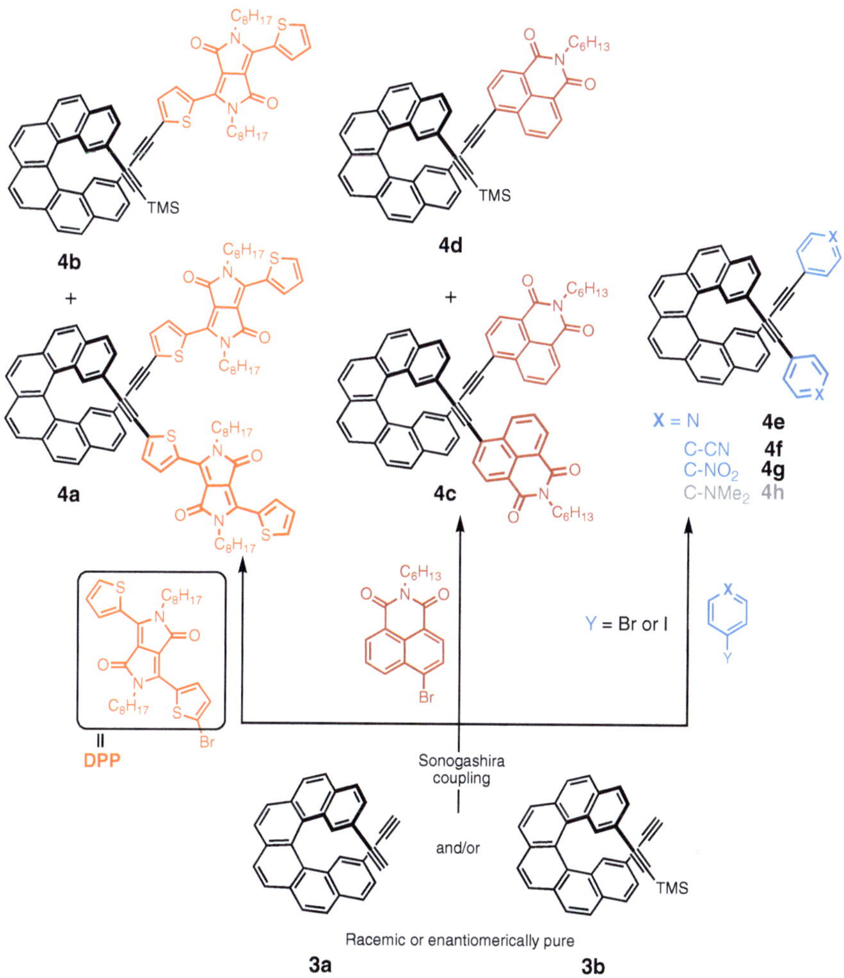

Figure 10.2 A new class of helicenes **4** with electron-accepting capabilities constructed via Sonogashira couplings from carbo[6]helicene derivatives **3**. Source: Based on [11].

of new chiral materials. Over the past few years, Crassous, Autschbach, Favereau, and coworkers have capitalized on this idea [11]. Namely, they developed and investigated a new family of helical π-conjugated systems **4** based on carbo[6]helicene with electron-withdrawing substituents (Figure 10.2). The synthesis of the family was efficient and divergent, in which the key step was a Sonogashira coupling of diacetylenic carbo[6]helicene precursors **3a** and/or **3b**, in either racemic or enantiomerically pure form, with a variety of electron-accepting aryl halides.

It should be noted that all of these molecules are technically better described as donor–acceptor helicenes, because the electron-withdrawing groups are spatially separated from the electron-donating carbo[6]helicene core through ethynyl bridges. Nevertheless, these helicenes can undergo reversible reduction process as demonstrated by cyclic voltammetry. For example, the reduction waves of **4a**

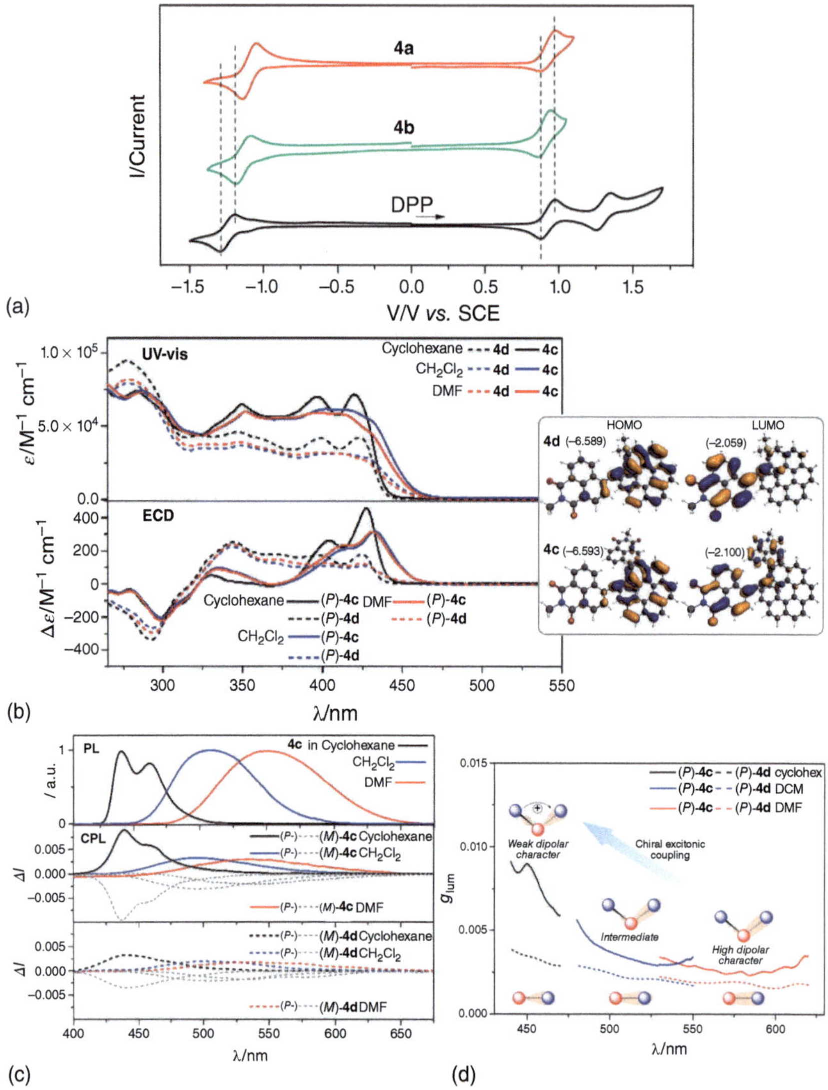

Figure 10.3 (a) Cyclic voltammograms of **DPPBr**, **4a** and **4b**. Source: [11a]/with permission of Royal Society of Chemistry. (b) UV-vis, ECD spectra, and calculated frontier orbitals of **4c** and **4d**. (c) CPL spectra and emission dissymmetry factors g_{lum} of **4c** and **4d**. (d) Observed solvent effects on dissymmetry factors g_{lum} of (*P*)-**4c** (solid lines) and (*P*)-**4d** (dashed lines) and the proposed mechanism. Source: [11b]/with permission of Royal Society of Chemistry.

and **4b** (ca. −1.1 V vs. SCE) are quite close to that of the DPP unit itself (−1.2 V vs. SCE), indicating only weak electronic interaction between DPP and the helicene (Figure 10.3a). Similar trends are also observed with **4c** and **4d**, suggesting that this relatively weak coupling between the terminal substituents is a universal feature for this class of ethynyl bridged molecules.

A hallmark of helicenes is the ability to selectively interact with one hand (left or right) of chiral light (i.e. circularly polarized light), which is also essential for applications in chiroptoelectronics. To interrogate such interaction of chiral light with chiral objects like helicenes, a wide range of spectroscopic techniques such as electronic circular dichroism (ECD)/vibrational circular dichroism (VCD), circularly polarized luminescence (CPL), and optical rotation/rotatory dispersion (ORD) arc routinely used [12]. However, the identification of the relationships between molecular structures and chiroptical properties remains a significant challenge.

To help address this challenge, helicenes **4** (Figure 10.2) serve as a suitable platform for studying the structure–function relationships. For instance, the UV-vis and ECD spectra of **4c** and **4d** were recorded and compared (Figure 10.3b). The absorption spectra of **4c** and **4d** share similar features, although the symmetric helicene **4c** shows universally higher absorptivities. Intramolecular charge transfer (ICT) character of **4c,d** can be visualized by the calculated frontier orbitals. Moreover, time-dependent density functional theory (TD-DFT) calculations suggest that the increased intensity of **4c** vs. **4d** is originated from exciton coupling between the ICT transitions from the π-orbitals of the [6]helicene core to the π^*-orbitals localized at either one of the naphthalimide units. Since **4d** only consists of one acceptor, such interactions are not viable.

The concept of excitonic coupling plays a crucial role in interpreting chiroptical properties of helicenes, as discussed here and in later sections of this chapter. The disparities between the ECD of (*P*)-**4c,d** and other carbo[6]helicene derivatives [13] demonstrate the non-innocent role of the naphthalimide moieties in the absorptions with chiral light (Figure 10.3b). The ECD comparison between optically pure **4c** and **4d** is also noteworthy, in which the biggest difference appears in the bands above 400 nm. TD-DFT calculations reveal that the ECD-intense excitations of (*M*)- or (*P*)-**4c** can be treated as an exciton couplet. Although the ECD spectra of (*M*)- and (*P*)-**4c,d** are clearly dependent on the polarity of the solvent, a more noticeable solvent effect is seen in their CPL spectra (Figure 10.3c). An up to threefold decrease of intensity is observed for (*M*)- and (*P*)-**4c** when switching from cyclohexane to DMF, whereas (*M*)- and (*P*)-**4d** exhibit similar CPL intensity in all solvents. This is consistent with the exciton coupling mechanism as described previously. In polar solvents, the more severe solvent and solute reorganization causes the reduction of symmetry, which renders excitations localized to a single naphthalimide group. This also creates a strong dipole moment in ICT state [14]. As a result, the electrostatic interaction between electric transition dipoles (i.e. exciton coupling) is suppressed, and thus the intensity of CPL is diminished (Figure 10.3d).

To demonstrate the practical application of these new helicenes, enantiomerically pure **4c,d** were used as non-fullerene electron acceptor materials in organic photovoltaic (OPV) devices [11c]. The helicenic electron acceptors, in either racemic or optically pure form, were blended with the commonly used wide bandgap donor polymer **P3HT** [**P3HT**=poly(3-hexylthiophene)]. The chirality of **4c** and **4d** in fact have a significant impact on the device performance. Namely, devices based on the racemic and the enantiopure forms of these π-helical non-fullerene acceptors have power conversion efficiencies of 0.4 and ca. 2.0%, respectively. This is attributed to a different solid-state organization of the pure enantiomers, thereby affecting the charge transport properties and the morphology of the active layers.

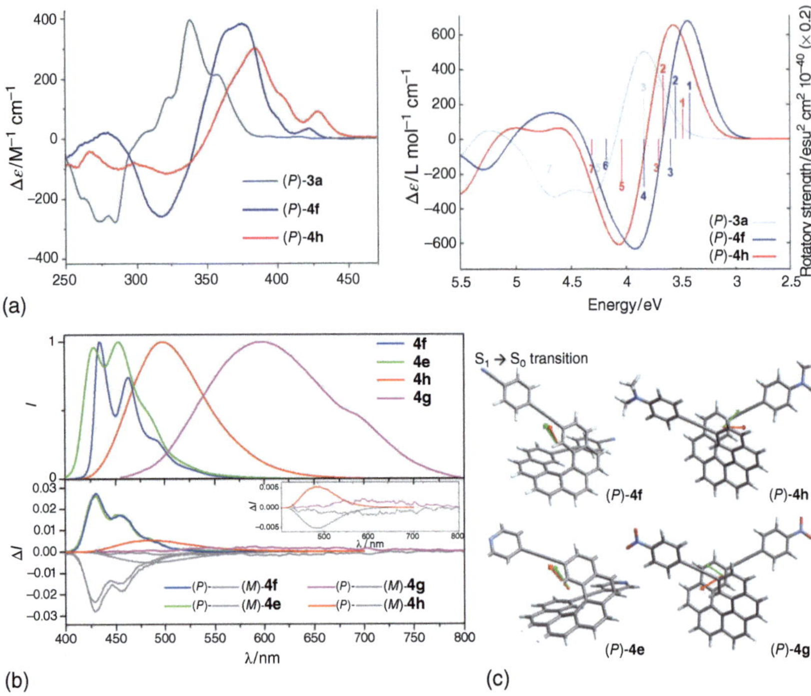

Figure 10.4 (a) Experimental (left) and computed (right) ECD spectra of (*P*)-**3a**, **4f**, and **4h**. (b) Unpolarized luminescence and CPL spectra of **4e–h**. (c) Visualization of the relevant transition dipoles for the S_1 to S_0 transition of (*P*)-**4e–h**. Source: [11d]/with permission of Royal Society of Chemistry.

The study of **4e–h** (cf. Figure 10.2) sheds additional light on the origin of the enhanced chiroptical properties observed for this new class of helicenes [11d]. Due to the similar spectroscopic properties of **4e** and **4f**, as well as of **4g** and **4h**, only the ECD spectra of (*P*)-**3a**, (*P*)-**4f**, and (*P*)-**4h** are presented (Figure 10.4a). The bisignate feature of exciton coupled ECD [15] (centered at ca. 350 nm) is even more evident compared to that of **4c,d**. TD-DFT calculations disclose the positive signal (excitations 1 and 2), and the adjacent negative bands both involve the π to π* and ICT transitions, reminiscent of the analysis of **4c,d**. Furthermore, the electric and magnetic transition dipole moments for the ECD couplet of each compound are inspected computationally. In theory, the anisotropy factor g is proportional to rotational strength (R), that is, the imaginary part of scalar product of the relevant electric (μ_e) and magnetic transition dipole (μ_m) moments ($R = ||\mu_e||\cdot||\mu_m||\cos\theta$) [16]. The two vectors turn out to orient nearly in parallel with angles of 5.9° and 7.4° for (*P*)-**4f** and (*P*)-**4h**, respectively, which give rise to the exceptional chiroptical properties, with a g_{abs} of 3.2 and 2.2% in dichloromethane. The g value difference also indicates a minor impact of the grafting substituents on the ECD of these functionalized helicene systems.

The effect of the substituents is more dramatic in the CPL spectra of (*M*)- and (*P*)-**4e–h** (Figure 10.4b). Namely, the CPL manifest maxima corresponding to that of the unpolarized spectra, with g_{lum} ranging from 0.15 for **4h** to 2.8% for **4e**. Notably, the impressive CPL intensities for **4e** and **4f** are among the highest recorded at the molecular level for organic molecules [17]. As in the case of the ECD, a helicene-mediated excitonic coupling of π transitions involving the substituents seems to be a major contributor to the strong CPL of **4e,f**, leading to large electric and magnetic transition dipole moments and a small angle between the vectors (Figure 10.4c). This exciton coupling mechanism that is also present in **4g,h**, however, promotes CPL to a less extent. Further analysis of the transition dipole moments reveal a larger angle between the two vectors (Figure 10.4c). In fact, the calculated rotational strengths of **4g,h** for S_1 to S_0 is higher than those of **4e,f**, but the concomitantly even larger dipole strengths (D) afford smaller dissymmetry factor ($g = 4R/D$, $D = ||\mu_e||^2$). Overall, these results provide new insights into how exciton coupling affect chiroptical properties and further reinforce the potential of π-conjugated helicenic acceptor molecules for chiral optoelectronic applications. They also build new opportunities for the design of novel CPL emitters and the development of more efficient CP-OLEDs, OPVs, and other devices.

10.3 Electron-accepting Heterohelicenes

The history of heterohelicenes can be traced back to more than a century ago, and in fact the first helicenes ever reported are heterohelicenes [18]; in 1903, Meisenheimer and Witte discovered the first azahelicene **5a** (Figure 10.5a). The electronegativity of N atoms in **5a** changes the inherent properties of the entire π system, making it an electron acceptor. Recently, the chemistry and applications of heterohelicenes have been rapidly growing. In this section, representative examples of electron-accepting heterohelicenes are discussed.

10.3.1 Azahelicenes

As demonstrated by **5a**, incorporation of N atoms is a convenient approach for modification of the electronic and chiroptical properties of helicenes. Azahelicenes with fused pyridines are arguably the most common electron poor azahelicenes or even heterohelicenes in general. Oxidative photocyclization is a classic and efficient way to construct such azahelicenes. Although most of the reported examples use electron-rich precursors [19], pyridines are compatible with the traditional Mallory reactions. For example, Srebro-Hooper, Crassous, and coworkers prepared a terpyridine ligand comprising two aza[6]helicenes (Figure 10.5b) [20]. The synthesis commenced with a double Wittig olefination between dialdehyde **5b** and phosphonium bromide **5c**, giving **5d** as a mixture of regioisomers. Subsequent photocyclization of **5d** with substoichiometric amount of iodine afforded double azahelicene **5e** as a statistical mixture of *meso* and racemic (*M,M* and *P,P*) stereoisomers, which were separated by chiral HPLC.

The use of transition metal-catalyzed transformations (e.g. cross couplings, [2+2+2] cycloisomerizations, etc.) represents another important synthetic strategy for accessing pyridohelicenes. A key advantage of this approach is the possible

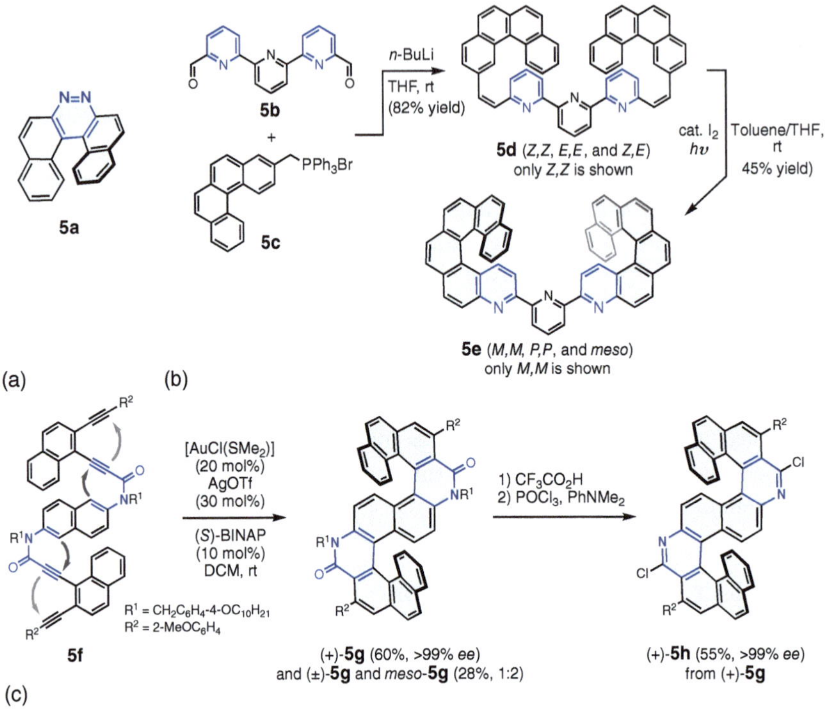

Figure 10.5 (a) The first reported electron-deficient azahelicene **5a**. (b) Pyridohelicene **5e** constructed by oxidative photocyclization. (c) An enantioselective synthesis of double azahelicenes **5h**.

formation of only one enantiomer from an achiral precursor in the presence of a chiral ligand. Indeed, Tanaka and coworkers discovered a landmark enantioselective synthesis of azahelicenes [9e], as demonstrated by the synthesis of S-shaped double azahelicenes **5g** and **5h** (Figure 10.5c). The key step involves a cascade enantioselective hydroarylation of alkyne **5f** by utilizing a single catalyst with a chiral ligand [AuCl(SMe$_2$)/(*S*)-BINAP]. As indicated by the arrows, the first hydroarylation gave an axially chiral biaryl intermediate, and the second event forged the azahelicene (+)-**5g**. The overall enantioselectivity and diastereoselectivity for this complex transformation was quite impressive with over 15:1 *e.r.* and 3.7:1 *d.r.* Moreover, the opposite enantiomer of (+)-**5g** was selectively obtained by using (*R*)-BINAP, and optically pure (+)-**5g** was simply purified from *rac*- and *meso*-**5g** via preparative TLC. After removal of the 4-alkoxybenzyl group on the nitrogen atoms, pyridohelicene (+)-**5h** was obtained by a subsequent chlorination of the imidic acid moieties. These S-shaped double azahelicenes exhibited enhanced CPL activities compared with that of the monoazahelicenes, with $g_{lum} = 2.8\%$ at 492 nm for (+)-**5g** and $g_{lum} = 1.1\%$ at 454 nm for (+)-**5h** in chloroform.

Recently, Audisio, Champagne, and coworkers reported a unique approach to synthesize a pyridine containing heterohelicene **6c** by 1,3-dipolar cycloaddition of sydnone **6a** with aryne **6b'** (Figure 10.6a) [21]. The electron acceptability of **6c** was

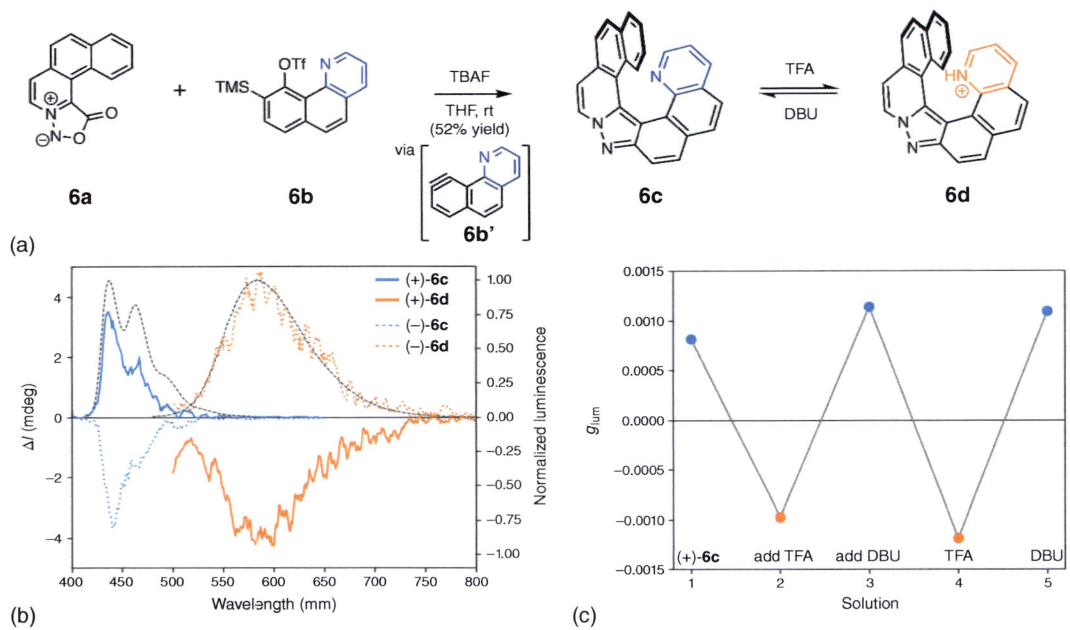

Figure 10.6 (a) A pyridine-containing helicene **6c** synthesized via 1,3-dipolar cycloaddition of sydnones with arynes. (b) pH-dependent chiroptical behaviors of **6c** and **6d**. (c) Reversible switching of g_{lum}. Source: [21]/with permission of American Chemical Society.

further increased in **6d** by protonation of the pyridyl nitrogen, which also affects the photophysical properties. Namely, the absorption and ECD spectra of **6d** is overall redshifted in comparison with those of **6c**. TD-DFT calculations unveil that protonation of the pyridine-like nitrogen significantly stabilizes the LUMO of **6d**, albeit the HOMO energy is nearly unchanged, resulting in a lower HOMO-LUMO gap. Remarkably, the CPL of **6d** not only is shifted to lower energy region with almost zero overlap with that of **6c**, but also the sign of **6d** is completely reversed compared with the original CPL of **6c** (Figure 10.6b). The pH-dependent switching behavior is also reversible, as shown by how g_{lum} responds to the sequential addition of trifluoroacetic acid (TFA) and 1,8-diazabicyclo[5.4.0]undec-7-ene (DBU) to a solution of (+)-**6c** (Figure 10.6c). The origin for this CPL sign reversal upon protonation can be rationalized by a slight change of the angle θ from 85.9° (**6c**) to 97.2° (**6d**) between the electric and magnetic transition dipoles (sign of cos θ is changed). The unusual and desirable switching properties of **6c,d** showcase the potential of electron-acceptor azahelicenes as promising chiroptical switches.

Electron-accepting azahelicenes can possess unique aggregation properties. In 2020, Qiu and coworkers discovered a spontaneous symmetry breaking phenomena in the aggregates of aza[4]helicenes [22]. As shown in Figure 10.7a, pyridohelicenes **7a** and **7b** rapidly racemize at ambient temperature [23], each of which was able to form nanoribbons (see the scanning electron microscope [SEM] image of aggregates from **7a**) in DMF/H_2O (1/99, v/v). Due to the fast enantioinversion process, these aggregates are supposed to be achiral. Indeed, aggregates of **7b** display neither CD nor CPL activities. However, interesting chiroptical properties were observed for the nanoribbons obtained from **7a** (Figure 10.7b), suggesting the atomic precision is crucial for the symmetry breaking process. The CD and CPL signals are well discernible with g_{lum} approaching 1%. Multiple runs of the experiment reveal that the overall chirality is totally batch dependent (Figure 10.7c). Single crystal structures of **7a** and **7b** were obtained to gain more insights into the solid-state packing mode: only one enantiomer (*P* or *M*) is found in an individual single crystal of **7a** (Figure 10.7d), while **7b** prefers a heterochiral stacking. Therefore, an initial formation of homochiral domains of **7a** could serve as the seed for the heterochiral packing to slowly become the thermodynamically favored homochiral form, thereby amplifying the chirality.

Since 2013, Fuchter and coworkers have demonstrated that active layers based on the blend of enantiopure aza[6]helicene (*M* or *P*)-**8a** with achiral polyfluorene-based polymers like **8b** are capable of generating CP emissions in devices (Figure 10.8a) [24]. In 2019, they further capitalized on the **8a**/**8b** blends reported in earlier preliminary studies, achieving incredible dissymmetry factors ($g_{lum} > 1$) and state-of-the-art device performance (4.0 cd A^{-1} with a brightness of 2054 cd m^{-2} at 10 V and a maximum brightness of >8000 cd m^{-2}) [24b]. The redshifted absorption of the thermally annealed thin film of the **8a**/**8b** blends (with 10% **8a**) compared with the as-cast films supports a larger distribution of conformational states (Figure 10.8b). The annealed films show strong ECD in spite of the as-cast films being essentially CD inactive, which also exhibits a linear increase of g_{abs} as a function of film thickness.

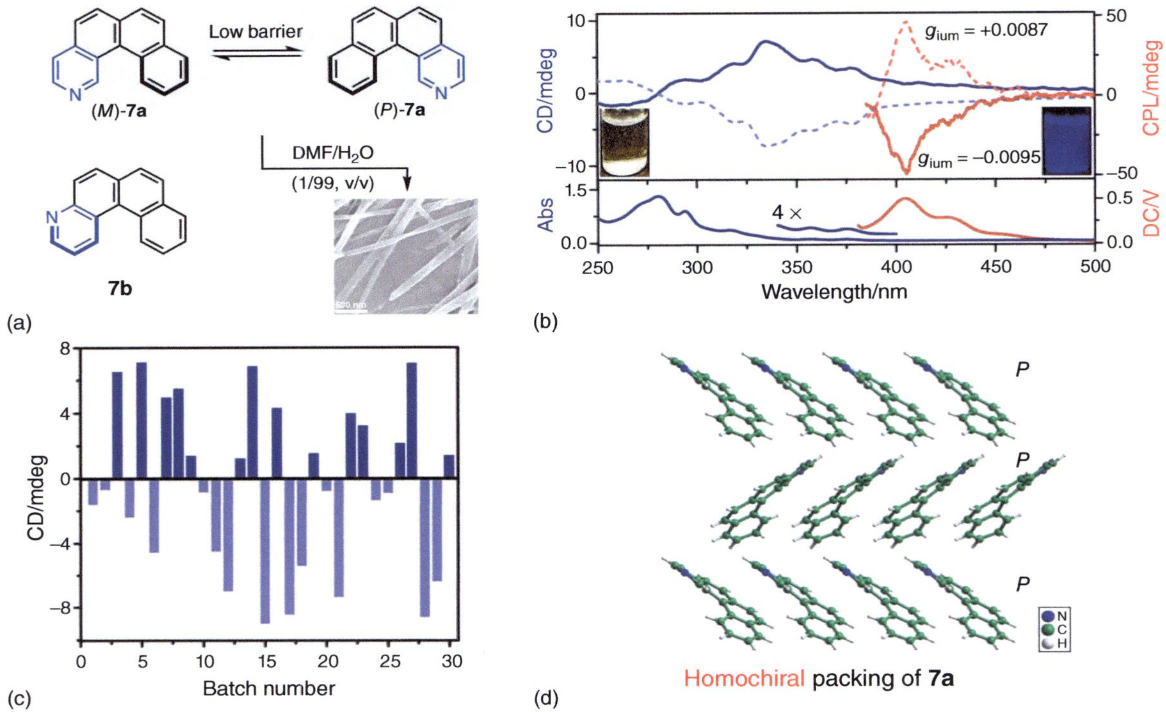

Figure 10.7 (a) The chemical structures of aza[4]helicenes **7a** and **7b**. (b) Aggregated nanoribbons of **7a** with CD and CPL activities. (c) Batch-dependent chirality of the aggregated nanoribbons. (d) Single crystal structure of **7a**. Source: [22], Figure 01, p. 03/with permission of American Chemical Society.

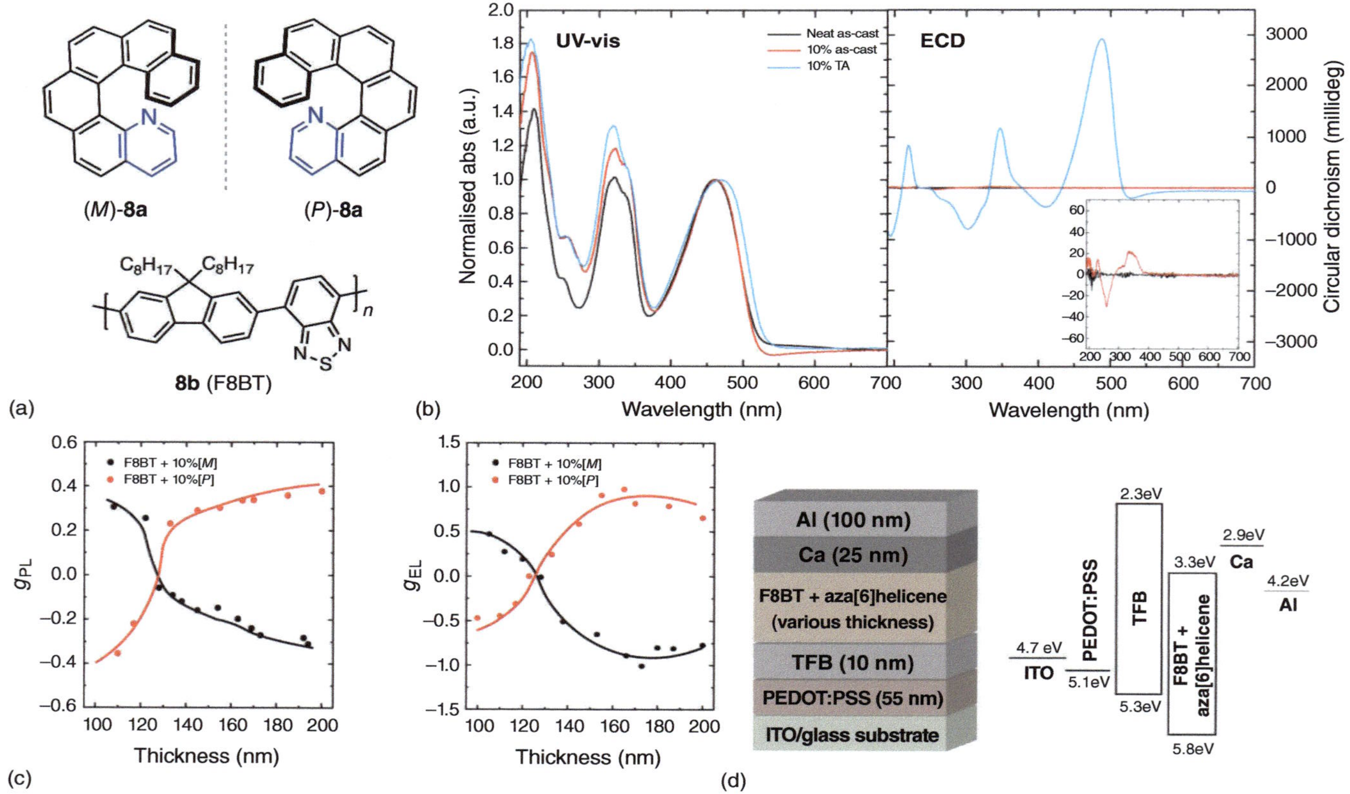

Figure 10.8 (a) Azahelicene **8a** and polyfluorene-based polymer **8b**. (b) Normalized absorption and CD spectra of thin films of as cast **8b**, as cast (see inset), and thermal annealed 10% **8a/8b** blend. (c) g_{pl} and g_{el} as a function of film thickness, respectively. (d) Device architecture and energy diagram. Source: [24b]/with permission of American Chemical Society.

The thickness-dependent photoluminescence and electroluminescence spectra are unprecedented (Figure 10.8c). The CPL first decreases (100–110 nm) to zero and then inverts sign (>120 nm) completely. Using the other antipode of the azahelicene **8a** gives an equal but opposite response. Moreover, when CP-OLEDs within the same thickness range were fabricated (see Figure 10.8d for device configuration), analogous trend was observed for the extracted g_{EL}: for the (*M*)-**8a** blends, g_{EL} varies from +0.51 (110 nm) to −1.05 (160 nm). An interplay between localized CP emission originating from molecular chirality (<120 nm thick) and CP light amplification or inversion through a chiral medium (> 120 nm) is proposed as the mechanism. Most recently, they discovered that the extraordinary chiroptical effects of these blends are elicited from short-range couplings between magnetic and electric dipoles, instead of long-range structural chirality as previously hypothesized [24c].

10.3.2 Incorporation of Oxygen Atoms in Helicenes

Introduction of quinone-like substructures into helicenes can greatly enhance the ability to undergo reversible reduction [25]. The synthesis of helicene quinones was first pioneered by Katz and coworkers in 1990 (Figure 10.9a) [27]. As mentioned earlier, the relatively well-developed oxidative photocyclization often suffers from functional group intolerance. Here, helicenebisquinone **9b** was created from an elegant double Diels–Alder reaction of divinyl benzene **9a** and excess 1,4-benzoquinone. In this transformation, 1,4-benzoquinone serves as both the dienophile and the oxidant.

Many other divinyl arenes can be employed to construct more structurally complex helicene quinones. Since then, helicene bis-*p*-quinones were explored in greater detail by Katz and coworkers in aspects such as radical anions [28] and supramolecular aggregation [2, 29]. They discovered that several chiral helicenequinones display enormous amplification of circular dichroism (CD) in their aggregates compared with their CD in dilute solution. Moreover, intense second-harmonic generation was observed in the Langmuir–Blodgett film of a nonracemic helicene bis-*p*-quinone **9c** (Figure 10.9a), providing a unique approach to second-order nonlinear optical materials [2].

In 2014, Diederich, Gescheidt, and coworkers studied the rather elusive *o*-quinone analog **9d** (Figure 10.9b) [26]. The synthesis began with a key nickel-catalyzed alkyne [2 + 2 + 2] cycloisomerization of triyne **9e**, followed by elimination of acetic acid to afford **9f** (Figure 10.9c). Helicene quinone **9d** was obtained after demethylation of **9f** and subsequent oxidation of the catechol moiety. The formation of a stable semiquinone radical anion $\mathbf{9d}^{\bullet-}$ was revealed by the initial reversible reduction wave observed in the cyclic voltammogram of **9d**. Therefore, spectroelectrochemical measurements were conducted for $\mathbf{9d}^{\bullet-}$. The comparison between the ECD spectra of **9d** and $\mathbf{9d}^{\bullet-}$ clearly unveils appreciable differences (Figure 10.9d). Namely, large $\Delta(\Delta\varepsilon)$ of up to 40 L mol^{-1} cm^{-1} are located at 357 nm, whereas smaller $\Delta(\Delta\varepsilon)$ are identified at 635 nm (4 L mol^{-1} cm^{-1}). In spite of that, the latter corresponds to a complete on–off for chiroptical switching. Besides, the switching of ECD signals

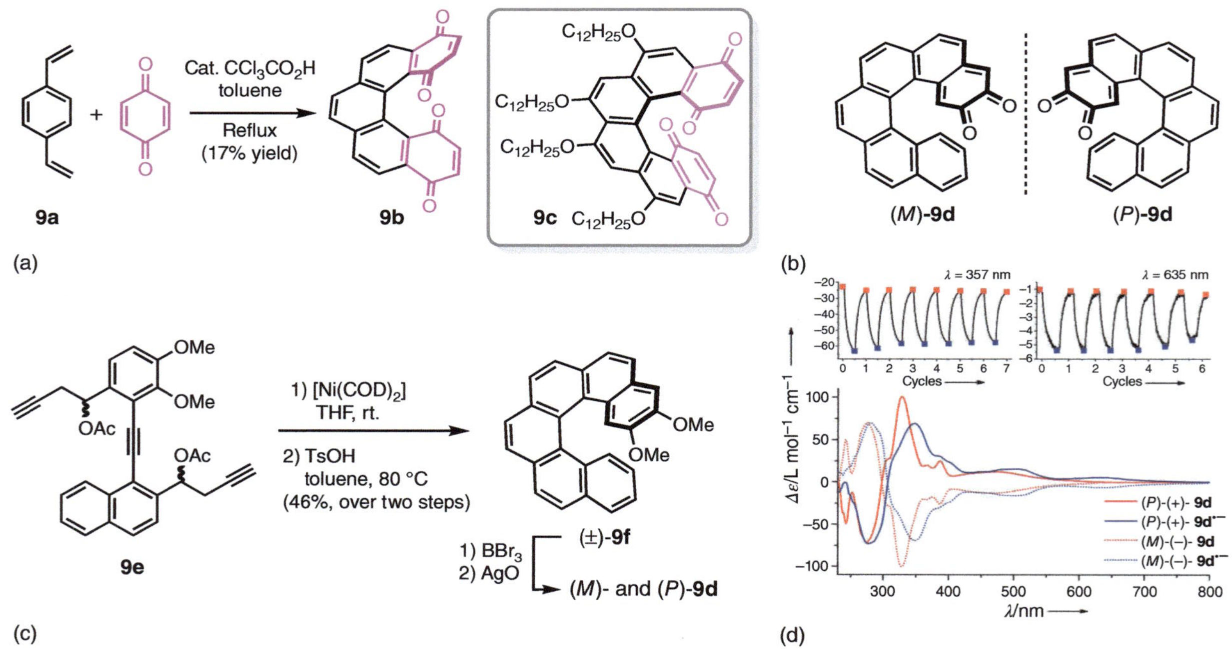

Figure 10.9 (a) Katz's Diels–Alder approach to helicenebisquinones. (b) Helicene *o*-quinone **9d**. (c) Synthetic route to **9d**. (d) ECD spectra and chiroptical switching behavior of **9d** and its radical anion **9d•−**. Source: [26]/with permission of American Chemical Society.

(Figure 10.9d, top) is still effective after multiple cycles. Such good recyclability and switching performance are rare for pure helicene-based chiroptical switches.

10.3.3 Borahelicenes

Boron incorporation endows helicenic scaffolds with electron-accepting character and Lewis acidic centers. However, the intrinsic reactivity of organoboranes with vacant p orbitals toward oxygen and moisture renders the synthesis of pure boron-incorporated helical PAHs challenging. To overcome this issue, N or O atoms are often concomitantly introduced as B–N or B–O units, and, therefore, examples of electron-accepting borahelicenes are scarce. Nevertheless, pyridohelicenes with tetravalent boron inclusion fall into this category. Extended π-conjugated molecules incorporating four-coordinate boron atoms have demonstrated potential in applications such as organic field-effect transistor (OFET)s, organic light-emitting diode (OLED)s, and sensors [30]. For example, Shen et al. prepared a series of cycloborylated hexa-, octa-, and decahelicenes (**10a–d**, Figure 10.10a), which were accessed via a general method as depicted in Figure 10.10b [31]. First, the pyridine ring reacts with BBr_3 followed by a net loss of HBr in the presence of an amine base afforded a *B*,*B*-dibromo-azaborole system. Then, the bromo atoms are replaced with methyl group upon the action of trimethylaluminum. Azaborahelicenes **10a–d** display blue fluorescence (λ_{em} ~ 420–450 nm) with strong quantum yields (0.21–0.49) for azaborahexahelicenes **10a,c** and only marginal values (~0.07) for the octa- and decahelicenes **10b,d**.

10.3.4 Cationic Heterohelicenes

Cationic helicenes are inherently prone to accept electron due to the positive charge delocalized across π system. The presence of heteroatoms is typically necessary to stabilize the cation. One good example is protonated azahelicenes, which are a subset of azahelicenes with one or more positively charged quaternary sp^2 nitrogen atoms (see **6d**, Figure 10.6). Fuchter and coworkers have achieved redox-driven

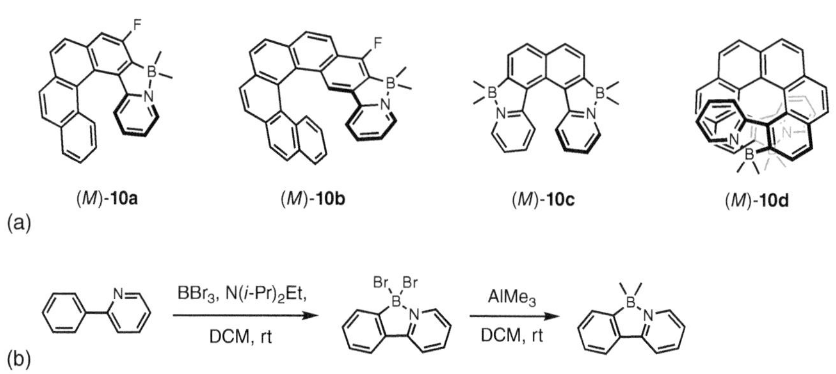

Figure 10.10 (a) Electron-accepting borahelicenes **10a–d**. (b) A generic synthetic route to *B*,*N*-heterocycles. Source: Modified from [31].

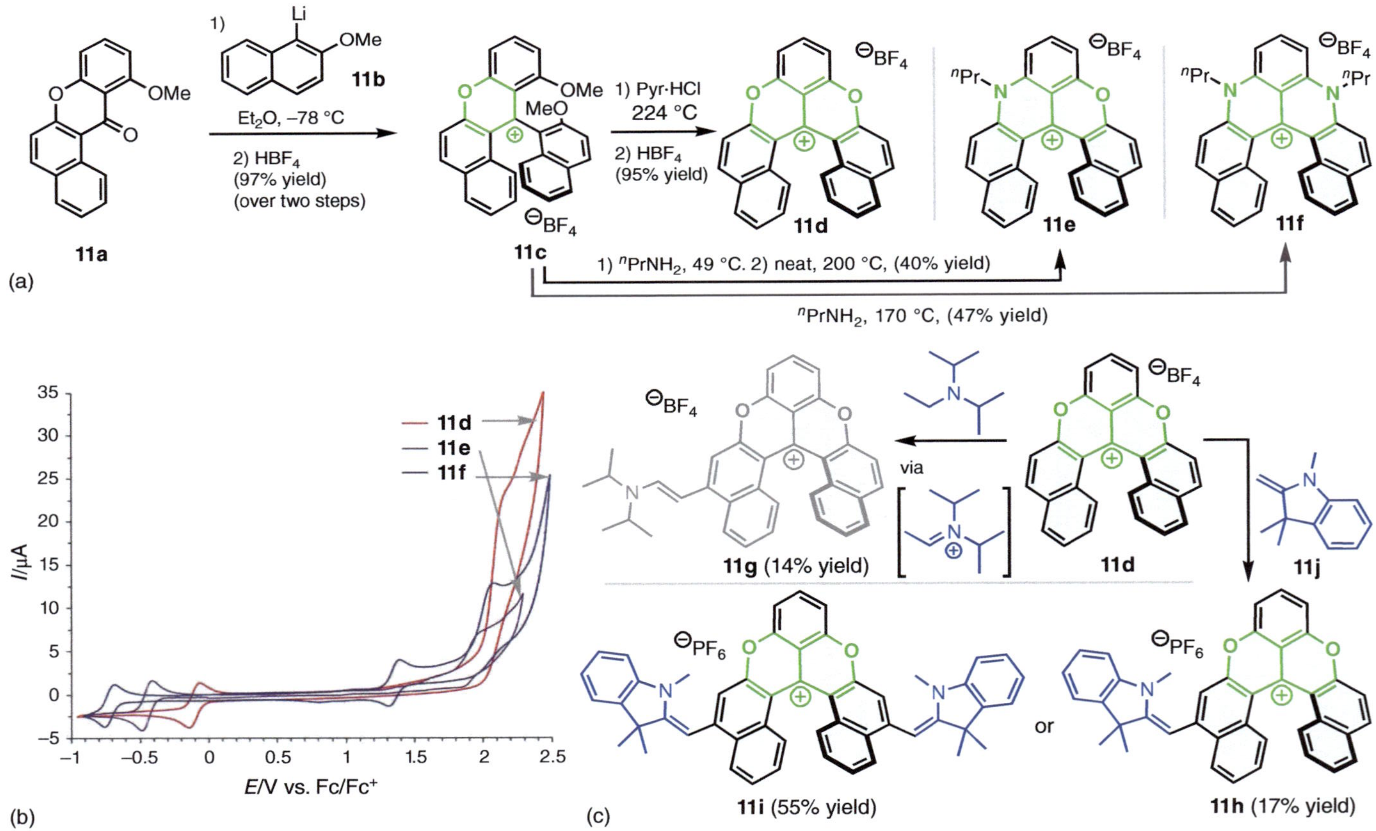

Figure 10.11 (a) A modular synthesis of cationic helicenes **11d–f**. (b) Cyclic voltammograms of 1 mM **11d–f**. (c) NIR-fluorophores **11g-i** derived from **11d**. Source: [33]/with permission of John Wiley & Sons.

chiroptical switching with an azoniahelicene [32]. The pyridinium moiety can readily and reversibly accept an electron to afford a neutral radical species, which rapidly undergoes dimerization. Therefore, robust switching of ECD signals over multiple reduction–oxidation cycles was observed by application of a series of potentiostatic pulses between reduction and oxidation.

On the other hand, carbocation-based heterohelicenes have also been studied. For instance, Lacour and coworkers reported a modular synthesis of carbocation-based helicenes [33a]. A common synthetic intermediate **11c** was employed in the synthesis of a series of cationic dioxa-, azaoxa-, and diaza[6]helicenes such as **11d–f**, which was prepared from **11a** via the addition of aryl lithium **11b** and a protonation/elimination sequence (Figure 10.11a). A number of other diaza[6]helicene derivatives were obtained by further functionalization of **11f**. Not surprisingly, cyclic voltammetry clearly shows a reversible single-electron reduction for each of **11d–f**, but the reduction potentials differ quite noticeably (Figure 10.11b) [33b]. Each replacement of an oxygen by a nitrogen induces a cathodic shift of ca. 0.3 V.

More recently, Lacour and coworkers have designed several chiral near-infrared (NIR) fluorophores **11g–i** based on cationic helicene **11d** that manifests the most positive reduction potential among **11d–f** [33c] (Figure 10.11c). The synthesis is based on an oxidative coupling with tertiary amines/enamines. Initially, **11d** easily abstracts an electron from the lone pair of a tertiary amine like Hünig's base, which undergoes deprotonation to give the related enamine. The carbon-centered cation of **11d** next directs the nucleophilic addition of the enamine to the *para* position. Subsequent single-electron oxidation by another molecule of **11d** and deprotonation furnishes the coupling product **11g**. Enamine **11j** can also be directly used as the coupling partner, giving rise to helicenes **11h** and **11i**. These chromophores exhibit large absorptivities (λ_{max} up to 791 nm), fluorescence (λ_{max} up to 887 nm), and intense ECD properties ($|\Delta\varepsilon|$ = ca. 60 M^{-1} cm^{-1} at 790 nm) in the far-red or NIR domains.

10.4 Electron-accepting π-extended Frameworks Embedded with Helicenes

Merging electron-accepting organic semiconductors with helicene architectures not only afford unique π-extended scaffolds but also impart great potential for device applications. This approach has attracted increasing attention in recent years, which has been facilitated by new synthetic development for accessing large π systems. Classic electron acceptors like naphthalene (di)imides, perylene diimides (PDIs), and higher rylene diimides are the most commonly used motifs in this class of helicenes. One of the reasons these subunits are used is that rylene diimides have shown promising properties such as large absorption at longer wavelengths, low lying LUMO, high photoluminescence quantum yields, and n-type redox behavior [34]. Moreover, they are readily available, and there is a rich library of reactions for chemical functionalizations of rylene diimides [35]. Finally, extension of the helical π-surface hampers the interconversion between each enantiomer, thereby increasing the thermal enantiostability. Numerous contorted rylene diimide molecules have proved to be relevant in applications such as organic field effect transistors, non-fullerene acceptors in solar cell, photodetectors, and organic pseudocapacitors [36]. Although in many cases, chiral helicene substructures are present, no characterization regarding the pure enantiomer has been discussed. There are a number of reasons why only racemic mixture was studied. First, in some cases the pure enantiomer is not enantiomerically stable. For example, most of the [4]helicene- and [5]helicene-based materials can only exist as a racemic mixture due to the fast interconversion [37]. Second, the chiral information is often insignificant for optoelectronics only involving nonpolarized light. Last, chiral resolution of the π-extended helicenes in relatively large quantities for device applications is challenging. This section only includes recent examples where π-extended helicenic acceptors in their optically pure form are explicitly studied.

10.4.1 Extended Helicenes with Naphthalimide Units

Naphthalene diimide (NDI) is the smallest homolog of the rylene diimide family. The first synthesis of an NDI-containing helicene was reported by Usui and coworkers [38]. Although NDI has been far less commonly incorporated in helicenes than PDIs, in the last few years more examples have emerged. For example, Hirsch group synthesized a diaza[7]helicene (**12b**, Figure 10.12) by applying NDIs into hexa-*peri*-hexabenzocoronenes (HBC) [39]. The imidazole moiety of **12a** was constructed by condensation of a naphthalimide monoanhydride with a diamino benzene derivative. The hexaphenylbenzene core was prepared via a well-developed Diels–Alder reaction/cheletropic extrusion of CO sequence. The diaza[7]helicene **12b** was obtained by a final Scholl oxidation of **12a**. The UV-vis spectrum of **12b** displays a characteristic absorption feature of HBC as well as a redshifted peak of NDI, compared with that of the precursor **12a**. The enantiomers of **12b** were separated via chiral HPLC, and their ECD spectra were measured. The mirror-imaged spectra show noticeable Cotton effects into the deep visible (730 nm), albeit with a modest maximum g_{abs} of 0.15%.

Recently, Ravat and coworkers designed and synthesized a new class of helicene diimides (Figure 10.13a) [40]. The synthesis started with Knoevenagel condensations between homophthalimide **13a** and different aromatic dialdehydes to yield photocyclodehydrogenation precursors **13b–d**. The following oxidative

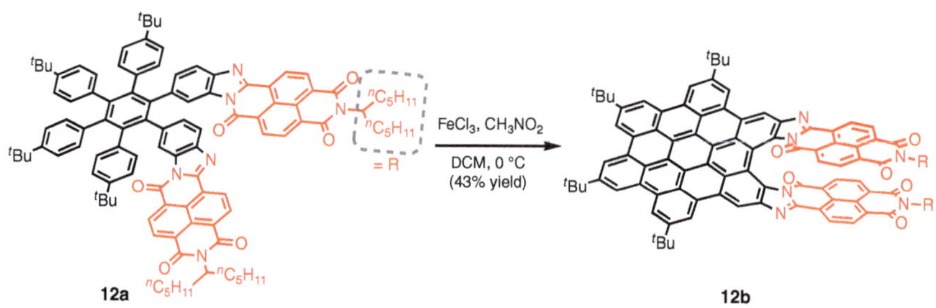

Figure 10.12 Synthesis of diaza[7]helicene **12b**. Source: Adapted from [39].

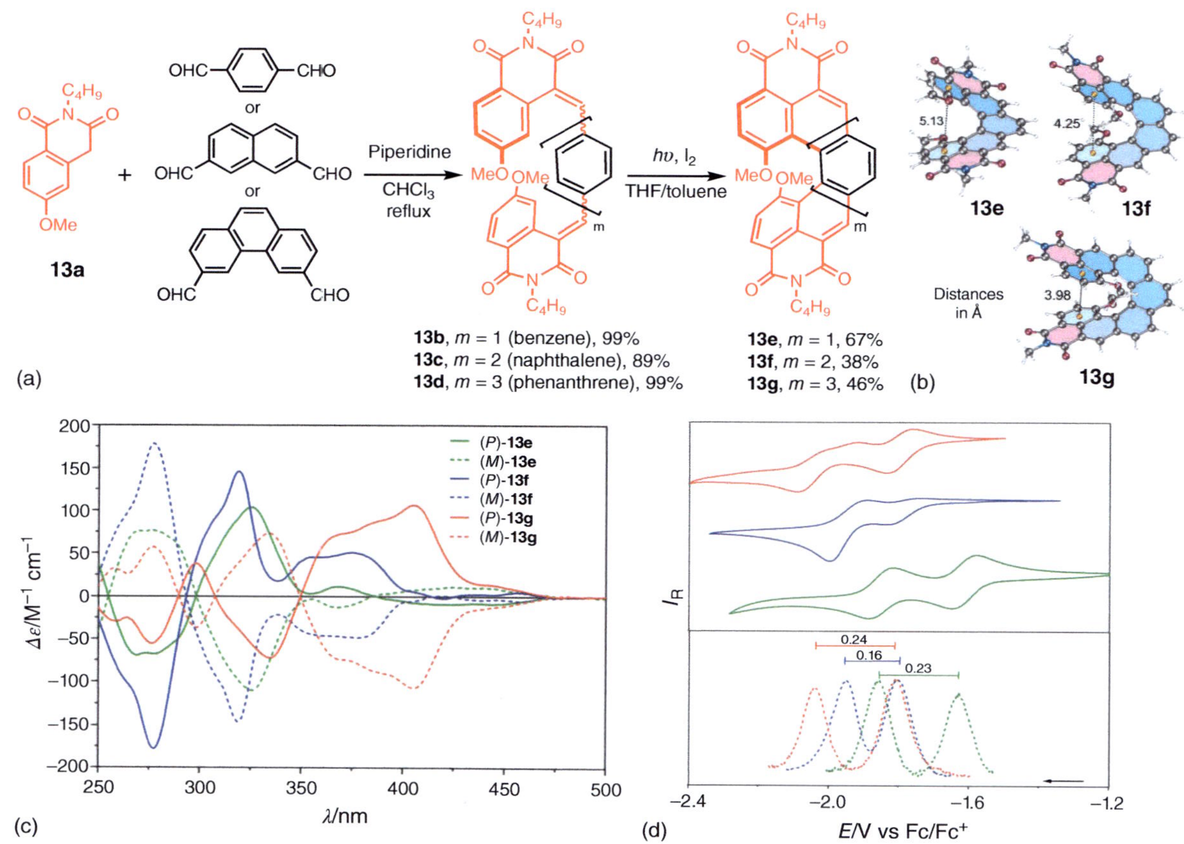

Figure 10.13 (a) Synthesis of [*n*]helicene diimides **13e–g**. (b) DFT-optimized structures of **13e–g**. (c) ECD spectra of (*M*)- and (*P*)-**13e–g**. (d) CV (top) and DPV (bottom) plots of **13e** (green), **13f** (blue), and **13g** (red). Source: [40]/with permission of American Chemical Society.

photocyclization furnished [*n*]helicene diimides **13e–g** (n = 5, 6, and 7). The optical spectroscopy of **13e–g** reveals the conjugation and electronic effects of naphthalimide moieties: they all show a large bathochromic shift in their absorption and emission spectra compared to those of corresponding [*n*]helicenes. This also translates to their chiroptical spectroscopies (Figure 10.13c). Furthermore, as the length of the helical axis increases, the longer helicene diimides **13f,g** exhibit enhanced chiroptical properties with g_{abs} value on the order of ~10^{-2}. Electrochemical analysis sheds additional light on the possible intramolecular through-space conjugation in these molecules (Figure 10.13d). The potential differences between the two reduction peaks are similar (ca. 0.24 V) for **13e** and **13g**, whereas that for **13f** is considerably smaller (0.16 V). The decreased peak gap in **13f** compared to **13e** can be explained by a reduced through-bond interaction between the two terminal imides as the helicene elongates. However, the trend does not continue in **13g**, because the [7]helicene backbone forces the two NDIs in close proximity (see Figure 10.13b for DFT-optimized structures), which provides additional through-space electronic couplings.

10.4.2 Perylene Diimide (PDI)-based Helicenes

PDI, on the other hand, has been utilized extensively to manipulate the electronic and chiroptical properties of helicenes in the past decade. In 2016, one of the initial studies was reported by our laboratory [41]. Our synthetic strategy stemmed from two previous findings: 1) selective bromination can occur at the bay position of *N,N′*-dialkylated PDI [42]; and 2) acene-linked PDIs preferentially fuse at the *peri*-positions of the acene during oxidative intramolecular photocyclizations [43]. As such, we prepared two PDI helicenes, **NPDH** and **APDH**, in two steps from **PDI-Br** (Figure 10.14a). The Suzuki–Miyaura cross coupling of 2,7-Bpin naphthalene or 2,7-Bpin anthracene with **PDI-Br** afforded the acene-bridged PDI dimers **14a** or **14b**, respectively. Next, the regioselective photocyclizations of **14a,b** furnished **NPDH** and **APDH**, respectively, as racemates in nearly quantitative yields. In either case, the alternative regioisomer from the cyclization occurring at the *ortho*-position of the acenes was not detected (● vs. ▲, Figure 10.14a).

The (*M*)- and (*P*)-isomers of **NPDH** and **APDH** are readily separable using chiral HPLC. However, the enantiomers of **APDH** interconvert at room temperature in solution. Conversely, the more severe steric repulsion of the [6]helicene backbone (highlighted in grey, Figure 10.14a) of **NPDH** impedes inversion: no isomerization was identified when a single enantiomer of **NPDH** heated at 250 °C for 1 hour. Notably, the parent [6]helicene racemizes completely in solution in ca. 13 minutes at 220 °C [44]. Accordingly, the chiroptical properties of **NPDH** was studied (Figure 10.14b). The pronounced Cotton effects in the visible region indicates the synergistic combination of the chirality of [6]helicene with the strong absorptivity of the PDIs.

Owing to the strong electron-accepting feature of **PDI**, **NPDH**, and **APDH** readily undergoes chemical or electrochemical reductions. Electrochemical reductions shed light on the electron delocalization in their anions after reductions (Figure 10.14c). The larger separation of the two reduction waves compared to the

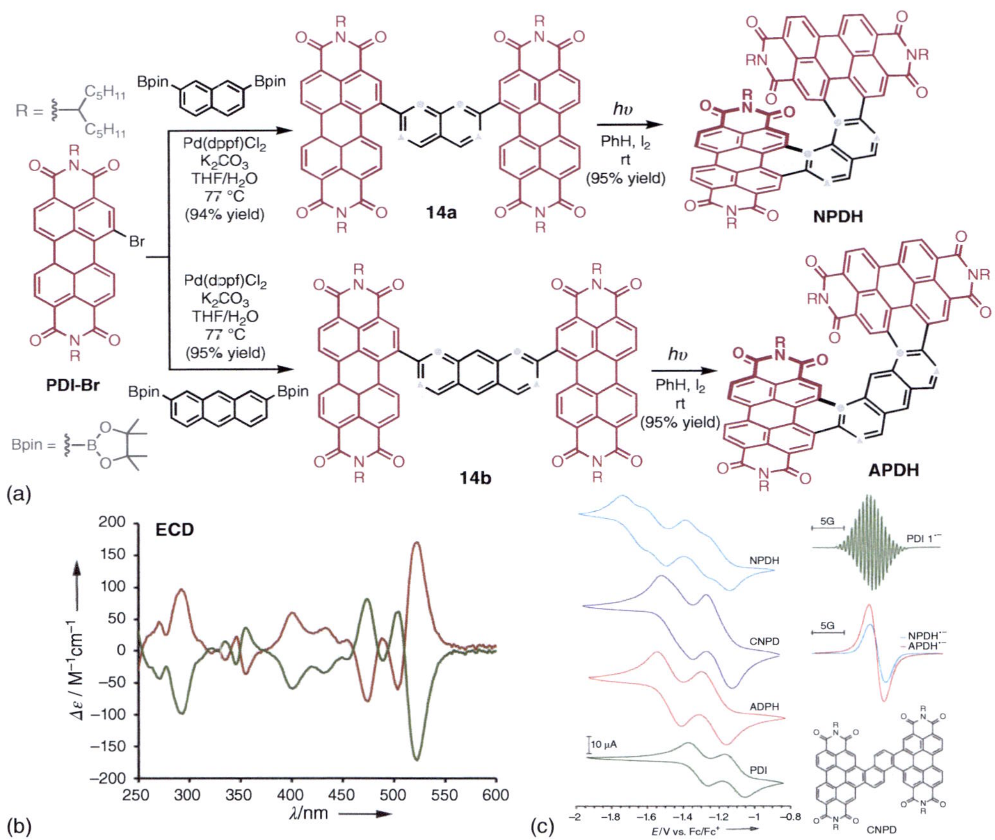

Figure 10.14 (a) Synthesis of **NPDH** and **APDH**. (b) ECD spectra of (*M*)- and (*P*)-**NPDH**. (c) Cyclic voltammograms and continuous-wave EPR spectra of **PDI** (green), **NPDH** (blue), **APDH** (red), and **CNPD** (purple). Source: [41]/with permission of John Wiley & Sons.

parent PDI suggests effective electronic coupling of PDI subunits in both dimers and more strongly in **NPDH** than in **APDH**. The helical structure is essential for such coupling: the CV of another dimer **CNPD** shows a diminished capacity to delocalize electrons relative to reduced **NPDH**. Finally, continuous-wave EPR spectroscopy reinforces that the PDIs in these helicenes share an added electron (Figure 10.14c): no hyperfine splitting is observed in the spectra of **NPDH**$^{\bullet-}$ and **APDH**$^{\bullet-}$ as compared with that of PDI$^{\bullet-}$.

After this initial study, we have demonstrated that [7]helicene-based PDI dimers can be constructed via a similar synthetic approach [45]. Here, a 3,6-disubstituted phenanthrene linker was utilized in the oxidative photocyclization (**15a,b**, Figure 10.15a). Although the regioselectivity was not as perfect as the cases in Figure 10.14, PDI-based [7]helicenes **PPDH** and **PPDH-OPe** were obtained in good yields. It should be noted that this Suzuki coupling/photocyclization sequence has been widely used in constructing PDI-derived nanographenes. In fact, it was used as the key step in the syntheses of essentially all the PDI helicenes that are discussed in this section. Therefore, the synthetic details of later examples are not elaborated. Single crystal X-ray diffraction (SCXRD) analysis uncovered that the racemate of **PPDH** assembles into columns of alternating *M*- and *P*-helices (Figure 10.15b). Besides, the extensive intramolecular overlap of PDI units is also evident.

The ECD spectra of the resolved enantiomers are depicted in Figure 10.15c. The largest transitions of (*M*)- or (*P*)-**PPDH** and -**PPDH-OPe** manifest as a bisignate pair centered at ca. 360 nm, whose intensity is nearly fourfold as that of **NPDH** at the same region. This difference from **NPDH** to **PPDH** deviates from the trend observed for the carbohelicenes: axial elongation of the carbohelicenes (e.g. [6]helicene to [7]helicene) does not significantly change the intensity of the largest Cotton effect [46]. The large increase in ECD from **NPDH** to **PPDH** in the 340–410 nm regime is not simply the result of an increased absorbance of the total light, because they display a considerable disparity in g_{abs} at these wavelengths (0.15% for **NPDH** and 0.68% for **PPDH**).

PDI-embedded [8]helicene and double[8]helicenes were recently disclosed by Wang and coworkers [47]. Besides the mono[8]helicene **P8H**, six stereoisomeric double[8]helicenes including two racemates (**PD8H-6R** and **PD8H-7R**) and two *meso* isomers (**PD8H-6M** and **PD8H-7M**) were accessed by using a [6]helicene precursor **16** as the template (Figure 10.16a). Each structure was unambiguously confirmed by SCXRD analysis of the achiral crystal. Remarkably, they also obtained single crystals of enantiopure (*P*,*P*)- and (*M*,*M*)-**PD8H-6R** (Figure 10.16b). Unlike the racemic crystals of **PD8H-6R**, where π–π stacking dominates, only C–H⋯π contacts are found in the solid-state packing of the two enantiomers. As a result, the two enantiomers stack via a loosely ordered pattern without any π–π interactions.

The optical resolution of **PD8H-6R** and **PD8H-7R** by chiral HPLC resulted in their corresponding enantiopure isomers, each of whose chiroptical properties was examined by CD and CPL spectroscopies (Figure 10.16c). The differences in ECD of **PD8H-6R** and **PD8H-7R** are relatively minor: the largest dissymmetry factor g_{abs} of (*M*,*M*)- or (*P*,*P*)-**PD8H-6R** is slightly higher than that of (*M*,*M*)- or (*P*,*P*)-**PD8H-7R** (1.2% at 411 nm vs. 0.76% at 403 nm). On the contrary, (*M*,*M*)- or (*P*,*P*)-**PD8H-6R**

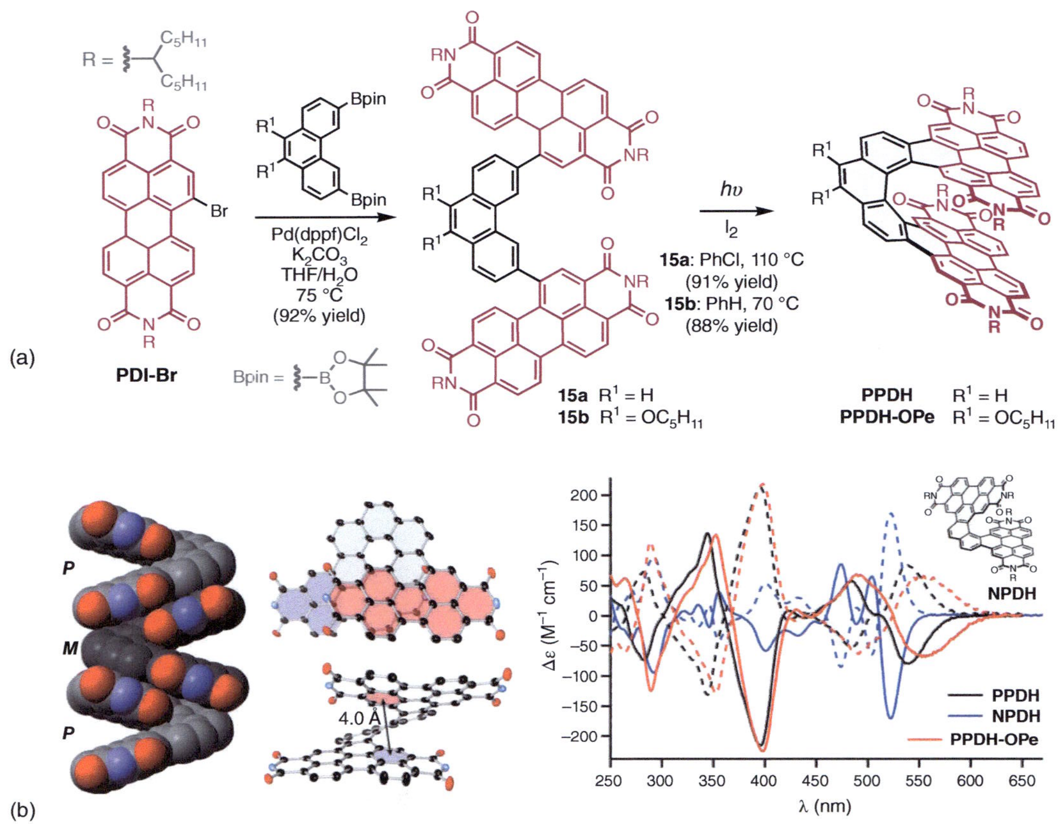

Figure 10.15 (a) Synthesis of **PPDH** and **PPDH-OPe**. (b) SCXRD structure of **PPDH**. (c) ECD comparison between **PPDH** (black), **PPDH-OPe** (red), and **NPDH** (blue); the spectra of the other enantiomer are shown in dashed lines. Source: [45]/Royal Society of Chemistry/CC BY 3.0.

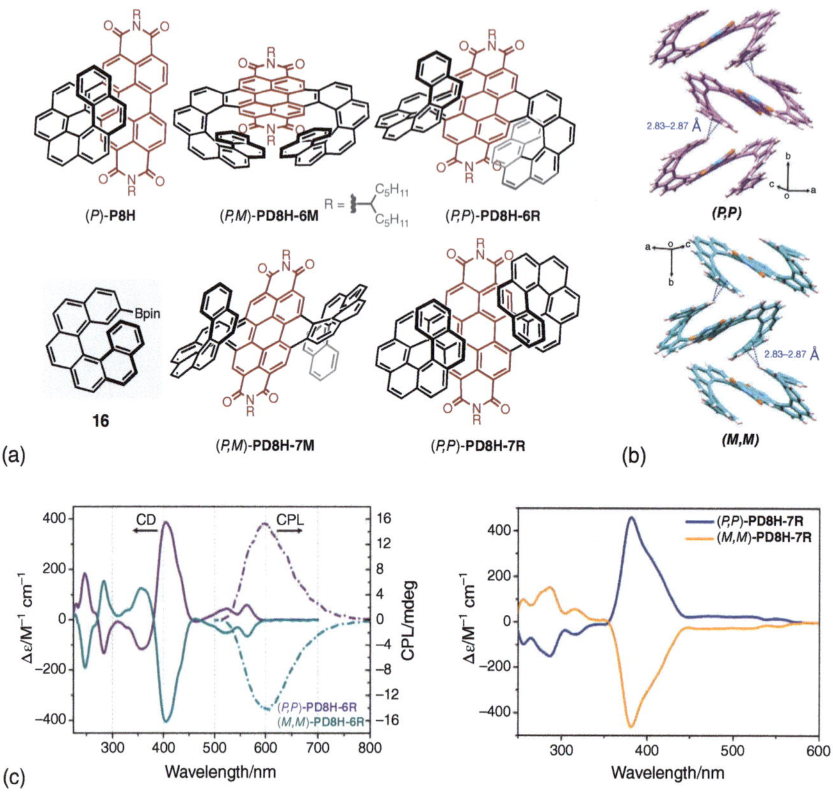

Figure 10.16 (a) The structures of PDI-embedded [8]helicene **P8H** and double[8]helicenes **PD8H**; only one enantiomer is shown for each structure. (b) Molecular packing of optically pure (*P*,*P*)- (top) and (*M*,*M*)-**PD8H-6R** (bottom). (c) Chiroptical properties of **PD8H-6R** (ECD and CPL) and **PD8H-7R** (ECD). Source: [47]/with permission of American Chemical Society.

exhibits a considerable fourfold increase in g_{lum} of CPL compared with that of (*M*,*M*)- or (*P*,*P*)-**PD8H-7R** (CPL not shown). All in all, (*M*,*M*)- or (*P*,*P*)-**PD8H-6R** is a superior candidate for applications in chiral optoelectronics.

The presence of PDI can even stabilize [5]helicene, which racemizes at room temperature. In 2020, Hirsch and coworkers described enantiomerically stable PDI-based [5]helicenes [48]. The structures are shown in Figure 10.17a. As expected, the [4]-helicene moiety of **17a** and **17b** isomerizes easily at room temperature. However, the inversion barrier of the [5]-helicene subunit in **17a** and **17b** is large enough to isolate the (*M*) and (*P*) isomers. Surprisingly, the relative position of the non-chiral [4]-helicene plays an important role in the ECD spectra (Figure 10.17b). The absolute configuration of each enantiomer is assigned by comparison with calculated ECD spectra. The ECD spectra of (*M*) and (*P*)-**17b** show only two major Cotton effects with $|\Delta\varepsilon|$ of only 88 and 36 M^{-1} cm^{-1}, respectively. On the other hand, (*M*) and (*P*)-**17a** exhibit five Cotton effects as well as increased $|\Delta\varepsilon|$ values of up to 148 M^{-1} cm^{-1}. The g_{abs} values reach up to 0.33% for (*M*) and (*P*)-**17a** and only 0.22% for (*M*) and (*P*)-**17b**.

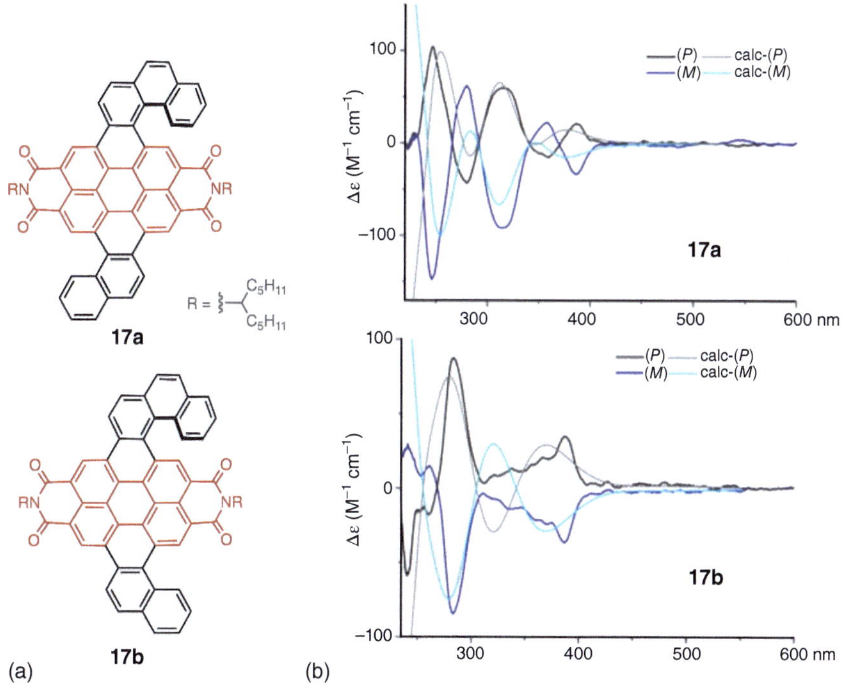

Figure 10.17 (a) The structures of **17a** and **17b**. (b) ECD spectra of **17a,b**. Source: [48]/with permission of John Wiley & Sons.

Another important strategy to construct complex helicene architectures is fusion of multiple helicenes into a nanographene core. The Wang lab has exploited this approach with PDI in several studies. For example, they attached four PDI derivatives (red) to a central dibenzocoronene diimide core (black) [49]. These extended nanographenes, **18a** and **18b,** feature six [5]helicene units (Figure 10.18a). Although there are overall 28 possible stereoisomers of **18a**, among which 14 pairs are enantiomers to one another, only one pair of enantiomers was formed selectively. Further annulation of the outer bay region of the peripheral four PDI units afforded **18b**, whose single crystal was obtained. SCXRD analysis of **18b** reveals that it adopts a D2-symmetric "four-bladed propeller" conformation consisting of another double "three-bladed propeller" (Figure 10.18b). In the solid-state, the two enantiomers, (*P*,*P*,*P*,*P*,*P*,*P*) and (*M*,*M*,*M*,*M*,*M*,*M*) shown in green and purple, respectively, are stacked via intermolecular π–π overlapping between the outer PDIs. This particular conformer is thermally stable, and no isomerization was seen when heated at 200 °C. As such, the chiroptical properties of **18a** were investigated (Figure 10.18c). Optically pure **18a** exhibits three predominant Cotton effects > 400 nm with g_{abs} of approximately 0.2%.

Likewise, Wang and coworkers utilized a nonplanar PAH core – corannulene (**19**) – to build penta-[6]helicene superstructures **CRP-1** and **CRP-2** [50]. These two pairs of enantiomers were formed in a ca. 1:1 ratio. The structures were assigned by SCXRD analysis. The isomer with higher symmetry, **CRP-1**, is calculated to be

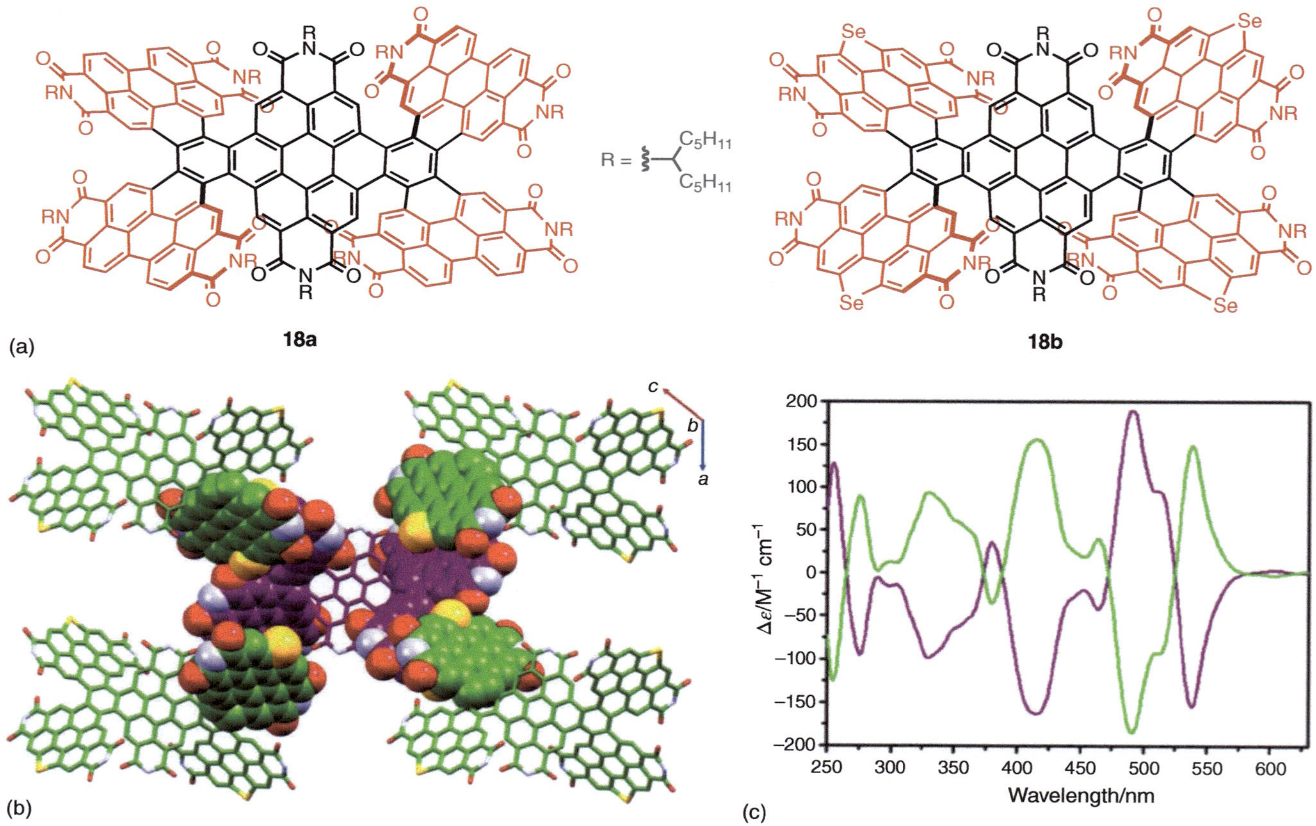

Figure 10.18 (a) Two nanographenes with dual-core, sixfold [5]helicenes **18a,b**. (b) Crystal packing of **18b**. (c) CD spectra of **18a** in $CHCl_3$ (the spectra of the two enantiomers are shown in green and violet). Source: [49]/with permission of John Wiley & Sons.

slightly more stable than **CRP-2** by 5.53 kcal mol^{-1}. However, a solution of **CRP-2** did not show any sign of change after heating at 220 °C. Thus, both **CRP-1** and **CRP-2** are enantiomerically stable, and the ECD of each enantiomer is examined (Figure 10.16b). The two diastereoisomers manifest nearly identical shape of the ECD, and, in fact, their ECD share the same spectral characteristics with **NPDH** (cf. Figure 10.14b). It should be noted that each PDI-[6]helicene unit in (*P*,*P*,*P*,*P*,*P*)-**CRP-1** is actually a (*P*)-**NPDH**. Additionally, the signal intensity for (+)- or (−)-**CRP-1** is 1.5-fold of that for (+)- or (−)-**CRP-2** across the entire spectra, indicating that the one **NPDH** subunit with a different chirality within **CRP-2** is cancelling the overall signal arose from other **NPDH** units.

The branched alkyl chains on the nitrogen atoms of PDIs support the tubular units and the lamellar arrangement in the honeycomb lattice of **CRP-1**, in which two chiral honeycomb layers alternately stacks. To investigate the charge transport properties of the honeycomb lattice, hexagonal microcrystals of **CRP-1** were prepared by *in situ* drop-coating method (Figure 10.19c). A transistor device based on individual hexagonal microcrystal with six electrodes and channels was fabricated to examine the charge transport properties of the crystal along different directions. The transistors along the six channels all show electron mobility at 10^{-4} cm^2 V^{-1} s, which implies that without any π–π stacking, the honeycomb lattice could also facilitate electron transport.

As demonstrated earlier, leveraging nanographene core in extended helicenes provided impressive scaffolds like **18a,b** and **CRP-1,2**, in which interesting electronic properties and molecular packing patterns do arise. However, this approach often leads to merely comparable chiroptical properties (e.g. ECD intensity) relative to that of simply one of the helicene components (cf. Figures 10.18c and 10.19b vs. Figures 10.14b, 10.15c, and 10.16c). In contrast, macromolecular helices assembled by multiple conjugated helicenes show enhanced ECD response [51]. Therefore, we suspected that nanoribbons adopting similar helix-of-helicenes structures would exhibit amplified ECD, as well as other properties befitting chiral electronic materials [52].

To this end, we developed a chiral and shape-persistent nanoribbon architecture **NP3H** (Figure 10.20a) [53]. Specifically, it is resulted from fusion of two naphthalene subunits with three PDI monomers, which can also be regarded as a homolog of **NPDH**. SCXRD verified the π-helix-of-helicenes structure of **NP3H**. The considerable compression of the helix results in extensive π-to-π overlap between the PDI subunits, preventing inversion to give the *meso* isomer. As anticipated, resolution of the enantiomers by preparative chiral HPLC showed no evidence for the presence of other stereoisomers. In addition, effective intermolecular π–π interactions mediate the formation of supramolecular columns of **NP3H** in the solid state (Figure 10.20b). These heterochiral columns consist of alternating (*M*)- and (*P*)-**NP3H**.

The considerable ECD of (*M*)- and (*P*)-**NP3H** distinguishes this π-helix from other helicenes as introduced previously (Figure 10.20c, top). For instance, the largest Cotton effects well exceed the maximum in ECD (ca. 260 M^{-1} cm^{-1}) encountered upon lengthening the carbo[n]helicenes [46]. This bisignate pair, centered at 390 nm, manifest $|\Delta\varepsilon|$ of 620 M^{-1} cm^{-1} and 820 M^{-1} cm^{-1} at 377 and 407 nm,

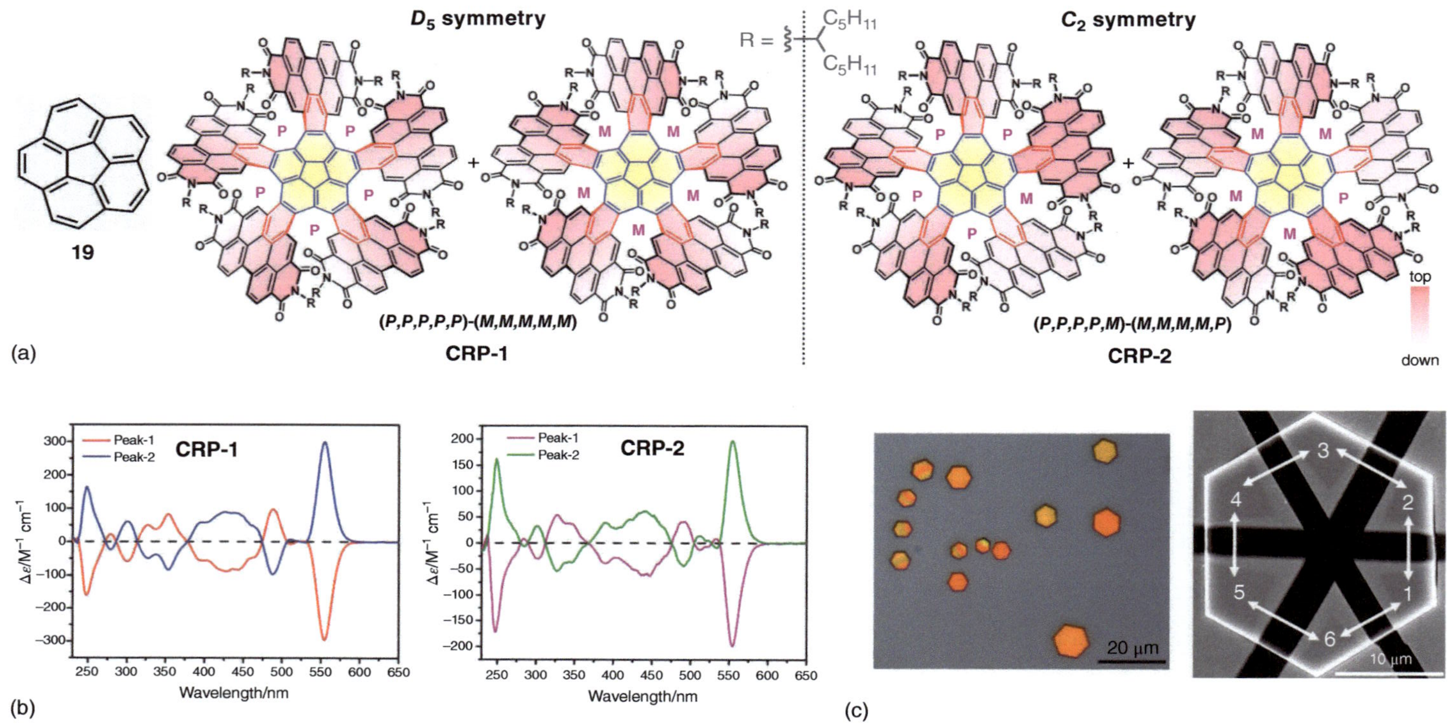

Figure 10.19 (a) Corannulene-based penta-[6]helicenes **CRP-1** and **CRP-2**. (b) ECD spectra of **CRP-1** (blue and red) and **CRP-2** (green and magenta). (c) Optical microscopy image of microcrystals of **CRP-1** and SEM image of a transistor device. Source: [50], Figure 03, p. 04/with permission of American Chemical Society.

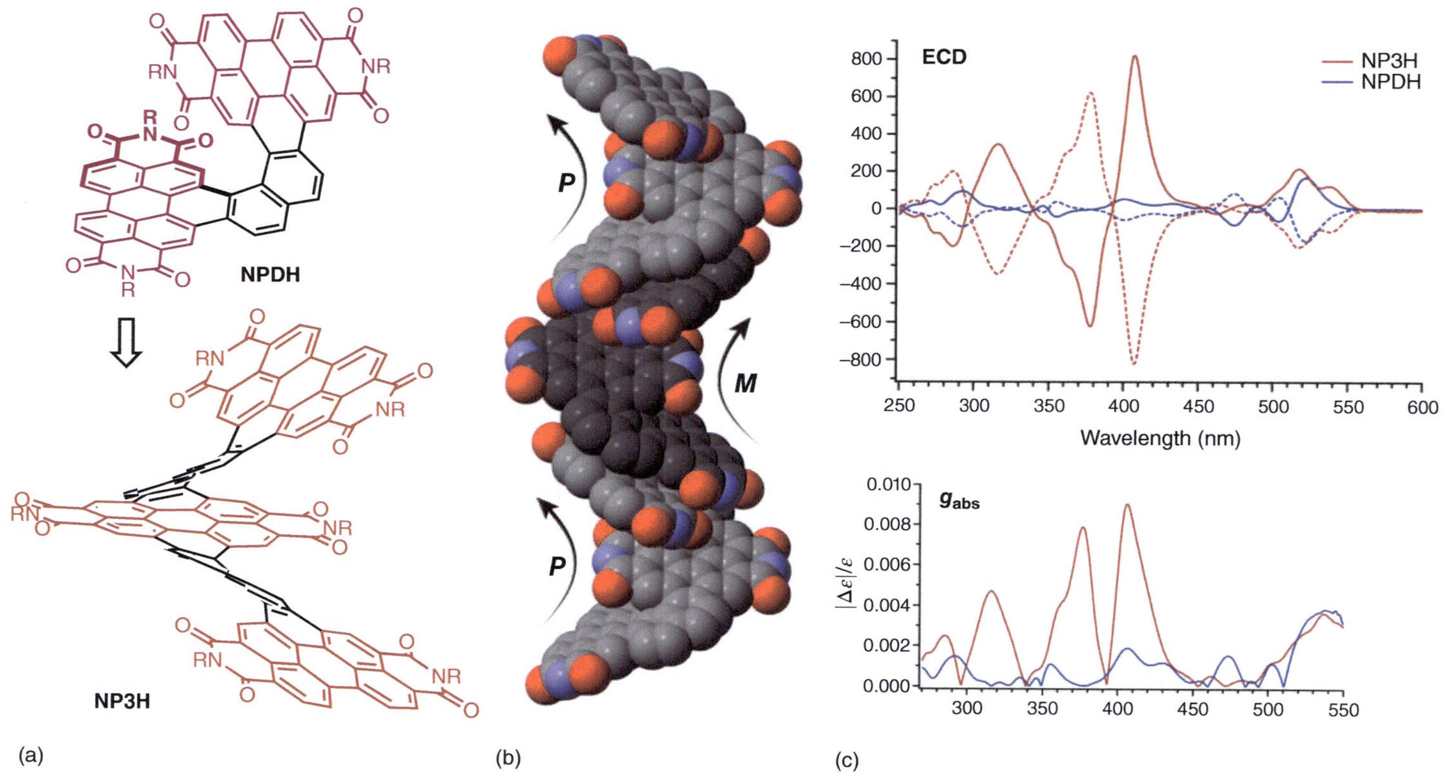

Figure 10.20 (a) **NP3H**, a chiral nanoribbon and homolog of **NPDH**. (b) Crystal packing of **NP3H**. (c) ECD spectra and dissymmetry factors of **NP3H** in comparison with **NPDH**. Source: [53]/with permission of American Chemical Society.

respectively. In stark contrast, (*M*)- or (*P*)-**NPDH** is significantly less ECD-active as exemplified by the corresponding couplet at 380 nm with $|\Delta\varepsilon|$ of only 41 M^{-1} cm^{-1} and 56 M^{-1} cm^{-1} at 355 and 401 nm, respectively. This 15-fold enhancement in $\Delta\varepsilon$ is not observed between the absorption spectra of **NPDH** and **NP3H**. Indeed, a comparison of the g_{abs} (Figure 10.20c, bottom) from 270–410 nm, once again, reveals significant amplification of ECD for **NP3H** relative to **NPDH**. For example, **NP3H** shows $|\Delta\varepsilon|/\varepsilon$ of 0.79% at 377 nm and 0.89% at 407 nm, a 7.2-fold and 5.9-fold increase over those transitions of **NPDH** at 355 and 401 nm. As such, elongating the helix from **NPDH** to **NP3H** increases both $|\Delta\varepsilon|$ and g_{abs}. This infers that the specific structural arrangements of PDIs and helicenes in **NP3H** productively augment the ability to selectively interact with only one hand of the chiral light.

Santoro and coworkers used several readily accessible computational tools to deconvolute the mechanism for the unusual chiral amplification from **NPDH** to **NP3H** [54]. First, TD-DFT, at the CAM-B3LYP/6-31G(d)/GD3BJ level of theory, calculations reproduce the experimental spectra of both **NPDH** and **NP3H**; thus the absolute configuration of each enantiomer is assigned (e.g. Figure 10.21a). Next, each major excitation in the ECD is carefully analyzed (see sticks in Figure 10.21a). Note that normally the TD-DFT electronic states are described by the excitations between Kohn–Sham (KS) molecular orbitals (MOs). However, the involved MOs for **NP3H** are numerous, and these states arise from different transitions with comparable weights: in many cases the major KS transition only contribute to ca. 30% of the total weight. Therefore, the relevant excited states are analyzed based on the difference in their electronic densities (DED) with respect to the ground state.

The plots of DED are two-color surfaces: red indicates a density increase, and blue means a density depletion. The DED maps of two major states for the largest ECD couplet of **NP3H** almost localize exclusively on the two carbo[6]helicene moieties (Figure 10.21b). This indicates the origin of the enhancement in (*M*)-**NP3H** is due to the coupling between the states on the two carbo[6]helicenes, because only a single [6]helicene motif, like (*M*)-**NPDH** along does not exhibit strong ECD signal. The calculated ECD of several model molecules, each of which is merely an isolated portion of (*M*)-**NP3H** without reoptimization of the structure (Figure 10.21c), reinforces the importance of exciton-like interactions between helicene subunits. Both **Fragment 1** and **Fragment 2** exhibit a similarly intense ECD couplet, whereas $\mathbf{PDI_3}$ is predicted to have only a very weak ECD response at $\lambda < 430$ nm. Moreover, the strong ECD of **Fragment 2** proves that the π-conjugation is not essential to the chiroptical enhancement, which further supports an exciton-like coupling mechanism.

Our laboratory further capitalized on the exciton-like coupling in **NP3H** to design even better chiral materials [55]. Two additional PDI units are installed on the original **NP3H** and **NPDH** core structures through an ethylene bridge, respectively (Figure 10.22a). The newly formed helicenes **WH[6][6]** and **WH[6]** features two outer double-[4]helicenes (highlighted in orange), which rapidly waggles in solution at ambient temperature. Once again, enantiomerically pure **WH[6][6]** – a structure that is reminiscent of **NP3H** – shows amplified ECD in virtually every transition with respect to chiral **WH[6]**, whose structure is based on **NPDH** (Figure 10.22b). More importantly, (*M*)- and (*P*)-**WH[6][6]** elicit superior

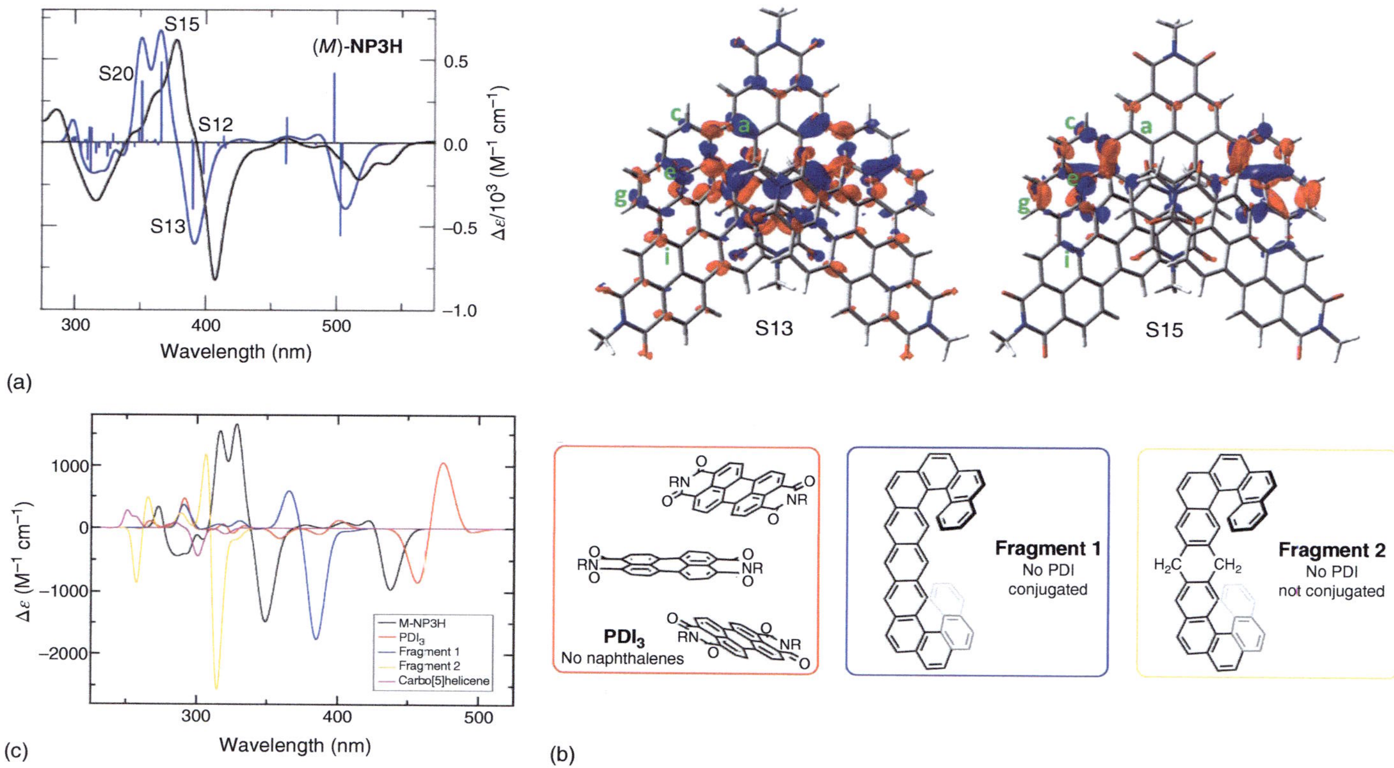

Figure 10.21 (a) Computed ECD spectra of (*M*)-**NP3H**. (b) Calculated difference in electronic density with respect to the ground state for S13 and S15 of (*M*)-**NP3H**. (c) Comparison of the ECD computed for (*M*)-**NP3H** and fragments of (*M*)-**NP3H**. Source: [54]/with permission of American Chemical Society.

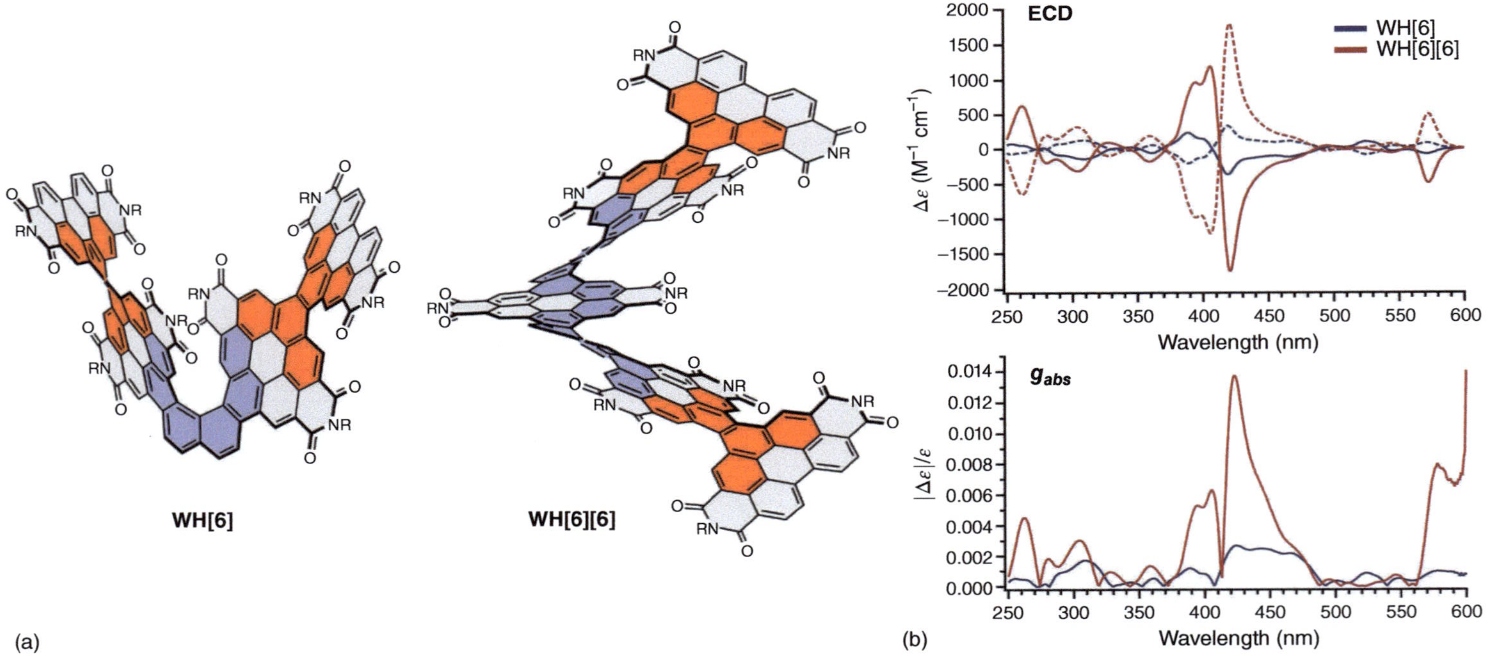

Figure 10.22 (a) Waggling helicenes **WH[6]** and **WH[6][6]**. (b) Comparison of the ECD spectra and g factors between **WH[6]** and **WH[6][6]**. Source: [55]/with permission of American Chemical Society.

chiroptical response in the visible region to the original (*M*)- and (*P*)-**NP3H** (cf. Figure 10.20c). Namely, the largest anisotropy factor of **WH[6][6]** is now over 1.3%, a nearly 1.5-fold increase from that of **NP3H**.

Computational conformation analysis by DFT identifies the dominant conformers of the waggling helicenes: only three most stable conformations represent effectively 100% of the solution ensemble at room temperature. The predicted ECD spectra for each conformer by TD-DFT calculations reveal that the ECD of **WH[6]** and **WH[6][6]** originates principally from the conformations whose [4]helicenes orient with the same handedness of the global helix – the chirality of the **NP3H** core. TD-DFT calculations also reproduce the increase in $|\Delta\varepsilon|$ from **WH[6]** to **WH[6][6]**, which implies the disparities in the intensity of their ECD do not stem from conformational differences. Instead, the simulated largest ECD bands in **WH[6][6]** (i.e. the bands from 380–430 nm) are insensitive upon conformational changes. This suggests that the rigid chiral **NP3H** substructure is the source of the observed chiroptical amplification.

In early 2021, Lin and coworkers demonstrated that π-extended PDI-heterohelicenes can be integrated into optoelectronic devices for broadband circularly polarized light detection [56]. The PDI-azahelicene **20** was designed by merging a distorted π-system – a prerequisite for strong coupling with chiral light – with a planar π-stacking motif, which is necessary for efficient charge transport (Figure 10.23a). Once each enantiomer was isolated by chiral HPLC, its photophysical properties were studied. It is noteworthy that **20** is CD active even in the NIR region (~780 nm). Besides, it exhibits high dissymmetry with the largest g_{abs} factor of 1.4% at 628 nm in solution. To test the applicability of **20** in devices, thin films of **20**, forged by a thermal evaporation process, were investigated using ECD spectroscopy. The thin films show similar CD behavior to the solution with slightly redshifted sharp peaks. Moreover, rotating and flipping the sample do not change CD spectra at all, suggesting that the chirality in the thin film stems from the intrinsic chirality of the helicenes, rather than from linear dichroism and birefringence [57].

To further investigate the potential of **20** as chiral organic semiconductors, thin-film-based OFETs were fabricated. Each enantiomer shows almost identical electrical properties. Notably, the optimum *P*/*M* films display ambipolar field-effect behaviors with typical V-shaped transfer curves, which exhibit high hole and electron mobility of 2.1×10^{-3} and 1.7×10^{-3} $cm^2\ V^{-1}\ s^{-1}$, respectively. Moreover, the photocurrent of phototransistor devices based on (*P*)-**20** upon NIR light irradiations were also measured: considerable changes in current response and threshold voltage (V_T) were observed in both p- and n-type operating transfer curves. Naturally, the photocurrent and V_T shift increase with more intense NIR light (Figure 10.23b). Finally, selective detection of CPL was achieved in OFETs of optically active **20** by recording the real-time photocurrent when illuminated by CPL. The p-type mode was found to give more sensitive CPL detection (Figure 10.23c). (*P*)-**20** based OFETs show higher photocurrent toward left-handed CPL, whereas (*M*)-isomer-based device preferentially interacts with right-handed CPL, consistent with the selectivity observed in the CD spectra of (*P*) or (*M*)-**20**.

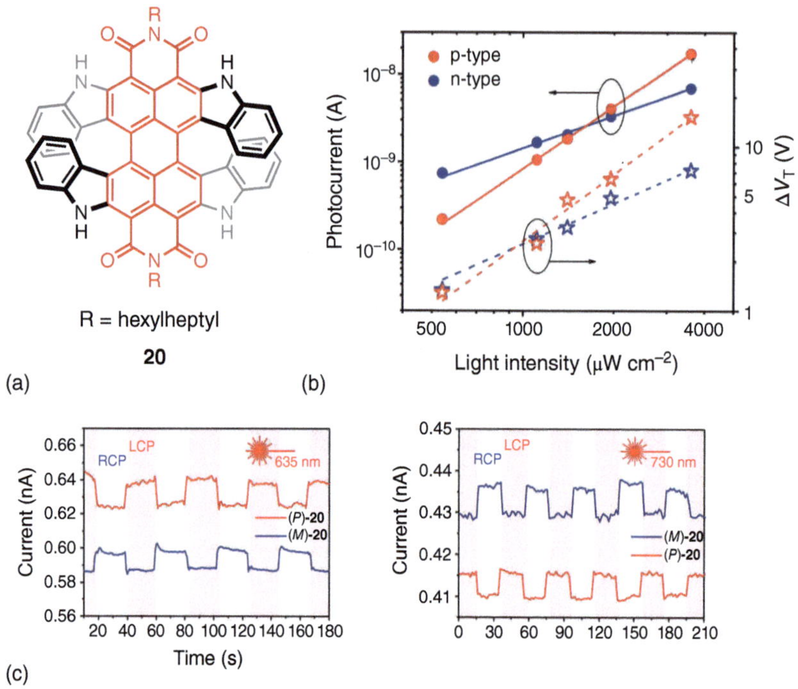

Figure 10.23 (a) PDI-embedded double-aza[7]helicene **20**. (b) Photocurrent and V_T of OFETs based on (*P*)-**20** under the NIR light irradiations. (c) Real-time photocurrent of (*P*)/(*M*)-**20**-based OFETs under CPL (λ = 635 or 730 nm) in p-type mode. Source: [56]/Springer Nature/CC BY 4.0.

Given the extraordinary chiroptical properties of **NP3H** and related structures (cf. Figures 10.20 and 10.22), we further pushed the length of the rigid helix. In our most recent study, three discrete, long, helicene nanoribbons **NP4H**, **NP5H**, and **NP5H-iso** were synthesized (Figure 10.24a) [58]. **NP5H** is also one of the longest discrete helicenoid structures ever reported [8a, 9c]: its helical backbone, from one end to the other, comprises four axially fused carbo[6]helicenes (i.e. 24 benzene rings). The structure of **NP5H** was also verified by SCXRD analysis. In the solid state, the same enantiomer constitutes a supramolecular chiral column with an intermolecular π–π stacking of ca. 3.4 Å.

As previously mentioned, multi-chromophores with chiral spatial interactions between each subunit (i.e. exciton couplings) often give rise to larger ECD, especially when the chromophores are degenerate. Accordingly, the many identical chromophores (e.g. PDI and [6]helicenes) in **NP4H** and **NP5H** enhance exciton-like couplings, which yield massive chiroptic amplification as compared with the shorter analogs (i.e. **NPDH** and **NP3H**, *vide supra*). Indeed, compared with (*P*)-**NPDH**, the chiroptical properties are greatly amplified in (*P*)-**NP4H** and (*P*)-**NP5H** (Figure 10.24b). In particular, (*P*)-**NP5H** exhibits the largest molar ECD (> 1900 M^{-1} cm^{-1}) in the visible spectrum ever recorded for a molecule, with g_{abs} up to 2.2% [59]. The difference in ECD intensity between (*P*)-**NP5H** and

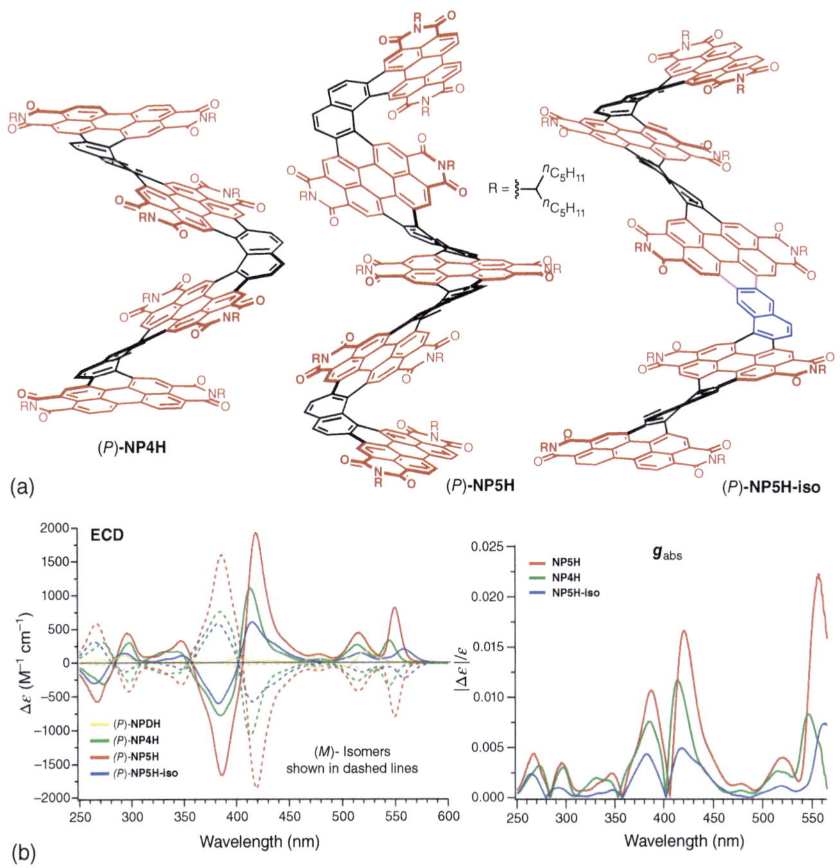

Figure 10.24 (a) Helicene nanoribbons (*P*)-**NP4H**, -**NP5H**, and -**NP5H-iso**. (b) Comparison of ECD spectra and g_{abs} of the helicene nanoribbons. Source: [58]/with permission of American Chemical Society.

(*P*)-**NP5H-iso** is also noteworthy. Although they both consist of five PDI units, perturbed symmetry in the latter structure disfavors exciton-like interactions. As a result, the ECD and g_{abs} of **NP5H** completely overshadows that of **NP5H-iso** (Figure 10.24b). The outstanding chiroptical properties, large molar absorptivity, and excellent fluorescence quantum yield (0.68) of **NP5H** makes it a promising material candidate for chiral optoelectronics. Finally, theoretical calculations reaffirm the role of exciton-like couplings in ECD enhancement, and even larger and more selective chiroptical responses are predicted to ensue from further axial elongation.

10.4.3 Terrylene-based Helicenes

As a result of the relatively challenging synthesis of longer rylene diimides, to the best of the authors' knowledge, there is only one report of terrylene diimide (TDI)-based helicenes. In 2011, Müllen and coworkers prepared

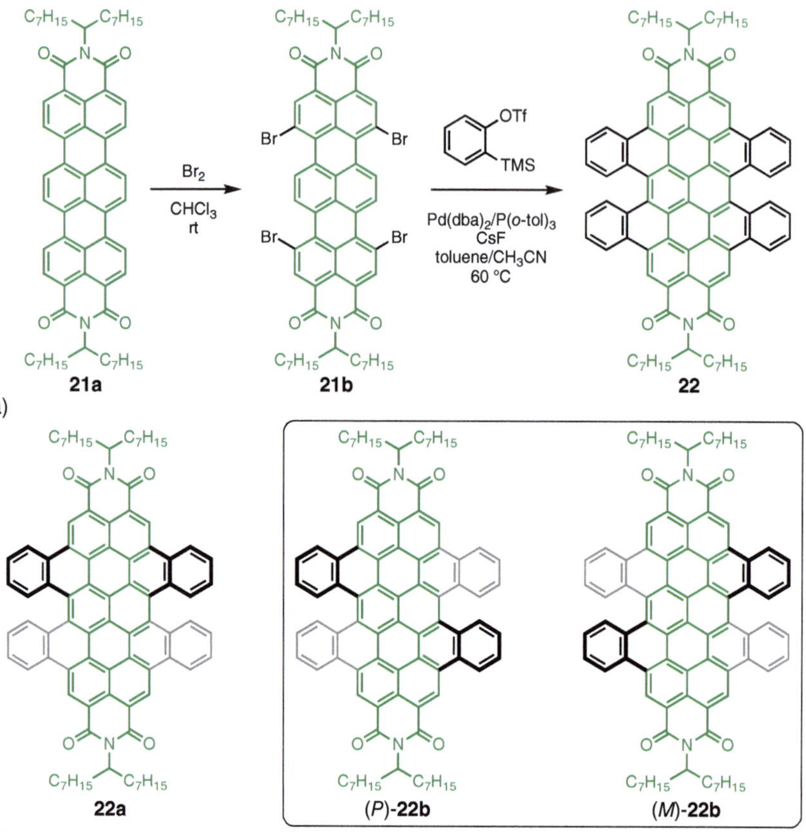

(a)

(b)

Figure 10.25 (a) Synthesis of terrylene diimide-based double[5]helicene **22**. (b) Three stereoisomers of **22**. Source: Lütke Eversloh et al. [60].

[5]helicene-containing TDI **22** [60]. The synthesis of **22** commenced with a selective bromination of the established terrylene diimide **21a** to give tetrabrominated TDI **21b** (Figure 10.25a). The subsequent key step was a fourfold Pd-catalyzed annulation of **21b** with *o*-trimethylsilyl-triflate benzene, which generated benzyne under the action of cesium fluoride. The mechanism of each benzannulation involves an oxidative addition of the TDI–Br bond with Pd (0), a migratory insertion to the benzyne, and the final cyclization to regenerate the catalyst.

Three possible stereoisomers were detected for **22** (Figure 10.25b). Among them, **22a** and **22b** are two diastereoisomers, which can be separated by standard chromatography on silica gel. Unlike **22a**, in which there is a mirror plane within the structure, the structure of **22b** is chiral: the two enantiomers, (*P*)- and (*M*)-**22b** were isolated via chiral HPLC, even though they are just [5]helicenes. A free activation enthalpy of $121.4 \pm 0.3\,\text{kJ mol}^{-1}$ is estimated at 80 °C. Despite a smaller optical bandgap of TDI compared with PDI [61], the UV-vis profile of **22b** does not show redshifted band compared with PDI-helicenes discussed earlier. In addition, the CD spectra (not shown here) of (*M*)- and (*P*)-**22b** only display one predominant

Cotton effect above 400 nm with only a weak molar ECD value of < 20 M^{-1} cm^{-1}. The weak ECD intensity of **22b** at longer wavelength (500–600 nm) may reflect the relatively planar nature of the TDI core.

10.5 Conclusion and Outlook

Since the discovery of helicene over a century ago, the rapid growth of synthetic methods has enabled the access to myriad helicene structures. As a particularly interesting type of helicenes, helicene-based electron acceptors invert the electronic nature of classic helicenes, which typically are electron donors. More recently, the structural complexity of helicene-based electron acceptors has been taken to the next level: many π-extended, helicenic frameworks of electron-accepting nature show promising chiroptical properties. This chapter covers some recent, representative examples of these unique helicenes, from the standpoint of materials synthesis, to fundamental understanding of chiral properties, and eventually to applications in chiral optoelectronic devices. Now, the ceiling of helicenic electron acceptors is no longer limited only by their synthetic inaccessibility: virtually any target molecule can be tackled by synthetic chemists. Instead, the current major challenge is how to (i) design molecular systems with large *g*-factors, (ii) integrate these impressive molecules into next-generation devices, and (iii) expand the breadth of applications that utilize chiral materials. Nonetheless, the increasing number of breakthroughs in the last decade suggests that the future of helicene-based electron acceptors has never looked brighter.

References

1 (a) Gingras, M. (2013). One hundred years of helicene chemistry. Part 1: non-stereoselective syntheses of carbohelicenes. *Chem. Soc. Rev.* 42 (3): 968–1006. (b) Gingras, M., Félix, G., and Peresutti, R. (2013). One hundred years of helicene chemistry. Part 2: stereoselective syntheses and chiral separations of carbohelicenes. *Chem. Soc. Rev.* 42 (3): 1007–1050. (c) Gingras, G. (2013). One hundred years of helicene chemistry. Part 3: applications and properties of carbohelicenes. *Chem. Soc. Rev.* 42 (3): 1051–1095. (d) Shen, Y. and Chen, C.F. (2012). Helicenes: synthesis and applications. *Chem. Rev.* 112 (3): 1463–1535. (e) Mori, T. (2021). Chiroptical properties of symmetric double, triple, and multiple helicenes. *Chem. Rev.* 121 (4): 2373–2412.

2 Verbiest, T., Van Elshocht, S., Kauranen, M. et al. (1998). Strong enhancement of nonlinear optical properties through supramolecular chirality. *Science* 282 (5390): 913–915.

3 (a) Isla, H. and Crassous, J. (2016). Helicene-based chiroptical switches. *Comptes Rendus Chim.* 19 (1–2): 39–49. (b) Kalachyova, Y., Guselnikova, O., Elashnikov, R. et al. (2019). Helicene-SPP-based chiral plasmonic hybrid structure: toward direct enantiomers SERS discrimination. *ACS Appl. Mater. Interfaces* 11 (1): 1555–1562.

4 Aillard, P., Voituriez, A., and Marinetti, A. (2014). Helicene-like chiral auxiliaries in asymmetric catalysis. *Dalt. Trans.* 43 (41): 15263–15278.

5 Kiran, V., Mathew, S.P., Cohen, S.R. et al. (2016). Helicenes - a new class of organic spin filter. *Adv. Mater.* 28 (10): 1957–1962.

6 (a) Brandt, J.R., Salerno, F., and Fuchter, M.J. (2017). The added value of small-molecule chirality in technological applications. *Nat. Rev. Chem.* 1: 45. (b) Pop, F., Zigon, N., and Avarvari, N. (2019). Main-group-based electro- and photoactive chiral materials. *Chem. Rev.* 119 (14): 8435–8478.

7 Majewski, M.A. and Stępień, M. (2019). Bowls, hoops, and saddles: synthetic approaches to curved aromatic molecules. *Angew. Chem. Int. Ed.* 58 (1): 86–116.

8 (a) Mori, K., Murase, T., and Fujita, M. (2015). One-step synthesis of [16]helicene. *Angew. Chem. Int. Ed.* 54 (23): 6847–6851. (b) Nakakuki, Y., Hirose, T., Sotome, H. et al. (2018). Hexa- *peri* -hexabenzo[7]helicene: homogeneously π-extended helicene as a primary substructure of helically twisted chiral graphenes. *J. Am. Chem. Soc.* 140 (12): 4317–4326. (c) Evans, P.J., Ouyang, J., Favereau, L. et al. (2018). Synthesis of a helical bilayer nanographene. *Angew. Chem. Int. Ed.* 57 (23): 6774–6779. (d) Fernández-García, J.M., Evans, P.J., Medina Rivero, S. et al. (2018). π-Extended corannulene-based nanographenes: selective formation of negative curvature. *J. Am. Chem. Soc.* 140 (49): 17188–17196. (e) Cruz, C.M., Castro-Fernández, S., Maçôas, E. et al. (2018). Undecabenzo[7]superhelicene: a helical nanographene ribbon as a circularly polarized luminescence emitter. *Angew. Chem. Int. Ed.* 57 (45): 14782–14786. (f) Kiel, G.R., Patel, S.C., Smith, P.W. et al. (2017). Expanded helicenes: a general synthetic strategy and remarkable supramolecular and solid-state behavior. *J. Am. Chem. Soc.* 139 (51): 18456–18459. (g) Nakakuki, Y., Hirose, T., and Matsuda, K. (2018). Synthesis of a helical analogue of kekulene: a flexible π-expanded helicene with large helical diameter acting as a soft molecular spring. *J. Am. Chem. Soc.* 140 (45): 15461–15469.

9 (a) OuYang, J. and Crassous, J. (2018). Chiral multifunctional molecules based on organometallic helicenes: recent advances. *Coord. Chem. Rev.* 376: 533–547. (b) Dhbaibi, K., Favereau, L., and Crassous, J. (2019). Enantioenriched helicenes and helicenoids containing main-group elements (B, Si, N, P). *Chem. Rev.* 119 (14): 8846–8953. (c) Nejedlý, J., Šámal, M., Rybáček, J. et al. (2017). Synthesis of long oxahelicenes by polycyclization in a flow reactor. *Angew. Chem. Int. Ed.* 56 (21): 5839–5843. (d) Fujikawa, T., Segawa, Y., and Itami, K. (2017). Laterally π-extended dithia[6]helicenes with heptagons: saddle-helix hybrid molecules. *J. Org. Chem.* 82 (15): 7745–7749. (e) Nakamura, K., Furumi, S., Takeuchi, M. et al. (2014). Enantioselective synthesis and enhanced circularly polarized luminescence of S-shaped double azahelicenes. *J. Am. Chem. Soc.* 136 (15): 5555–5558. (f) Kato, K., Furukawa, K., Mori, T., and Osuka, A. (2018). Porphyrin-based air-stable helical radicals. *Chem. - A Eur. J.* 24 (3): 572–575.

10 Hrbac, J., Pavelka, V., Crassous, J. et al. (2020). Redox and optically active carbohelicene layers prepared by potentiodynamic polymerization. *Electrochem. Commun.* 113: 106689.

11 (a) Dhbaibi, K., Favereau, L., Srebro-Hooper, M. et al. (2018). Exciton coupling in diketopyrrolopyrrole-helicene derivatives leads to red and near-infrared circularly polarized luminescence. *Chem. Sci.* 9 (3): 735–742. (b) Dhbaibi, K., Favereau, L., Srebro-Hooper, M. et al. (2020). Modulation of circularly polarized luminescence through excited-state symmetry breaking and interbranched exciton coupling in helical push-pull organic systems. *Chem. Sci.* 11 (2): 567–576. (c) Josse, P., Favereau, L., Shen, C. et al. (2017). Enantiopure versus racemic naphthalimide end-capped helicenic non-fullerene electron acceptors: impact on organic photovoltaics performance. *Chem. Eur. J.* 23 (26): 6277–6281. (d) Dhbaibi, K., Abella, L., Meunier-Della-Gatta, S. et al. (2021). Achieving high circularly polarized luminescence with push-pull helicenic systems: from rationalized design to top-emission CP-OLED applications. *Chem. Sci.* 12 (15): 5522–5533.

12 Zhu, S. and Sun, M. (2020). Electronic circular dichroism and Raman optical activity: principle and applications. *Appl. Spectrosc. Rev.* 56 (7): 553–587.

13 Nakai, Y., Mori, T., and Inoue, Y. (2013). Circular dichroism of (Di)methyl- and diaza[6]helicenes. A combined theoretical and experimental study. *J. Phys. Chem. A* 117 (1): 83–93.

14 Li, Y., Zhou, M., Niu, Y. et al. (2015). Solvent-dependent intramolecular charge transfer delocalization/localization in multibranched push-pull chromophores. *J. Chem. Phys.* 143 (3): 034309.

15 Pescitelli, G., Di Bari, L., and Berova, N. (2014). Application of electronic circular dichroism in the study of supramolecular systems. *Chem. Soc. Rev.* 43 (15): 5211–5233.

16 Schellman, J.A. (1975). Circular dichroism and optical rotation. *Chem. Rev.* 75 (3): 323–331.

17 Li, M., Lin, W.B., Fang, L., and Chen, C.F. (2017). Recent progress on circularly polarized luminescence of chiral organic small molecules. *Acta Chim. Sin.* 75 (12): 1150–1163.

18 Meisenheimer, J. and Witte, K. (1903). Reduction von 2-Nitronaphtalin. *Ber. Dtsch. Chem. Ges.* 36 (4): 4153–4164.

19 Jørgensen, K.B. (2010). Photochemical oxidative cyclisation of stilbenes and stilbenoids-the Mallory-reaction. *Molecules* 15 (6): 4334–4358.

20 Isla, H., Srebro-Hooper, M., Jean, M. et al. (2016). Conformational changes and chiroptical switching of enantiopure bis-helicenic terpyridine upon Zn^{2+} binding. *Chem. Commun.* 52 (35): 5932–5935.

21 Yen-Pon, E., Buttard, F., Frédéric, L. et al. (2021). Heterohelicenes through 1,3-dipolar cycloaddition of sydnones with arynes: synthesis, origins of selectivity, and application to pH-triggered chiroptical switch with CPL sign reversal. *JACS Au* 1 (6): 807–818.

22 Wang, F., Gan, F., Shen, C., and Qiu, H. (2020). Amplifiable symmetry breaking in aggregates of vibrating helical molecules. *J. Am. Chem. Soc.* 142 (38): 16167–16172.

23 Zheng, Y.H., Lu, H.Y., Li, M., and Chen, C.F. (2013). Synthesis, structures, and optical properties of aza[4]helicenes. *Eur. J. Org. Chem.* 2013 (15): 3059–3066.

24 (a) Yang, Y., Da Costa, R.C., Smilgies, D.M. et al. (2013). Induction of circularly polarized electroluminescence from an achiral light-emitting polymer via a chiral small-molecule dopant. *Adv. Mater.* 25 (18): 2624–2628. (b) Wan, L., Wade, J., Salerno, F. et al. (2019). Inverting the handedness of circularly polarized luminescence from light-emitting polymers using film thickness. *ACS Nano* 13 (7): 8099–8105. (c) Wade, J., Hilfiker, J.N., Brandt, J.R. et al. (2020). Natural optical activity as the origin of the large chiroptical properties in π-conjugated polymer thin films. *Nat. Commun.* 11 (1): 6137. (d) Greenfield, J.L., Wade, J., Brandt, J.R. et al. (2021). Pathways to increase the dissymmetry in the interaction of chiral light and chiral molecules. *Chem. Sci.* 12 (25): 8589–8602.

25 (a) Urbano, A. and Carreño, M.C. (2013). Enantioselective synthesis of helicenequinones and -bisquinones. *Org. Biomol. Chem.* 11 (5): 699–708. (b) Tani, F., Narita, M., and Murafuji, T. (2020). Helicene radicals: molecules bearing a combination of helical chirality and unpaired electron spin. *ChemPlusChem* 85 (9): 2093–2104.

26 Schweinfurth, D., Zalibera, M., Kathan, M. et al. (2014). Helicene quinones: redox-triggered chiroptical switching and chiral recognition of the semiquinone radical anion lithium salt by electron nuclear double resonance spectroscopy. *J. Am. Chem. Soc.* 136 (37): 13045–13052.

27 Liu, L. and Katz, T.J. (1990). Simple preparation of a helical quinone. *Tetrahedron Lett.* 31 (28): 3983–3986.

28 (a) Yang, B., Liu, L., Katz, T.J. et al. (1991). Electron delocalization in helical quinone anion radicals. *J. Am. Chem. Soc.* 113 (23): 8993–8994. (b) Liberko, C.A., Miller, L.L., Katz, T.J., and Liu, L. (1993). The electronic structure of helicene-bisquinone anion radicals. *J. Am. Chem. Soc.* 115 (6): 2478–2482.

29 (a) Nuckolls, C., Katz, T.J., and Castellanos, L. (1996). Aggregation of conjugated helical molecules. *J. Am. Chem. Soc.* 118 (15): 3767–3768. (b) Nuckolls, C., Katz, T.J., Katz, G. et al. (1999). Synthesis and aggregation of a conjugated helical molecule. *J. Am. Chem. Soc.* 121 (1): 79–88.

30 Li, D., Zhang, H., and Wang, Y. (2013). Four-coordinate organoboron compounds for organic light-emitting diodes (OLEDs). *Chem. Soc. Rev.* 42 (21): 8416–8433.

31 Shen, C., Srebro-Hooper, M., Jean, M. et al. (2017). Synthesis and chiroptical properties of hexa-, octa-, and deca-azaborahelicenes: influence of helicene size and of the number of boron atoms. *Chem. Eur. J.* 23 (2): 407–418.

32 Brandt, J.R., Pospíšil, L., Bednárová, L. et al. (2017). Intense redox-driven chiroptical switching with a 580 mV hysteresis actuated through reversible dimerization of an azoniahelicene. *Chem. Commun.* 53 (65): 9059–9062.

33 (a) Torricelli, F., Bosson, J., Besnard, C. et al. (2013). Modular synthesis, orthogonal post-functionalization, absorption, and chiroptical properties of cationic [6]helicenes. *Angew. Chem. Int. Ed.* 52 (6): 1796–1800. (b) Bosson, J., Labrador, G.M., Pascal, S. et al. (2016). Physicochemical and electronic properties of cationic [6]helicenes: from chemical and electrochemical stabilities to far-red (polarized) luminescence. *Chem. Eur. J.* 22 (51): 18394–18403. (c) Bosson, J., Labrador, G.M., Besnard, C. et al. (2021). Chiral near-infrared fluorophores by self-promoted oxidative coupling of cationic helicenes with amines/enamines. *Angew. Chem. Int. Ed.* 60 (16): 8733–8738.

34 (a) Kobaisi, M.A., Bhosale, S.V., Latham, K. et al. (2016). Functional naphthalene diimides: synthesis, properties, and applications. *Chem. Rev.* 116 (19): 11685–11796. (b) Würthner, F., Saha-Möller, C.R., Fimmel, B. et al. (2016). Perylene bisimide dye assemblies as archetype functional supramolecular materials. *Chem. Rev.* 116 (3): 962–1052.

35 Nowak-Król, A. and Würthner, F. (2019). Progress in the synthesis of perylene bisimide dyes. *Org. Chem. Front.* 6 (8): 1272–1318.

36 (a) Hartnett, P.E., Ramakrishna Matte, H.S.S., Eastham, N.D. et al. (2016). Ring-fusion as a perylenediimide dimer design concept for high-performance non-fullerene organic photovoltaic acceptors. *Chem. Sci.* 7 (6): 3543–3555. (b) Ohtsuka, N., Nakano, M., Nakagawa, S. et al. (2020). Naphthalene diimide-incorporated helical thienoacene: a helical molecule with high electron mobility, good solubility, and thermally stable solid phase. *Chem. Commun.* 56 (82): 12343–12346. (c) Liang, N., Meng, D., and Wang, Z. (2021). Giant rylene imide-based electron acceptors for organic photovoltaics. *Acc. Chem. Res.* 54 (4): 961–975. (d) Ball, M., Zhong, Y., Wu, Y. et al. (2015). Contorted polycyclic aromatics. *Acc. Chem. Res.* 48 (2): 267–276. (e) Peurifoy, S.R., Sisto, T.J., Ng, F. et al. (2019). Dimensional control in contorted aromatic materials. *Chem. Rec.* 19 (6): 1050–1061.

37 Kato, K., Segawa, Y., and Itami, K. (2019). Symmetric multiple carbohelicenes. *Synlett* 30 (4): 370–377.

38 Kogiso, T., Yamamoto, K., Suemune, H., and Usui, K. (2012). Synthesis and characterization of 1,8-naphthalimide with [6]helicene skeleton. *Org. Biomol. Chem.* 10 (15): 2934–2936.

39 Dusold, C., Sharapa, D.I., Hampel, F., and Hirsch, A. (2021). π-Extended diaza[7]helicenes by hybridization of naphthalene diimides and hexa-*peri*-hexabenzocoronenes. *Chem. Eur. J.* 27 (7): 2332–2341.

40 Saal, F., Zhang, F., Holzapfel, M. et al. (2020). Helicene diimides (n = 5, 6, and 7): through-bond versus through-space conjugation. *J. Am. Chem. Soc.* 142 (51): 21298–21303.

41 Schuster, N.J., Paley, D.W., Jockusch, S. et al. (2016). Electron delocalization in perylene diimide helicenes. *Angew. Chem. Int. Ed.* 55 (43): 13519–13523.

42 Rajasingh, P., Cohen, R., Shirman, E. et al. (2007). Selective bromination of perylene diimides under mild conditions. *J. Org. Chem.* 72 (16): 5973–5979.

43 Li, Y., Xu, L., Liu, T. et al. (2011). Anthraceno-perylene bisimides: the precursor of a new acene. *Org. Lett.* 13 (20): 5692–5695.

44 Martin, R.H. and Marchant, M.J. (1972). Thermal racemisation of [6], [7], [8] and [9] helicene. *Tetrahedron Lett.* 13 (35): 3707–3708.

45 Milton, M., Schuster, N.J., Paley, D.W. et al. (2019). Defying strain in the synthesis of an electroactive bilayer helicene. *Chem. Sci.* 10 (4): 1029–1034.

46 Nakai, Y., Mori, T., and Inoue, Y. (2012). Theoretical and experimental studies on circular dichroism of carbo[n]helicenes. *J. Phys. Chem. A* 116 (27): 7372–7385.

47 Liu, B., Böckmann, M., Jiang, W. et al. (2020). Perylene diimide-embedded double [8]helicenes. *J. Am. Chem. Soc.* 142 (15): 7092–7099.

48 Weiss, C., Sharapa, D.I., and Hirsch, A. (2020). Coronenohelicenes with dynamic chirality. *Chem. Eur. J.* 26 (62): 14100–14108.

49 Liu, G., Koch, T., Li, Y. et al. (2019). Nanographene imides featuring dual-core sixfold [5]helicenes. *Angew. Chem. Int. Ed.* 58 (1): 178–183.

50 Meng, D., Liu, G., Xiao, C. et al. (2019). Corannurylene pentapetalae. *J. Am. Chem. Soc.* 141 (13): 5402–5408.

51 (a) Sugiura, H., Nigorikawa, Y., Saiki, Y. et al. (2004). Marked effect of aromatic solvent on unfolding rate of helical ethynylhelicene oligomer. *J. Am. Chem. Soc.* 126 (45): 14858–14864. (b) Saito, N., Terakawa, R., Shigeno, M. et al. (2011). Side chain effect on the double helix formation of ethynylhelicene oligomers. *J. Org. Chem.* 76 (12): 4841–4858. (c) Schaack, C., Sidler, E., Trapp, N., and Diederich, F. (2017). Helical threads: enantiomerically pure carbo[6]helicene oligomers. *Chem. Eur. J.* 23 (57): 14153–14157. (d) Roose, J., Achermann, S., Dumele, O., and Diederich, F. (2013). Electronically connected [n]helicenes: synthesis and chiroptical properties of enantiomerically pure (E)-1,2-di([6]helicen-2-yl)ethenes. *Eur. J. Org. Chem.* 2013 (16): 3223–3231.

52 Yang, Y., Da Costa, R.C., Fuchter, M.J., and Campbell, A.J. (2013). Circularly polarized light detection by a chiral organic semiconductor transistor. *Nat. Photonics* 7 (8): 634–638.

53 Schuster, N.J., Hernández Sánchez, R., Bukharina, D. et al. (2018). A helicene nanoribbon with greatly amplified chirality. *J. Am. Chem. Soc.* 140 (20): 6235–6239.

54 Aranda, D., Schuster, N.J., Xiao, X. et al. (2021). Origin of chiroptic amplification in perylene-diimide helicenes. *J. Phys. Chem. C* 125 (4): 2554–2564.

55 Schuster, N.J., Joyce, L.A., Paley, D.W. et al. (2020). The structural origins of intense circular dichroism in a waggling helicene nanoribbon. *J. Am. Chem. Soc.* 142 (15): 7066–7074.

56 Zhang, L., Song, I., Ahn, J. et al. (2021). π-Extended perylene diimide double-heterohelicenes as ambipolar organic semiconductors for broadband circularly polarized light detection. *Nat. Commun.* 12 (1): 142.

57 Albano, G., Lissia, M., Pescitelli, G. et al. (2017). Chiroptical response inversion upon sample flipping in thin films of a chiral benzo[1,2-b:4,5-b′]dithiophene-based oligothiophene. *Mater. Chem. Front.* 1 (10): 2047–2056.

58 Xiao, X., Pedersen, S.K., Aranda, D. et al. (2021). Chirality amplified: long, discrete helicene nanoribbons. *J. Am. Chem. Soc.* 143 (2): 983–991.

59 (a) Zhu, Y., Guo, X., Li, Y., and Wang, J. (2019). Fusing of seven HBCs toward a green nanographene propeller. *J. Am. Chem. Soc.* 141 (13): 5511–5517. (b) Werner, A., Michels, M., Zander, L. et al. (1999). “Figure eight” cyclooctapyrroles: enantiomeric separation and determination of the absolute configuration of a binuclear metal complex. *Angew. Chem. Int. Ed.* 38 (24): 3650–3653. (c) Brahma, S., Ikbal, S.A., Dhamija, A., and Rath, S.P. (2014). Highly enhanced bisignate circular dichroism of ferrocene-bridged Zn(II) bisporphyrin tweezer with extended chiral substrates due to well-matched host-guest system. *Inorg.*

Chem. 53 (5): 2381–2395. (d) Sato, S., Yoshii, A., Takahashi, S. et al. (2017). Chiral intertwined spirals and magnetic transition dipole moments dictated by cylinder helicity. *Proc. Natl. Acad. Sci. U. S. A.* 114 (50): 13097–13101. (e) Wang, J., Zhuang, G., Chen, M. et al. (2020). Selective synthesis of conjugated chiral macrocycles: sidewall segments of (−)/(+)-(12,4) carbon nanotubes with strong circularly polarized luminescence. *Angew. Chem. Int. Ed.* 59 (4): 1619–1626.

60 Lütke Eversloh, C., Liu, Z., Müller, B. et al. (2011). Core-extended terrylene tetracarboxdiimide: synthesis and chiroptical characterization. *Org. Lett.* 13 (20): 5528–5531.

61 Lee, S.K., Zu, Y., Herrmann, A. et al. (1999). Electrochemistry, spectroscopy and electrogenerated chemiluminescence of perylene, terrylene, and quaterrylene diimides in aprotic solution. *J. Am. Chem. Soc.* 121 (14): 3513–3520.

11

Chiroptical Properties of Helicenes

Historical Perspective and Structure–Property Relationships

Sergio Abbate[1], *Giovanna Longhi*[1], *and Tadashi Mori*[2]

[1] *Dipartimento di Medicina Molecolare e Traslazionale (DMMT), Università degli Studi di Brescia, Brescia, Italy*
[2] *Department of Applied Chemistry, Graduate School of Engineering, Osaka University, Osaka, Japan*

11.1 Introduction

Helicenes are highly conjugated polycyclic aromatic molecules that are *ortho*-fused to exhibit unique three-dimensional spiral structure twisted either clockwise or counterclockwise, that is either *P* or *M* configuration. Owing to their uniquely distorted helical geometry and extended π-conjugation, optical and chiroptical properties of helicenes have been subjected to extensive and continued studies both in experimental and theoretical research fields. Recently, symmetrical and multiple helicenes that contain two or more helicene units within the molecule have been attracting considerable attention due to their more fascinating chiroptical responses. In this chapter, chiroptical properties of helicene derivatives are summarized by historical and current perspectives.

11.2 Chiroptical Spectroscopies of Model Hexahelicenes: A Brief History

Hexahelicene, or [6]helicene, is a molecule, whose study has served very much not only for the successive applications in technology, chemistry, and biochemistry but also for the advancement of chiroptical and theoretical studies; that is why we think opportune to retrace the first chiroptical investigations on hexahelicene and why we are confident that this will be useful for the understanding of the more complicated and newest helicene-based molecules discussed in this book. Hexahelicene was first synthesized by Newman and Lednicer in 1956 [1], and its specific optical rotation (OR) was measured for the separate enantiomers. Observed OR was huge ($[\alpha]_{589}$ = −3640 for (−)-hexahelicene in chloroform [1]); this raised hopes that the absolute configuration be easily assigned by chiroptical methods in the absence of X-ray diffraction studies. Albert Moscowitz [2, 3], who at the time

Helicenes: Synthesis, Properties, and Applications, First Edition.
Edited by Jeanne Crassous, Irena G. Stará, and Ivo Starý.

was a Harvard PhD student with W. Moffitt, ran the first *ab initio* study of the OR of hexahelicene. (Moffitt first introduced the linear combination of atomic orbitals (LCAO) method to cata-condensed systems [4].) Moscowitz indeed applied the Rosenfeld formula [5] for the calculation of $[\alpha]_{589}$ in a way, which we may call now as sum-over-states method. Within the Hückel approximation, he arrived at the configuration assignment $(-) \leftrightarrow P$ and $(+) \leftrightarrow M$. Other, more primitive, approaches arrived at the opposite configuration assignment [6, 7]. The correct assignment was indeed the latter, as demonstrated later by a former postdoc of Moscowitz, David Lightner, [8] who succeeded in synthesizing 2-bromohexahelicene, exhibiting $[\alpha]_{589} = -3560$ for the (−) enantiomer in chloroform, and was also able to get a crystal out of it. In this way he could unequivocally demonstrate that the correct assignment is $(+) \leftrightarrow P$ and $(-) \leftrightarrow M$. The reason for the failure of Moscowitz treatment was recognized in the use of too small a number of excited states treating configuration interaction while applying the Hückel approximation, inevitable at the time, due to limited computational resources. A more complete set of orbitals while applying the same Moscowitz approach led to the correct assignment also theoretically [9]. Indeed, several years later Grimme [10, 11] arrived at the correct assignment by density functional theory (DFT) calculations, based on the generally practiced linear response approach. A worthwhile general reference for theoretical aspects is the work of Autschbach [12] (see also references therein), where full account is given on the "renaissance" in the theoretical treatment of chiroptical properties, after more than 20 years of stagnation.

However, the study of Moscowitz was important for the subsequent theoretical developments and the introduction of several concepts still used nowadays in chiroptical studies, like the application of the Kramers–Kronig relation [13] to relate optical rotatory dispersion (ORD) and electronic circular dichroism (ECD), as a generalization of the Rosenfeld equation [5], and the concept of inherently dissymmetric chromophore [3, 14] to describe the relevant aspect of this distorted, π-conjugated long molecular structure. Other chiroptical spectroscopies may be able to evidence in a more useful way the subtle differences between hexahelicene and substituted hexahelicenes, like vibrational circular dichroism (VCD) [15], Raman optical activity (ROA) [16], and vibronically resolved ECD and circularly polarized luminescence (CPL) spectra [17]. These data are still scarce: in the case of 2-bromohexahelicene, we are able to show all these data in a single figure (Figure 11.1). We notice that the presence of bromine quenches fluorescence, giving a hardly detectable CPL signal, however, permitting one to record the ROA spectrum.

All the data of Figure 11.1 were correctly interpreted by the use of DFT and TD-DFT calculations; illuminating insight into the mechanism governing ECD was provided by Nakai, Mori, and Inoue through the definition of substituent-sensitive and helical sense-responsive transitions [18]; such idea has been extended to CPL phenomena [17]. A few comments are worthwhile making. (i) By far the most common situation is when the sign of CPL is the same as the sign of the lowest energy ECD band [19], since the structure of the excited and ground electronic state is often similar; however, in the case of hexahelicenes, the lowest energy ECD

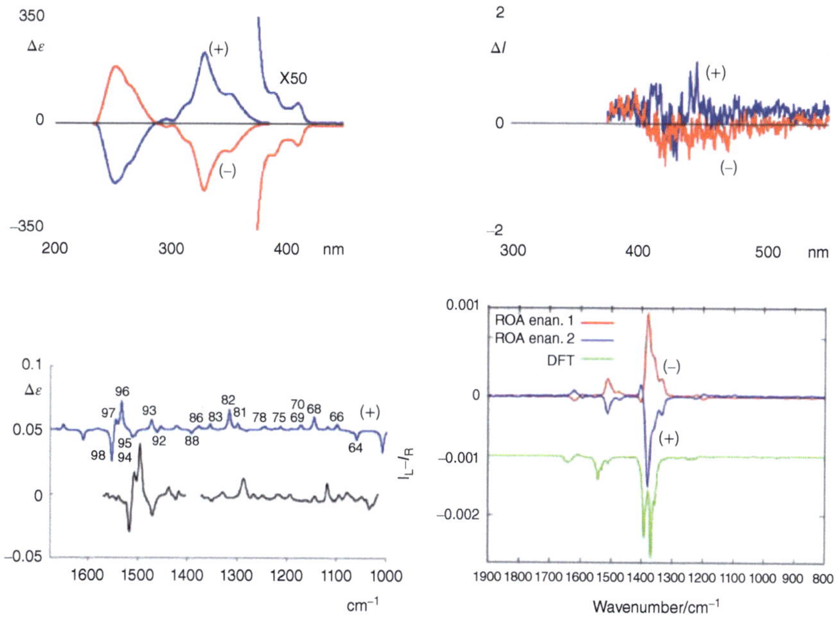

Figure 11.1 Experimental ECD (top left), CPL (top right), VCD (bottom left), and ROA (bottom right) spectra of (−)-2-bromohexahelicene (red curves) and (+)-2-bromohexahelicene (blue curves). Source: Data taken from the papers [15–17]. For VCD only the data for the dextrorotatory (+) enantiomer is given. (Calculated VCD and ROA spectra for the (+) enantiomer in black and green, respectively, are also reported.)

band and the CPL band do not necessarily reflect the helical sense but show a substituent-sensitive character, since the signal is dominated by vibronic effects [20]. (ii) Vibrational normal modes are thus important for this kind of systems, and the study of the VCD and ROA spectra is quite useful: the VCD spectrum is dominated by the band at ca. 1500 cm^{-1}, which is called G-band for polycyclic aromatic hydrocarbons (PAH) [15, 21], while the ROA spectrum is dominated by the band at ca. 1350 cm^{-1}, which is called D-band for PAH [16, 21]; the normal modes underlying both vibrational bands are well-defined linear combinations of C=C/C–C stretchings and C=C–H bendings, able to influence the conjugation of the PAH-chain. In publication [20] it is indeed shown that these modes participate in the shaping of vibronic aspect of CPL and ECD bands of the four hexahelicenes studied there, even though together with a number of other normal modes. (iii) Finally, according to the single-excited state (SES) interpretation by Nafie [22, 23] of resonance and pre-resonance ROA spectra, the sign of all ROA bands is the same and is the opposite of the ECD resonant transition. This is indeed the case for 2-bromohexahelicene: most of the ROA bands have the same sign, equal to that for the largest ROA D-band, the latter being opposite to the most intense ECD band at ~330 nm and of the bands at higher wavelengths (Figure 11.1); few minor observed ROA bands escape this simple and efficacious rule.

11.3 Helical Length Dependence of Chiroptical Properties of Helicenes

One of the interesting arguments in the optical and chiroptical properties of helicenes is on how they depend on number of annulated benzene rings *n* or on helical elongation (Figure 11.2). Such a study, however, is relied on the preparation of a series of [*n*]helicenes, particularly in optically pure forms. As stated in the previous section, synthesis and optical resolution of [6]helicene had been first achieved in 1956 by Newman and Lednicer [1], which was followed by Goedicke and Stegemeyer for [5]helicene in 1970 [24]. The optical resolution in these experiments was conducted by a partial crystallization of a diastereomeric complex between the helicene and optically active 2-[[(2,4,5,7-tetranitro-9-fluorenylidene)amino]oxy]propionic acid, but the methods later turned out to be less effective in terms of optical purity because they often were only partially resolved and the latter helicene is susceptible to the thermal racemization [18]. In 1974, the group of Martin reported the chiroptical properties of [7]-, [8]-, and [9]helicenes [25]. It is to note that the optically pure [7]helicene was obtained by crystal picking of the conglomerate, while partial resolution by spontaneous resolution was followed by repeated crystallizations to obtain

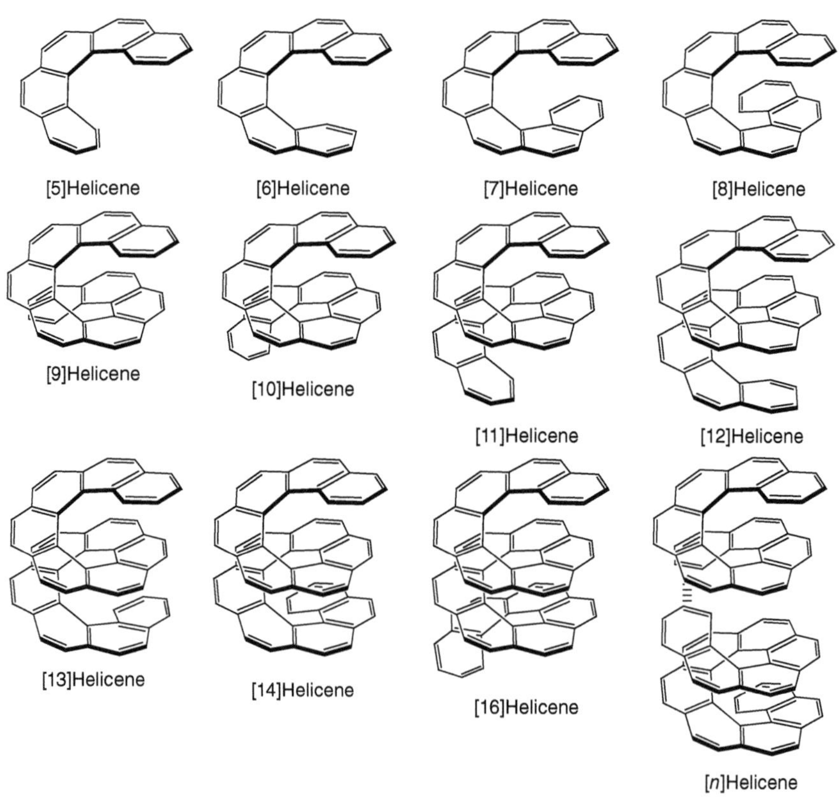

Figure 11.2 Structure of carbo[*n*]helicenes.

improved optical purity for others. Larger helicenes such as [10]–[14]helicenes [26–28] as well as [16]helicene [29] were also prepared in racemic forms, but the chiroptical properties have not been reported. Remarkably, enantiomer separation of up to [14]helicenes by chiral high performance liquid chromatography (HPLC) has been also documented [30]. In brief, the chiroptical studies on the dependence of molecular size has not been much explored experimentally so far, and the optical purity was not fully confirmed in most of the classical reports, hindering the systematic investigations of chiroptical properties of [n]helicenes.

In such a limited situation in experiment, theoretical investigation has provided substantial insights on the length dependence of chiroptical properties such as ECD of [n]helicenes. Indeed, a convergence behavior of the physicochemical properties of parent carbo[n]helicenes up to $n = 14$ has been extensively discussed with an aid from DFT calculation [31]. Accordingly, the equilibrium structure and stabilization energy difference are converged at or before [14]helicene with a helical pitch of 3.75 Å. In contrast, ECD and absorption spectra are converged more rapidly with size to the value of the extrapolated bandgap of ca. 2.5 eV. Remarkably, the rotational strengths R are roughly conserved over the carbo[n]helicene series, while the excitation energies and intensities are decreasing upon lengthening of [n]helicenes. Such analysis has been more sensibly performed by one of us using the approximate second-order coupled cluster calculations [18]. Consequently, comparable observations were found in the excitation energies and intensities, while the rotational strength R were found somewhat dispersed with molecular size. In the same study, experimental chiroptical properties of [5]- and [6]helicenes were corrected by employing optically pure samples resolved and checked by chiral HPLC. More careful analysis revealed that the transition energies at the ^1B_b and ^1B_a bands are found linearly dependent on the reciprocal number n in carbo[n]helicenes. More interestingly, the dissymmetry (g) factor, which is determined by differential molar extinction coefficient ($\Delta\varepsilon$) divided by total molar extinction coefficient (ε), was also found to be linear to $1/n$ for carbo[n]helicenes larger than [6]helicene, where the helix starts to overlap (Figure 11.3). By these theoretical calculations at the RI-CC2/TZVPP level, all the characteristic experimental features of chiroptical properties of up to [9]helicenes were accurately reproduced. On the contrary, semiempirical intermediate neglect of differential overlap for spectroscopic modification (INDO/S) method has been successfully applied to predict the chiroptical property of much larger helicenes up to [36]helicene with a modest compromise in accuracy [32].

In a solenoid coil, magnetic flux density is linearly dependent on the number of turns. Thus, the observed correlation in g value against $1/n$ is somewhat puzzling, given that the g can be approximated to $4R/D$ in isotropic solutions, where D denotes the dipole strength and the rotational strength R is the dot product between the electric and magnetic dipole transition moments. Consequently, the convergence behavior of g values in [n]helicenes is, at least in part, attributed to the limited electronic interaction between the helices, which is experimentally determined as around 50 benzene rings for the ^1B_b band. Extrapolation of the linear correlation between g and $1/n$ to infinite n affords an ultimate g value as large as 0.11 for [∞]helicene

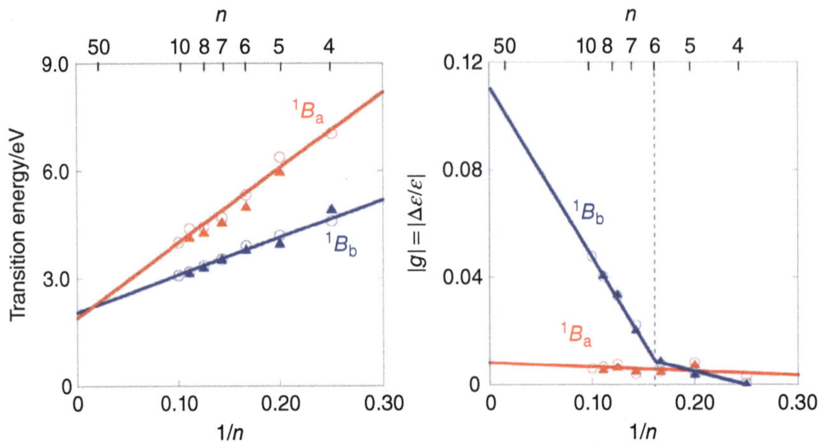

Figure 11.3 Plots of experimental (triangle) and theoretical (open circle) transition energies (left) and dissymmetry (*g*) factors (right) against the reciprocal number of rings (1/*n*) in carbo[*n*]helicenes at the ^1B_b (blue) and ^1B_a (red) bands. The lines are the least squares fit for the theoretical values. Source: Reproduced with permission from [18]. Copyright 2012, American Chemical Society.

or 0.09 for [50]helicene, in which the effective conjugation length is taken into account.

CPL is an additional chiroptical property of interest for helicenes [17, 33]. However, less has been explored for the size dependence primarily due to the facile intersystem crossing and smaller fluorescence quantum yield of larger helicenes [34].

Recently, the size dependence on the chiroptical properties of polyaza- [35], azabora- [36], and azaoxahelicenes [37] has been reported. However, related studies on hetero[*n*]helicenes are still restricted due to the experimental limitations. More investigations are desired in both experimental and theoretical perspectives in order to better understand the *n*-dependence of electric and magnetic dipole transition moments and eventually optical and chiroptical properties of [*n*]helicenes to facilitate the advancement of molecular design of helicene-based nanostructures.

11.4 Impact of Symmetry in Multiple Helicenes on Their Chiroptical Properties

All pristine carbo[*n*]helicenes bear C_2-symmetry element perpendicular to their helical axis. In theory, absence of improper rotations in these chiral helicenes is an origin of the natural optical activity, and the symmetry provides one cause for the intense chiroptical responses of helically chiral helicene [38]. Substitution at outer helices, replacement of carbon(s) by heteroatom(s), and insertion of and/or replacement by other cyclic moieties have been successfully applied to rationally modify the chiroptical property of helicenes. In all these modifications, however,

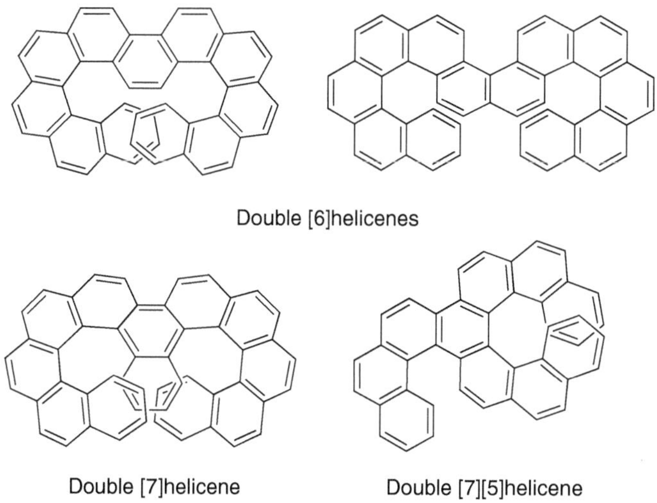

Figure 11.4 Classical examples of fused and annulated double helicenes. Note that chiroptical properties were not reported for these molecules.

structural alterations in helix are not much expected to afford divergent chiroptical responses; also chemical substitution providing exciton coupling to optical activity has been achieved [39]. On the other hand, placing chromophores in chiral environment or in relative chiral arrangement may provide strong chiroptical responses due to their supramolecular structures and intermolecular interactions [40]. Alternatively, proper arrangement of two or more helicenes in various symmetries has been explored with great success to improve the chiroptical property [41]. For instance, the fusion of two, three, and more helicene moieties into common aromatic framework leads to double, triple, and multiple helicenes, respectively. In classical examples of C_2-symmetric fused and annulated double [6]helicenes, fused double [7]helicene, and asymmetrically fused double [7,5]helicene, none of chiroptical properties such as ECD were reported (Figure 11.4) [42–44]. The multiple helicenes have received considerable attention in recent years, and more variety of derivatives become experimentally available [45, 46]. Some of them have been optically resolved, and their chiroptical properties are also reported recently. In these molecules, symmetry plays key roles in remarkably improved chiroptical responses. In this section, we showcase several recent examples discussing the chiroptical properties of multiple helicenes. There are not a few reports on the synthesis, racemization behavior, and electronic and chiroptical properties of multiple helicenes, but we illustrate herein the properties of relatively simple pristine (unsubstituted and hydrocarbon) helicenes alone, in order to stress the symmetry effect on the chiroptical property of helicenes (Figure 11.5), followed by some π-extended analogs.

As the simplest model for property study of multiple helicenes, X-shaped D_2-symmetric and S-shaped C_2-symmetric fused double [6]helicenes (i.e. dinaphtho[2,1-*i*:1′,2′-*l*]hexahelicene and diphenanthro[3,4-*c*:3′,4′-*l*]chrysene) were examined [47]. Crystal structural investigations as well as theoretical calculations

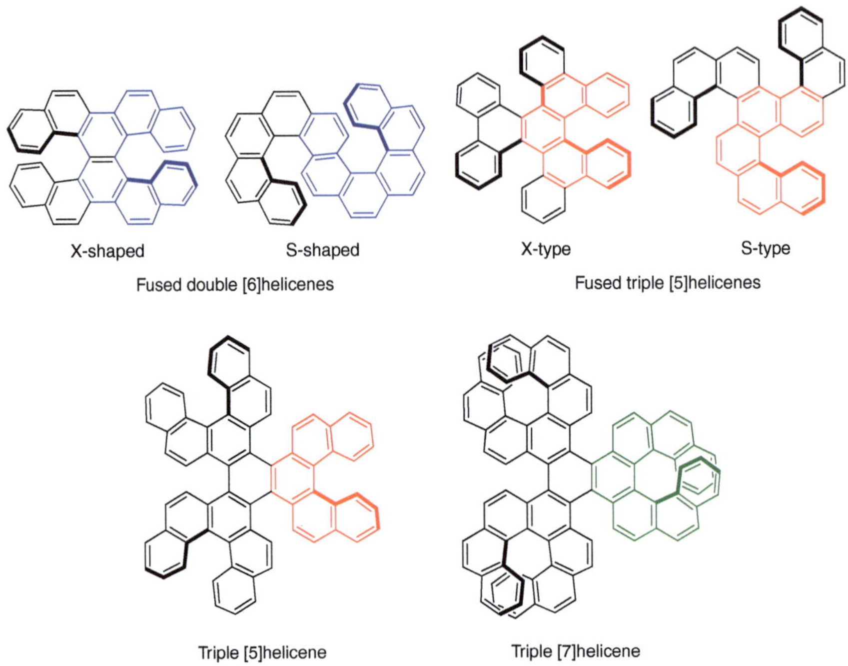

Figure 11.5 Fused (top) and benzene-annulated (bottom) double and triple helicenes discussed in this section. (*P,P*)- and (*P,P,P*)-enantiomers are shown for clarity.

revealed the insignificant structural changes in helicene structure upon fusion of two [6]helicene units. Nevertheless, the observed dissymmetry (*g*) factor, particularly at the ^1B_b band, became considerably enhanced to 0.016 (X-shaped) and 0.022 (S-shaped), respectively, as compared to that of parent [6]helicene (0.009) (Figure 11.6). Cotton effects at the ^1L_b and ^1L_a bands, which are essentially hidden in the intense ^1B_b band in parent [6]helicene, were more apparent in the double helicenes. These observations are explained by the symmetrical placement of two helicene units that allows to align the corresponding electric and magnetic dipole transition moments, the angle between which is the crucial factor to determine the rotational strength *R* in the ECD, increasing the intensity of dissymmetry (*g*) factor.

The parent [5]helicene promptly racemizes at ambient temperature. In contrast, fused triple [5]helicenes with D_3- and C_3-symmetry are more stable due to the presence of diastereomeric isomers on the racemization potential surfaces. Accordingly, ECD spectra of X- and S-type triple [5]helicenes were investigated (Figure 11.7) [48, 49]. Likewise, the observed dissymmetry (*g*) factor at the ^1B_b band was substantially enhanced to 0.010 for X-type triple [5]helicene as compared with that of parent [5]helicene (0.004), due to the proper alignment of the dipole transition moments.

Besides the fused [6]- and [5]helicenes mentioned previously, triple helicenes annulated to benzene core that contain six helical units have been reported. The preparation of such triple [5]helicene has been reported almost simultaneously by two independent groups [50, 51]. They provided qualitatively comparable ECD spectra in pattern (Figure 11.8). The analogous triple [7]helicene has been also

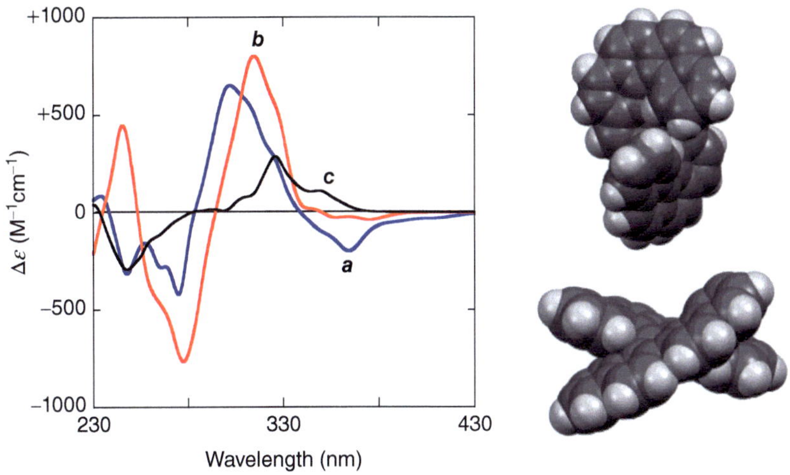

Figure 11.6 Left: experimental ECD spectra of X-shaped (blue, *a*) and S-shaped (red, *b*) fused double [6]helicenes. For comparison, spectrum for parent [6]helicene is also shown (black, *c*). Right: crystal structures of fused double [6]helicenes. Source: [47]/CC BY-4.0.

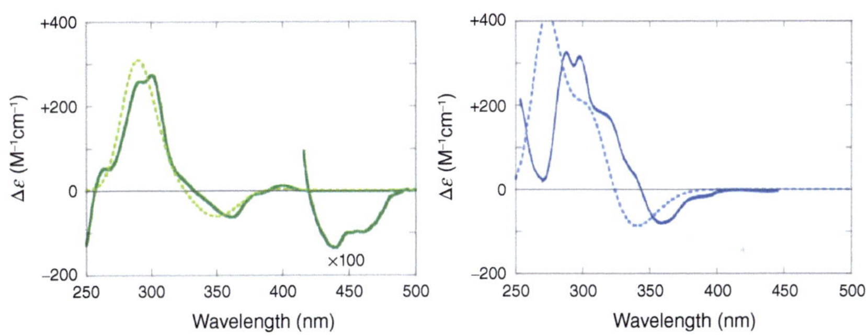

Figure 11.7 Experimental (solid) and calculated (dashed) ECD spectra of X-type (left) and S-type (right) fused triple [5]helicenes. Source: Reproduced with permission from [48]. Copyright 2018, American Chemical Society.

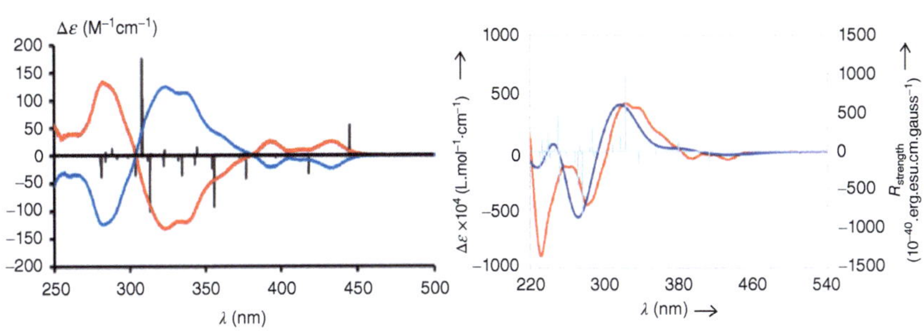

Figure 11.8 ECD spectra of triple [5]helicene annulated to single benzene core from two different sources. Bars and blue line on right are theoretical spectra, while others are experimental spectra. Source: Reproduced with permission from [50, 51]. Copyright 2017, American Chemical Society.

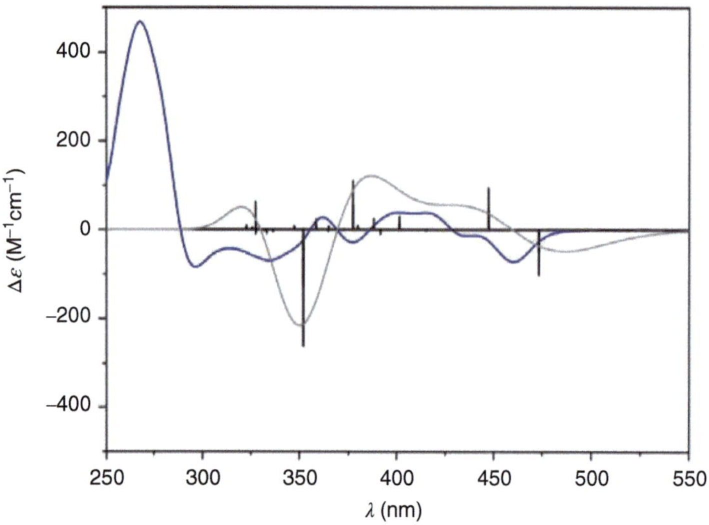

Figure 11.9 Experimental (blue) and theoretical (gray and bar spectra) ECD spectra of triple [7]helicene annulated to single benzene core. Source: Reproduced with permission from [52]. Copyright 2019, Wiley-VCH Verlag GmbH.

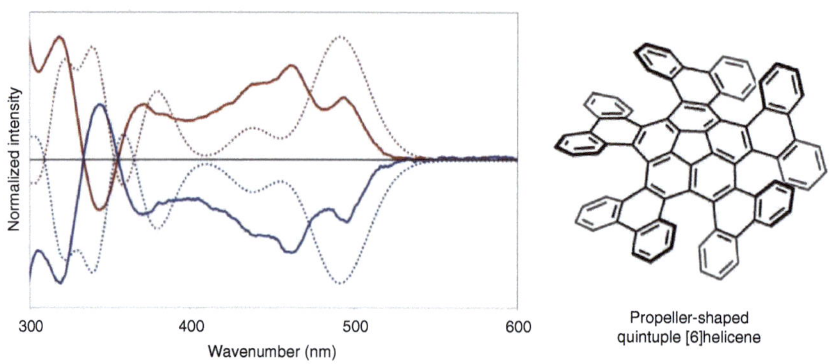

Figure 11.10 Experimental (solid) and theoretical (dotted) ECD spectra of propeller-shaped quintuple [6]helicene. Source: Reproduced with permission from [53]. Copyright 2018, Wiley-VCH Verlag GmbH.

reported, but the corresponding ECD spectrum was found more complicated (Figure 11.9) [52]. More complex helical molecule containing multiple helicene substructures has been also reported. Thus, propeller-shaped quintuple [6]helicene fused with a bowl-shaped corannulene core was prepared, and the enantiomers were partially resolved by chiral HPLC, providing complex signals in the ECD spectrum (Figure 11.10) [53].

The extension of π-system in helicenes would be also accessible by annulation of additional benzene rings in examples such as follows. As for a primary substructure of helically twisted graphenes, hexa-*peri*-hexabenzo[7]helicene was prepared [54]. A small bandgap of 1.55 eV and excellent chiroptical property with a considerably

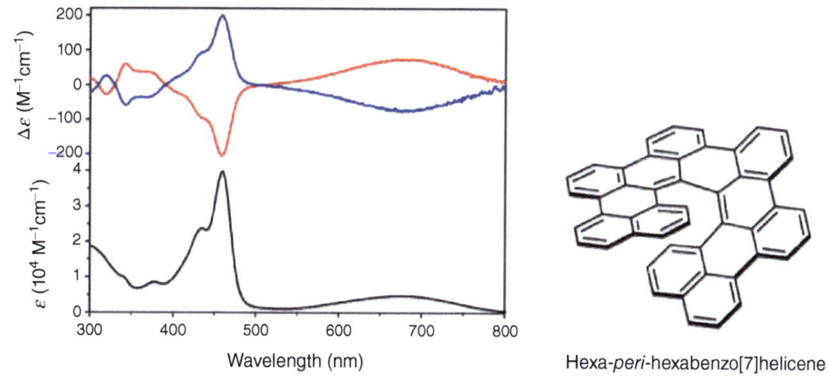

Figure 11.11 ECD (top) and absorption (bottom) spectra of hexa-*peri*-hexabenzo[7]helicene. Source: Reproduced with permission from [54]. Copyright 2018, American Chemical Society.

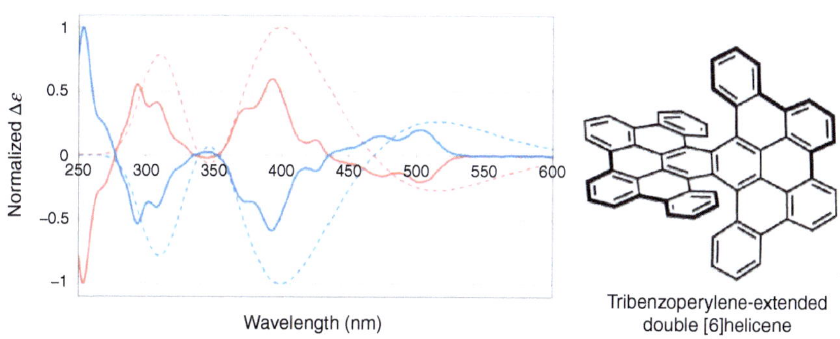

Figure 11.12 Experimental (solid) and calculated (dashed) ECD spectra of tribenzoperylene-extended double [6]helicene. Source: Reproduced with permission from [55]. Copyright 2015, American Chemical Society.

large dissymmetry (g) factor of 0.016 at 680 nm was reported for this symmetrically extended helical molecule, ascribable to the effective exciton delocalization (Figure 11.11).

The π-extension strategy has been also applied in the multiple helicenes. Thus, the X-shaped double [6]helicene was annulated by two tribenzo[*b,n,p,qr*]perylene substructures to provide unique helical structure with a combination of planar and helical π-systems [55]. The diastereomeric and enantiomeric separation were successfully performed by chiral HPLC, and the experimental ECD spectrum of twisted chiral isomers was compared with the calculated ECD pattern for an assignment of absolute configuration (Figure 11.12). Due to the limited solubility, unfortunately, quantitative analyses on the chiroptical property were not provided.

Finally, CPL of multiple helicenes is briefly described. Although CPL should reflect the chirality in the excited state, a direct correlation between the experimental dissymmetry (g) factors of luminescence and lowest energy absorption (i.e. CPL and ECD) was found to fit to a linear expression of $|g_{CPL}| = 0.81 \times |g_{ECD}|$ for all the available CPL and ECD data of the electronically allowed π–π* transition of

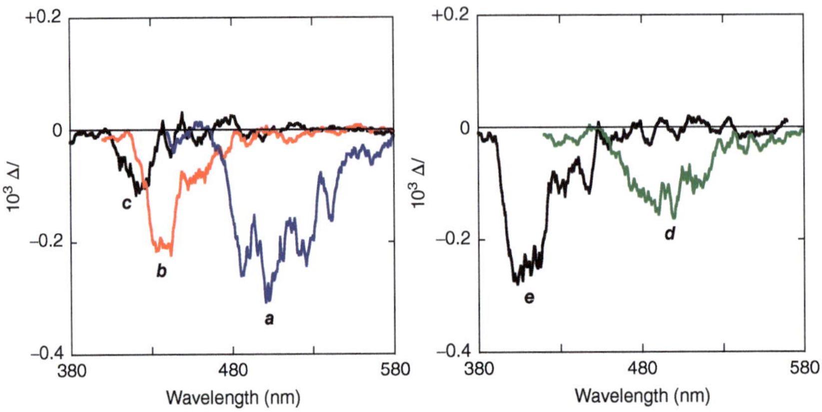

Figure 11.13 CPL spectra of fused double [6]helicenes (left) and triple [5]helicene (right). Blue, (a): X-shaped double [6]helicene. Red, (b): S-shaped double [6]helicene. Green, (d): triple [5]helicene. For comparison, spectra for parent [6]- and [5]helicenes are also shown (black, (c and e, respectively)). Source: [47]/CC BY-4.0.

rigid aromatic systems [56], suggesting relatively minor (and somewhat systematic) structural relaxation of such π-systems in the excited state. Because of the fast intersystem crossing and reduced fluorescence quantum yield for larger helicenes [34], the CPL property has been less explored for multiple helicenes. Nevertheless, increased dissymmetry (g) factors of 0.0025 and 0.0021 for X- and S-shaped fused double [6]helicenes, respectively, have been reported with a large red shift of emission wavelength, as compared with the $g_{CPL} = 0.001$ for parent [6]helicene (Figure 11.13) [47]. Again, a proper alignment of two helicene units allows to make parallel the relevant electric and magnetic dipole transition moments, their relative angle being the crucial factor to improve the rotational strength R and dissymmetry (g) factor in these double helicenes. A similar enhancement has been also reported in double azahelicenes [57]. In the triple [5]helicene, the dissymmetry (g) factor was comparable or slightly reduced as compared with the parent [5]helicene, but the undesired racemization was entirely avoided.

To summarize this section, symmetry plays a key role in improving the chiroptical responses both in ECD and CPL in multiple helicenes where more than two helicene units are arranged into a single molecule. Such structural modification allows to align the relevant electric and magnetic dipole transition moments in parallel (or antiparallel) fashion, maximizing the angle factor of the rotational strength R, thus the dissymmetry (g) factor therefrom. As demonstrated herein in hydrocarbon helicenes, symmetrically expanded helicene derivatives such as multiple and π-expanded helicenes are envisaged as a promising molecular motif for superior chiroptical materials. In the future, this kind of strategy should be more exploited also in substituted heterohelicenes and related helically chiral molecules.

11.5 Relevance of Substituents, of Vibronic Couplings, and of Charge Transfer Effects in the ECD Spectra of Helicene Systems

With the aim of increasing optical, optoelectronic, and nonlinear optics properties, the roles of substituent effects with increasing complexity have been examined. In Figure 11.14 it is first of all interesting to examine simple substituents in position 2 of carbohexahelicene considering absorption, fluorescence, ECD, and CPL spectra. The presence of an electron-rich atom like nitrogen, while leaving practically unchanged the intense ECD features (due to magnetically and electrically allowed transitions involving excited ^1B_b and ^1B_a states), has a strong influence on the ECD intensity of the ^1L_b transition, which gains electric dipole activity and exhibits higher

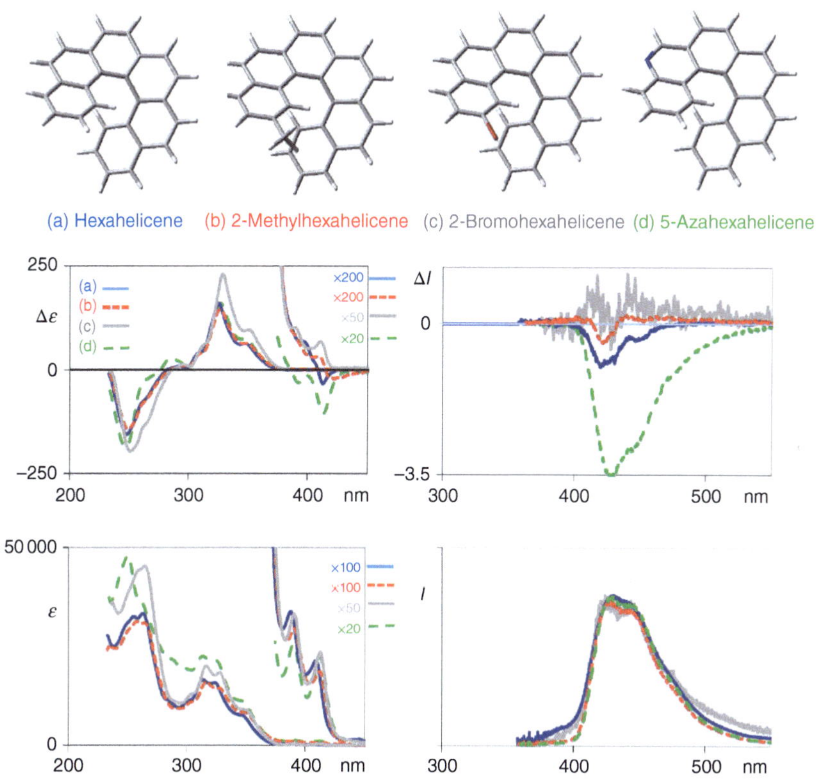

Figure 11.14 Top: three-dimensional structures of (a) hexahelicene, (b) 2-methylhexahelicene, (c) 2-bromohexahelicene, and (d) 5-azahexahelicene. Middle: superposed experimental ECD (left) and CPL (right) spectra of hexahelicenes (a)–(d). Bottom: superposed experimental absorption (left) and fluorescence (right) spectra of hexahelicenes (a)–(d).

optical activity also due to favorable electric dipole–magnetic dipole reciprocal orientation, namely, larger value for the angle between the two [17, 18]. For the 2-bromo derivative, the emission behavior is affected also by heavy atom effect quenching fluorescence activity [58]. On the contrary the simple pristine and methyl-helicenes are of theoretical interest for their simplicity. The lowest energy ^1L_a and ^1L_b transitions, common to PAH, are not easily accounted for by the TD-DFT calculations [59, 60]. Moreover, the contribution of vibronic terms to the ECD intensity of carbohexahelicene is of major importance, as already suggested by Weigang et al. [61], with vibronic components, which are not even constant in sign. This fact has been analyzed in detail adopting either the Adiabatic Hessian or the Vertical Hessian model [20]: both the Duschinsky and Herzberg–Teller effects need to be considered to account for the observation that absorption and ECD spectral shapes are different, and do not show mirror symmetry with respect to their emission counterparts.

Obviously, having in mind optical applications, enhancement of long wavelength absorption and ECD bands and of fluorescence and CPL bands due to heteroatoms is quite relevant and leads one to consider other helicene systems: among them we mention thiahelicenes due to their advantageous electronic properties [62]. Considering simple thiahelicene, ECD features resemble the ones of carbohelicene, as shown in Figure 11.15 [63] with improved optical response. On top of the experimental spectra of (+) and (−)-pentathia[9]heterohelicene, which were recorded a while ago, we report the ECD spectrum of (*P*)-enantiomer, computed *ex novo* by the TD-DFT method at the M06/TZVP level, in the form of sticks proportional to transition intensities as well as in the form of convoluted bands. The agreement is quite acceptable, confirming also in this case the correlation $P \leftrightarrow (+)$; we also notice that the longest wavelength transitions toward the ^1L_b and ^1L_a states (in order of decreasing wavelength) are quite intense, as expected for heterohelicenes, and are of similar magnitude as the next transition in wavelength, which we checked, by looking at the localized MOs and their symmetry, to be toward the ^1B_b state. Finally, we notice that the ^1L_b and ^1L_a transitions exhibit opposite sign, having opposite C_2-symmetry-species character. Other interesting examples of opportunely substituted thiahelicenes, designed so as to enhance optical properties, are given in Refs. [64, 65].

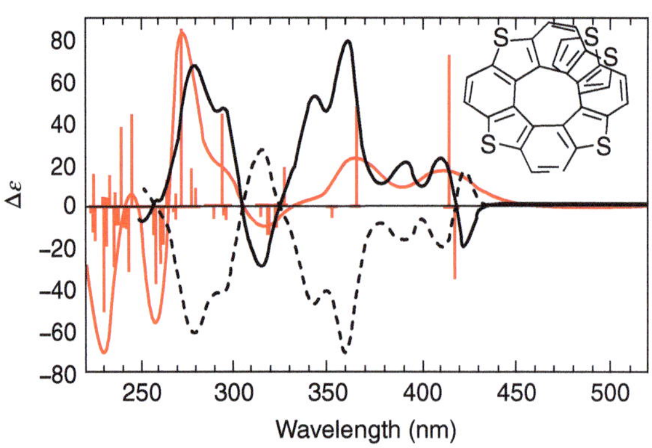

Figure 11.15 Superposed experimental ECD spectra of (+) and (−)-pentathia [9]heterohelicene [63] and of the calculated ECD spectrum of (*P*)-enantiomer at the M06/TZVP level with Gaussian16 package [66]. Gaussian band shapes with 0.2 eV bandwidth for computed features.

Playing with different substituent chemical groups on helicene type molecules may allow one to monitor a great role in photophysical responses from the material. In particular, researchers aim at preparing versatile compounds devised to exhibit charge transfer transitions with the possibility of tuning absorption and emission wavelengths to extend the spectroscopic range toward lower energy values. As an instructive example we considered three similar molecules (Figure 11.16) that are highly fluorescent azabora[5]helicene dyes bearing groups with increasing electron donor properties: with different X-peripheral substituents, one may tune absorption and emission wavelength and also sign and intensity of ECD and CPL bands [67]. TD-DFT calculations show that ground state structures are superimposable, while HOMO and LUMO orbitals are quite different for the three molecules. Consequently, differences in the magnitude and sign of rotational strengths (well accounted for by the calculations) are of pure electronic origin and are reflected also in the different angle between electric and magnetic dipole transition moments. In brief, in

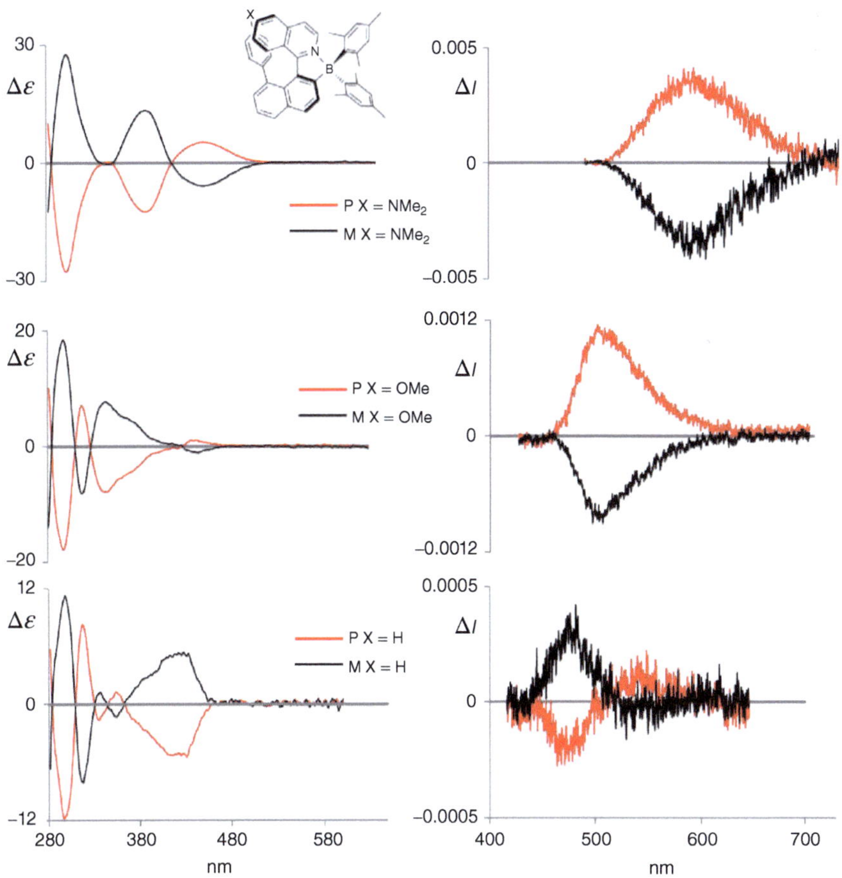

Figure 11.16 ECD (left) and CPL (right) of both enantiomers of three borylated arylisoquinoline skeleton-based helicenes bearing three different substituents. Source: Redrawn from Ref. [67].

Figure 11.16 one notices that not only wavelengths but also ECD and CPL intensities are sensitive markers of different electronic responses. Numerous other examples with charge transfer transitions are present in the literature and are worth consulting [68–70], the last reference dealing with important push–pull effects associated with chiral structures.

Finally, as extensively illustrated previously, chiral PAHs were synthesized either by considering helically elongated helicenes or by extending sideways the π-system of the helical motif. In this paragraph we focus on the role of substituent groups, and we consider examples for which the distortion of nanographene-like moieties is accomplished with bulky groups introduced in correspondence of a bay, cove, or fjord region of a PAH [71]. Helicenes with $n \geq 4$ are the limiting example for fjord regions. The role of the substituents is to promote a chiral structure; at the same time pendant groups can confer solubility to PAH [72–74]. The steric hindrance thus generated is also large enough to provide the possibility of enantiomer resolution [74]; however, also in this case some conformational mobility of the substituents may be present and may influence the spectroscopic response [73]. In Figure 11.17 we report

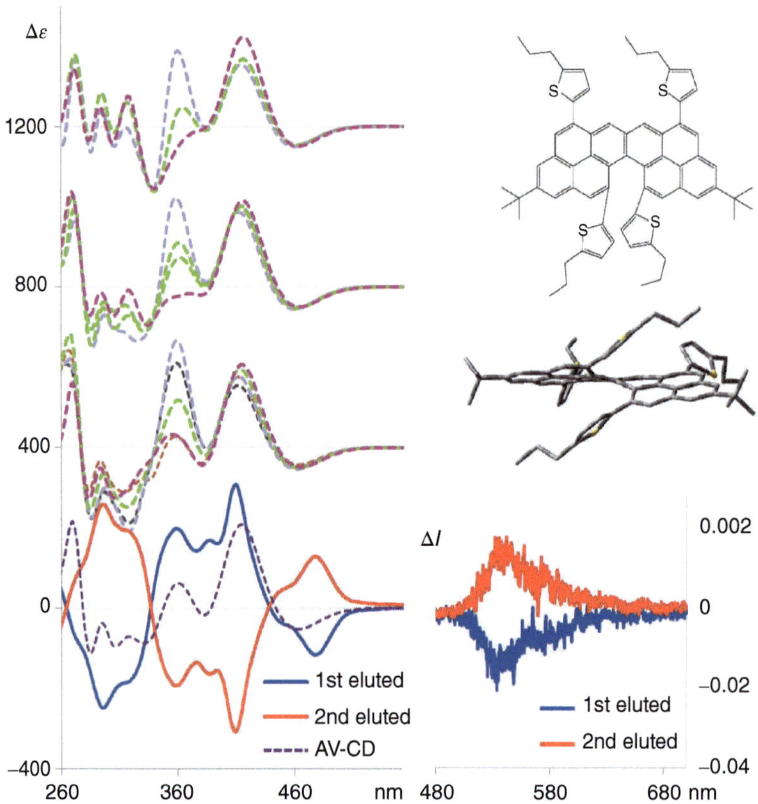

Figure 11.17 Bottom: experimental ECD (left) and CPL (right) spectra of HPLC separated enantiomers of pyreno[a]pyrene helicene system with two thiophene groups in the cove area. Top: calculated ECD spectra of the pyreno[a]pyrene system structures on the right. Source: Based on [73].

one example from a recent study [72]: a pyreno[a]pyrene-based helicene hybrid in which aryl rings present in the cove region give a chiral distortion to the extended π-structure with good chiroptical response, both in absorption and emission. After careful conformational search one notices how the higher energy bands are largely influenced by the conformational landscape; on the contrary the HOMO–LUMO transition located in the extended aromatic part is not influenced by the pendant group orientation. Another observation can be made on these systems [72–74]: the shape of the first absorption/ECD band and of the fluorescence/CPL is the same with mirror inversion. Finally, the fact that the dissymmetry (g) factor is constant throughout the vibronic structures suggests that Franck–Condon terms dominate.

11.6 Concluding Remarks

In this chapter, chiroptical properties such as ORD (or OR), ECD, VCD, ROA, and CPL of helicenes are briefly discussed. Owing to the distinctively distorted three-dimensional structure and extended π-conjugation, helicenes as prototype examples of helical chirality have been and are the target of extensive investigations for their optical and chiroptical features both experimentally and theoretically. The chiroptical investigations of hexahelicene are of historical importance in the advancement of theoretical concepts such as the sum-over-states method, the Kramers–Kronig relation, and the Rosenfeld equation, which are fundamental in the current chiroptical studies. The ECD and CPL of model helicenes have been more thoroughly investigated to increase the knowledge of chiroptical property–structure relations and to develop a design principle for superior and/or desired chiroptical molecules and materials. Thus, the effects of helix size (number of rings), substitution, and substituents, as well as symmetrical fusion of multiple helicene moieties on both optical and chiroptical properties have been expressly explored thus far. Although we see in this chapter the substantial advancement of chiroptical studies on helicenes and derivatives, more examples and studies are certainly needed to comprehensively understand and to eventually control and fine-tune the chiroptical responses of helicenes. Still, it is confirmed in this chapter that the favorable chiroptical properties based on the helicene structures are of central importance in chiral materials science.

References

1 Newman, M.S. and Lednicer, D. (1956). The synthesis and resolution of hexahelicene. *J. Am. Chem. Soc.* 78: 4765–4770. https://doi.org/10.1021/ja01599a060.
2 Moscowitz, A. (1957). On optical activity – hexahelicene. PhD thesis. Harvard University.
3 Moscowitz, A. (1961). Some applications of the Kronig-Kramers theorem to optical activity. *Tetrahedron* 13: 48–56. https://doi.org/10.1016/S0040-4020(01)92204-5.

4 Moffitt, W. (1954). The electronic spectra of cata-condensed hydrocarbons. *J. Chem. Phys.* 22: 320–333. https://doi.org/10.1063/1.1740059.
5 Rosenfeld, L. (1929). Quantenmechanische Theorie der Natürlichen Optischen Aktivität von Flüissigkeiten und Gasen. *Z. Phys.* 52: 161–174. https://doi.org/10.1007/BF01342393.
6 Fitts, D.D. and Kirkwood, J.G.J. (1955). The theoretical optical rotation of phenanthro[3,4-c]phenanthrene. *J. Am. Chem. Soc.* 77: 4940–4941. https://doi.org/10.1021/ja01623a085.
7 Tinoco, I. and Woody, R.W. (1964). Optical rotation of oriented helices. IV. A free electron on a helix. *J. Chem. Phys.* 40: 160–165. https://doi.org/10.1063/1.1724854.
8 Lightner, D.A., Hefelfinger, D.T., Frank, G.W. et al. (1972). Hexahelicene. Absolute configuration. *J. Am. Chem. Soc.* 94: 3492–3497. https://doi.org/10.1021/ja00765a041.
9 Brickell, W.S., Brown, A., Kemp, C.M., and Mason, S.F. (1971). π-Electron absorption and circular dichroism spectra of [6]- and [7]-helicene. *J. Chem. Soc. A*: 756–760. https://doi.org/10.1039/J19710000756.
10 Grimme, S. (2001). Calculation of frequency dependent optical rotation using density functional response theory. *Chem. Phys. Lett.* 339: 380–388. https://doi.org/10.1016/S0009-2614(01)00356-6.
11 Furche, E., Ahlrichs, R., Wachsmann, C. et al. (2000). Circular dichroism of helicenes investigated by time-dependent density functional theory. *J. Am. Chem. Soc.* 122: 1717–1724. https://doi.org/10.1021/ja991960s.
12 Autschbach, J. (2009). Computing properties from first principles theoretical methods: background and illustrative examples. *Chirality* 21: E116–E152. https://doi.org/10.1002/chir.20789.
13 Kramers, M.H.A. (1927). La Diffusion de la Lumière par les Atomes. *Atti Cong. Intern. Fisicia (Transactions of Volta Centenary Congress) Como* 2: 545–557.
14 Moscowitz, A., Krueger, W.C., Kay, I.T. et al. (1964). On the origin of the optical activity in the urobilins. *Proc. Natl. Acad. Sci. USA* 52: 1190–1194. https://doi.org/10.1073/pnas.52.5.1190.
15 Abbate, S., Lebon, F., Longhi, G. et al. (2009). Experimental and calculated vibrational and electronic circular dichroism spectra of 2-Br-hexahelicene. *Phys. Chem. Chem. Phys.* 11: 9039–9043. https://doi.org/10.1039/B909991C.
16 Johannessen, C., Blanch, E.W., Villani, C. et al. (2013). Raman and ROA spectra of (−)- and (+)-2-Br-hexahelicene: experimental and DFT studies of a π-conjugated chiral system. *J. Phys. Chem. B* 117: 2221–2230. https://doi.org/10.1021/jp312425m.
17 Abbate, S., Longhi, G., Lebon, F. et al. (2014). Helical sense-responsive and substituent-sensitive features in vibrational and electronic circular dichroism, in circularly polarized luminescence, and in Raman spectra of some simple optically active hexahelicenes. *J. Phys. Chem. C* 118: 1682–1695. https://doi.org/10.1021/jp4105695.
18 Nakai, Y., Mori, T., and Inoue, Y. (2012). Theoretical and experimental studies on circular dichroism of carbo [*n*]helicenes. *J. Phys. Chem. A* 116: 7372–7385. https://doi.org/10.1021/jp304576g.

19 Longhi, G., Castiglioni, E., Abbate, S. et al. (2013). Experimental and calculated CPL spectra and related spectroscopic data of camphor and other simple chiral bicyclic ketones. *Chirality* 25: 589–599. https://doi.org/10.1002/chir.22176.

20 Liu, Y., Cerezo, J., Mazzeo, G. et al. (2016). Vibronic coupling explains the different shape of electronic circular dichroism and of circularly polarized luminescence spectra of hexahelicenes. *J. Chem. Theory Comput.* 12: 2799–2819. https://doi.org/10.1021/acs.jctc.6b00109.

21 Negri, F., Castiglioni, C., Tommasini, M., and Zerbi, G. (2002). A computational study of the Raman spectra of large polycyclic aromatic hydrocarbons: toward molecularly defined subunits of graphite. *J. Phys. Chem. A* 106 (106): 3306–3317. https://doi.org/10.1021/jp0128473.

22 Nafie, L.A. (1996). Theory of resonance Raman optical activity: the single electronic state limit. *Chem. Phys. Lett.* 205: 309–322. https://doi.org/10.1016/0301-0104(95)00400-9.

23 Nafie, L.A. (2008). Theory of Raman scattering and Raman optical activity: near resonance theory and levels of approximation. *Theor. Chem. Account* 119: 39–55. https://doi.org/10.1007/s00214-007-0267-9.

24 Goedicke, C. and Stegemeyer, H. (1970). Resolution and racemization of pentahelicene. *Tetrahedron Lett.* 11: 937–940. https://doi.org/10.1016/S0040-4039(01)97871-2.

25 Martin, R.H. and Marchant, M.J. (1974). Resolution and optical properties ($[\alpha]_{max}$, ORD and CD) of hepta-, octa- and nonahelicene. *Tetrahedron* 30: 343–345. https://doi.org/10.1016/S0040-4020(01)91468-1.

26 Martin, R.H. and Baes, M. (1975). Helicenes: photosyntheses of [11], [12] and [14]helicene. *Tetrahedron* 31: 2135–2137. https://doi.org/10.1016/0040-4020(75)80208-0.

27 Martin, R. (1974). The Helicenes. *Angew. Chem. Int. Ed.* 13: 649–660. https://doi.org/10.1002/anie.197406491.

28 Martin, R.H., Morren, G., and Schurter, J.J. (1969). [13] helicene and [13] helicene-10,21-d_2. *Tetrahedron Lett* 10: 3683–3688. https://doi.org/10.1016/S0040-4039(01)88487-2.

29 Mori, K., Murase, T., and Fujita, M. (2015). One-step synthesis of [16]helicene. *Angew. Chem. Int. Ed.* 54: 6847–6851. https://doi.org/10.1002/anie.201502436.

30 Kim, Y.H., Tishbee, A., and Gil-Av, E. (1980). Chiral recognition by small biological molecules. Resolution of helicenes on silica gel coated with riboflavin. *J. Am. Chem. Soc.* 102: 5915–5917. https://doi.org/10.1021/ja00538a040.

31 Rulíšek, L., Exner, O., Cwiklik, L. et al. (2007). On the convergence of the physicochemical properties of [*n*]helicenes. *J. Phys. Chem. C* 111: 14948–14955. https://doi.org/10.1021/jp075129a.

32 Botek, E. and Champagne, B. (2007). Circular dichroism of helical structures using semiempirical methods. *J. Chem. Phys.* 127: 204101. https://doi.org/10.1063/1.2805395.

33 Zhao, W.-L., Li, M., Lu, H.-Y., and Chen, C.-F. (2019). Advances in helicene derivatives with circularly polarized luminescence. *Chem. Commun.* 55: 13793–13803. https://doi.org/10.1039/C9CC06861A.

34 Birks, J.B., Birch, D.J.S., Cordemans, E., and Donckt, E.V. (1976). Fluorescence of the higher helicenes. *Chem. Phys. Lett.* 43: 33–36. https://doi.org/10.1016/0009-2614(76)80750-6.

35 Otani, T., Sasayama, T., Iwashimizu, C. et al. (2020). Short-step synthesis and chiroptical properties of polyaza[5]–[9]helicenes with blue to green-colour emission. *Chem. Commun.* 56: 4484–4487. https://doi.org/10.1039/D0CC01194K.

36 Pedersen, S.K., Eriksen, K., and Pittelkow, M. (2019). Symmetric, unsymmetrical, and asymmetric [7]-, [10]-, and [13]helicenes. *Angew. Chem. Int. Ed.* 58: 18419–18423. https://doi.org/10.1002/anie.201910214.

37 Shen, C., Srebro-Hooper, M., Jean, M. et al. (2017). Synthesis and chiroptical properties of hexa-, octa-, and deca-azaborahelicenes: influence of helicene size and of the number of boron atoms. *Chem. Eur. J.* 23: 407–418. https://doi.org/10.1002/chem.201604398.

38 Murphy, V.I. and Khar, B. (2011). Planar hydrocarbons more optically active than their isomeric helicenes. *J. Am. Chem. Soc.* 133: 12918–12921.

39 Dhbaibi, K., Favereau, L., Srebro-Hooper, M. et al. (2018). Exciton coupling in diketopyrrolopyrrole-helicene derivatives leads to red and near-infrared circularly polarized luminescence. *Chem. Sci.* 9: 735–742. https://doi.org/10.1039/c7sc04312k.

40 Albano, G., Pescitelli, G., and Di Bari, L. (2020). Chiroptical properties in thin films of π-conjugated systems. *Chem. Rev.* 120: 10145–10243. https://doi.org/10.1021/acs.chemrev.0c00195.

41 Mori, T. (2021). Chiroptical properties of symmetric double, triple, and multiple helicenes. *Chem. Rev.* 121: 2373–2412. https://doi.org/10.1021/acs.chemrev.0c01017.

42 Peña, D., Cobas, A., Pérez, D. et al. (2003). Dibenzo[*a,o*]phenanthro[3,4-*s*]pycene, a configurationally stable double helicene: synthesis and determination of its conformation by NMR and GIAO calculations. *Org. Lett.* 5: 1863–1866. https://doi.org/10.1021/ol034433t.

43 Martin, R.H., Eyndels, C., and Defay, N. (1974). Double helicenes: diphenanthro[4,3-a; 3′,4′-o]picene and benzo[s]diphenanthro[4,3-a; 3′,4′-o]picene. *Tetrahedron* 30: 3339–3342. https://doi.org/10.1016/S0040-4020(01)97510-6.

44 Laarhoven, W.H. and de Jong, M.H. (1973). Photodehydrocyclizations of stilbene-like compounds VIII: synthesis of hexaheliceno[3,4-c]hexahelicene. *Recl. Trav. Chim. Pays-Bas* 92: 651–657. https://doi.org/10.1002/recl.19730920612.

45 Li, C., Yang, Y., and Miao, Q. (2018). Recent progress in chemistry of multiple hHelicenes. *Chem. Asian J.* 13: 884–894. https://doi.org/10.1002/asia.201800073.

46 Kato, K., Segawa, Y., and Itami, K. (2019). Symmetric multiple carbohelicenes. *Synlett* 30: 370–377. https://doi.org/10.1055/s-0037-1610283.

47 Tanaka, H., Ikenosako, M., Kato, Y. et al. (2018). Symmetry-based rational design for boosting chiroptical responses. *Comms. Chem.* 1: 38. https://doi.org/10.1038/s42004-018-0035-x.

48 Tanaka, H., Kato, Y., Fujiki, M. et al. (2018). Combined experimental and theoretical study on circular dichroism and circularly polarized luminescence of configurationally robust D_3-symmetric triple pentahelicene. *J. Phys. Chem. A* 122: 7378–7384. https://doi.org/10.1021/acs.jpca.8b05247.

49 Saito, H., Uchida, A., and Watanabe, S. (2017). Synthesis of a three-bladed propeller-shaped triple [5]helicene. *J. Org. Chem.* 82: 5663–5668. https://doi.org/10.1021/acs.joc.7b00486.

50 Hosokawa, T., Takahashi, Y., Matsushima, T. et al. (2017). Synthesis, structures, and properties of hexapole helicenes: assembling six [5]helicene substructures into highly twisted aromatic systems. *J. Am. Chem. Soc.* 139: 18512–18521. https://doi.org/10.1021/jacs.7b07113.

51 Berezhnaia, V., Roy, M., Vanthuyne, N. et al. (2017). Chiral nanographene propeller embedding six enantiomerically stable [5]helicene units. *J. Am. Chem. Soc.* 139: 18508–18511. https://doi.org/10.1021/jacs.7b07622.

52 Zhang, F., Michail, E., Saal, F. et al. (2019). Stereospecific synthesis and photophysical properties of propeller-shaped $C_{90}H_{48}$ PAH. *Chem. Eur. J.* 25: 16241–16245. https://doi.org/10.1002/chem.201904962.

53 Kato, K., Segawa, Y., Scott, L.T., and Itami, K. (2018). A quintuple [6]helicene with a corannulene core as a C_5-symmetric propeller-shaped π-system. *Angew. Chem. Int. Ed.* 57: 1337–1341. https://doi.org/10.1002/anie.201711985.

54 Nakakuki, Y., Hirose, T., Sotome, H. et al. (2018). Hexa-peri-hexabenzo [7]helicene: homogeneously π-extended helicene as a primary substructure of helically twisted chiral graphenes. *J. Am. Chem. Soc.* 140: 4317–4326. https://doi.org/10.1021/jacs.7b13412.

55 Fujikawa, T., Segawa, Y., and Itami, K. (2015). Synthesis, structures, and properties of π-extended double helicene: a combination of planar and nonplanar π-systems. *J. Am. Chem. Soc.* 137: 7763–7768. https://doi.org/10.1021/jacs.5b03118.

56 Tanaka, H., Inoue, Y., and Mori, T. (2018). Circularly polarized luminescence and circular dichroisms in small organic molecules: correlation between excitation and emission dissymmetry factors. *ChemPhotoChem* 2: 386–402. https://doi.org/10.1002/cptc.201800015.

57 Nakamura, K., Furumi, S., Takeuchi, M. et al. (2014). Enantioselective synthesis and enhanced circularly polarized luminescence of S-shaped double azahelicenes. *J. Am. Chem. Soc.* 136: 5555–5558. https://doi.org/10.1021/ja500841f.

58 Lakowicz, J.R. (2006). *Principles of Fluorescence Spectroscopy*, 3rde. New York: Springer.

59 Krykunov, M., Grimme, S., and Ziegler, T. (2012). Accurate theoretical description of the ^1L_a and ^1L_b excitedsStates in acenes using the all order constricted variational density functional theory method and the local density approximation. *J. Chem. Theory. Comput.* 8: 4434–4440. https://doi.org/10.1021/ct300372x.

60 Shi, B., Nachtigallová, D., Aquino, A.J.A. et al. (2019). Emission energies and stokes shifts for single polycyclic aromatic hydrocarbon sheets in comparison to the effect of excimer formation. *J. Phys. Chem. Lett.* 10 (18): 5592–5597. https://doi.org/10.1021/acs.jpclett.9b02214.

61 Weigang, O.E. Jr., Turner, J.A., and Trouard, P.A. (1966). Emission polarization and circular dichroism of hexahelicene. *J. Chem. Phys.* 45: 1126–1134. https://doi.org/10.1063/1.1727727.

62 Arnaboldi, S., Cauteruccio, S., Grecchi, S. et al. (2019). Thiahelicene-based inherently chiral films for enantioselective electroanalysis. *Chem. Sci.* 10: 1539–1548. https://doi.org/10.1039/C8SC03337D.

63 Caronna, T., Sinisi, R., Catellani, M. et al. (2001). Photochemical synthesis and optical properties of high membered thiohelicenes. *Synthetic Metals* 119: 79–80. https://doi.org/10.1016/S0379-6779(00)00749-9.

64 Yamamoto, Y., Sakai, H., Yuasa, J. et al. (2016). Controlled excited-state dynamics and enhanced fluorescence property of tetrasulfone[9]helicene by a simple synthetic process. *J. Phys. Chem. C* 120: 7421–7427. https://doi.org/10.1021/acs.jpcc.6b01123.

65 Longhi, G., Castiglioni, E., Villani, C. et al. (2016). Chiroptical properties of the ground and excited states of two thia-bridged triarylamine heterohelicenes. *J. Photochem. Photobiol. A* 331: 138–145. https://doi.org/10.1016/j.jphotochem.2015.12.011.

66 Frisch, M.J., Trucks, G.W., Schlegel, H.B. et al. (2016). *Gaussian 16, Revision C.01*. Wallingford, CT: Gaussian, Inc.

67 Domínguez, Z., López-Rodríguez, R., Álvarez, E. et al. (2018). Azabora[5]helicene charge-transfer dyes show efficient and spectrally variable circularly polarized luminescence. *Chem. Eur. J.* 24: 12660–12668. https://doi.org/10.1002/chem.201801908.

68 Moussa, M.E.S., Srebro, M., Anger, E. et al. (2013). Chiroptical properties of carbo[6]helicene derivatives bearing extended π-conjugated cyano substituents. *Chirality* 25: 455–465. https://doi.org/10.1002/chir.22201.

69 Zhao, Z.-H., Liang, X., He, M.-X. et al. (2019). Triarylborane-based [5]helicenes with full-color circularly polarized luminescence. *Org. Lett.* 21: 9569–9573. https://doi.org/10.1021/acs.orglett.9b03734.

70 Yamamoto, Y., Sakai, H., Yuasa, J. et al. (2016). Synthetic control of the excited-state dynamics and circularly polarized luminescence of fluorescent "push–pull" tetrathia[9]helicenes. *Chem. Eur. J.* 22: 4263–4273. https://doi.org/10.1002/chem.201504048.

71 Rieger, R. and Müllen, K. (2010). Forever young: polycyclic aromatic hydrocarbons as model cases for structural and optical studies. *J. Phys. Org. Chem.* 23: 315–325. https://doi.org/10.1002/poc.1644.

72 Yang, W., Longhi, G., Abbate, S. et al. (2017). Chiral peropyrene: synthesis, structure, and properties. *J. Am. Chem. Soc.* 139: 13102–13109. https://doi.org/10.1021/jacs.7b06848.

73 Bam, R., Yang, W., Longhi, G. et al. (2019). Four-fold alkyne benzannulation: synthesis, properties, and structure of pyreno[a]pyrene-based helicene hybrids. *Org. Lett.* 21: 8652–8656. https://doi.org/10.1021/acs.orglett.9b03273.

74 Sitaula, P., Malone, R.J., Longhi, G. et al. (2020). π-Extended helical nanographenes: synthesis and photophysical properties of naphtho[1,2-a]pyrenes. *ChemRxiv* https://doi.org/10.26434/chemrxiv.12931703.v1.

12

Photophysical and Chiroptical Properties of Metal–Organic Helicenic Systems: Experiment vs. Theory

Monika Srebro-Hooper[1], Jeanne Crassous[2], and Jochen Autschbach[3]

[1] *Department of Theoretical Chemistry, Faculty of Chemistry, Jagiellonian University, Krakow, Poland*
[2] *Institut des Sciences Chimiques de Rennes, CNRS, ISCR – UMR 6226, Université de Rennes 1, Rennes, France*
[3] *Department of Chemistry, University at Buffalo, State University of New York, Buffalo, NY, USA*

12.1 General Aspects

12.1.1 Introduction

The inherent helical chirality of helicenes combined with their π-delocalized electronic structure provides their derivatives with high optical activity (OA). This is reflected in large-magnitude values of optical rotation (OR) parameters and intense signals in electronic circular dichroism (ECD) spectra. Helicenes are also optically active in the infrared spectral region, therefore exhibiting vibrational circular dichroism (VCD) or Raman optical activity (ROA), and, in addition, they demonstrate emission properties, including circularly polarized luminescence (CPL). These unique photophysical and chiroptical properties along with a number of other interesting features, such as electrical conductivity, self-organization, catalytic activity, and biological activity, have allowed helicenes to find numerous applications in various areas of chemistry and materials engineering, including nanotechnology, catalysis, biological chemistry, supramolecular chemistry, and polymer chemistry. Of particular interest, especially recently, is the possibility of utilizing helicenes in optoelectronics – as, for example, molecular switches, molecular motors, logic gates, nonlinear optical materials, and emissive and semiconducting materials for organic light-emitting diode (OLED) technology [1–5]; regarding the latter, see also Chapter 15 by Fuchter and coworkers.

The unique OA of helicenes is inextricably linked to their geometric (helical twist) and electronic structure (π-conjugation). Therefore, the conventional strategy to develop new helicene-based functional materials is to modify this structure chemically. This can be efficiently achieved by, for example: increasing the number of fused aromatic rings, incorporating heteroatoms into the

Helicenes: Synthesis, Properties, and Applications, First Edition.
Edited by Jeanne Crassous, Irena G. Stará, and Ivo Starý.

π-skeleton, or functionalizing the helical core with external organic substituents. Another appealing approach to modify the electronic and (chir)optical properties of helicenes is to generate multihelicenic or metal–organic architectures incorporating helicenic units. Accordingly, in recent years, metal-based helicene derivatives have been attracting growing attention. The metal ions may furnish helical ligands with new properties, such as redox activity and phosphorescence [1–3, 6], among others. See also Chapter 5.

Rational engineering of helicene-based materials requires identifying and fully understanding the relationships between the structure of the system and its properties. Here, theoretical studies employing quantum-chemical methods and tools have proved to be extremely useful. In-depth analysis of spectroscopic measurements with the help of quantum-chemical calculations often not only affords the basis for interpreting experimental data and understanding the factors responsible for any observed trends but also provides direct information about how a measured physicochemical property depends on the atomic (geometric and electronic) structure of a given system. The identification of these relations is the first step in being able to predict experimental results qualitatively, and to design new systems with desired properties.

In recent years, a variety of metal–organic helicenic structures have been prepared and characterized in detail in our laboratories via joint experimental and computational studies. In these works, simulations of UV-vis absorption, ECD, VCD, and ROA spectra, calculations of specific $[\alpha]$ and molar $[\varphi] = [\alpha]{\cdot}M/100$ (here, M is the molecular weight) OR parameters, along with modeling of emission properties (fluorescence, phosphorescence, and CPL: wavelengths, lifetimes, and dissymmetry factors g_{lum}) were performed, and their results were analyzed in comparison with experiments to rationalize spectroscopic features of compounds of interest. While details on the synthesis and measured properties of many of these systems are provided by one of us in Chapter 5, in this contribution, we focus mainly on the corresponding theoretical research, and, by presenting representative case studies, we demonstrate the importance of computations in providing insights into structure–properties relationships in metal–organic helicenes.

12.1.2 Theory Developments in the Context of Helicene Optical Activity

The discovery of OA, both natural and magnetic field-induced, predates the development of quantum mechanics. However, a thorough and quantitative understanding of OA requires a quantum mechanics framework. We focus on the natural OA of isotropic samples. The first complete quantum-mechanical account of natural electronic OA, viz. OR, its dispersion (ORD), and ECD, is credited to L. Rosenfeld who published on the topic in 1929 [7]. An early account of the classical and quantum-theoretical treatment of OR and ECD, including Rosenfeld's theory, was provided by Condon [8]. See also more recent reviews by some of us [9, 10], monographs [11–13], and comprehensive collections [14, 15], regarding details of the formalism not covered here, unit conversions, and overviews of available computational methods and tools.

For an excitation number j out of the ground state, the isotropic rotatory strength is given by

$$R_j = \mathrm{Im}\left[\mathbf{d}_j \cdot \mathbf{m}_j^*\right] \tag{12.1}$$

Here, $\mathbf{d}_j$ and $\mathbf{m}_j$ are the electric and magnetic transition dipole vectors, respectively, for the transition. The pseudo-scalar R_j has the opposite sign for the two enantiomers of a chiral molecule and can be obtained experimentally from the integrated intensity of the ECD band for the j-th transition. Conversely, a calculated R_j can be converted to a broadened $\Delta\varepsilon$ or ellipticity of the ECD spectrum. The relationship between R_j and $\Delta\varepsilon$ is the same as that between the dipole strength D_j of a transition and the absorption coefficient ε [16]. The isotropic OR parameter $\beta(\omega)$ as a function of the circular frequency of a mono-chromatic light beam is given in terms of the rotatory strengths as

$$\beta(\omega) = \frac{2c}{3\hbar}\sum_j \frac{R_j}{\omega_j^2 - \omega^2} \tag{12.2}$$

for frequencies that do not coincide with the excitation frequencies ω_j. The previous equation (12.2) is in cgs-based Gaussian units, and the constants have their usual meaning. The OR(D) for a collection of identical molecules at a given concentration is proportional to $\omega^2\beta(\omega)$. In emission spectra, instead of differential absorption of circularly polarized light one is dealing with CPL [17, 18]. For electric dipole-allowed transitions, the associated dissymmetry factors are given by

$$g_j = \frac{4R_j}{D_j} \tag{12.3}$$

Numerous semiempirical theoretical models (vide infra) for calculating ORD and ECD for molecules were developed in the decades following Rosenfeld's seminal work, but they were not sufficiently accurate for quantitative predictions or reliable analyses of the cause of the observed OA. Albert Moscowitz summarized the state-of-the-art in the early 1960s [19, 20], see also Ref. [21]. First-principles calculations were not possible until much later. In the context of this chapter, there is notable evidence of a strong interest in the OA of helicenes in the early to mid-1960s literature on OA theory. For example, Tinoco and Woody (TW) [22] developed an "electron on a helix" model, i.e. a helical analog of the quantum-mechanical particle-in-a-box taught to every chemistry student, to assign the absolute configuration of [6]helicene by comparing its observed and calculated ORD. We mention for the interested reader here that TW also discussed a variety of the aforementioned semiempirical methods in some detail. The absolute configuration of [6]helicene was solved in 1972 [23], showing that the helix model predicted the sign of the ORD correctly, but even so TW were rightfully cautious in noting the "agreement between theory and experiment is only fair and we are suspicious even of this agreement."

First-principles calculations of natural electronic OA became available eventually, in the form of time-dependent Hartree–Fock (HF) perturbation theory (also often referred to as random-phase approximation). A review article [24] from 1980 surveyed the underlying theory, and the somewhat lamentable status quo as far as computations on real and interesting molecules were concerned. OA calculations embedded in electronic structure methods based on semiempirical fitting of parameters to avoid the costly evaluation of electron repulsion integrals were also not particularly successful. A "renaissance" in chiroptical quantum chemistry [25] took place with the development of correlated yet efficient first-principles methods that can be used to predict, confirm, and assign experimental data. At the forefront was the development of time-dependent density functional theory (TD-DFT) response methods for electronic OA. It is no surprise that some of the pioneering TD-DFT studies [26–33] showcase the applicability to helicene OR, ORD, and ECD. Coupled-cluster wavefunction methods for electronic OA were also developed around the same time [30, 34], although the scaling of their computational effort with the system size continues to limit their range of applications, unless further approximations are applied [35].

Natural OA also manifests itself in circular dichroism of the absorption spectrum for vibrational transitions (VCD) [36–41]. Similar to ECD, VCD is characterized in terms of the rotatory strengths of the vibrational transitions. Computational VCD methodology has been around for quite some time [40–45]. HF calculations were reasonably successful in predicting VCD spectra, but the field also has benefited tremendously from the development of corresponding DFT methods [46, 47]. The VCD effect is rather subtle, and therefore the calculations must include terms that arise from the vibronic coupling. For brevity, we leave a discussion of the VCD theory to the literature cited already. Likewise, we refer the reader to the relevant literature regarding the theoretical and computational framework for calculating Raman OA (ROA) intensities [48–56]. In a nutshell, in the usually adopted short-time approximation, the differential Raman scattering of circularly polarized light is determined from invariants formed by derivatives of the electric dipole–dipole or dipole–quadrupole polarizability tensor and the derivatives of the mixed electric dipole–magnetic dipole polarizability tensor. The derivatives are taken with respect to the vibrational normal modes. The latter tensor has an isotropic average of $-\omega\beta(\omega)$ and is therefore seen to be related closely to the electronic OA of the molecule. In particular, when the circular frequency ω of the incident light in the ROA experiment becomes close to one of the electronic resonances ω_j, there is the possibility of a strong resonance enhancement, along with other effects.

Due to the sizes of helicene derivatives, and metal–organic helicenic systems in particular, that are typically encountered in contemporary research, the only practical computational methods for describing their ground- and excited-state properties and OA are presently DFT and TD-DFT (and, with limitations, certain coupled-cluster approximations). In the remainder of this section, some remarks regarding the state-of-the-art of DFT-based calculations are provided.

Satisfactory reproduction of the experimental chiroptical properties of helicenes often requires the use of exchange-correlation functionals with a high contribution of exact exchange (eX) [35, 57–59], such as the global hybrid BHLYP (eX: 50%) [60, 61] or range-separated functionals (RSFs) [62], in which the amount of eX depends on the interelectronic distances and changes from a functional-specific minimum threshold up to a maximum of 100%. Further improvements between calculated and experimental results may be obtained by using RSFs with a system-specific parameterization that is derived non-empirically by minimizing the electron delocalization error and/or enforcing the vertical ionization potential (IP) condition, HOMO energy = –IP [58, 63–65]. Note that in computational studies of some metal–organic helicenes presented next, global hybrids with a lower fraction of eX (B3LYP: 20% [61, 66, 67]; PBE0: 25% [68]) were sometimes found better suited. For π-chromophores such as those discussed herein, and calculations at the (TD-)DFT level, the use of standard double-zeta valence-polarized basis sets, without polarization functions for hydrogen atoms and diffuse functions, such as SV(P) [69], in the calculations of chiroptical properties does not significantly deteriorate the quality of the obtained results [59].

The results of OA calculations for helicene derivatives are typically very sensitive to the calculated parameters of the molecular structure of the system (in particular, the distance between the helicene terminal rings, i.e. helical pitch). Those parameters are strongly dependent on the inclusion of dispersion effects (for example, via the popular semiempirical dispersion corrections D3 [70]) in the geometry optimizations [35, 57–59].

Because the measurements of photophysical and chiroptical properties for helicenic systems are usually carried out in aprotic, nonpolar solvents (typically dichloromethane), the inclusion of solvation effects via continuum models (such as conductor-like screening model (COSMO) or polarizable continuum model (PCM)) [71] in the calculations of these properties (with the exception of specific cases such as, for example, modeling of protonated helicene derivatives [72], or calculations of phosphorescence lifetimes [73]) has shown no great impact on the obtained agreement with experimental data.

For closed-shell systems, triplet (near-)instability of the ground-state is sometimes encountered in (TD-)DFT calculations that employ functionals with large fractions of eX. More reliable results for modeling of triplet excited states in organometallic complexes with helicene ligands are thus often obtained by using the Tamm-Dancoff approximation [74, 75] in TD-DFT [76–79].

Purely electronic descriptions of photophysical and chiroptical properties of helicene derivatives are often sufficient to rationalize experimentally observed trends, although to reproduce the fine structures present in the measured spectra, vibronic contributions need to be taken into account [80–85]. Nonequilibrium structure effects might also be of importance [86].

Finally, we note recent developments in so-called simplified TD-DFT methods for calculations of OR and ECD for large molecular systems that were shown to provide a satisfactory agreement between simulated and experimental results also for helicene-based systems, with a tremendous reduction in the computational cost [86–89]. This may pave the way for routine modeling of large helicenic compounds.

12.2 Case Studies

12.2.1 Azahelicene-Phosphole Pd(II) and Cu(I) Complexes: OR and ECD Analyses

12.2.1.1 Complexes with Extended π-Conjugated Helicenic P,N Bidentate Ligands

As seen in Chapter 5, heteroditopic aza[6]helicene-phosphole **1** is a 1,4-chelating ligand [90, 91]. It possesses two coordination centers with different stereo-electronic properties ("hard" nitrogen and "soft" phosphorus) and upon coordination to square-planar d^8-metal center Pd(II) and tetrahedral d^{10}-metal center Cu(I), it gives access to C_2-symmetrical complexes in a controlled manner. Namely, the complexation of **1** with Pd(II) and Cu(I) yielded stable metal-bis(helicene) complexes **2** and **3** (Figure 12.1a).

It is striking that **2** has a molar OR (MR) that is almost twice the MR of **3** (see values in Figure 12.1b). This immediately raises the question of whether the metal orbitals are somehow responsible for this difference, by participating in the relevant transitions that cause the OR according to Equation (12.2). Indeed, the ECD spectra for the two systems show a much higher ECD intensity at low photon frequencies (long wavelengths, energies below ca. 4 eV) for **2** (Figure 12.1c), which is expected to translate into a larger OR because of the denominators in Equation (12.2).

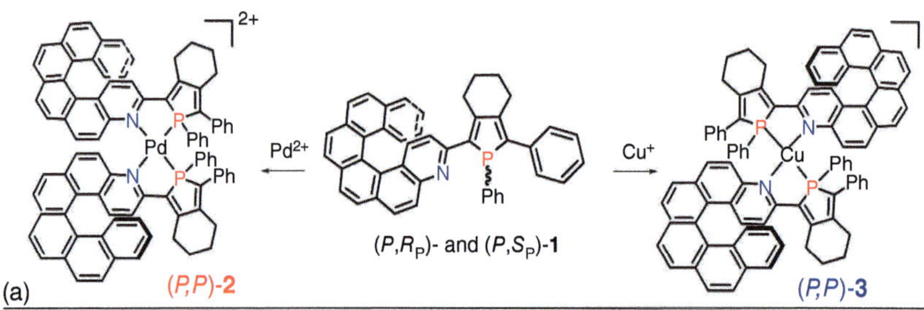

Expt. OR	$[\alpha]_D^{23}$	$[\phi]_D^{23}$
(P,R_P^*)-**1**	+1350	+8200
(P,P)-**2**	+1275	+23080
(P,P)-**3**	+910	+13120

(b)

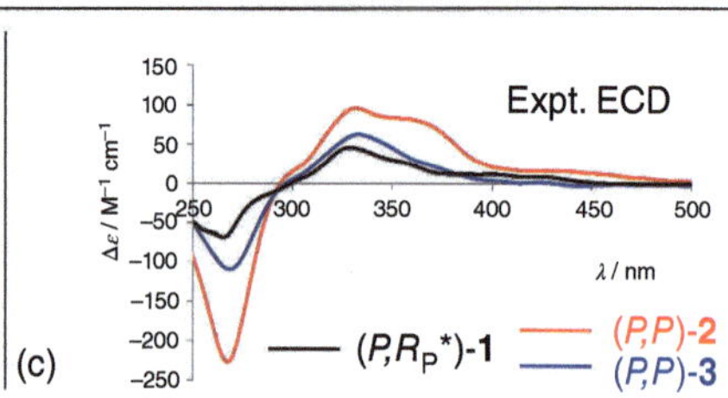

Figure 12.1 (a) Stereoselective complexation process of aza[6]helicene-phosphole **1** acting as a 1,4-P,N-chelate ligand to Pd(II) and Cu(I) complexes **2** and **3**. (b) Experimental (dichloromethane) specific and molar rotation values, in respectively 10^{-1} deg cm^2 g^{-1} and deg cm^2 dmol^{-1} units, and (c) the corresponding ECD spectra for compounds **1** (black), **2** (red), and **3** (blue). Source: Based on Ref. [90].

However, a painstaking analysis of the ECD and MR of the two complexes painted an unexpected picture as far as the role of the metal centers is concerned [91].

First, calculated and experimental MRs and ECD spectra were in good agreement, allowing for reliable assignment of the observed effects of the metal substitution by computations. Second, "sum-over-states" (SOS) analyses, based on Equation (12.2), confirmed that the large difference in the MR between **2** and **3** is mainly due to the large intensity differences in the ECD bands below ca. 4 eV (above ca. 300 nm). See Figure 12.2a and b for representative results. However, although the metal orbitals were found to be involved in numerous excitations (see Figure 12.2c), their overall participation in the OA was not decisive. Instead, the critical factor appeared to be the relative positions of the phenyl groups of one azahelicene-phosphole ligand to the azahelicene moiety of the other ligand. In a nutshell, the preferred coordination of d^8 Pd(II), square-planar, leads to distorted square-planar coordination of Pd in **2**. The preferred coordination of d^{10} Cu(I), tetrahedral, leads to distorted tetrahedral coordination of Cu in **3**. These different metal coordination geometries result in large differences in the distances and orientations of the phenyl groups of the phosphole ligands relative to the terminal rings of the azahelicene moieties. The theoretical analysis showed that the MR caused by the azahelicene becomes much smaller when another phenyl group is in proximity, likely due to excitonic effects. The flexibility of Cu to adopt a distorted tetrahedral geometry allows the

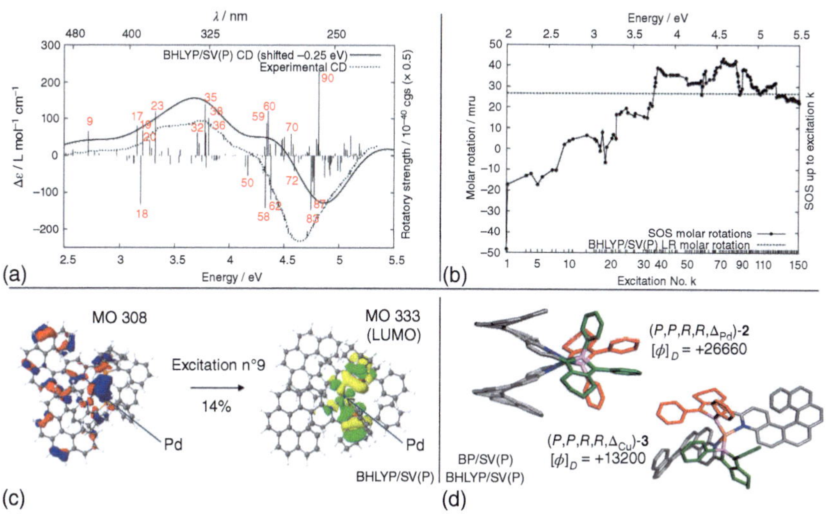

Figure 12.2 (a) Simulated (redshifted by 0.25 eV) ECD spectrum of $(P,P,R_P,R_P,\Delta_{Pd})$-**2** compared with experiment. Numbered excitations are those selected for MO-pair contributions and "sum-over-states" (SOS) analyses [90, 91]. (b) Calculated SOS molar rotation for $(P,P,R_P,R_P,\Delta_{Pd})$-**2**. $\omega^2\beta(\omega)$, in 10^3 deg cm^2 dmol^{-1} units abbreviated as mru, plotted as a function of k, the number of excitations included in the SOS. The converged linear-response result is indicated by a dotted line. (c) Metal-centered MOs (±0.03 au) of $(P,P,R_P,R_P,\Delta_{Pd})$-**2** involved (14%) in its excitation n°9. (d) Optimized structures for **2** and **3** along with the corresponding calculated MR values, in deg cm^2 dmol^{-1} units. Source: Panels a-c: Adapted from Refs. [90, 91]. Copyright 2009, ACS and 2010, Wiley. Panel d: Based on Ref. [91].

phenyl groups to crowd next to the helicenes (in the Cu Δ configuration) and cause a lower MR than would be possible from a distorted square-planar geometry (see Figure 12.2d). This highlights how the nature of the metal (coordination geometry and electronic interaction) can impact the amplitude of the chiroptical properties in metal–organic helicenic systems.

12.2.1.2 Assemblies with Multitopic Helicenic Ligands

Chapter 5 also presented multitopic 2,6-bis(2-pyridyl-phosphole)-pyridine and 2,6-bis(aza[6]helicene-phosphole)-pyridine, **4a** and **4b** in Figure 12.3a and c, respectively, which appear well-suited to generate helicates by acting as N,P,N,P,N-helicands upon an original coordination mode on Cu(I)-dimers [92]. The enantiopure *M*- or *P*-azahelicenes yielded enantiopure helicates (−)- and (+)-**5b**, which closely assembled two helicene-capped helicands **4b** around four Cu(I) centers, and resulted in strong chiroptical activity. Indeed, as seen in Figure 12.4a, complex (+)-**5b** displays more intense ECD bands than those of (+)-**4b** and demonstrates the additional intensity of moderate magnitude between 400–600 nm.

The Cu(I)-helicate core present in **5b** was characterized in detail via tetrametallic complex $[Cu_4(\mathbf{4a})_2](BF_4)_4$ (**5a**) that was prepared from **4a** (see Figure 12.3a), and for which X-ray crystallography revealed the double-stranded helicate structure built on two heterotopic N,P-helicands, as illustrated in Figure 12.3b. Calculations of the chiroptical properties for **5a** (note that enantiopure **4a** and **5a** were not obtained, so they could not be studied experimentally) enabled to elucidate the ECD signature for the N,P,N,P,N-helicand/Cu(I)-helicate core in **5b** and to shed some light on the origin of the low-energy ECD bands appearing upon complexation of **4b** into **5b**. The simulated ECD spectrum of (+)-**5a** (Figure 12.4b) displays a positive band at ca. 375 nm that corresponds well to the experimental positive band around 400 nm observed for **5b**. This suggests an efficient chiral transfer from the azahelicene moieties incorporated within large π-conjugated ligands in **4b** to the helicate metal-ligand core in **5b**. Furthermore, molecular orbital (MO)-pair analysis of dominant excitations demonstrated that the overall involvement of the Cu orbitals

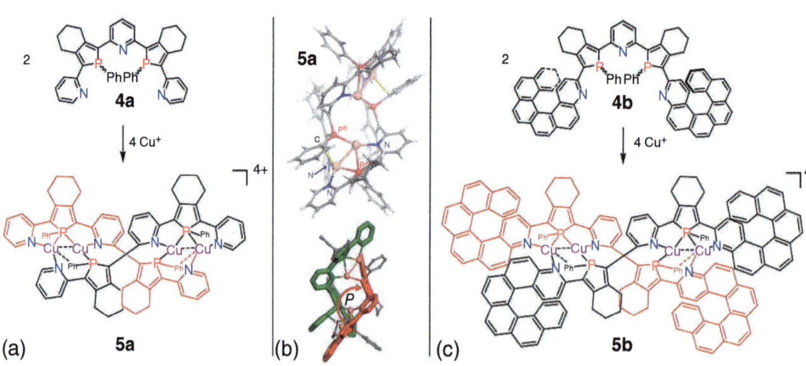

Figure 12.3 (a) and (c) Coordination-driven synthesis of helicates **5a,b** from helicands **4a,b**. (b) X-ray structure of helicate **5a** and the two highlighted *P*-(+)-helicands. Source: Panels a and c: Based on Ref. [92]. Panel b: Adapted from Ref. [92]. Copyright 2013, Wiley.

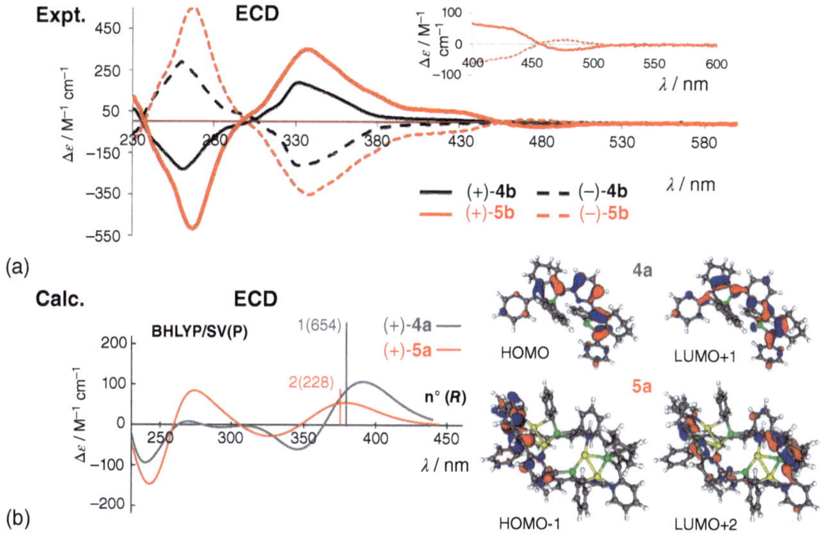

Figure 12.4 (a) Experimental (dichloromethane) ECD spectra of **4b** (black) and **5b** (red) with enlargement of the low-energy ECD bands of **5b** shown in the inset. (b) Left: Simulated ECD spectra of **4a** (grey) and **5a** (orange) along with the corresponding dominant low-energy excitations (with rotatory strength *R*, in 10^{-40} cgs units, values given in parentheses). Right: Selected MOs (±0.04 au) involved in the numbered low-energy excitations for **4a** and **5a**. Source: Based on Ref. [92].

in the ECD-active bands is negligible, indicating that the Cu(I)-dimers do not electronically determine the chiroptical property of helicates **5a** and **5b**. Indeed, the low-energy ECD band in **5a** is due to π-to-π* transitions within π-conjugated helicand moieties that show signatures of intraligand charge transfer (ILCT) and, quite surprisingly, strong ligand-to-ligand CT (LLCT) character, while no metal-to-ligand CT (MLCT) takes place. See, for example, excitation n°2 for **5a** that displays high rotatory strength and involves HOMO-1 and nearly-degenerate LUMO+2 and LUMO+3 as well as LUMO and LUMO+1 (Figure 12.4b). LLCTs were also found to contribute to many other transitions with high rotatory strengths, such as the ones underlying the simulated band centered around 275 nm. Notably, the computed ECD spectral envelopes for the helicand structure of **4a** in its frozen (+) helicity appears to be overall similar to that for its corresponding Cu complex **5a** although with visibly decreased intensity at wavelengths shorter than 300 nm that highlights a key role of helicand-to-helicand LLCT transitions in determining strong OA of the Cu(I)-helicates. Accordingly, the theoretical studies clearly showed that the role of metal centers is to assemble ligands **4a,b** and to freeze their chirality upon an original coordination mode involving both their N,μ-P,N- and P,N-moieties, and, by holding two helicands closely together, to allow both ILCT and LLCT to take place resulting in high chiroptical activity.

To conclude this section, we mention briefly bimetallic gold(I)-alkynyl-helicene architectures obtained via coordination to bis-phosphole ligands [93]; see Figure 5.10 for structure visualizations. Here again, detailed experimental and computational studies showed that Au(I) ions play rather an indirect role in shaping chiroptical properties of the complexes, by assembling bis-phosphole bidentate ligands with ethynyl-helicene monodentate ones. Furthermore, calculations enabled to highlight the importance of (i) the presence or absence of Au(I)–Au(I) interactions, (ii) helicene-to-bis-phosphole LLCTs, and (iii) positions (distance and orientation) of the helicenic units with respect to each other in determining the ECD and OR responses of these metal–organic helicenes. Interestingly, effects of both exciton coupling between the chromophore moieties and strong LLCT-type transitions, described earlier (vide supra) for azahelicene-phosphole Pd(II) and Cu(I) complexes, seem to reinforce each other for these Au(I)-helicenic derivatives to provide them with strong chiroptical properties.

12.2.2 Ru- and Fe-[6]helicene Redox-Triggered Chiroptical Switches: UV-vis, OR, ECD, VCD, and ROA Analyses

12.2.2.1 Ruthenium-vinyl-helicenes

References [94, 95] present mono- and bis-ethylene-[6]helicene derivatives **6a** and **6b** that are functionalized with one or two ruthenium(II)-based organometallic fragment(s) to generate **7a** and **7b** complexes, in which the Ru^{2+} ion(s) is (are) linked to the helicene via a vinyl bridge (see Figure 12.5a and Chapter 5). This was demonstrated to have a great impact on physicochemical and (chir)optical properties of the systems. Calculations showed that the unsaturated carbon linker allows for effective electronic interaction between the metal d orbitals and the delocalized helicene π-system resulting in the presence of π-electron density that extends from the metal through the vinyl bridge to the helicene group; this is clearly visible, for example, in the canonical orbitals (see HOMO for **7a** in Figure 12.5b). The extensive π-conjugation leads to the involvement of Ru orbitals in the low-energy electronic excitations of the complexes, which accounts for the differences in the observed chiroptical properties of these compounds (in particular: the presence of additional intensity in the low-energy region of the ECD spectra and the corresponding strong increase in OR values) when compared to their parent helicenes (see Figure 12.5c for a comparison of the spectra of **6a** and **7a**). The presence of the metal fragment(s) also results in intense IR and VCD spectra of **7a,b** that demonstrate typical helicene signatures with additional features originating from the substituent, as confirmed via computations (see Figure 12.5d for representative data) [95].

More interestingly, in combination with the advantageous redox properties of ruthenium, the Ru–[6]helicene π-electron coupling was also shown to be responsible for the effective, reversible oxidation/reduction processes of the helicene at fairly low potentials (Figure 12.5a). These processes are accompanied by pronounced changes in the compounds' ECD spectra (see Figure 12.5c for **7a** / $[\mathbf{7a}]^{\cdot+}$), which established the first electrochemical redox chiral switch that was based on a pure [6]helicene moiety [4, 5]. The calculated electron spin density of the oxidized complexes clearly shows, although the oxidation formally takes place on the ruthenium ion ($Ru^{2+} \rightarrow Ru^{3+}$), the spin density created in this process is not only localized on the metal ion but also spread out over the helical π-ligand (Figure 12.5b). Therefore,

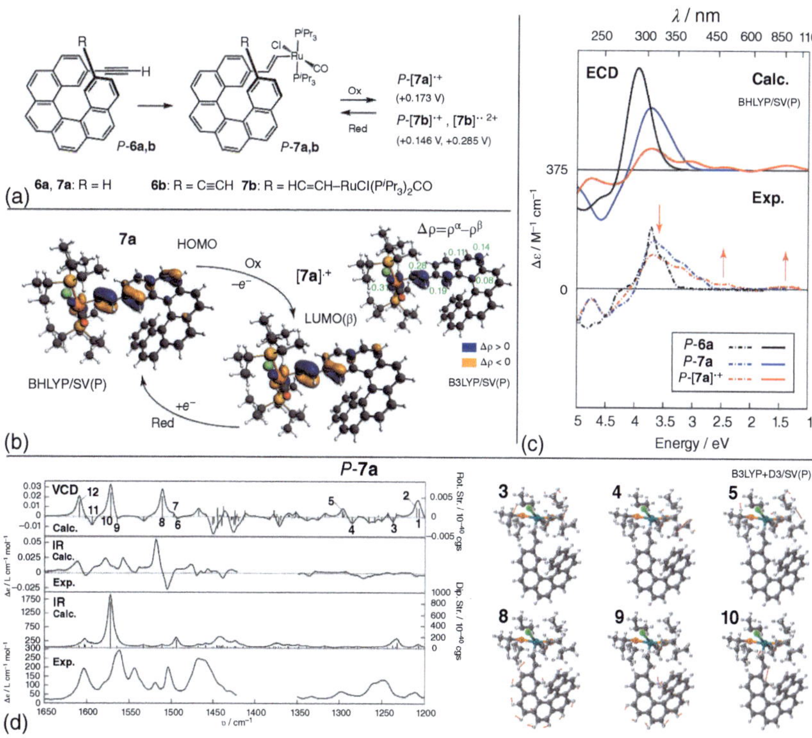

Figure 12.5 (a) Mono- and bis-ethynyl-helicenes **6a** and **6b** along with the corresponding mono- and bis-ruthenium(II)-vinyl-helicene complexes **7a** and **7b** and their reversible oxidation (Ox)/reduction (Red) processes. The numbers listed are oxidation potentials versus ferrocene/ferrocenium standard. (b) Isosurfaces (±0.05 au) of frontier orbitals, HOMO for **7a** and LUMO in a set of orbitals describing β-spin electrons (representing the "electron hole" created upon oxidation) for $[\mathbf{7a}]^{\cdot+}$, and isosurface (±0.005 au) of electron spin density $\Delta\rho$ for $[\mathbf{7a}]^{\cdot+}$ with fractions of the total integrated spin density obtained from Mulliken decomposition. (c) Experimental (dichloromethane and dichloroethane) and simulated ECD spectra for **6a**, **7a** and $[\mathbf{7a}]^{\cdot+}$. Arrows indicate changes in the spectrum of **7a** occurring as a result of its oxidation. (d) Experimental (dichloromethane) and simulated IR and VCD spectra for **7a**. Numbered vibrations correspond to those analyzed in detail; atomic displacements for selected normal modes are shown on the right. Source: Panels a-c: Based on Refs. [94, 95]. Panel d: Adapted from Ref. [95]. Copyright 2015, Wiley.

the oxidation/reduction necessarily affects the electronic and chiroptical properties of this chromophore. The corresponding computational analysis of the ECD spectra for oxidized species provided a detailed interpretation of the observed spectral changes in terms of underlying excitations. It was determined that an increase in the involvement of ruthenium orbitals in electronic transitions for Ru^{3+}-based complexes is the main origin of the observed changes [95]. In particular, the presence of orbitals that represent the "electron hole" created upon the oxidation of the compound (see LUMO for $[\mathbf{7a}]^{\cdot+}$ in Figure 12.5b) leads to low-energy CT transitions (ligand-to-metal CT (LMCT), MLCT, as well as ILCT, ligand = helicene fragment).

12.2.2.2 Iron-Ethynyl-Helicenes

Ligands **6a,b** were also used to generate metal–organic systems based on Fe(II) ions bound to [6]helicene via an alkynyl linker (**8a,b** in Figure 12.6a, see Chapter 5 for synthetic details), which demonstrated that the idea of creating a redox-triggered chiroptical switch based on a [6]helicene moiety can be successfully implemented by using metal centers other than Ru, and unsaturated carbon bridges other than vinyl fragments [96, 97]. As in the case of the Ru-based systems **7a,b**, the Fe compounds also exhibit an effective π-electron metal–helicene conjugation, thereby enabling delocalization of the electron spin density over part of the helicene upon oxidation at the metal. Compared to **7a,b**, the amount of unpaired electronic charge on the [6]helicene fragment, generated via the oxidation ($Fe^{2+} \rightarrow Fe^{3+}$), is much lower here (see, for example, electron spin density for $[\mathbf{8a}]^{\cdot+}$ in Figure 12.6b), which results in higher stability of oxidized species of these compounds and enables their operation as chiroptical molecular switches not only in the visible and ultraviolet spectral regions (changes in UV-vis and ECD spectra, and in OR (D-line) parameters), but also in the near- and far-infrared spectral range (changes in VCD and ROA spectra, and in OR (1.54 μm) parameters).

The systems were thoroughly studied via calculations that enabled their full characterization: (i) A detailed interpretation of the changes in UV-vis and ECD spectra observed for these complexes upon oxidation emerges in terms of underlying

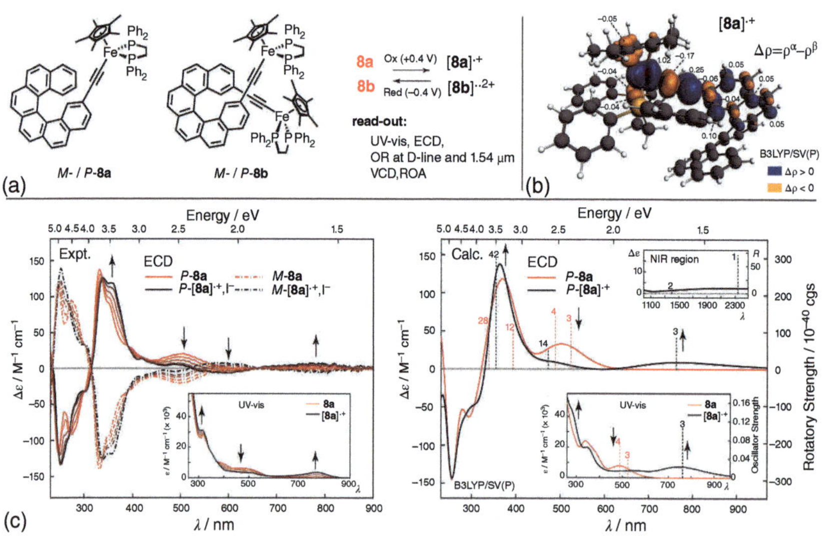

Figure 12.6 (a) Mono- and bis-iron(II)-ethynyl-helicene complexes **8a** and **8b** and their redox-triggered chiroptical switching (at ± 0.4 V versus a Pt pseudopotential reference electrode). (b) Isosurface (±0.0015 au) of electron spin density $\Delta\rho$ for $[\mathbf{8a}]^{\cdot+}$ with fractions of the total integrated spin density obtained from Mulliken decomposition. (c) Experimental (dichloromethane) and simulated (with the corresponding dominant excitations) UV-vis and ECD spectra showing changes (highlighted by arrows) in the spectral properties of **8a** upon its chemical oxidation. Source: Panel a: Based on Refs. [96, 97]. Panels b and c: Adapted from Ref. [97]. Copyright 2018, Wiley.

excitations (see Figure 12.6c for **8a**/[**8a**]$^{\cdot+}$). For example, modifications in the low-energy spectral regions (as in the case of the aforementioned Ru systems) are due to electronic transitions that involve orbitals representing the "electron hole" generated upon oxidation, while the increase in ECD intensity at higher energy that is observed for the oxidized species originates from the stronger helicene character of excitations occurring in this range. (ii) An increase and sign inversion of OR parameters that were observed at 1.54 μm upon oxidation (for example: from MR of 2370 to −4380 deg cm^2 $dmol^{-1}$ for **8a**/[**8a**]$^{\cdot+}$) are associated with the presence of ECD-active electronic excitations around this wavelength for the oxidized species (see Figure 12.6c), which causes anomalous OR dispersion that can change the sign of the OR. (iii) Experimental interpretation of the VCD spectra was confirmed, highlighting the strong signatures originating from alkynyl and helicene moieties accompanied by strong contribution of the metal(s) to the vibration intensities. (iv) The calculated ROA spectra were analyzed in terms of electronic excitations, investigating potential deviations from the two-state model of resonance ROA (commonly used in the literature). (It has been confirmed in the meantime that near- and at-resonance ROA of metal complexes is not necessarily monosignate [98].)

12.2.3 Platinahelicenes: UV-vis, OR, ECD, Non-polarized Emission and CPL Analyses

12.2.3.1 Mono- and Bis-cycloplatinated Helicenes

The synthesis of the first cycloplatinated helicene **12** in 2010 [99] (see Figure 12.7a) paved the way for straightforward modification of electronic and spectroscopic properties of helicenic systems via incorporation of metallic ions into a helical π-conjugated backbone. As elaborated in Chapter 5, numerous metallahelicenes have been developed since then, and among them, several other Pt(II)-based complexes, also including Pt-[8]helicene **10** and Pt_2-[6]helicene **11** [100] shown in Figure 12.7a. Experimentally, the considered series of systems demonstrates intense ECD signatures and red phosphorescence at room temperature with efficient CPL, see Figure 12.7b and c, and first-principles modeling was used to analyze effects of the size of the helicene, the helicity, and the role of the metal center(s) in shaping their properties [99–101].

Theoretical simulations, correctly reproducing important experimentally observed trends (albeit with some discrepancies as far as relative peak intensities for the different compounds are concerned) as illustrated in Figure 12.7b, revealed the crucial contribution of the Pt ion(s) to the chiroptical properties of these metallahelicenes via (i) creation or enlargement of the helicoidal framework due to the ring closure that accompanies orthometallation, and (ii) affecting electronic properties of the π-conjugated system by efficient interaction of the Pt d orbitals with those of the ligand (see Figure 12.7d). In particular, the increased π-conjugation path accounts for the redshift and enhanced intensity of the ECD spectrum for Pt-[8]helicene **10** vs. its parent ligand pyridine-[6]helicene **9**, while pronounced involvement of metal orbitals in the intense (and primarily of π-to-π* nature) excitations of Pt_2-[6]helicene **11** and Pt-[6]helicene **12** explains differences in their ECD envelopes as compared

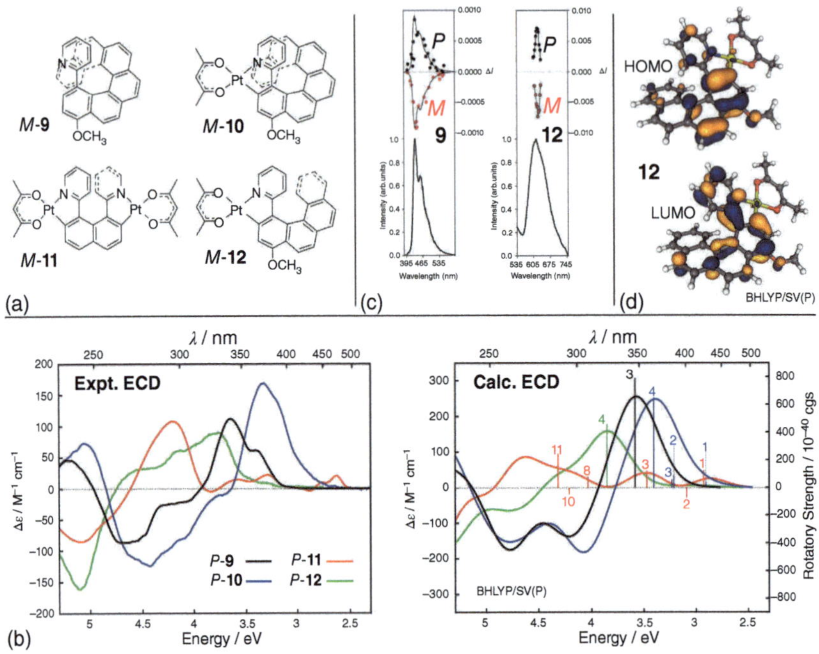

Figure 12.7 (a) Example cycloplatinated complexes **10-12** studied in Refs. [99–101] along with reference helicene ligand **9**. (b) Experimental (dichloromethane) and simulated (redshifted by 0.25 eV) ECD spectra of **9-12**. Numbered excitations were analyzed in detail. (c) CPL and total luminescence spectra (respectively top and bottom in each panel, dichloromethane) of ligand **9** and complex **12**. (d) Isosurfaces (±0.04 au) of frontier MOs for **12**. Source: Panels a, b and d: Based on Refs. [99, 100]. Panel c: Adapted from Ref. [100]. Copyright 2014, RSC.

to **9**. Visible participation of the Pt d orbitals was also established for T_1-to-S_0 phosphorescence transitions in **10-12** and high structural similarity between their S_1- and T_1-excited states was demonstrated that may indeed facilitate a rapid intersystem crossing responsible for the elimination of the fluorescence processes in these compounds. Calculations with multireference wavefunction-based methods including spin-orbit coupling (SOC) effects predicted nonzero positive rotatory strengths for the T_1-to-S_0 emission for **12** that are consistent with the sign of the CPL in the experiment [102]. Very recently, TD-DFT calculations for the complex confirmed the previous wavefunction-based calculations and helped to reconcile previously noted discrepancies between the calculated and observed g_{lum}.

12.2.3.2 Bipyridine-Helicene-Based Multifunctional pH-Switchable Platform

As mentioned in Chapter 5, bipyridine-helicene **13** can act as a C,N chelate leading to cycloplatinated complex **14** (Figure 12.8a), and both systems, thanks to the presence of non-coordinated nitrogen atom(s), established unprecedented multi-functional pH-triggered switches [103]. Unlike for Pt(II)-based systems described earlier, a computational analysis did not indicate a strong involvement of the metal d orbitals in the lower-energy excitations for **14**, assigning them as predominantly

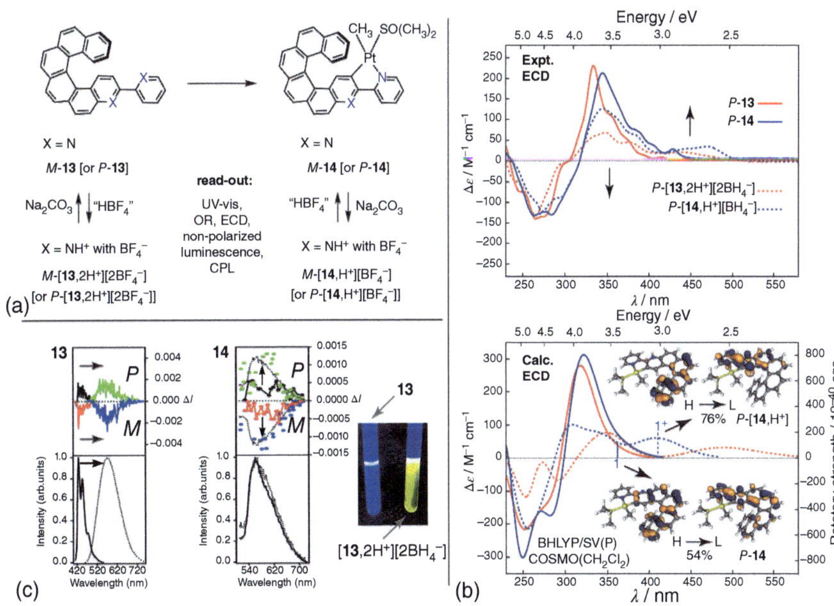

Figure 12.8 (a) Multifunctional acid-/base-triggered switches based on organic and cycloplatinated helicenes **13** and **14**. (b) Experimental (dichloromethane) and simulated ECD spectra of **13** and **14** and of their corresponding protonated species. Arrows in the top panel indicate changes in the spectra of **13** and **14** occurring as a result of their protonation. Isosurfaces (±0.04 au) of HOMO (H) and LUMO (L) of **14** and [**14**,H$^+$] that contribute to the lowest-energy excitations. (c) CPL and total luminescence spectra (respectively top and bottom in each panel, dichloromethane/acetone) of **13** and **14** (black and red/black lines) and of their protonated species (green and blue/dashed lines; changes in the spectra occurring upon protonation indicated additionally by arrows) along with visual fluorescence switching of **13**. Source: Panels a and b: Based on Ref. [103]. Panel c: Adapted from Ref. [103]. Copyright 2015, Wiley.

π-to-π* within bipyridine ligand with partial ILCT from helicene moiety to pyridine fragment (see, for example, visualization of a dominant MOs pair contributing to the lowest-energy excitation presented in Figure 12.8b), thus showing their akin character to that of low-energy excitations in **13**. This explains overall similar ECD spectral envelopes obtained for both systems (Figure 12.8b). Accordingly, the main effect of the metal was recognized to be to rigidify the structure of the chromophore and to promote effective intersystem crossing that results in red phosphorescence of the compound at room temperature with no fluorescence band (see Figure 12.8c).

More importantly, calculations enabled also rationalizing pronounced modifications of photophysical and chiroptical properties occurring for both systems upon their protonation (see Figure 12.8b, c). For example, it was demonstrated that the appearance of low-energy ECD bands for the protonated species is due to the stabilization of the LUMO, reducing the HOMO-LUMO gap, and increase in the helicene-to-bipyridine CT component in the predominantly HOMO-to-LUMO lowest-energy excitations. The stronger CT character emerges as an effect of (i) a loss of π-conjugation in the frontier orbitals caused by a reduction in aromaticity of the pyridine moiety, as evidenced by nucleus-independent chemical shift calculations, and (ii) an increase of the overall contribution of the HOMO-LUMO pair to the transition upon protonation (see Figure 12.8b for illustrative results for **14**/ [**14**,H^+]). The interplay between effects of HOMO-LUMO character of electronic transition and LUMO stabilization also explains drastically different changes in luminescence properties of **13** and **14** occurring due to their protonation. Namely, for the ligand, the S_1-to-S_0 transition affords a slightly larger contribution from the HOMO-LUMO pair, and accordingly, the strong stabilization of the LUMO results in a large redshift of the emission upon protonation (see Figure 12.8c). The decreased HOMO-LUMO character of the T_1-to-S_0 transition for the complex diminishes the effect of the LUMO-level stabilization, leading to only a very subtle decrease in the phosphorescence energy for the corresponding protonated species.

12.2.4 Re(I) and Ir(III) Helicenic Phosphors: OR, ECD and Non-polarized Emission Analyses

12.2.4.1 Bipyridine-Helicene-Rhenium Complexes

N,N-bidentate coordination mode of bipyridine-helicene ligand **13** was further utilized to obtain the first examples of Re(I)-based phosphors **15** and **16** (see Figure 12.9a and Chapter 5) that exhibit CPL [104]. Here again, calculations showed no strong involvement of the metal d orbitals in the helicenic π-electron system, but indicated that the main role of Re is to help increase the π-conjugation pathway and to polarize π-electron density within the ligand compared to the parent helicene. This polarization is visible in the corresponding canonical orbitals (i.e. in both **15** and **16**, the π-density shifts toward the helicene moieties' terminal rings in the HOMO and toward the bipyridine fragment in the LUMO, see Figure 12.9b) and appears similar to the effect of the protonation of **14** with quite analogous consequences for ECD spectral trends (vide supra). Namely, the theoretical analysis demonstrated that the visible differences in low-energy ECD between the complexes **15** and **16** and the ligand **13** arise predominantly from ILCT electronic transitions within bipyridine-helicene (helicene-to-bipyridine) and whose quantitative degree of CT strongly depends on the nature of the ligand bound to Re in the axial position and, accordingly, the neutral versus cationic character of the complex (see Figure 12.9b). For example, the lowest-energy excitation shows a higher contribution from the HOMO-to-LUMO transition for cationic complexes with dimethylphenyl isocyanide (**16**^{1,2}) than for neutral systems with chloride in the axial position (**15**^{1,2}); this effectively increases the CT character of the excitation and leads to its shift toward lower energy values and, thus, to the emergence of additional intensity in the low-energy region of the ECD spectrum for **16**^{1,2}.

It was also shown that the type of ligand bound to Re in the axial position is of key importance for the observed emission properties of the complexes, in particular, the phosphorescence lifetime, the value of which is at least an order of magnitude higher for **16**^{1,2} than for **15**^{1,2} (see Figure 12.9c). Although the results of calculations of the emissive T_1 excited-state properties of **15** and **16** deviated numerically from

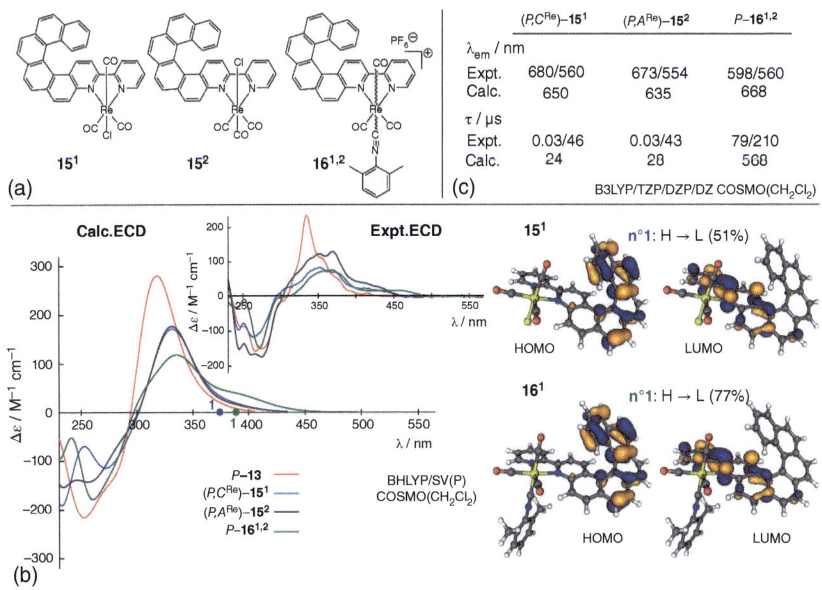

Figure 12.9 (a) Rhenium(I) complexes with bipyridine-helicene ligand **15** and **16**. Superscripts 1 and 2 differentiate diastereoisomers that differ in absolute configuration at the metal center (*C*, *A*) but demonstrate the same stereochemistry of the helicene moiety (either *P* or *M*). (b) Left: Simulated and experimental (dichloromethane) ECD spectra of complexes **15** and **16** and ligand **13**. First excitation energies for **15** and **16** are indicated by dots on the abscissa. Right: Isosurfaces (±0.04 au) of HOMO (H) and LUMO (L) involved in the lowest-energy electronic excitation for **15**^{1} and **16**^{1}. (c) Experimental (298 K, dichloromethane/77 K, diethyl ether/isopentane/ethanol) and computed photophysical properties of **15** and **16**. Source: Based on Ref. [104].

the experiments, the trend was correctly reproduced. A detailed analysis of the computed (involving relativistic SOC effects) T_1 excited states showed that the observed differences in the phosphorescence lifetimes of the systems are due to the different nature of this state – predominantly ^{3}MLCT in the case of **15**^{1,2} and predominantly ^{3}ILCT (within the helicenic ligand) in the case of **16**^{1,2}. As the involvement of Re orbitals facilitates the formally spin-forbidden T_1-to-S_0 phosphorescence transitions via SOC, decreasing the corresponding lifetimes, their diminished participation (lesser MLCT character of T_1) leads to longer lifetimes of triplet excited state, i.e. longer phosphorescence decay times.

12.2.4.2 Cycloiridiated Complexes with Helicenic N-Heterocyclic Carbene

In Chapter 5, several chiral Ir(III) and Re(I) complexes bearing helicenic N-heterocyclic carbene (NHC) ligands are presented; here we focus on cycloiridiated octahedral complexes **17**^{1} and **17**^{2} (Figure 12.10a) constituting the first example of systems with purely π-helical NHC [105].

Computational analysis of the electronic structures of these compounds evidenced extended π-conjugation between the NHC unit and the other fused phenyl rings within the π-helical carbenic ligand accompanied by efficient electronic interaction

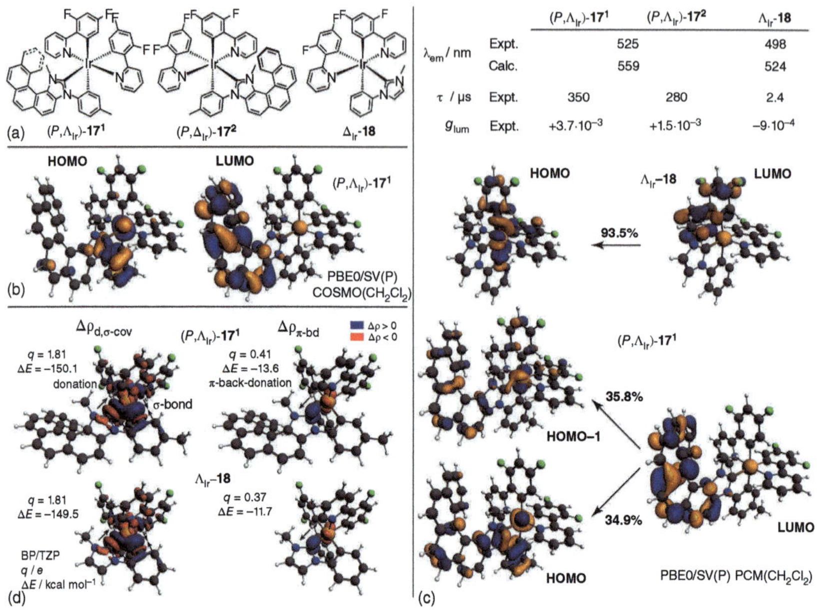

Figure 12.10 (a) Iridium(III) complexes with the purely π-helical NHC **17** along with reference non-helicenic system **18**. Superscripts 1 and 2 differentiate diastereoisomers differing in absolute configuration at the metal center (Λ, Δ). (b) Isosurfaces ($\pm$0.04 au) of HOMO and LUMO of **17**^{1}. (c) Isosurfaces ($\pm$0.0025 au) of dominant contributions to the differential electron density, $\Delta\rho$, describing the interaction between Ir and NHC ligand in **17**^{1} and **18** along with the corresponding electron-charge flow measure q and orbital energy stabilization ΔE obtained using the ETS-NOCV method. The blue/red color corresponds to the accumulation/loss of electron density. Arrows indicate direction of electron-density flow. (d) Top: Experimental (298 K, dichloromethane) and computed photophysical properties of **17** and **18**. Bottom: Isosurfaces ($\pm$0.04 au) of MOs involved in the phosphorescence for **18** and **17**^{1} along with the percentage contribution of a given MOs pair to this transition. Source: Based on Ref. [105].

between the NHC-[5]helicene and the Ir center. Both electronic features are visible in isosurfaces of the canonical frontier orbitals (see Figure 12.10b) and in the results of extended transition state – natural orbitals for chemical valence (ETS-NOCV) charge and bonding-energy decomposition analyses [106]. As can be seen in Figure 12.10c, the Ir–NHC-helicene bond reveals a typical donor–acceptor character with σ-donation from the ligand to the metal center and π-back-donation from the metal to the ligand, clearly enhanced – compared to the model complex **18** without helicenic NHC – due to the extended π-electron system of the helicene moiety. The strong interaction between the helicenic NHC and Ir has also recently been confirmed by Barone and coworkers, who additionally performed simulations of vibronic (chir)optical spectra of **17** and **18** [85], complementing our studies based on purely electronic features.

As revealed by calculations, both of the aforementioned factors (strongly delocalized electronic structure of NHC-helicene and effective electronic interaction between this ligand and Ir) are responsible for the observed chiroptical and photophysical properties of the compounds **17**^{1,2}, in particular: (i) for the increase in the intensity of ECD spectra and the strong dependence of the OA on the absolute configuration at the iridium, with the Λ_{Ir} (Δ_{Ir}) center enhancing the effect of *P* (*M*) stereochemistry of the helicenic ligand ("match/mismatch" effect), observed for **17**^{1,2} when compared to the system **18** (with non-helicenic NHC), (ii) for the redshifted phosphorescence of **17**^{1,2} and their extremely long lifetimes compared to **18**, and (iii) for the independence of the sign of the CPL dissymmetry parameter, g_{lum}, on the absolute configuration at the metal fragment. These unprecedented emission features result from the character of the T_1 excited state, which, in the case of **17**^{1,2}, is strongly delocalized across the NHC-helicene ligand (predominantly ^{3}ILCT and ^{3}MLCT) and demonstrates the reduced contribution of Ir orbitals in emission transitions in comparison to **18** (^{3}MLCT), see Figure 12.10d.

12.3 Concluding Remarks

First-principles theoretical calculations are now commonly used in aiding the analysis of spectroscopic measurements and providing meaningful and intuitive interpretations. Herein, we have presented representative cases of metal–organic helicenic systems that were thoroughly characterized via joint experimental and computational research. Calculations reproduced the experimental data well and greatly contributed to the understanding of the factors responsible for the observed experimental trends and features, linking them with the atomic-scale structure and bonding. While most of the standard photophysical and chiroptical phenomena (UV-vis, OR, ECD, VCD, ROA, non-polarized and circularly polarized fluorescence, non-polarized phosphorescence) appear to be routinely accessed via TD-DFT, further developments are needed to enable a better theoretical description of circularly polarized phosphorescence. Much work is also needed to develop low-scaling wavefunction theory methods that could be used for such metal–organic architectures.

One of the main advantages of computational modeling is the possibility to examine the compounds in terms of optical and conformational isomers that are often inaccessible or not separable for experimental techniques and to assess the effects of various factors (solvation, dynamics of the system, and vibrational contributions) on the observed signals. On the other hand, calculations are usually restricted to an a priori chosen model with a specific treatment of solvent, dynamics, vibrations, etc., while in reality, all these factors contribute to the measured property. It is thus evident that experimental and theoretical studies complement each other, and only in combination do they enable to gain broad knowledge on the relationships between structure and photophysical and chiroptical properties in a given system that may be useful in the developments of new compounds with targeted features.

Abbreviations

au	atomic units
CD	circular dichroism
COSMO	conductor-like screening model
CPL	circularly polarized luminescence
CT	charge transfer
DFT	density functional theory
ECD	electronic CD
ETS-NOCV	extended transition state – natural orbitals for chemical valence
eX	exact exchange
HF	Hartree–Fock
HOMO	highest occupied MO
ILCT	intraligand CT
IP	ionization potential
LLCT	ligand-to-ligand CT
LMCT	ligand-to-metal CT
LUMO	lowest unoccupied MO
MLCT	metal-to-ligand CT
MO	molecular orbital
MR	molar OR
NHC	N-heterocyclic carbene
OA	optical activity
OLED	organic light-emitting diode
OR	optical rotation
ORD	OR dispersion
PCM	polarizable continuum model
ROA	Raman OA
RSFs	range-separated functionals
SOC	spin-orbit coupling
SOS	sum-over-states
SV(P)	split-valence basis set with one set of polarization functions for non-hydrogen atoms
TD-DFT	time-dependent DFT
VCD	vibrational CD

References

1 Chen, C.-F. and Shen, Y. (2017). *Helicene Chemistry: From Synthesis to Applications*. Berlin Heidelberg: Springer-Verlag.

2 Shen, Y. and Chen, C.-F. (2012). Helicenes: synthesis and applications. *Chem. Rev.* 112 (3): 1463–1535.

3 Gingras, M. (2013). One hundred years of helicene chemistry. Part 3: applications and properties of carbohelicenes. *Chem. Soc. Rev.* 42 (3): 1051–1095.

4 Isla, H. and Crassous, J. (2016). Helicene-based chiroptical switches. *C. R. Chimie* 19 (1): 39–49.
5 Ravat, P., Šolomek, T., and Juríček, M. (2019). Helicenes as chiroptical photoswitches. *ChemPhotoChem* 3 (4): 180–186.
6 OuYang, J. and Crassous, J. (2018). Chiral multifunctional molecules based on organometallic helicenes: recent advances. *Coord. Chem. Rev.* 376: 533–547.
7 Rosenfeld, L. (1929). Quantenmechanische Theorie der natürlichen optischen Aktivität von Flüssigkeiten und Gasen. *Z. Physik* 52 (3): 161–174.
8 Condon, E.U. (1937). Theories of optical rotatory power. *Rev. Mod. Phys.* 9 (4): 432–457.
9 Autschbach, J. (2009). Computing chiroptical properties with first-principles theoretical methods: background and illustrative examples. *Chirality* 21 (1E): E116–E152.
10 Srebro-Hooper, M. and Autschbach, J. (2017). Calculating natural optical activity of molecules from first principles. *Annu. Rev. Phys. Chem.* 68 (1): 399–420.
11 Barron, L.D. (2004). *Molecular Light Scattering and Optical Activity*. Cambridge: Cambridge University Press.
12 Polavarapu, P.L. (2016). *Chiroptical Spectroscopy: Fundamentals and Applications*. Boca Raton: CRC Press.
13 Autschbach, J. (2020). *Quantum Theory for Chemical Applications: From Basic Concepts to Advanced Topics*. New York, NY: Oxford University Press.
14 Berova, N., Polavarapu, P.L., Nakanishi, K., and Woody, R.W. (ed.) (2012). *Comprehensive Chiroptical Spectroscopy, Volume 1: Instrumentation, Methodologies, and Theoretical Simulations*. New York, NY: John Wiley & Sons, Ltd.
15 Berova, N., Polavarapu, P.L., Nakanishi, K., and Woody, R.W. (ed.) (2012). *Comprehensive Chiroptical Spectroscopy, Volume 2: Applications in Stereochemical Analysis of Synthetic Compounds, Natural Products, and Biomolecules*. New York, NY: John Wiley & Sons, Ltd.
16 Schellman, J.A. (1975). Circular dichroism and optical rotation. *Chem. Rev.* 75 (3): 323–331.
17 Luk, C.K. and Richardson, F.S. (1974). Circularly polarized luminescence spectrum of camphorquinone. *J. Am. Chem. Soc.* 96 (7): 2006–2009.
18 Pritchard, B. and Autschbach, J. (2010). Calculation of the vibrationally resolved, circularly polarized luminescence of *d*-camphorquinone and (*S*,*S*)-*trans*-β-hydrindanone. *ChemPhysChem* 11 (11): 2409–2415.
19 Moscowitz, A. (1962). Theoretical aspects of optical activity part one: small molecules. *Adv. Chem. Phys.* 4: 67–112.
20 Djerassi, C. (1960). *Optical Rotatory Dispersion; Applications to Organic Chemistry*. New York: McGraw-Hill.
21 Tinoco, I. Jr., (1962). Theoretical aspects of optical activity part two: polymers. *Adv. Chem. Phys.* 4: 113–160.
22 Tinoco, I. Jr., and Woody, R.W. (1964). Optical rotation of oriented helices. IV. A free electron on a helix. *J. Chem. Phys.* 40 (1): 160–165.
23 Lightner, D.A., Hefelfinger, D.T., Powers, T.W. et al. (1972). Hexahelicene. Absolute configuration. *J. Am. Chem. Soc.* 94 (10): 3492–3497.

24 Hansen, A.E. and Bouman, T.D. (1980). Natural chiroptical spectroscopy: theory and computations. *Adv. Chem. Phys.* 44: 545–644.
25 Polavarapu, P.L. (2007). Renaissance in chiroptical spectroscopic methods for molecular structure determination. *Chem. Rec.* 7 (2): 125–136.
26 Furche, F., Ahlrichs, R., Wachsmann, C. et al. (2000). Circular dichroism of helicenes investigated by time-dependent density functional theory. *J. Am. Chem. Soc.* 122 (8): 1717–1724.
27 Grimme, S. (2001). Calculation of frequency dependent optical rotation using density functional response theory. *Chem. Phys. Lett.* 339 (5–6): 380–388.
28 Grimme, S., Furche, F., and Ahlrichs, R. (2002). An improved method for density functional calculations of the frequency-dependent optical rotation. *Chem. Phys. Lett.* 361 (3–4): 321–328.
29 Stephens, P.J., Devlin, F.J., Cheeseman, J.R., and Frisch, M.J. (2001). Calculation of optical rotation using density functional theory. *J. Phys. Chem. A* 105 (22): 5356–5371.
30 Ruud, K. and Helgaker, T. (2002). Optical rotation studied by density-functional and coupled-cluster methods. *Chem. Phys. Lett.* 352 (5–6): 533–539.
31 Autschbach, J., Ziegler, T., van Gisbergen, S.J.A., and Baerends, E.J. (2002). Chiroptical properties from time-dependent density functional theory. I. Circular dichroism spectra of organic molecules. *J. Chem. Phys.* 116 (16): 6930–6940.
32 Autschbach, J., Patchkovskii, S., Ziegler, T. et al. (2002). Chiroptical properties from time-dependent density functional theory. II. Optical rotations of small to medium sized organic molecules. *J. Chem. Phys.* 117 (2): 581–592.
33 Pecul, M., Ruud, K., and Helgaker, T. (2004). Density functional theory calculation of electronic circular dichroism using London orbitals. *Chem. Phys. Lett.* 388 (1–3): 110–119.
34 Tam, M.C. and Crawford, T.D. (2006). Ab initio determination of optical rotatory dispersion in the conformationally flexible molecule (*R*)-epichlorohydrin. *J. Phys. Chem. A* 110 (6): 2290–2298.
35 Friese, D.H. and Hättig, C. (2014). Optical rotation calculations on large molecules using the approximate coupled cluster model CC2 and the resolution-of-the-identity approximation. *Phys. Chem. Chem. Phys.* 16 (13): 5942–5951.
36 Deutsche, C.W. and Moscowitz, A. (1968). Optical activity of vibrational origin. I. A model helical polymer. *J. Chem. Phys.* 49 (7): 3257–3272.
37 Deutsche, C.W. and Moscowitz, A. (1970). Optical activity of vibrational origin. II. Consequences of polymer conformation. *J. Chem. Phys.* 53 (7): 2630–2644.
38 Craig, D.P. and Thirunamachandran, T. (1978). A theory of vibrational circular dichroism in terms of vibronic interactions. *Mol. Phys.* 35 (3): 825–840.
39 Nafie, L.A. and Freedman, T.B. (1983). Vibronic coupling theory of infrared vibrational transitions. *J. Chem. Phys.* 78 (12): 7108–7116.
40 Stephens, P.J. (1985). Theory of vibrational circular dichroism. *J. Phys. Chem.* 89 (5): 748–752.

41 Rauk, A. and Yang, D. (1992). Vibrational circular dichroism and infrared spectra of 2-methyloxirane and *trans*-2,3-dimethyloxirane: ab initio vibronic coupling theory with the 6-31G*(0.3) basis set. *J. Phys. Chem.* 96 (1): 437–446.

42 Nafie, L.A. (1992). Velocity-gauge formalism in the theory of vibrational circular dichroism and infrared absorption. *J. Chem. Phys.* 96 (8): 5687–5702.

43 Yang, D. and Rauk, A. (1994). Vibrational circular dichroism intensities by ab initio second-order Møller–Plesset vibronic coupling theory. *J. Chem. Phys.* 100 (11): 7995–8002.

44 Bour, P. and Keiderling, T.A. (1993). Ab initio simulations of the vibrational circular dichroism of coupled peptides. *J. Am. Chem. Soc.* 115 (21): 9602–9607.

45 Tam, C.N., Bour, P., and Keiderling, T.A. (1996). Vibrational optical activity of (3*S*,6*S*)-3,6-dimethyl-1,4-dioxane-2,5-dione. *J. Am. Chem. Soc.* 118 (42): 10285–10293.

46 Devlin, F.J., Stephens, P.J., Cheeseman, J.R., and Frisch, M.J. (1996). Prediction of vibrational circular dichroism spectra using density functional theory: camphor and fenchone. *J. Am. Chem. Soc.* 118 (26): 6327–6328.

47 Nicu, V.P., Autschbach, J., and Baerends, E.J. (2009). Enhancement of IR and VCD intensities due to charge transfer. *Phys. Chem. Chem. Phys.* 11 (10): 1526–1538.

48 Polavarapu, P.L., Hecht, L., and Barron, L.D. (1993). Vibrational Raman optical activity in substituted oxiranes. *J. Phys. Chem.* 97 (9): 1793–1799.

49 Helgaker, T., Ruud, K., Bak, K.L. et al. (1994). Vibrational Raman optical activity calculations using London atomic orbitals. *Faraday Discuss.* 99: 165–180.

50 Costante, J., Hecht, L., Polavarapu, P.L. et al. (1997). Absolute configuration of bromochlorofluoromethane from experimental and ab initio theoretical vibrational Raman optical activity. *Angew. Chem. Int. Ed.* 36 (8): 885–887.

51 Nafie, L.A. (1996). Theory of resonance Raman optical activity: the single electronic state limit. *Chem. Phys.* 205 (3): 309–322.

52 Vargek, M., Freedman, T.B., Lee, E., and Nafie, L.A. (1998). Experimental observation of resonance Raman optical activity. *Chem. Phys. Lett.* 287 (3–4): 359–364.

53 Jensen, L., Autschbach, J., Krykunov, M., and Schatz, G.C. (2007). Resonance vibrational Raman optical activity: a time-dependent density functional theory approach. *J. Chem. Phys.* 127 (13): 134101.

54 Crawford, T.D. and Ruud, K. (2011). Coupled-cluster calculations of vibrational Raman optical activity spectra. *ChemPhysChem* 12 (17): 3442–3448.

55 Cheeseman, J.R. and Frisch, M.J. (2011). Basis set dependence of vibrational Raman and Raman optical activity intensities. *J. Chem. Theory Comput.* 7 (10): 3323–3334.

56 Johannessen, C., Blanch, E.W., Villani, C. et al. (2013). Raman and ROA spectra of (−)- and (+)-2-Br-hexahelicene: experimental and DFT studies of a π-conjugated chiral system. *J. Phys. Chem. B* 117 (7): 2221–2230.

57 Srebro, M., Govind, N., de Jong, W.A., and Autschbach, J. (2011). Optical rotation calculated with time-dependent density functional theory: the OR45 benchmark. *J. Phys. Chem. A* 115 (40): 10930–10949.

58 Srebro, M. and Autschbach, J. (2012). Tuned range-separated time-dependent density functional theory applied to optical rotation. *J. Chem. Theory Comput.* 8 (1): 245–256.

59 El Sayed Moussa, M., Srebro, M., Anger, E. et al. (2013). Chiroptical properties of carbo[6]helicene derivatives bearing extended π-conjugated cyano substituents. *Chirality* 25 (8): 455–465.

60 Becke, A.D. (1993). A new mixing of Hartree–Fock and local density-functional theories. *J. Chem. Phys.* 98 (2): 1372–1377.

61 Lee, C., Yang, W., and Parr, R.G. (1988). Development of the Colle-Salvetti correlation-energy formula into a functional of the electron density. *Phys. Rev. B* 37 (2): 785–789.

62 Yanai, T., Tew, D.P., and Handy, N.C. (2004). A new hybrid exchange–correlation functional using the Coulomb-attenuating method (CAM-B3LYP). *Chem. Phys. Lett.* 393 (1): 51–57.

63 Autschbach, J. and Srebro, M. (2014). Delocalization error and "Functional Tuning" in Kohn–Sham calculations of molecular properties. *Acc. Chem. Res.* 47 (8): 2592–2602.

64 Bensalah-Ledoux, A., Pitrat, D., Reynaldo, T. et al. (2016). Large-scale synthesis of helicene-like molecules for the design of enantiopure thin films with strong chiroptical activity. *Chem. Eur. J.* 22 (10): 3333–3346.

65 Isla, H., Srebro-Hooper, M., Jean, M. et al. (2016). Conformational changes and chiroptical switching of enantiopure bis-helicenic terpyridine upon Zn^{2+} binding. *Chem. Commun.* 52 (35): 5932–5935.

66 Becke, A.D. (1993). Density-functional thermochemistry. III. The role of exact exchange. *J. Chem. Phys.* 98 (7): 5648–5652.

67 Stephens, P.J., Devlin, F.J., Chabalowski, C.F., and Frisch, M.J. (1994). Ab initio calculation of vibrational absorption and circular dichroism spectra using density functional force fields. *J. Phys. Chem.* 98 (45): 11623–11627.

68 Adamo, C. and Barone, V. (1999). Toward reliable density functional methods without adjustable parameters: the PBE0 model. *J. Chem. Phys.* 110 (13): 6158–6170.

69 Weigend, F. and Ahlrichs, R. (2005). Balanced basis sets of split valence, triple zeta valence and quadruple zeta valence quality for H to Rn: design and assessment of accuracy. *Phys. Chem. Chem. Phys.* 7 (18): 3297–3305.

70 Ehrlich, S., Moellmann, J., and Grimme, S. (2013). Dispersion-corrected density functional theory for aromatic interactions in complex systems. *Acc. Chem. Res.* 46 (4): 916–926.

71 Tomasi, J., Mennucci, B., and Cammi, R. (2005). Quantum mechanical continuum solvation models. *Chem. Rev.* 105 (8): 2999–3094.

72 Nakai, Y., Mori, T., Sato, K., and Inoue, Y. (2013). Theoretical and experimental studies of circular dichroism of mono- and diazonia[6]helicenes. *J. Phys. Chem. A* 117 (24): 5082–5092.

73 Mori, K., Goumans, T.P.M., van Lenthe, E., and Wang, F. (2014). Predicting phosphorescent lifetimes and zero-field splitting of organometallic complexes with time-dependent density functional theory including spin–orbit coupling. *Phys. Chem. Chem. Phys.* 16 (28): 14523–14530.

74 Tamm, I. (1945). Relativistic interaction of elementary particles. *J. Phys. USSR* 9: 449–460.

75 Dancoff, S.M. (1950). Non-adiabatic meson theory of nuclear forces. *Phys. Rev.* 78 (4): 382–385.

76 Hirata, S. and Head-Gordon, M. (1999). Time-dependent density functional theory within the Tamm–Dancoff approximation. *Chem. Phys. Lett.* 314 (3): 291–299.

77 Peach, M.J.G. and Tozer, D.J. (2012). Overcoming low orbital overlap and triplet instability problems in TDDFT. *J. Phys. Chem. A* 116 (39): 9783–9789.

78 Peach, M.J.G., Warner, N., and Tozer, D.J. (2013). On the triplet instability in TDDFT. *Mol. Phys.* 111 (9–11): 1271–1274.

79 Anger, E., Srebro, M., Vanthuyne, N. et al. (2014). Helicene-grafted vinyl- and carbene-osmium complexes: an example of acid–base chiroptical switching. *Chem. Commun.* 50 (22): 2854–2856.

80 Weigang, O.E., Turner, J.A., and Trouard, P.A. (1966). Emission polarization and circular dichroism of hexahelicene. *J. Chem. Phys.* 45 (4): 1126–1134.

81 Liu, Y., Cerezo, J., Mazzeo, G. et al. (2016). Vibronic coupling explains the different shape of electronic circular dichroism and of circularly polarized luminescence spectra of hexahelicenes. *J. Chem. Theory Comput.* 12 (6): 2799–2819.

82 Xu, Q., Liu, Y., Zhao, X. et al. (2020). Vibrationally resolved electronic circular dichroism and circularly polarized luminescence spectra of a boron-fused double helicene: a theoretical study. *Spectrochim. Acta Part A: Mol. Spectrosc.* 231: 118132.

83 Liu, Y., Xu, Q., Sun, J. et al. (2020). Insights for vibronic effects on spectral shapes of electronic circular dichroism and circularly polarized luminescence of aza[7]helicene. *Spectrochim. Acta Part A: Mol. Spectrosc.* 239: 118475.

84 Liu, Y., Xu, Q., Liu, L. et al. (2021). The spectral-shapes of absorption, emission, ECD and CPL of a fluorene-fused [7]helicene: vibronic effect and solvent inhomogenous broadening. *Spectrochim. Acta Part A: Mol. Spectrosc.* 249: 119293.

85 Yang, Q., Fusè, M., Bloino, J., and Barone, V. (2021). Interplay of stereo-electronic, vibronic and environmental effects in tuning the chiroptical properties of an Ir(III) cyclometalated N-heterocyclic carbene. *Spectrochim. Acta Part A: Mol. Spectrosc.* 254: 119631.

86 Bannwarth, C., Seibert, J., and Grimme, S. (2016). Electronic circular dichroism of [16]helicene with simplified TD-DFT: beyond the single structure approach. *Chirality* 28 (5): 365–369.

87 Bannwarth, C. and Grimme, S. (2014). A simplified time-dependent density functional theory approach for electronic ultraviolet and circular dichroism spectra of very large molecules. *Comput. Theor. Chem.* 1040–1041: 45–53.

88 Bannwarth, C. and Grimme, S. (2015). Electronic circular dichroism of highly conjugated π-systems: breakdown of the Tamm–Dancoff/configuration interaction singles approximation. *J. Phys. Chem. A* 119 (15): 3653–3662.

89 de Wergifosse, M., Seibert, J., and Grimme, S. (2020). Simplified time-dependent density functional theory (sTD-DFT) for molecular optical rotation. *J. Chem. Phys.* 153 (8): 084116.

90 Graule, S., Rudolph, M., Vanthuyne, N. et al. (2009). Metal–bis(helicene) assemblies incorporating π-conjugated phosphole-azahelicene ligands: impacting chiroptical properties by metal variation. *J. Am. Chem. Soc.* 131 (9): 3183–3185.

91 Graule, S., Rudolph, M., Shen, W. et al. (2010). Assembly of π-conjugated phosphole azahelicene derivatives into chiral coordination complexes: an experimental and theoretical study. *Chem. Eur. J.* 16 (20): 5976–6005.

92 Vreshch, V., El Sayed Moussa, M., Nohra, B. et al. (2013). Assembly of helicene-capped N,P,N,P,N-helicands within Cu^{I} helicates: impacting chiroptical properties by ligand–ligand charge transfer. *Angew. Chem. Int. Ed.* 52 (7): 1968–1972.

93 El Sayed Moussa, M., Chen, H., Wang, Z. et al. (2016). Bimetallic gold(I) complexes with ethynyl-helicene and bis-phosphole ligands: understanding the role of aurophilic interactions in their chiroptical properties. *Chem. Eur. J.* 22 (17): 6075–6086.

94 Anger, E., Srebro, M., Vanthuyne, N. et al. (2012). Ruthenium-vinylhelicenes: remote metal-based enhancement and redox switching of the chiroptical properties of a helicene core. *J. Am. Chem. Soc.* 134 (38): 15628–15631.

95 Srebro, M., Anger, E., Moore, B. II, et al. (2015). Ruthenium-grafted vinylhelicenes: chiroptical properties and redox switching. *Chem. Eur. J.* 21 (47): 17100–17115.

96 Shen, C., Loas, G., Srebro-Hooper, M. et al. (2016). Iron alkynyl helicenes: redox-triggered chiroptical tuning in the IR and near-IR spectral regions and suitable for telecommunications applications. *Angew. Chem. Int. Ed.* 55 (28): 8062–8066.

97 Shen, C., Srebro-Hooper, M., Weymuth, T. et al. (2018). Redox-active chiroptical switching in mono- and bis-iron ethynylcarbo[6]helicenes studied by electronic and vibrational circular dichroism and resonance Raman optical activity. *Chem. Eur. J.* 24 (56): 15067–15079.

98 Abella, L., Ludowieg, H.D., and Autschbach, J. (2020). Theoretical study of the Raman optical activity spectra of $[M(en)_3]^{3+}$ with M = Co, Rh. *Chirality* 32 (6): 741–752.

99 Norel, L., Rudolph, M., Vanthuyne, N. et al. (2010). Metallahelicenes: easily accessible helicene derivatives with large and tunable chiroptical properties. *Angew. Chem. Int. Ed.* 49 (1): 99–102.

100 Shen, C., Anger, E., Srebro, M. et al. (2014). Straightforward access to mono- and bis-cycloplatinated helicenes displaying circularly polarized phosphorescence by using crystallization resolution methods. *Chem. Sci.* 5 (5): 1915–1927.

101 Anger, E., Rudolph, M., Norel, L. et al. (2011). Multifunctional and reactive enantiopure organometallic helicenes: tuning chiroptical properties by structural variations of mono- and bis(platinahelicene)s. *Chem. Eur. J.* 17 (50): 14178–14198.

102 Ludowieg, H.D., Srebro-Hooper, M., Crassous, J., and Autschbach, J. (2022). Optical activity of spin-forbidden electronic transitions in metal complexes from Time-Dependent Density Functional Theory with Spin-Orbit Coupling. ChemistryOpen: in press. DOI 10.1002/open.202200020.

103 Saleh, N., Moore, B., Srebro, M. et al. (2015). Acid/base-triggered switching of circularly polarized luminescence and electronic circular dichroism in organic and organometallic helicenes. *Chem. Eur. J.* 21 (4): 1673–1681.

104 Saleh, N., Srebro, M., Reynaldo, T. et al. (2015). Enantio-enriched CPL-active helicene–bipyridine–rhenium complexes. *Chem. Commun.* 51 (18): 3754–3757.

105 Hellou, N., Srebro-Hooper, M., Favereau, L. et al. (2017). Enantiopure cycloiridiated complexes bearing a pentahelicenic N-heterocyclic carbene and displaying long-lived circularly polarized phosphorescence. *Angew. Chem. Int. Ed.* 56 (28): 8236–8239.

106 Mitoraj, M.P., Michalak, A., and Ziegler, T. (2009). A combined charge and energy decomposition scheme for bond analysis. *J. Chem. Theory Comput.* 5 (4): 962–975.

13

Helicene Derivatives with Circularly Polarized Luminescence

Wen-Long Zhao[1], Meng Li[2], Hai-Yan Lu[1], and Chuan-Feng Chen[1,2]

[1] *School of Chemical Sciences, University of Chinese Academy of Sciences, Beijing, China*
[2] *Beijing National Laboratory for Molecular Sciences, CAS Key Laboratory of Molecular Recognition and Function, Institute of Chemistry Chinese Academy of Sciences, Beijing, China*

13.1 Introduction

Light is a kind of electromagnetic wave, which can be described by oscillating electric fields and perpendicular magnetic fields. Circularly polarized light is the light with a circular track at the end of the electric vector. The magnitude of the vector does not change, but the direction of the vector changes with time and shows a fixed phase difference. When a beam of polarized light comes in, if the corresponding electromagnetic vector is a counterclockwise spiral, then the beam is left circularly polarized light, and if it is a clockwise spiral, the beam is right circularly polarized light. The left or right circular polarization endows light with chiral characteristics [1]. The phenomenon that the luminophore emits different left-handed or right-handed circularly polarized light is called circularly polarized luminescence (CPL). The CPL spectrum can directly reflect the excited state structural information of the chiral luminescence system, and it can provide a detection method for the mechanism of generation, transmission, and amplification of chirality [2]. CPL materials have attracted great attention and interest in recent years for their wide application prospects in 3D information displays [3], spin information communication [4], information storage [5], CPL lasers [6], biological probes [7], CPL split sensors [8], and promotion of asymmetric photochemistry [9].

Fluorescence quantum yield (Φ_F) is an important parameter reflecting the luminescence ability of the illuminating system, which is the ratio of the number of emitted photons to the number of absorbed photons, the magnitude of which ranges from 0 to 100%. Another important parameter that can be reflected in the CPL spectrum is the luminescence dissymmetric factor (g_{lum}), which is measured by the following formula: $g_{lum} = 2(I_L - I_R)/(I_L + I_R)$, where I_L and I_R refer to left-handed and right-handed polarized light intensities, respectively. The value of g_{lum} is between −2 and +2, in which a negative value is right-handed, and a positive value is left-handed. Moreover, g_{lum} values of 0 and 2 indicate nonpolarized light and the completely

Helicenes: Synthesis, Properties, and Applications, First Edition.
Edited by Jeanne Crassous, Irena G. Stará, and Ivo Starý.

circularly polarized light, respectively. In addition, g_{lum} can also be theoretically defined as $4|m||\mu|\cos\theta/(|m|^2 + |\mu|^2)$, where m and μ are the magnetic and electrical transition dipole moments, respectively. The θ is the angle between m and μ [10]. To date, molecular systems with high CPL g_{lum} values mainly come from the chiral lanthanide complexes, which showed the highest g_{lum} value of 1.38 [11].

Generally, chiral lanthanide complexes [12] and chiral polymers [13] are the hot compounds in the early studies of CPL. The f–f transitions of the internal configuration allowed by the lanthanide are electric dipole forbidden but magnetically allowed, which can potentially increase the g_{lum} value. Chiral polymers have relatively high Φ_F and g_{lum} values. However, the chiral properties of the polymers are unstable due to the change of their conformations, making it difficult to accurately study their "structure–CPL activity" relationships as well. Moreover, limited types of the chiral monomers and difficulty in the material synthesis and purification could also restrict the research progress of CPL of chiral polymers [14].

Compared with chiral polymers, organic small molecules have not only simple and clear molecular structures and high fluorescence quantum yields but also wide variety and easy derivatization. These advantages make them hotspot materials for the exploration of CPL [15, 16]. Helicenes [17] are aromatic ortho-fused polycyclic compounds with helical structures, and their molecular skeletons can be distorted by the steric hindrance between the terminal aromatic rings. Although such molecules do not have any asymmetric carbon centers, their twisted and extended π-conjugated molecular structures exhibit special helical chirality, and chiral helix has also proven to be a valuable structural design for exploring CPL of small organic molecules. Consequently, helicenes and their derivatives have widely potential applications in optoelectronic devices, nonlinear optical materials, asymmetric organic synthesis, and so on [18]. This chapter presents recent advances in helicenes and their derivatives with CPL properties.

13.2 Carbohelicenes with CPL Properties

The carbohelicene molecule typically contains a helically twisted π-conjugated ortho-fused benzene ring system, which was designated as [*n*]helicene, where n represents the number of the rings forming a helix in the ortho-fusion. In 2003, Venkataraman et al. [19] first introduced a chiral auxiliary group into the helicene skeleton to obtain optically pure helicene derivatives (*M*)-**1a**/(*P*)-**1a** and (*M*)-**1b**/(*P*)-**1b** (Figure 13.1) and found that they showed CPL properties with the g_{lum} of $-1.1\times10^{-3}/+0.9\times10^{-3}$ (λ_{em} = 453 nm) and $-7\times10^{-4}/+8\times10^{-4}$ (λ_{em} = 478 nm), respectively. This work opened the door to the research of helicene derivatives with CPL properties. In 2012, Tanaka et al. [20, 21] synthesized a series of helicene derivatives **2–5** (Figure 13.1) with CPL properties. The $|g_{lum}|$ values of helicene derivatives **2** and **3** in chloroform were 3.0×10^{-2} and 3.2×10^{-2}, respectively. For **4**, its $|g_{lum}|$ value in chloroform was less than 1×10^{-3}. In comparison, the $|g_{lum}|$ value of double azahelicene **5** in chloroform was up to 1.1×10^{-2}, which might be due to the synergistic effect of the relatively rigid chiral double helix structure

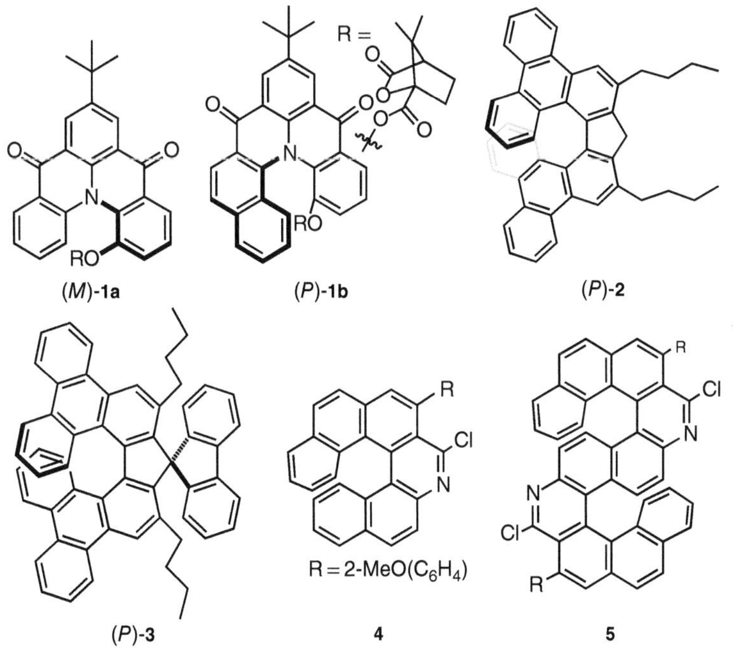

Figure 13.1 Molecular structures of helicene derivatives (*M*)-**1a**, (*P*)-**1b**, (*P*)-**2**, (*P*)-**3**, **4**, and **5**.

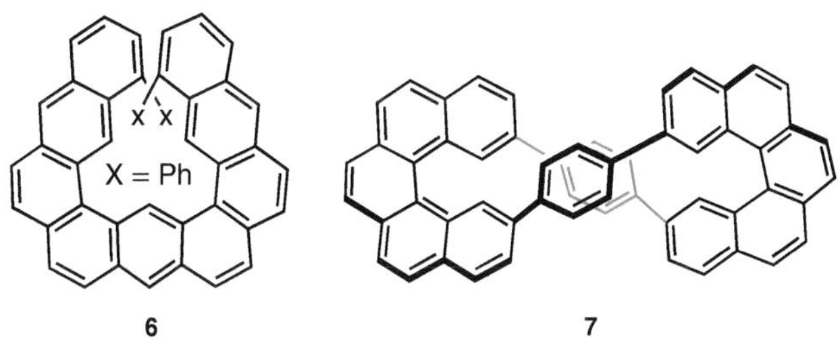

Figure 13.2 Molecular structures of helicene derivatives **6** and **7**.

of **5** [21]. Anthracene can also be used as the building element of helicene such as **6** (Figure 13.2). The benzene ring at the end of the triangular skeleton increased the molecular twisted degree, which promoted the $|g_{lum}|$ value of **6** reached 1.3×10^{-2} [22].

Since g_{lum} can be theoretically defined as $4|m||\mu|\cos\theta/(|m|^2 + |\mu|^2)$, where m and μ are the magnetic and electrical transition dipole moments, respectively, the purpose of increasing the $|g_{lum}|$ value can be achieved by rationally tuning the m and μ of the molecule itself and increasing the $|\cos\theta|$ value. Recently, an 8-shaped helicene dimer **7** with an obvious D_2 symmetry (parallel m and μ) has been reported [23]. Here, the $|\cos\theta|$ was constantly at its maximum. Additionally, two [5]helicenes were connected to form large rings and improved the $|m|$ value. Consequently, the $|g_{lum}|$ value of **7** reached 1.5×10^{-2}. This work proved that the design of D_2 symmetry molecule structure was a good strategy for increasing the $|g_{lum}|$ value.

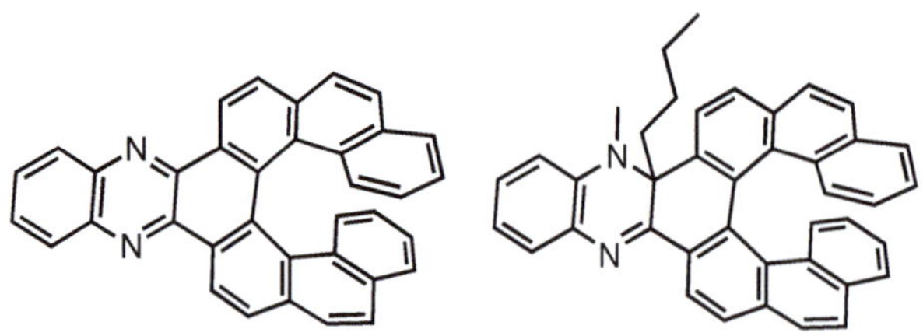

(*P*)-, (*M*)-**8a** (*P*)-, (*M*)-**8b**

Figure 13.3 Molecular structures of helicene derivatives **8a** and **8b**.

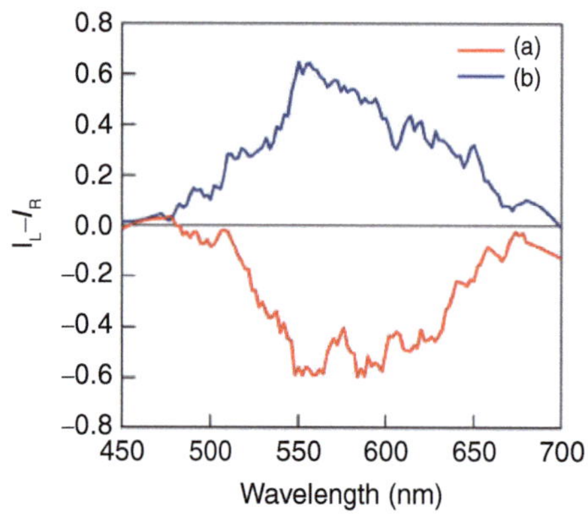

Figure 13.4 CPL spectra of (a) (*P*)-**8b** and (b) (*M*)-**8b** in THF. Source: Reproduced with permission from Ref. [24]. Copyright 2015, American Chemical Society.

By introducing a quinoline structure into [7]helicene, Sakai et al. [24] synthesized **8a**. They also obtained 1,2-dialkylquinoxalinefused [7]helicene **8b** by introducing two alkyl chains (Figure 13.3). It was found that the twist angle between the quinoxaline and the [7]helicene subunit in **8a** was 159°, and its Φ_F was 0.05. When **8a** was alkylated to produce **8b**, the quinoxaline fragment was almost coplanar with the [7]helicene skeleton, and the Φ_F value was increased to 0.25. Moreover, (*M*)-**8b**/(*P*)-**8b** also exhibited significant CPL properties with the $|g_{lum}|$ value of 4.0×10^{-3} (Figure 13.4).

When a substituent with excellent photoelectric properties was introduced into a conventional carbohelicene skeleton, the luminescence efficiency and CPL property of the carbohelicene derivatives could be obviously improved. For example, when a maleimide unit was introduced into the skeleton of simple [5]helicene ($\Phi_F = 3\%$) to obtain helicene derivatives (*P*)-**9**/(*M*)-**9** (Figure 13.5) [25], their Φ_F increased significantly to 37%. Moreover, the isolated enantiomers also exhibited CPL properties with a $|g_{lum}|$ of 2.4×10^{-3}.

By a platinum-catalyzed double-cyclization reaction, Nozaki et al. [26] synthesized a blue CPL [7]helicene-like compound **10** ($\Phi_F = 40\%$) with one fluorene moiety (Figure 13.5). The enantiomers of **10** also showed good CPL properties (Figure 13.6), and the g_{lum} values were about 3×10^{-3}. Introduction of a fluorene subunit into the framework of the helicene was the key step to this high fluorescence property, which could provide a promising molecular design for luminescent helicene derivatives.

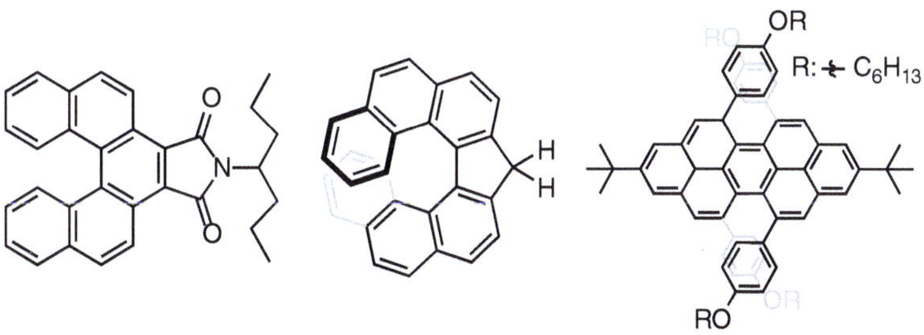

Figure 13.5 Molecular structures of helicene derivatives **9**, **10**, and **11**.

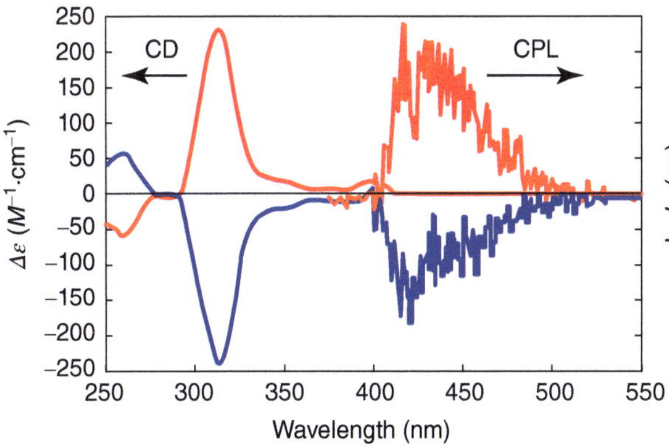

Figure 13.6 CD and CPL spectra of **10** in CH_2Cl_2. Source: Reproduced with permission from Ref. [26]. Copyright 2016, American Chemical Society.

With a *p*-terphenyl-2,2″,6,6″-tetrayne derivative as precursor, Chalifoux et al. [27] carried out a fourfold alkyne cyclization reaction to form chiral peropyrene enantiomers (*P,P*)-**11** and (*M,M*)-**11** (Figure 13.5). Due to the repulsion of the two benzene ring substituents, the torsion angle between the chiral helicene segments was 28°, and the distance between the two benzene rings on the same side of the molecule was 2.9 Å, while (*P,P*)-**11** exhibited a certain degree of CPL property and g_{lum} value of about 7.7×10^{-4}.

In 2018, Crassous et al. [28] reported the synthesis and optical properties of the π-conjugated diketopyrrolopyrrole helicene derivatives **12a–c** (Figure 13.7) with CPL properties. The Φ_F values of **12a–c** were 41, 41, and 35%, respectively. The estimated $|g_{lum}|$ values of the optically pure **12a–c** were 1.0×10^{-4}, 7.0×10^{-4}, and 9.0×10^{-4}, respectively. Due to the potential applications of diketopyrrolopyrrole and helicenes as optoelectronic materials, this work may open up a new perspective for chiral organic materials.

Recently, Qiu et al. combined [6]helicene skeleton with tetraphenyl ethylene (TPE) and obtained carbohelicene derivatives **13a** and **13b** (Figure 13.8) with aggregation-induced emission (AIE) properties [29]. Compared with **13b**, the two TPE units constituted large steric hindrance at both terminals of **13a**, which

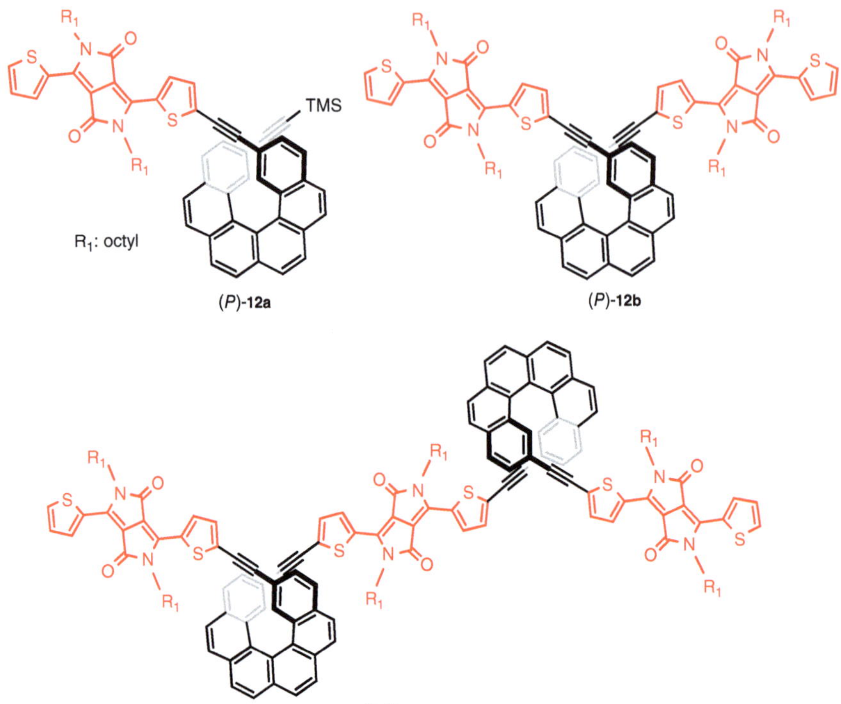

Figure 13.7 Molecular structures of helicene derivatives (*P*)-**12a**, (*P*)-**12b**, and (*P,P*)-**12c**.

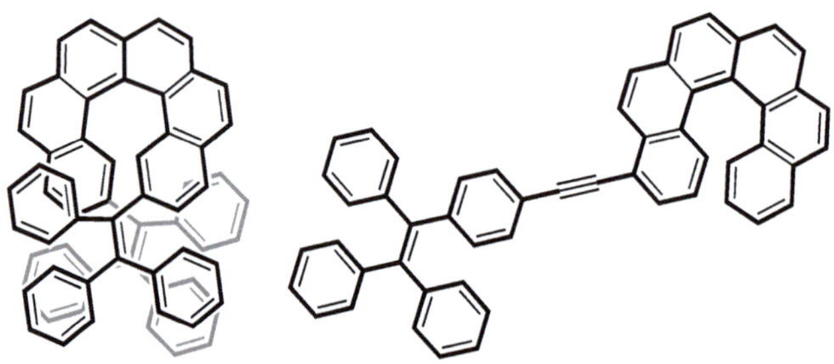

Figure 13.8 Molecular structures of helicene derivatives **13a** and **13b**.

increased the distance between the two ends of the helical skeleton and then improved the molecular rigidity, thus effectively transferring the helical chirality to the whole molecule and promoting the $|g_{lum}|$ value of **13a** to 1.5×10^{-2}. On the other hand, they utilized carbon–carbon triple bond as a linker to connect [6]helicene and TPE subunit, which made the molecule more flexible and obviously weakened the whole helical chirality of **13b**, but the existence of TPE unit made **13b** exhibit typical AIE properties. Consequently, although the $|g_{lum}|$ value of **13b** was only 1.0×10^{-3}, its Φ_F could reach 37%.

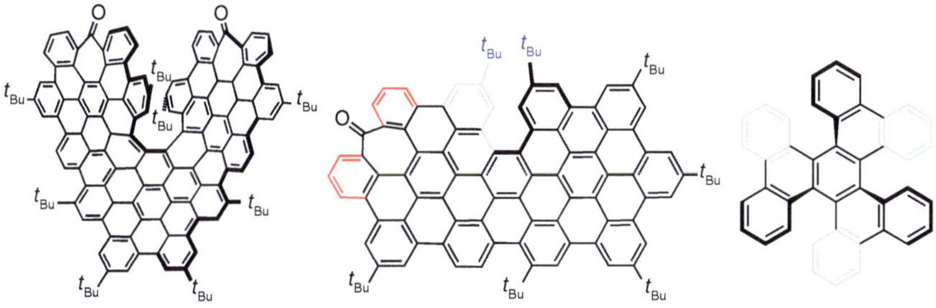

Figure 13.9 Molecular structures of helicene derivatives (*M*)-**14**, (*P*)-**15**, and (*P,P,P*)-**16**.

Recently, Campaña et al. [30] reported the synthesis and optical properties of the first complete π-conjugated extension [7]helicene derivative **14** (Figure 13.9), in which the [7]helicene was completely surrounded by ortho-fused benzene rings to form an extended π-conjugate twisted structure. **14** could emit orange-red fluorescence and CPL under ultraviolet light irradiation, and its Φ_F in CH_2Cl_2 was 9.8%, which was 4.7 times greater than that of [7]helicene ($\Phi_F = 2.1\%$). The $|g_{lum}|$ reached 2×10^{-3} at 610 nm. The design and synthesis of **14** provided a new perspective for the study of π-extended nanohelicene with two-photon absorption (TPA) and CPL properties. In the same year, they also reported a kind of nanographene **15** containing a helical structure (Figure 13.9) [31]. The intramolecular [5]helicene structure in **15** induced a π-extended aromatic backbone, which made it exhibit CPL properties with the $|g_{lum}|$ value of 2.3×10^{-4}. Moreover, **15** also displayed TPA and two-photon emission (TPE) effects through the introduction of an electron push–pull group on the molecular structure, and its Φ_F in CH_2Cl_2 was 13%.

By combination of three pentahelicene (PH) units, Mori et al. [32] synthesized (*P,P,P*)-**16** (hexabenzotriphenylene) (Figure 13.9) with stable conformation and chemical properties in 2018. This compound had no undesirable racemization and degradation relative to pH. Its Φ_F in CH_2Cl_2 was 0.018, and its $|g_{lum}|$ value was 1.3×10^{-3} at 480 nm.

In 2014, Chen et al. [33] reported a new synthetic method to obtain optically stable benzo[5]helicene derivatives by introducing two methyl groups at the 1,10-position of the [5]helicene skeleton, which showed no racemization under 120 °C for 12 hours. According to a similar approach, they also conveniently obtained a series of enantiomeric helical aromatic imides (*P*)- and (*M*)-**17a–e** (Figure 13.10) based on the tetrahydro[5]helicene skeleton, in 2016 [34]. Due to the different electron push–pull ability of the aromatic groups, the enantiomers not only exhibited large Stokes shifts and high quantum yields but also showed mirror-image and full-color CPL properties with the g_{lum} values at -0.2×10^{-3} to -1.5×10^{-3} for the *P* configuration enantiomers, and $+0.4 \times 10^{-3}$ to $+1.4 \times 10^{-3}$ for the *M* configuration enantiomers (Figure 13.11). Recently, Chen et al. [35] utilized the reprecipitation method to prepare the first organic chiral nanoparticles based on the enantiomers (*P*)- and (*M*)-**17a–e**, which exhibited full color and white CPL properties. It was further found that these water-soluble nanoparticles showed strong emission

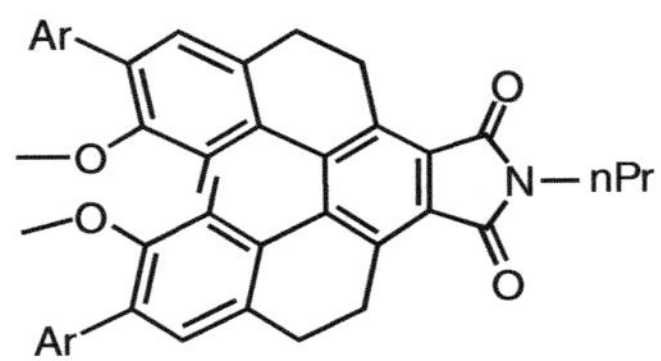

(P)-, (M)-**17a:** Ar = p-$F_3CC_6H_4$
(P)-, (M)-**17b:** Ar = Ph
(P)-, (M)-**17c:** Ar = p-$MeOC_6H_4$
(P)-, (M)-**17d:** Ar = p-$Ph_2NC_6H_4$
(P)-, (M)-**17e:** Ar = 5-Ph_2N-2-thienyl

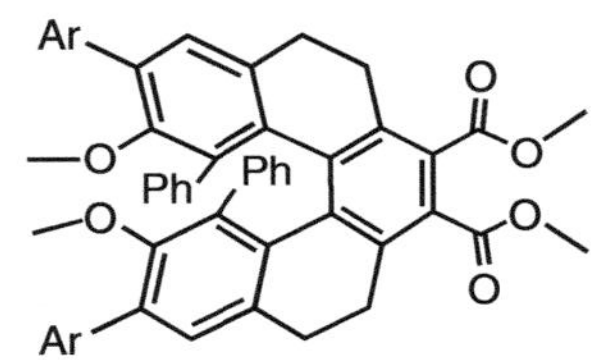

(P)-, (M)-**18a:** Ar = p-CNC_6H_4
(P)-, (M)-**18b:** Ar = Ph
(P)-, (M)-**18c:** Ar = p-$MeOC_6H_4$
(P)-, (M)-**18d:** Ar = p-FC_6H_4
(P)-, (M)-**18e:** Ar = p-$CH_3C_6H_4$

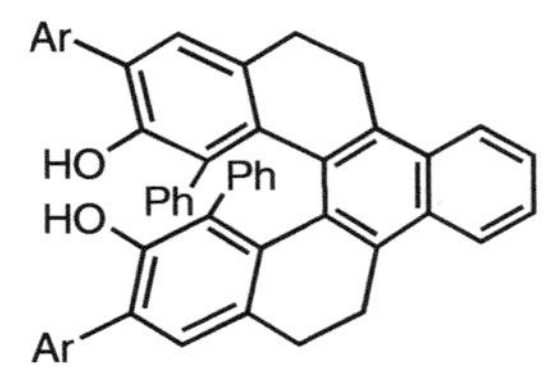

(P)-, (M)-**19a:** Ar = p-$MeOC_6H_4$
(P)-, (M)-**19b:** Ar = p-MePH
(P)-, (M)-**19c:** Ar = C_6H_5
(P)-, (M)-**19d:** Ar = p-FC_6H_4
(P)-, (M)-**19e:** Ar = p-NCC_6H_4
(P)-, (M)-**19f:** Ar = C_6F_5

Figure 13.10 Molecular structures of enantiomeric hydro[5]helicene derivatives (P)-, (M)-**17a–e**; (P)-, (M)-**18a–e**; and (P)-, (M)-**19a–f**.

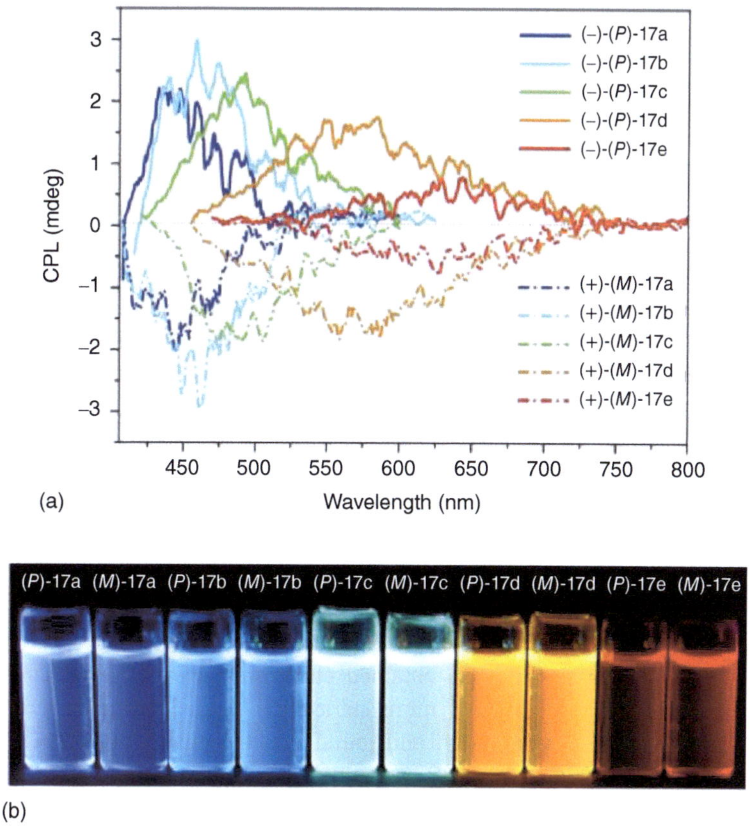

Figure 13.11 (a) CPL spectra of (−)-(*P*)-**17a–e** and (+)-(*M*)-**17a–e** in THF (c = 5×10^{-5} M). (b) Photographs of enantiomers (−)-(*P*)-**17a–e** and (+)-(*M*)-**17a–e** in THF under irradiation at 365 nm.

intensities, high Φ_F values and large Stokes shifts (4521–6405 cm^{-1}). In particular, the chiral organic nanoparticles also exhibited mirror-image and full-color CPL properties in water. The maximum emission wavelengths of the chiral nanoparticles in water ranged from 463 to 603 nm; their Φ_F values increased from 3 to 12%; and their $|g_{lum}|$ values ranged from 1.0×10^{-4} to 1.7×10^{-3}. This work could provide an opportunity for further exploration of the chiral organic nanoparticle's applications in chiral bioanalysis and bioimaging.

By introducing phenyl groups with greater steric hindrance than methyl groups at the 1,10-position of helicene derivatives, Chen and coworkers also synthesized a series of optically stable enantiomeric tetrahydro[5]helicene derivatives (*P*)- and (*M*)-**18a–e** containing two ester groups (Figure 13.10) [36]. It was found that the maximum emission wavelengths of **18a–e** in THF ranged from 438 to 442 nm, and their Φ_F values ranged from 40 to 59%. Moreover, the g_{lum} values were in the region of $+3.66 \times 10^{-3}$ to $+6.41 \times 10^{-3}$ for (*P*)-**18a–e** and 3.40×10^{-3} to 6.59×10^{-3}

for (*M*)-**18a–e**, respectively. Thus, a kind of helicene derivative with blue CPL properties was obtained. Similarly, Chen et al. [37] also obtained a series of optically stable enantiomeric tetrahydro-benzo[5]helicene derivatives (*P*)- and (*M*)-**19a–f** (Figure 13.10) and found that the maximum emission wavelengths of **19a–f** in dichloromethane ranged from 406 to 416 nm, the Φ_F of **19a–f** ranged from 22 to 35%, and their $|g_{lum}|$ values were about 1×10^{-4}. It was further found that **19f** displayed significant enantioselective recognition toward the enantiomers of tryptophan methyl esters.

In 2018, Chen et al. [38] designed and synthesized **20a–d** (Figure 13.12). It was found that compared with **20a–c**, **20d** with more rigid structure showed not only redshift in emission spectrum but also stronger CPL with amplified luminescence dissymmetric factor. However, the Φ_F values of **20a–c** were greater than that of **20d**, among which the Φ_F of **20c** was as high as 41%, while the Φ_F of **20d** was only 19% in dichloromethane. The enhanced π–π stacking interaction and intermolecular electron transfer of **20d** might lead to its partial fluorescence quenching. It was further found that enantiomers **20a–20c** displayed CPL properties, and the enantiomers with *P* configuration exhibited g_{lum} values of -2.5×10^{-4} to -4.1×10^{-4}, and the *M*-enantiomers exhibited g_{lum} values of $+2.6\times10^{-4}$ to $+4.2\times10^{-4}$. Moreover, the g_{lum} values of (*P*)-**20d** and (*M*)-**20d** at 455 nm were -4.52×10^{-3} and $+4.43\times10^{-3}$, respectively, which was nearly 20 times higher than that of **20a** (Figure 13.13). Recently, they also obtained enantiomeric benzo[5]helicene derivatives **21a–g** (Figure 13.12) with more sterically hindered aromatic substituent groups at the 1,10-position [39] and found that the enantiomers **21a–g** not only showed high racemization barriers and significant thermal stabilities but also exhibited significant CPL properties. Moreover, enantiomer (*M*)-**21f** could also self-assemble into a helical nanotube by halogen bond interaction in the solid state. Recently, they reported enantiomeric helicene derivatives (*P*)- and (*M*)-**21h** (Figure 13.12), both of which also showed significant mirror-image CPL properties with the $|g_{lum}|$ values of about 6×10^{-3} [40]. They also fabricated circularly polarized organic light-emitting diodes (CP-OLEDs) based on (*P*)- and (*M*)-**21h**, respectively. Notably, the CP-OLEDs using the (*P*)- and (*M*)-**21h** as emitters and a TADF molecule as sensitizer exhibited well circularly polarized electro-luminescence (CPEL) performance with the electroluminescent dissymmetric factor ($|g_{EL}|$) of -2.2×10^{-3} and $+2.3\times10^{-3}$ for (*P*)- and (*M*)-**21h**-based devices, respectively.

More recently, Crassous et al. [41] reported phosphole helicene derivatives (*P*)-**22**/(*M*)-**22** (Figure 13.14) with the Φ_F of 10% in dichloromethane, which exhibited blue mirror-image CPL at 450 nm, and the g_{lum} values were $+8\times10^{-4}$ and 7×10^{-4} at 452 nm, respectively. Moreover, the enantiomers (*P*)-**22**/(*M*)-**22** were also used as blue-emitting chiral dopants to fabricate OLEDs. Unfortunately, no circularly polarized electroluminescence properties of the obtained OLEDs were detected. Generally, the twist degree of helical framework has a significant influence on its CPL properties. The more ortho-fused benzene subunits arranged

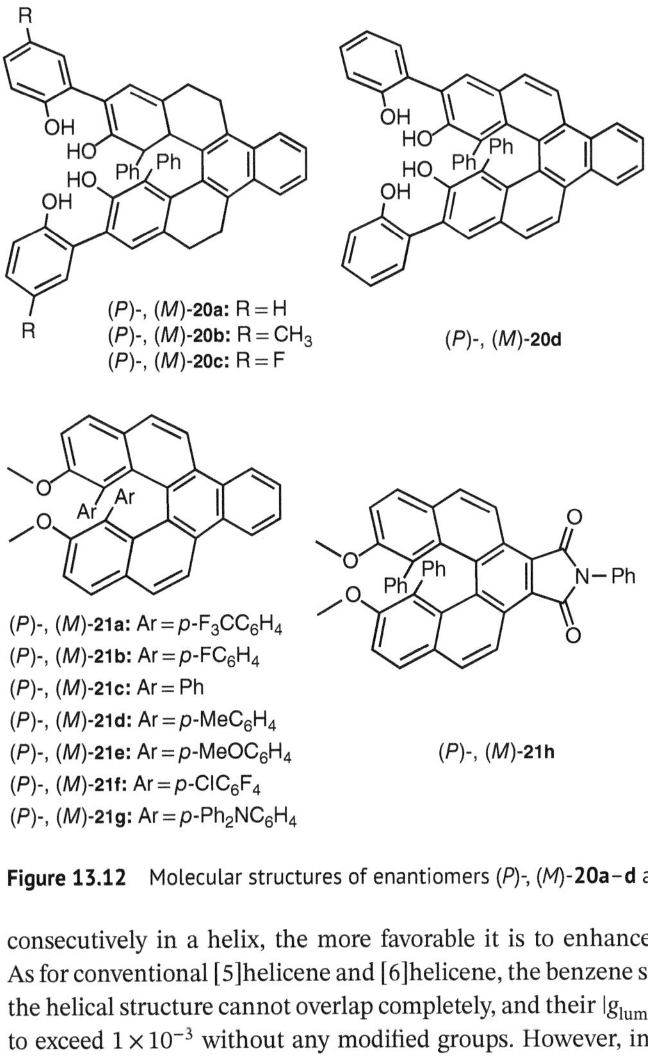

Figure 13.12 Molecular structures of enantiomers (*P*)-, (*M*)-**20a–d** and (*P*)-, (*M*)-**21a–h**.

consecutively in a helix, the more favorable it is to enhance the CPL properties. As for conventional [5]helicene and [6]helicene, the benzene subunits at the ends of the helical structure cannot overlap completely, and their $|g_{lum}|$ values were difficult to exceed 1×10^{-3} without any modified groups. However, introducing more than six ortho-fused benzene subunits into the helix structure face the challenge of difficult synthesis and low yield. Therefore, it is a compromise proposal to improve the CPL properties of helicene derivatives by modifying the conventional [6]helicene and increasing the twist degree of helical framework. Wang et al. [42] obtained **23a** (Figure 13.14) by introducing [6]helicene into the framework of perylene imide. The distance between the terminal helical skeleton and the parent structure of perylene imide was about 3.3Å, thus the $|g_{lum}|$ values of **23a** was increased to 2×10^{-3}. Compared with **23a** (Figure 13.14), the $|g_{lum}|$ values of **23b** ($\Phi_F = 10.9\%$) was only 1.2×10^{-3} [42], this result can be attributed to the fact that the three perylene imide units with high steric resistance reduced the torsion degree of the helical structure formed at their edges.

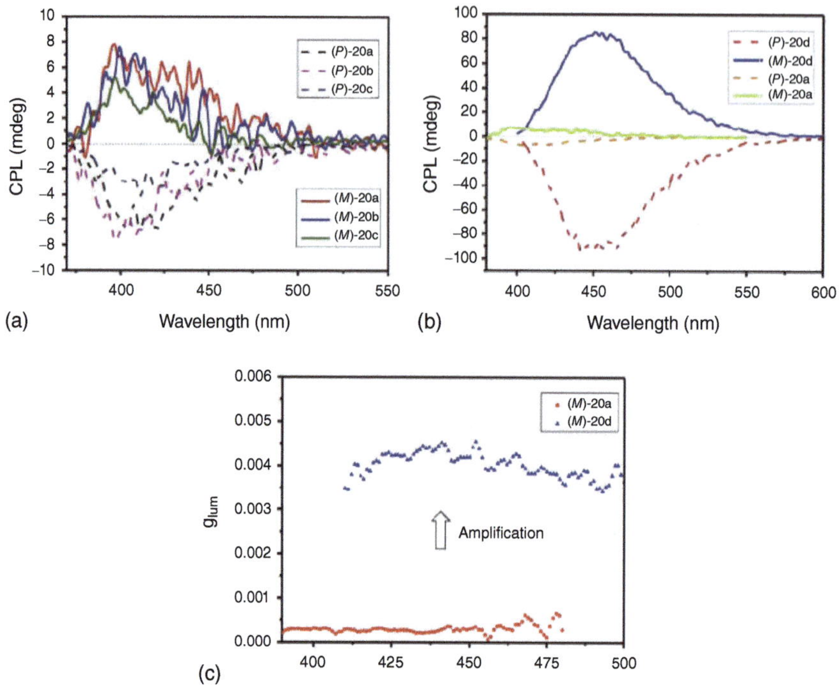

Figure 13.13 (a) Mirror-image CPL spectra of (*P*)- and (*M*)-**20a–c** in DCM, (b) mirror-image CPL spectra of (*P*)- and (*M*)-**20a** and **20d** in DCM, and (c) dissymmetry factor g_{lum} versus wavelength of (*M*)-**20a** and (*M*)-**20d** in DCM ($c = 8.0 \times 10^{-5}$ M).

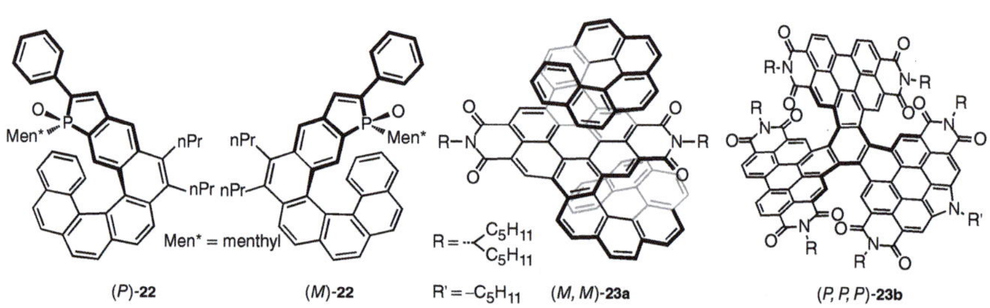

Figure 13.14 Molecular structures of enantiomers (*P*)-**22**, (*M*)-**22**, (*M,M*)-**23a**, and (*P,P,P*)-**23b**.

13.3 Heterohelicenes with CPL Properties

Heterohelicenes are important kinds of helicene derivatives, in which the heteroatoms (such as *O*, *N*, *S*, *Si*, etc.) are inserted into the ortho-fused ring systems [43]. Thus, the heterohelicenes generally contain a five-membered ring such as furan, thiophene, or pyrrole and six-membered ring pyridine. Heterohelicene derivatives also exhibited excellent CPL properties, and they have attracted much attention for their potential applications in materials science, asymmetric catalysis, and supramolecular chemistry.

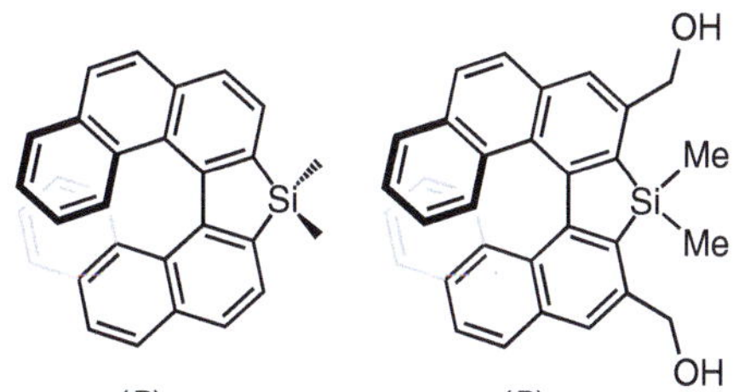

Figure 13.15 Molecular structures of sila[7]helicene derivatives (*P*)-**24** and (*P*)-**25**.

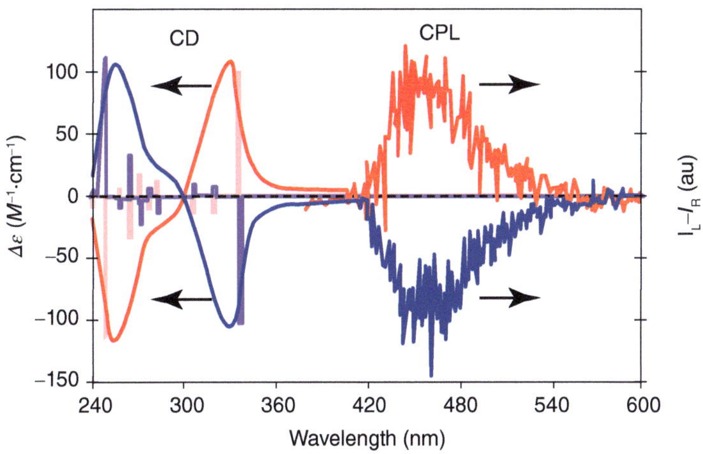

Figure 13.16 CD (c = 4.8×10^{-6} M) and CPL (c = 6.8×10^{-6} M) spectra of (*M*)-**24** (blue) and (*P*)-**24** (red) in chloromethane. The blue and red bars show the calculated CD spectra. Source: Reproduced with permission from Ref. [44]. Copyright 2013, American Chemical Society.

13.3.1 Silahelicenes with CPL Properties

In 2013, Nozaki et al. [44] synthesized chiral sila[7]helicene **24** (Figure 13.15) and found that its quantum yield was up to 17% in the solid state. It was also found that the enantiomers of **24** showed mirror-imaged CD and CPL spectra in CH_2Cl_2 (Figure 13.16) and the g_{lum} value of the enantiomer (*P*)-**24** was $+3.5 \times 10^{-3}$. By the double [2+2+2] cycloaddition reaction of diaryltetrayne and silicon-linked bis(propargyl alcohol), Tanaka and coworkers [45] obtained sila[7]helicene **25** (Figure 13.15) with the Φ_F of 15% in chloroform. They found that (*P*)-**25** also showed excellent CPL properties, and its g_{lum} value was 1.6×10^{-2}.

13.3.2 Borohelicenes with CPL Properties

In 2016, Hatakeyama et al. [46] reported boron-fused double [5]helicene **26** (Figure 13.17) and found that its Φ_F was 65% and its $|g_{lum}|$ value in chloromethane was 1.7×10^{-3}. In the same year, Knight and coworkers [47] reported another kind of borohelicene, helical chiral *N,N,O,O*-boron-chelated dipyrromethenes **27a–d** (Figure 13.17). They found that **27a–d** exhibited CPL activities in acetonitrile as

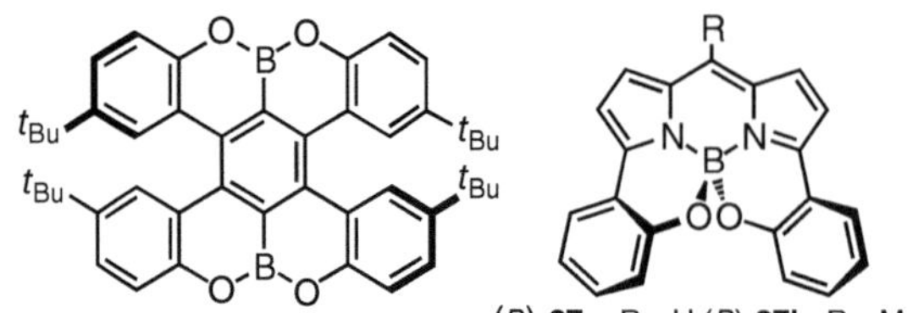

(*P*,*P*)-**26** (*P*)-**27a:** R = H (*P*)-**27b:** R = Me
(*M*)-**27c:** R = 4-Me(C_6H_4) (*P*)-**27d:** R = 2-pyridyl

Figure 13.17 Molecular structures of (*P*,*P*)-**26** and (*P*)-**26a–d**.

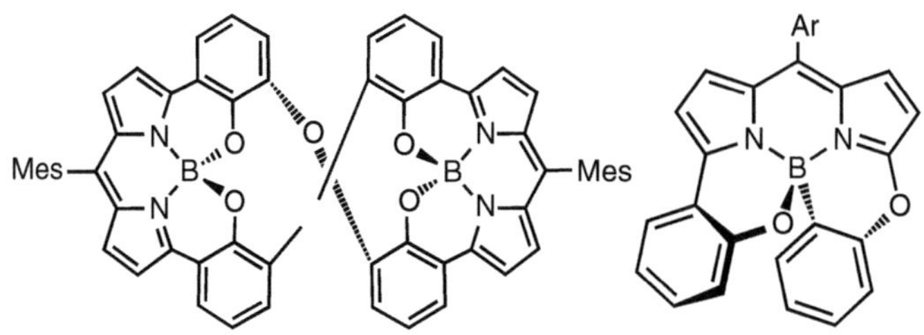

(*P*, *P*)-**28** (*P*)-**29**

Figure 13.18 Molecular structures of (*P*,*P*)-**28** and (*P*)-**29**.

well, and the $|g_{lum}|$ values of **27a–d** were 4.7×10^{-3} (623 nm), 3.3×10^{-3} (635 nm), 4.3×10^{-3} (637 nm), and 4.2×10^{-3} (675 nm), respectively.

Recently, Nabeshima et al. [48] synthesized double BODIPY (bis(boron-dipyrromethene)) complex **28** with a twisted figure-of-eight structure (Figure 13.18). It was found that (*P*,*P*)-**28** exhibited strong CPL activity at 663 nm in chloroform, and its Φ_F and g_{lum} values were 58% and 9×10^{-3}, respectively. Soon after, they also reported a helically chiral BODIPY **29** with a *N,N,O,C*-boron chelation structure (Figure 13.18) [49]. The enantiomers (*P*)- and (*M*)-**29** exhibited significant mirror-image CPL spectra at 622 nm in hexane, and the $|g_{lum}|$ and Φ_F were 3.7×10^{-3} and 49%, respectively.

By embedding bis(haloBODIPY) into an unstable helical chiral structure, De la Moya et al. [50] reported a simple method for the synthesis of chiral BODIPY derivatives **30a** and **30b** (Figure 13.19). It was found that the Φ_F values of **30a** and **30b** were 15 and 16%, respectively, and their $|g_{lum}|$ values were estimated to be both 1.0×10^{-3}. It was worth noting that the introduction of the BODIPY unit into the helical framework made it a feasible strategy to enhance the CPL property. As shown in Figure 13.20, by embedding the BODIPY unit into [7]helicene, the $|g_{lum}|$ value of **31a** reached 1.7×10^{-3}, which was significantly higher than **30a–b**. Moreover, rigid structure decreased the energy loss caused by the molecular vibration, thus the Φ_F value of **31a** reached 21.8%. Although the $|g_{lum}|$ value of **31b** (1.0×10^{-3}) was significantly smaller than that of **31a**, when the benzene subunits were introduced into the periphery of helical segment, the Φ_F value of **31b** (33%) was higher than that of **31a**. Interestingly, the chiroptical properties of the helical framework was enhanced by the attachment of axial chirality. Consequently, **31c** ($|g_{lum}| = 1.5 \times 10^{-3}$, $\Phi_F = 29.6\%$)

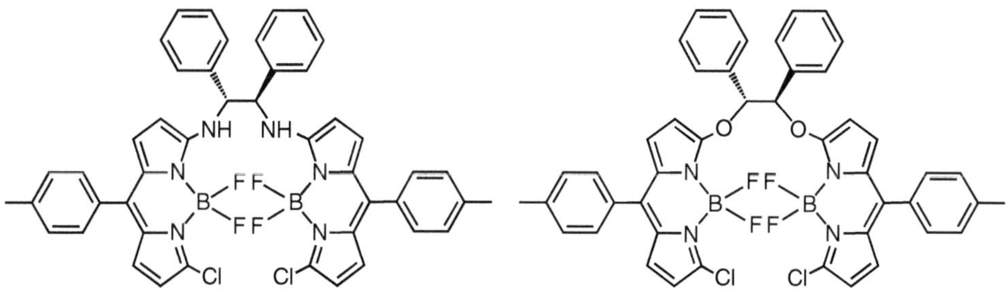

Figure 13.19 Molecular structures of (*R*,*R*)-**30a** and (*R*,*R*)-**30b**.

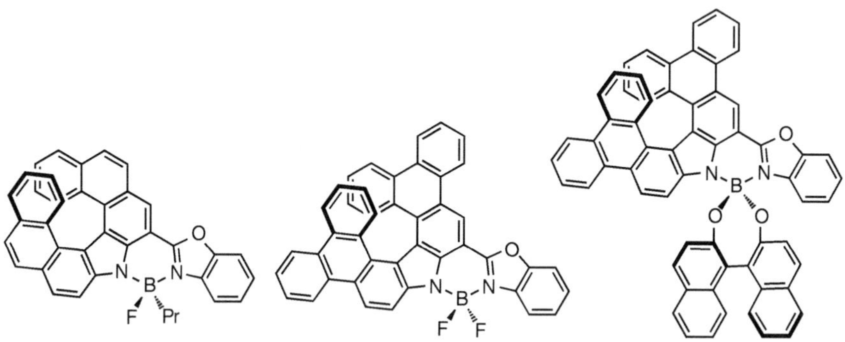

Figure 13.20 Molecular structures of (*M*)-**31a**, (*M*)-**31b**, and (*R*,*M*)-**31c**.

exhibited more intense CD and CPL than **31b** because of the existence of BANOL segment [51].

The boron–nitrogen bond has a certain polarity with respect to the carbon–carbon double bond, and it can also be used to construct helicene derivatives with CPL characteristics. In 2017, Crassous et al. [52] reported four boron-azahelicenes **32a–d** (Figure 13.21). The enantiomers of the boron-azahelicenes also showed significant CPL properties, and the $|g_{lum}|$ values of **32a–d** were 9.0×10^{-4}, 7.0×10^{-4}, 2.3×10^{-3}, and 1.0×10^{-3}, respectively. The introduction of boron nitrogen heteroatoms can effectively change the energy level of this kind of heterohelicenes, which might make them become a new kind of organic photoelectric material.

More recently, Ros and coworkers prepared three pairs of enantiomers (*P*)-**33a–c** and (*M*)-**33a–c** based on the boronated arylquinoline skeleton (Figure 13.22) [53]. It was found that the Φ_F values of both **33a** and **33b** were about 30% in toluene and acetonitrile, but the Φ_F of **33c** (13%) was slightly lower than those of **33a–b**. The enantiomers of **33a–c** exhibited good CPL properties, and the $|g_{lum}|$ values of **33a** and **33b** were 2.5×10^{-4} (480 nm) and 9.5×10^{-4} (500 nm), respectively. Interestingly, **33c** exhibited a strong redshifted CPL signal, and its g_{lum} (3.5×10^{-3}) was larger than those of **33a–b**.

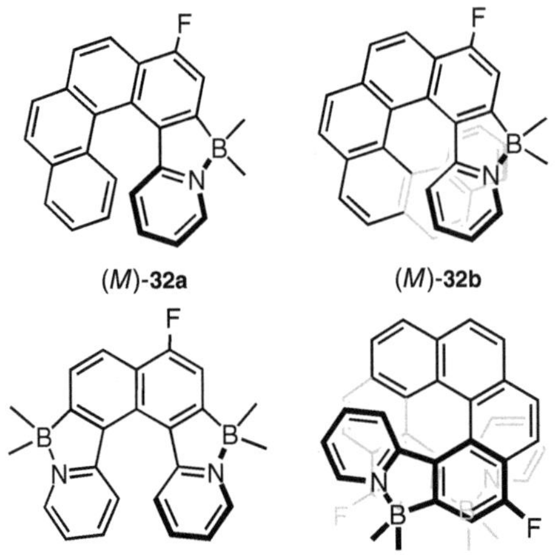

(*M*)-**32a** (*M*)-**32b**

(*M*)-**32c** (*M*)-**32d**

Figure 13.21 Molecular structures of boron-azahelicenes (*M*)-**32a–d**.

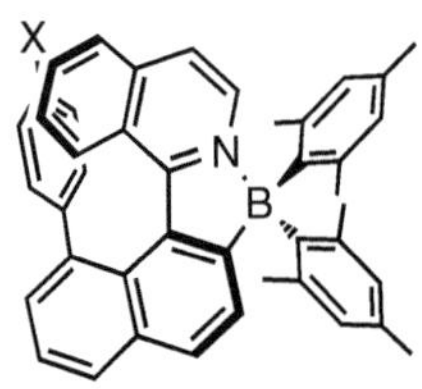

(*P*)-**33a**: X = H
(*P*)-**33b**: X = OMe
(*P*)-**33c**: X = NMe_2

Figure 13.22 Molecular structures of (*P*)-**33a–c**.

13.3.3 Azahelicenes with CPL Properties

Since Venkataraman et al. [19] first reported azahelicenes (*M*)-**1a** and (*P*)-**1b** with CPL properties in 2003, azahelicenes have become one of the most important heterohelicenes. In 2012, Shinokubo et al. [54] reported a pair of enantiomeric azahelicene analogs (*P*)- and (*M*)-**34a** (Figure 13.23) and found that they had high racemization barriers and stable chiral properties. The $|g_{lum}|$ value of **34a** was 3×10^{-3}, and its Φ_F was 36%. Furthermore, Shinokubo et al. [55] also obtained an eight-shaped azahelicene dimer **34b** (Figure 13.23). They found that compared with **34a**, the Φ_F of **34b**

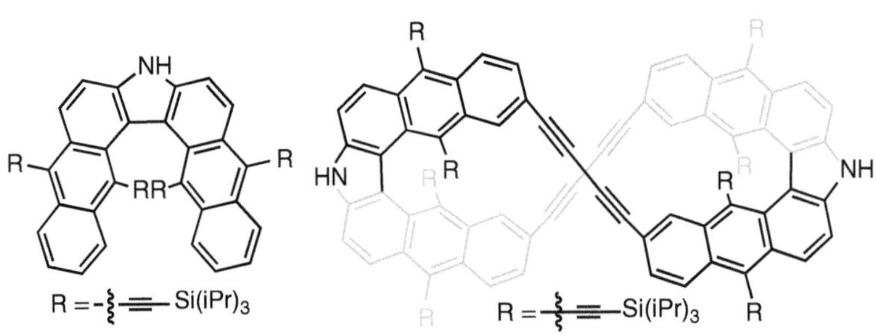

(*P*)-, (*M*)-**34a** (*M*,*M*)-**34b**

Figure 13.23 Molecular structures of azahelicene analogs (*P*)-**34a**, (*M*)-**34a**, and eight-shaped azahelicene dimer (*M*,*M*)-**34b**.

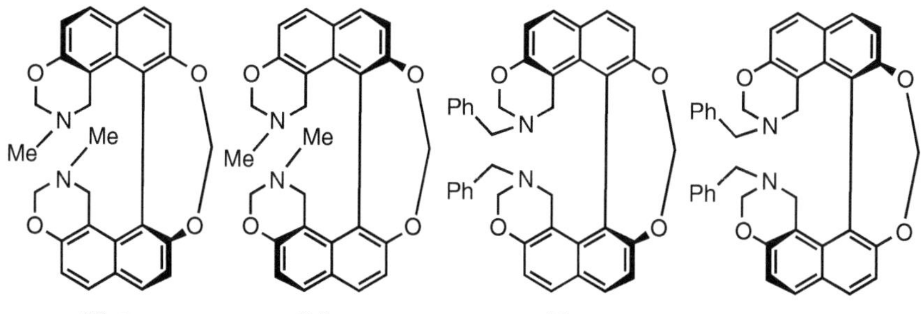

Figure 13.24 Molecular structures of helicene derivatives (*P*)-**35a,** (*M*)-**35a**, (*P*)-**35a,** and (*M*)-**35b**.

was enhanced to 55%. Moreover, good CPL properties for the enantiomers of **34b** could also be observed, and the $|g_{lum}|$ value was found to be 8.5×10^{-3}, which was probably because of their rigid molecular structure.

In 2014, Muller and coworkers [56] synthesized bisoxazine-based helicene derivatives **35a** and **35b** (Figure 13.24) and found that both **35a** and **35b** showed CPL properties. The g_{lum} values of (*P*)-**35a**/(*M*)-**35a** and (*P*)-**35b**/(*M*)-**35b** were $+1.5 \times 10^{-3}/-9.0 \times 10^{-4}$ and $+1.4 \times 10^{-3}/-1.3 \times 10^{-3}$, respectively. This work can provide a new option for the synthesis of structurally rigid and optically pure helicene-like molecules with CPL properties.

To investigate the influence of different heteroatoms on the emission and CPL properties of heterohelicene, Lacour et al. [57] synthesized cationic dioxa[6]helicene **36a**, azaoxa[6]helicene **36b**, and diaza[6]helicene **36c** (Figure 13.25) in 2016. They studied the photophysical properties of the cationic heterohelicenes and found that their fluorescence emissions range from orange to near infrared, and the Φ_F values of **36a–c** were 12, 22, and 31%, respectively. Moreover, the enantiomers (*P*)-**36a–c**/(*M*)-**36a–c** also showed CPL properties, and their g_{lum} values were estimated to be $-3.2 \times 10^{-4}/+4.1 \times 10^{-4}$, $-2.1 \times 10^{-3}/+2.0 \times 10^{-3}$, and $-1.2 \times 10^{-3}/+1.0 \times 10^{-3}$ at 595, 614, and 658 nm, respectively.

Compared with **37a** ($\Phi_F = 13\%$), **37b** ($\Phi_F = 35\%$) and **37c** ($\Phi_F = 35\%$) (Figure 13.26) have higher luminescence quantum efficiencies. It was found that diaza[4]helicene derivatives (*P*)-**37a–c**/(*M*)-**37a–c** also showed CPL properties, and their g_{lum} values were $+1.3 \times 10^{-3}/-1.0 \times 10^{-3}$, $+1.6 \times 10^{-3}/-1.7 \times 10^{-3}$, and $+9.0 \times 10^{-4}/-8.0 \times 10^{-4}$, respectively. This work indicated that different substituents could affect the photophysical properties of the chiral cationic diaza[4]helicene derivatives [58], such as

Figure 13.25 Molecular structures of cationic heterohelicenes (*P*)-, (*M*)-**36a–c**.

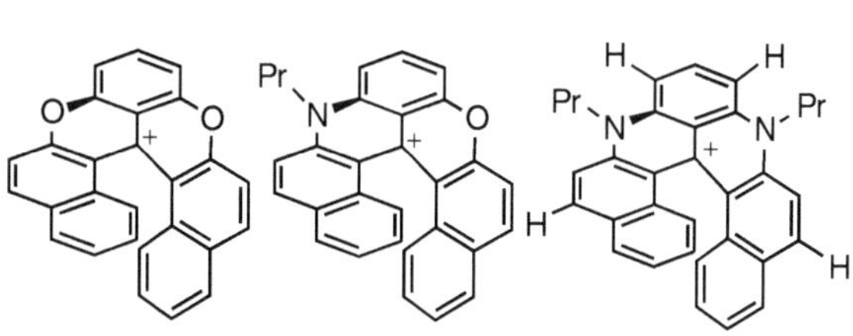

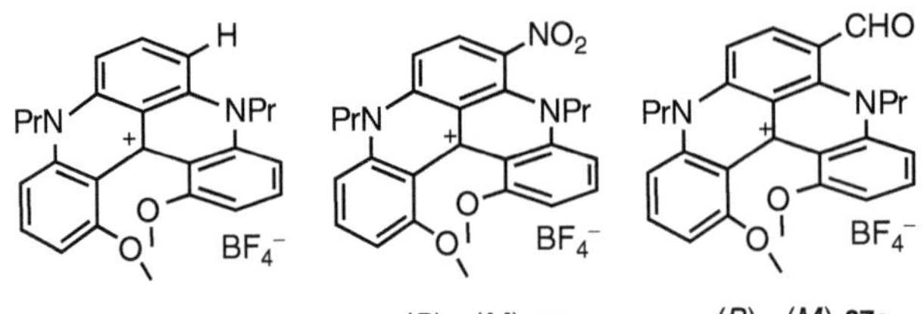

Figure 13.26 Molecular structures of cationic diaza[4]helicene derivatives **37a**, **37b**, and **37c**.

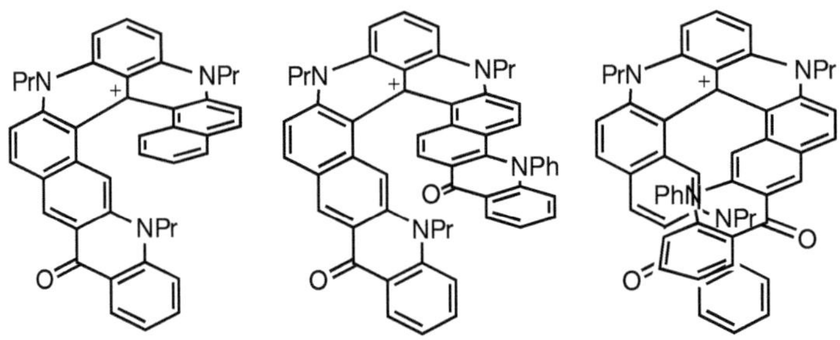

Figure 13.27 Molecular structures of cationic diaza[4]helicene derivatives **38a**, **38b**, and **38c**.

they showed little effect on the g_{lum} values of their CPL. Based on the structure of **37a–c**, **38a–c** (Figure 13.27) were obtained by introducing tetracene with longer conjugated structure into the helical skeleton [59]. The extended conjugated structure increased the absorption of excited photons by helicene derivatives, resulting in the Φ_F values of **38a–c** (29, 25, and 21% for **38a**, **38b**, and **38–c**) smaller than **37b–c**. In addition, the number of tetracene subunits and their arrangement on the helical skeleton affected the CPL properties of **38a–c**; the mono-tetracene **38a** presented a $|g_{lum}|$ value of 1×10^{-4}; the non-symmetrical **38b** exhibited a $|g_{lum}|$ value of 6×10^{-4}, and the bis-tetracene **38c** showed a $|g_{lum}|$ value of 1.4×10^{-3}.

13.3.4 Oxalohelicenes with CPL Properties

Due to the nonplanar structures, helical-like molecules could still exhibit strong fluorescence in the aggregation state. Moreover, their assemblies could also show enhanced CPL properties. The g_{lum} values of the assemblies were usually one or two orders higher than those of helicene monomers [60–62]. In 2011, Takeuchi et al. [63] designed and synthesized multi-hydrogen-functionalized helicene derivatives (*P*)-**39** and (*M*)-**39** (Figure 13.28), which could form a helical fiber structure in a nonpolar solvent such as toluene or chloroform. The fibrous self-assemblies obtained from monomers (*P*)-**39** and (*M*)-**39** showed mirror-image CPL signal with the $|g_{lum}|$ value of 3.5×10^{-2} at 476 nm. Such chiral assemblies could be used to construct chiral materials with g_{lum} values comparable to those of lanthanide complexes.

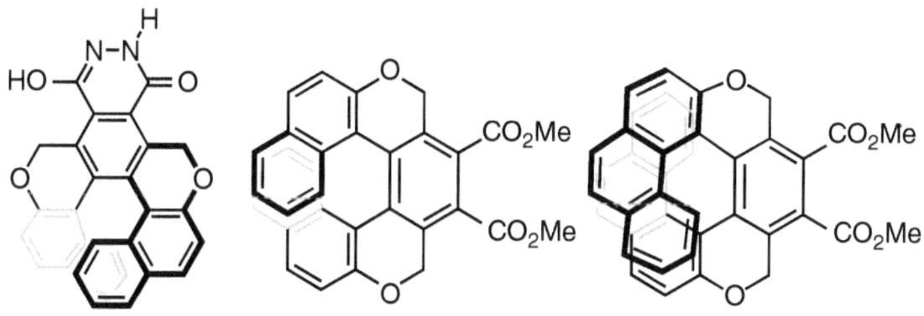

(*M*)-**39** (*P*)-**40a** (*P*)-**40b**

Figure 13.28 Molecular structures of helicene derivatives (*M*)-**39**, (*P*)-**40a**, and (*P*)-**40b**.

Recently, Tanaka and coworkers [64] synthesized [9]helicene-like molecules **40a** and **40b** based on the phenanthrenol group by the rhodium-mediated intramolecular [2 + 2 + 2] cycloaddition of 3-phenanthrenol-linked triynes (Figure 13.28). The Φ_F values of **40a** and **40b** in dichloromethane were 23 and 18%, respectively, and their $|g_{lum}|$ values at 473 and 547 nm were 9.5×10^{-4} and 1.1×10^{-3}, respectively. **40b** had more extended π-conjugated structure, so its g_{lum} was slightly larger than that of **40a**.

13.3.5 Thiohelicenes with CPL Properties

In 2016, Villani and coworkers [65] synthesized the optically pure disulfide [6]helicene derivative (*P*)-**41** (Figure 13.29) and its enantiomer (*M*)-**41** and studied their CPL properties based on the DFT calculation and experiments. It was found that the Φ_F value of **41** in chloroform was only 1.6%, but the $|g_{lum}|$ value of the enantiomers was 9×10^{-3}. This work indicated that although the thiohelicene derivatives showed CPL properties, their Φ_F values could be significantly decreased by sulfur atoms.

The similar phenomenon was also found by Hasobe and coworkers [66]. As shown in Figure 13.29, the luminescence quantum efficiencies of enantiomers (*P*)-**42a**/(*M*)-**42a** were just only 3%. However, when they were oxidized to sulfone derivatives (*P*)-**42b**/(*M*)-**42b**, the Φ_F values of (*P*)-**42b**/(*M*)-**42b** significantly increased to 27%. Moreover, enantiomers (*P*)-**42b** and (*M*)-**42b** also showed significant mirror-image CPL spectra ($|g_{lum}| = 8.3 \times 10^{-4}$) in THF.

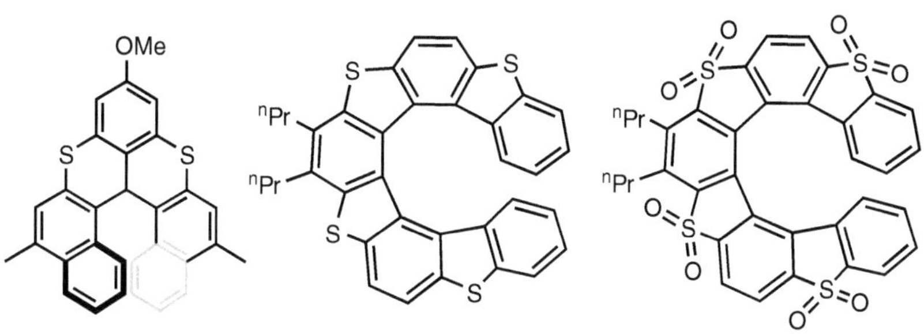

(*P*)-**41** (*P*)-, (*M*)-**42a** (*P*)-, (*M*)-**42b**

Figure 13.29 Molecular structures of thiohelicenes (*P*)-**41**; (*P*)-, (*M*)-**42a**; and (*P*)-, (*M*)-**42b**.

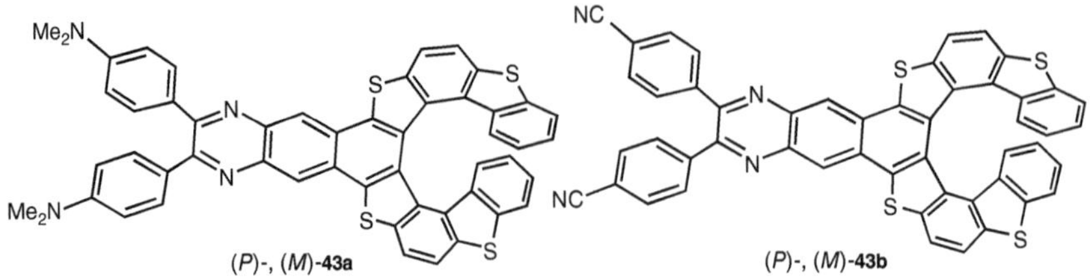

Figure 13.30 Structures of tetrathia[9]helicene derivatives (*P*)-**43a**/(*M*)-**43a** and (*P*)-**43b**/(*M*)-**43b**.

By introducing a quinoxaline structure with different groups on the tetrathia[9] helicene skeleton, Hasobe et al. successfully synthesized tetrathia[9]helicene derivatives **43a** and **43b** (Figure 13.30) [67]. It was found that the Φ_F values of **43a** and **43b** in toluene were 39% and 26%, respectively, which were significantly greater than those of the original tetrathia[9]helicene (Φ_F = 2%) and the quinoxaline-tetrathia[9]helicene (Φ_F = 4%). Moreover, (*P*)-**43b** and (*M*)-**43b** also showed CPL properties, and their $|g_{lum}|$ value was 3.0×10^{-3}. This strategy of introducing electronic push–pull groups on the helical framework to enhance its fluorescence and CPL performance could provide a new perspective for the future development of optically pure heterohelicene derivatives with high luminescence quantum efficiencies.

13.4 Metal Helicene Derivatives with CPL Properties

By introducing transition metals on the molecular skeleton or using helicene as a metal ligand, helicene derivatives could efficiently emit circularly polarized phosphorescence for the heavy atom effect of transition metals. In 2014, Crassous et al. [68] introduced a transition metal Pt on the helicene skeleton and obtained enantiopure mono-cycloplatinated-[8]helicene **44a** and bis-cycloplatinated-[6]helicene derivatives **44b** (Figure 13.31). The luminescence quantum yield of **44b** (Φ_F = 10%) was higher than that of **44a** (Φ_F = 5.6%). Moreover, **44a** and **44b** exhibited obvious CPL properties, and the g_{lum} values of (*P*)-**44a**/(*M*)-**44a** and (*P*)-**44b**/(*M*)-**44b** were $+4.0 \times 10^{-3}/-5.0 \times 10^{-3}$ and $+5.0 \times 10^{-3}/-5.0 \times 10^{-3}$, respectively. As a comparison, (*P*)-**44c**/(*M*)-**44c** ($g_{lum} = +1.3 \times 10^{-3}/-1.1 \times 10^{-3}$) exhibited higher g_{lum} than those of **44a–b**. (*P*)-**44c**/(*M*)-**44c** were also used as emitting layers to construct a CP-OLED

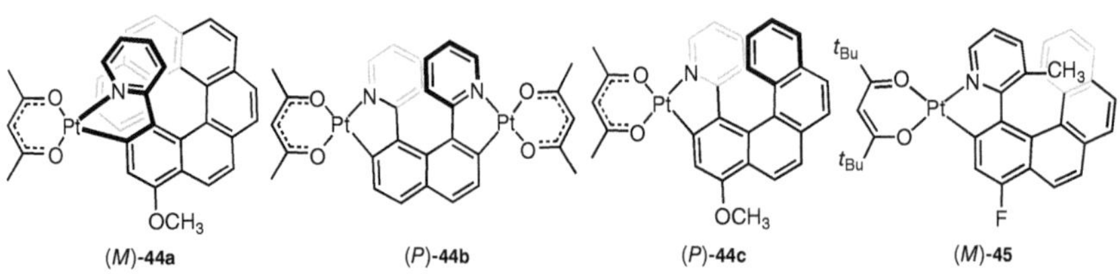

Figure 13.31 Molecular structures of helicene derivatives (*M*)-**44a**, (*P*)-**44b**, (*P*)-**44c**, and (*M*)-**45**.

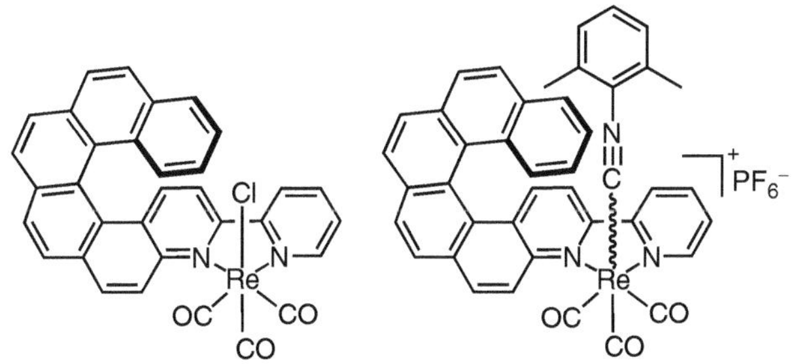

Figure 13.32 Molecular structures of (*M*)-**46a** and (*M*)-**46b**.

by Fuchter's group, and the $|g_{EL}|$ value was up to 0.38 [69], which was meaningful for the research of next-generation OLED displays. Recently, Zheng et al. [70] reported a new metal helicene derivative **45** (Figure 13.32) based on **44c**. The introduction of the large hindered groups (-CF_3, ^{t}Bu) endowed **45** with a more twisted structure, which in turn induced it to exhibit more obvious CPL properties than **44c**, the g_{lum} values of (*P*)-**45** and (*M*)-**45** were -5.9×10^{-3} and $+6.0 \times 10^{-3}$, respectively. They also fabricated two CP-OLEDs by utilizing (*P*)-**45** and (*M*)-**45** as emitters, both devices exhibited obvious CP electroluminescence properties, and the $|g_{EL}|$ values of (*P*)- and (*M*)-**45** were 1×10^{-3} at 653 nm.

In 2015, Crassous et al. [71] synthesized the rhenium-based enantiomers (*P*)-**46a**/(*M*)-**46a** and (*P*)-**46b**/(*M*)-**46b** with red CPL properties by using helicene derivatives containing a dipyridine structure as ligands (Figure 13.32). The g_{lum} values of (*P*)-**46a**/(*M*)-**46a** and (*P*)-**46b**/(*M*)-**46b** were the same order of magnitude, which were $+3.1 \times 10^{-3}/-2.8 \times 10^{-3}$ and $+1.3 \times 10^{-3}/-1.5 \times 10^{-3}$, respectively. But the luminescence quantum yield of **46b** reached 6% by changing the π-conjugated ligand, which was higher than that of **46a** (0.16%). This work not only provided the first rhenium-based helicene derivatives with CPL properties but also realized the increase of quantum yields by changing the ligand, which proved that adjusting the interaction between helical skeleton and metal ligand was an effective strategy to optimize the properties of CPL. Based on this work, they reported two pairs of enantiomers **47a$^{1-2}$ and 47b$^{1-2}$** (Figure 13.33) [72]. When the large CO ligand was located above the Re atom, the steric hindrance effect was more obvious at the end of helical skeleton than that of halogens' ligand; consequently, helicene derivatives (*P*,*A*$_{Re}$)-**47a^2**/**47b^2** showed obvious CPL properties with g_{lum} values of $+5.3 \times 10^{-3}$ and $+5.6 \times 10^{-3}$ at 525 nm, respectively. In contrast, the g_{lum} values of (*P*,*C*$_{Re}$)-**47a^1**/**47b^1** were smaller and negative (-4.4×10^{-4} and -1.4×10^{-3}). The g_{lum} of epimer **47a^2** was 12 times larger than that of **47a^1**, revealing a strong "match/mismatch" effect that existed between the helix and metal stereochemistries, which indicated that CPL response could be optimized by manipulating Re(I) stereochemistry. Notably, although the phosphorescence quantum yield of **47b^1** was only 9%, its phosphorescence lifetime was up to 0.7 ms.

Moreover, Crassous and coworkers also reported a bis-helicenic terpyridine ligand **48a** and its zinc complex **48b** (Figure 13.34) [73]. It was found that **48a** exhibited

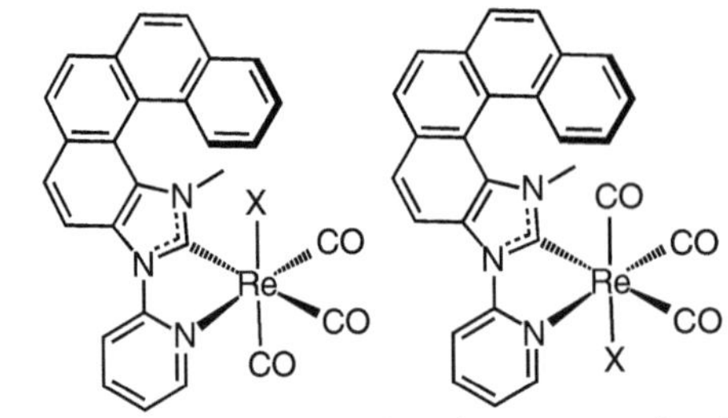

Figure 13.33 Molecular structures of $\mathbf{47a^{1-2}}$ and (M)-$\mathbf{47b^{1-2}}$.

Figure 13.34 Molecular structures of (P,P)-**48a** and (P,P)-**48b**.

strong blue fluorescence (λ_{max} = 421 nm, Φ_F = 8.4%) in dichloromethane. Moreover, the enantiomers (M,M)-**48a** (-8.4×10^{-3}) and (P,P)-**48a** ($+8.6 \times 10^{-3}$) also displayed remarkable CPL properties. When zinc ions were combined with the ligands, the obtained complexes (P,P)- and (M,M)-**48b** exhibited obvious redshift of their emission bands (λ_{max} = 480 nm, Φ_F = 19%), as well as CPL properties with g_{lum} values of $+1.2 \times 10^{-3}$ and -1.4×10^{-3}, respectively. The trans-(W-form) conformation of ligand **48a** could be conveniently transferred to U-shaped conformation of the Zn complex, which could provide a new perspective for the multifunctional chiral complexes and chiral switches by double-helical ligands.

Ir–NHC complexes by using helicene-based azacarbene as a ligand were also reported by Crassous and coworkers [74a]. As shown in Figure 13.35, the enantiomer without helicene ligand Δ-**49a** exhibited CPL properties, and its g_{lum} value was just $+9 \times 10^{-4}$. When a helicene ligand was introduced, P,Δ-**49b** showed higher g_{lum} ($+1.5 \times 10^{-3}$ at 530 nm) than that of Δ-**49a**. More recently, they further reported new Ir–NHC–helicene derivatives Λ/Δ-**49c** (Figure 13.35) [74b], which emitted long-lived green circularly polarized phosphorescence (g_{lum}=3.1×10^{-3}) in both solution and the solid state. This work proved that helicene as a ligand could enhance the CPL properties of Ir–NHC complexes.

In 2017, Avarvari et al. [75] reported a chiral metal diamine dithiolene complex (M)-**50** (Figure 13.36). The platinum(II) center was combined with 2,20-bipyridine and dithia[6]helicene ligand to form a metal helicene-like compound **50**. At room temperature, the enantiomer **50** exhibited CPL properties in acetonitrile with a $|g_{lum}|$ of 3×10^{-4}. Its emission band was centered at 715 nm, and the Φ_F was 0.15%.

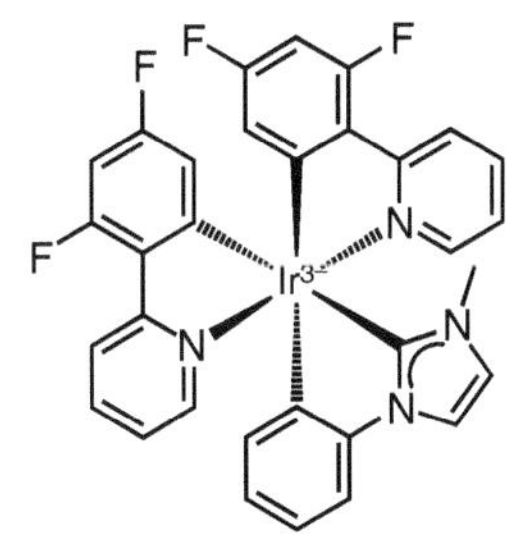

(Δ_{Ir})-, (Λ_{Ir})-**49a**

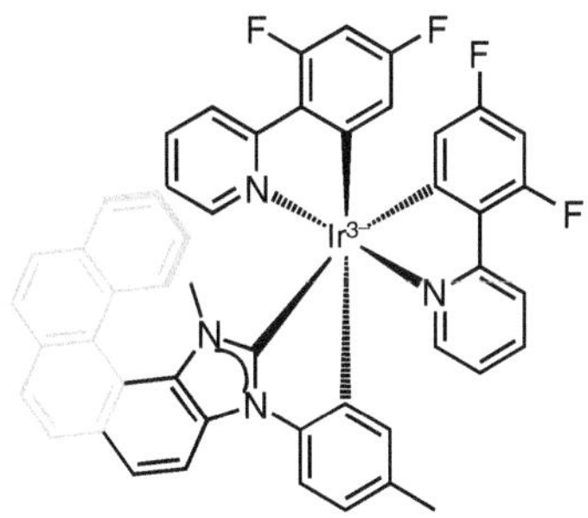

(P, Δ_{Ir})-, (M, Λ_{Ir})-**49b**^{1}

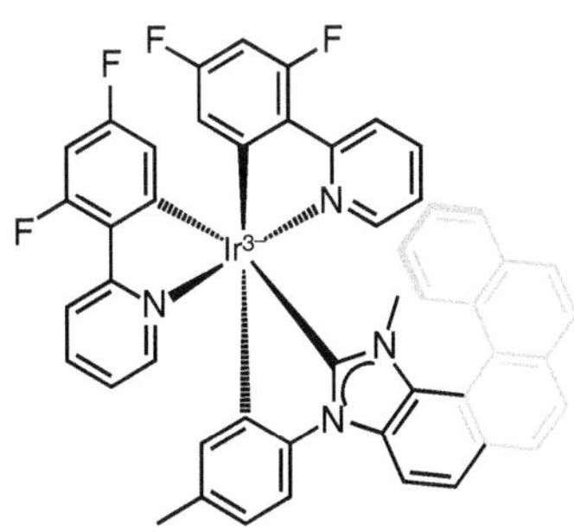

(P, Δ_{Ir})-, (M, Λ_{Ir})-**49b**^{2}

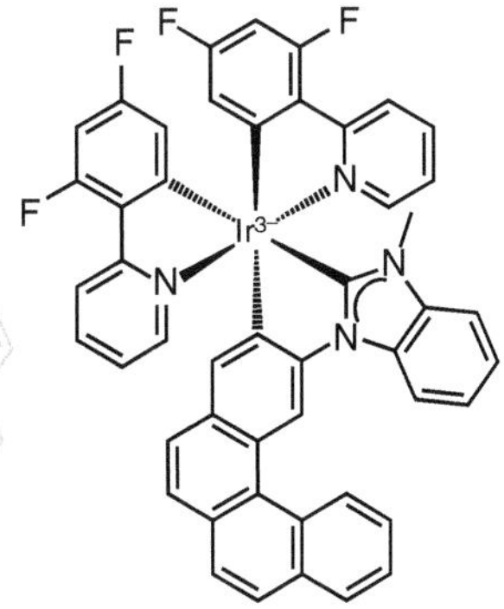

(*P*,)-, (*M*)-**49c**

Figure 13.35 Molecular structures of helicene derivatives (Δ_{Ir})-, (Λ_{Ir})-**49a**, (P, Λ_{Ir})-, (M, Δ_{Ir})-**49b**^{1}, (P, Δ_{Ir})-, (M, Λ_{Ir})-**49b**^{2}, and (*Delta*)-, (*Lambda*)-**49c**.

Figure 13.36 Molecular structures of (*M*)-**50** and (*M,M*)-**51**.

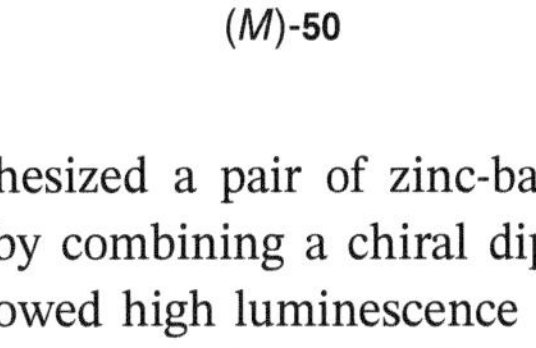

(*M*)-**50**

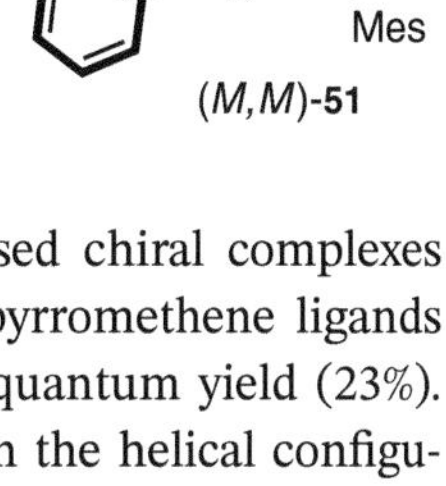

(*M,M*)-**51**

Recently, Hasobe et al. [75] synthesized a pair of zinc-based chiral complexes (*M,M*)-**51**/(*P,P*)-**51** (Figure 13.36) by combining a chiral dipyrromethene ligands with zinc(II) and found that it showed high luminescence quantum yield (23%). Due to the optical response of the strong exciton coupling in the helical configuration, the complexes (*M,M*)-**51**/(*P,P*)-**51** showed strong CPL properties, and their $|g_{lum}|$ value was calculated to be 2.2×10^{-2} at 660 nm. This dual ligand synergistic complexation can provide a new way for the design of chiral complexes with high g_{lum} values.

13.5 Regulation of CPL for Helicene Derivatives

Helicene derivatives not only showed CPL properties, but the regulation of their CPL could also be efficiently achieved by changing the external environmental conditions. For example, it was found that (*P*)-**52**/(*M*)-**52** exhibited blue fluorescence and CPL properties in CH_2Cl_2 [76]. When **52** (Figure 13.37) was protonated with HBF_4, **52-2H** with orange color emission and CPL in CH_2Cl_2 solution could be produced. Based on the helicene–bipyridine proligand **52**, Crassous and coworkers also obtained a "rollover" cycloplatinated derivative **53a** (Figure 13.37). It was found that **53a** showed CPL properties with a $|g_{lum}|$ value of 1.1×10^{-3}. When **53a** was protonated by HBF_4 to form **53b**, an enhanced CPL with a $|g_{lum}|$ of 2.2×10^{-3} was found. **53a** could also be deprotonated to **53a** in the presence of Na_2CO_3, which realized acid/base-triggered regulation of CPL in helicene derivatives [77].

In 2016, Sakai et al. [78] designed and synthesized helicene derivative **54** (Figure 13.38) by introducing a benzimidazole structure into the [5]helicene skeleton. Both (*P*)-**54** and (*M*)-**54** presented yellow CPL properties, and the $|g_{lum}|$ value was 9.45×10^{-3}. After **54** was protonated to form **54-H** (Figure 13.38), the enantiomers of **54-H** showed red CPL properties with a $|g_{lum}|$ value of 5.92×10^{-3}. Besides, the enantiomers of **54** with yellow CPL properties could be recovered after **54-H** reacted with pyridine. This work indicated that the protonation-induced CPL color regulation of a helicene derivative could be efficiently achieved.

The Φ_F (2% in water, $\lambda_{em} = 612$ nm) of **55** (Figure 13.39) was very low. As a result, its CPL property was difficult to detect. After **55** was protonated by HCl to form **55-H**,

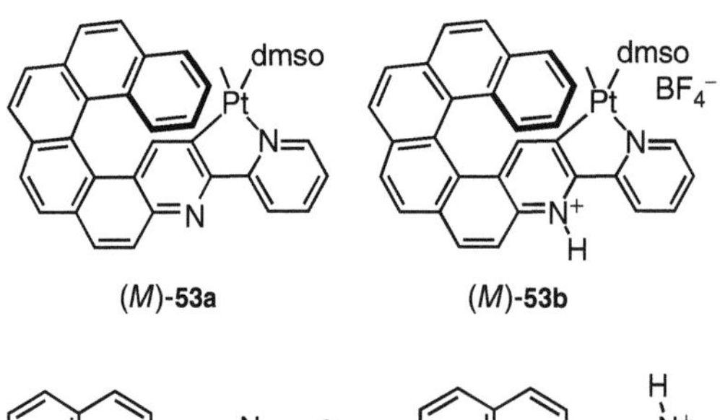

Figure 13.37 Molecular structures of **52**, **52-2H**, **53a**, and **53b**.

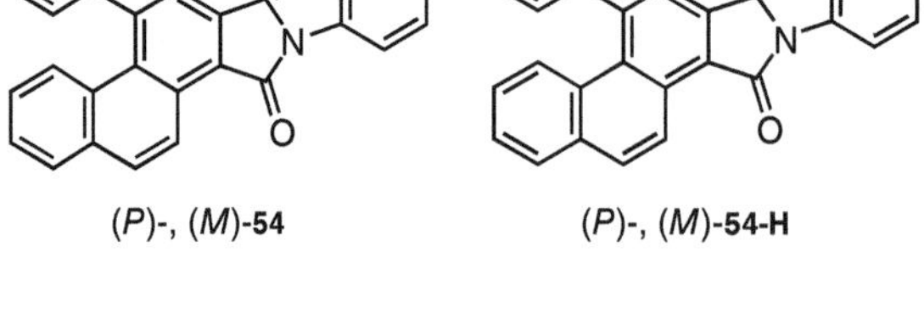

Figure 13.38 Molecular structures of **54** and **54-H**.

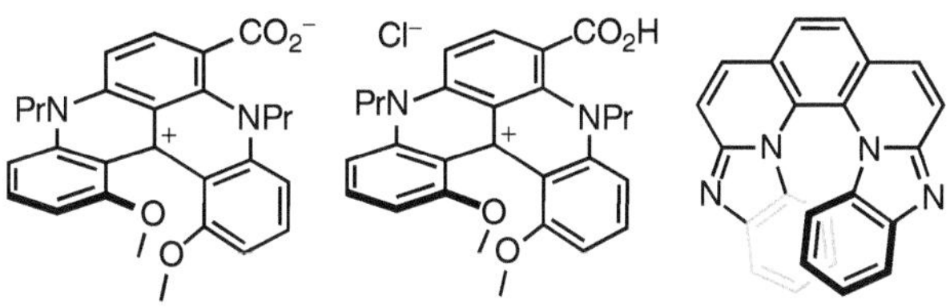

Figure 13.39 Molecular structures of (*P*)-**55**, (*P*)-**55-H**, and (*M*)-**56**.

the Φ_F (10% in water, λ_{em} = 587 nm) was significantly improved. Simultaneously, the CPL signals for the enantiomers of **55-H** could also be detected in the acidic acetonitrile solution, and the $|g_{lum}|$ value was about 5×10^{-4}. Moreover, **55-H** could return to **55** upon the addition of NaOH. This work realized a water-soluble pH-triggered CPL chiroptical switch in red spectral regions for the zwitterionic [4]helicene derivatives [79].

Tetraaza[7]helicene **56** [80] (Figure 13.39) with the Φ_F of 39% could be efficiently synthesized by a two-step reaction. It was found that the Φ_F of **56** could obviously increase to 80% under acidic conditions, and the protonated enantiomers (*P*)-**56** and (*M*)-**56** showed significant CPL properties ($|g_{lum}|$ = 9×10^{-3}). Moreover, the CPL properties of (*P*)-**56**/(*M*)-**56** could also be reversibly regulated by the protonation–deprotonation of nitrogen atoms, which could make them find potential applications in the recognition and sensing of stimulus responsiveness. Moreover, CPL properties of helicene derivatives can be modulated by the oxidation aromatization reaction. For example, helicenes (*P*)-**58a–f** could be obtained by one-step oxidation reaction from hydrohelicenes (*P*)-**57a–f** (Figure 13.40) [81]. The g_{lum} values of (*P*)-**57a–f** were generally in the magnitude of 10^{-4}, and the g_{lum} values of (*P*)-**58a–f** were at the range of -5.3×10^{-3} to -4.9×10^{-3}. The CPL signal inversion of (*P*)-**58a–f** relative to (*P*)-**57a–f** could be attributed to the difference of their efficient chiral fluorophores between axial chirality and helical chirality. In addition,

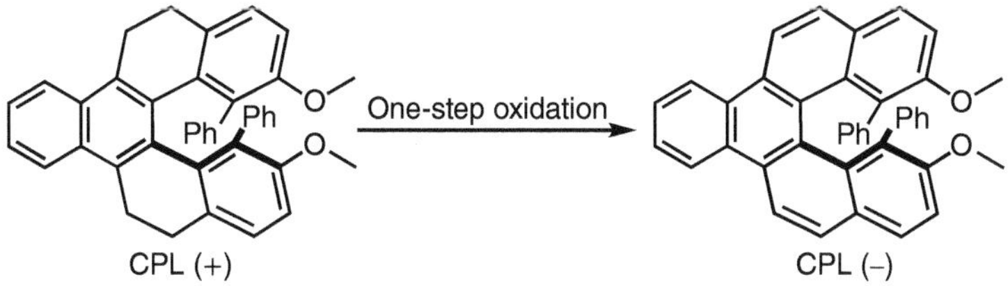

(*P*)-**57a:** Ar = *p*-$F_3CC_6H_4$
(*P*)-**57b:** Ar = *p*-C_6H_4
(*P*)-**57c:** Ar = C_6H_5
(*P*)-**57d:** Ar = *p*-MeC_6H_4
(*P*)-**57e:** Ar = *p*-$MeOC_6H_4$
(*P*)-**57f:** Ar = ClC_6H_4

(*P*)-**58a:** Ar = *p*-$F_3CC_6H_4$
(*P*)-**58b:** Ar = *p*-C_6H_4
(*P*)-**58c:** Ar = C_6H_5
(*P*)-**58d:** Ar = *p*-MeC_6H_4
(*P*)-**58e:** Ar = *p*-$MeOC_6H_4$
(*P*)-**58f:** Ar = ClC_6H_4

Figure 13.40 Molecular structures of (*P*)-**57a–f** and (*P*)-**58a–f**.

the extended π-conjugated structure increased the intramolecular charge transfer effect and emission characteristics, thus the $|g_{lum}|$ values of (*P*)-**58a–f** were all greater than those of (*P*)-**57a–f**.

13.6 Summary and Outlook

In summary, various helicenes and their derivatives including carbohelicenes, heterohelicenes, and metal helicene derivatives with circularly polarized luminescent properties have been discovered. In particular, the regulation of CPL for some helicene derivatives with *N*-containing heterocycles or carboxylate ions could be efficiently achieved. However, it was still difficult to obtain helicene derivatives with high Φ_F as well as large g_{lum} at the same time. The CPL regulation methods and modes for the molecules with helical chirality were also very limited. Moreover, we are not at the stage of practical applications of the helicene derivatives with CPL properties in materials science and devices. Therefore, the particular focus of this research area in the next few years will be (i) how to design and synthesize helicene derivatives with stable chiral structures, large g_{lum}, and high fluorescence quantum yields; (ii) how to control the CPL properties of helicene derivatives in different modes; (iii) how to expand the applications of the molecules with helical chirality in circularly polarized electroluminescence; and (iv) how to achieve the applications of helicene derivatives with CPL properties in sensors, biological imaging, and other research fields. It is believable that with the development of CPL materials science, the helicene derivatives with CPL properties will draw increasing interest and become one of the most attractive and important projects in this research area.

Acknowledgments

We thank the National Natural Science Foundation of China (21672211, 21572233, 21871272), Beijing National Laboratory for Molecular Sciences (BNLMS-CXXM-202105), and the Youth Innovation Promotion Association CAS (2019034) for financial supports.

References

1 (a) Brittain, H.G. (1996). *Chirality* 8: 357–363. (b) Riehl, J.P. and Richardson, F.S. (1986). *Chem. Rev.* 86: 1–16.
2 Muller, G. (2014). *Luminescence of Lanthanide Ions in Coordination Compounds and Nanomaterials* (ed. A. de Bettencourt-Dias), 77–124. Hoboken: Wiley.
3 Yang, Y., Costa, R.C.D., Smilgies, D.M. et al. (2013). *Adv. Mater.* 25: 2624–2628.
4 Farshchi, R., Ramsteiner, M., Herfort, J. et al. (2011). *Appl. Phys. Lett.* 98: 162508.
5 Yang, Y., Costa, R.C.D., Fuchter, M.J., and Campbell, A.J. (2013). *Nat. Photon.* 7: 634–638.

6 Jiménez, J., Cerdán, L., Moreno, F. et al. (2017). *J. Phys. Chem. C* 121: 5287–5292.

7 (a) Muller, G. (2009). *Dalton Trans.* 44: 9692–9707. (b) Hassey, R., Swain, E.J., Hammer, N.I. et al. (2006). *Science* 314: 1437–1439.

8 Song, F., Wei, G., Jiang, X. et al. (2013). *Chem. Commun.* 49: 5772–5774.

9 Sato, I., Yamashima, R., Kadowaki, K. et al. (2001). *Angew. Chem. Int. Ed.* 40: 1096–1098.

10 N. Berova, K. Nakanishi and R. W. Woody, *Circular Dichroism: Principles and Applications*, Wiley-VCH, New York, NY, USA, 2e, 2000, pp. 196–197.

11 Lunkley, J.L., Shirotani, D., Yamanari, K. et al. (2008). *J. Am. Chem. Soc.* 130: 13814–13815.

12 (a) Carr, R., Evans, N.H., and Parker, D. (2012). *Chem. Soc. Rev.* 41: 7673–7686. (b) Aspinall, H.C. (2002). *Chem. Rev.* 102: 1807–1850. (c) Bunzli, J.C.G. and Piguet, C. (2002). *Chem. Rev.* 102: 1897–1928. (d) Heffern, M.C., Matosziuk, L.M., and Meade, T.J. (2014). *Chem. Rev.* 114: 4496–4539.

13 (a) Yashima, E., Ousaka, N., Taura, D. et al. (2016). *Chem. Rev.* 116: 13752–13990. (b) Watanabe, K. and Akagi, K. (2014). *Sci. Technol. Adv. Mater.* 15: 044203.

14 Jin, X., Sang, Y., Shi, Y. et al. (2019). *ACS Nano.* 13: 2804–2811.

15 Ma, K., Chen, W., Jiao, T. et al. (2019). *Chem. Sci.* 10: 6821–6827.

16 Sánchez-Carnerero, E.M., Agarrabeitia, A.R., Moreno, F. et al. (2015). *Chem. -Eur. J.* 21: 13488–13500.

17 (a) Shen, Y. and Chen, C.-F. (2012). *Chem. Rev.* 112: 1463–1535. (b) Lin, W.-B., Li, M., Fang, L., and Chen, C.-F. (2018). *Chin. Chem. Lett.* 29: 40–46.

18 (a) Zhao, Z.-H., Zhang, M.-Y., Liu, D.-H., and Zhao, C.-H. (2018). *Org. Lett.* 20: 7590–7593. (b) Fang, L., Lin, W.-B., Shen, Y., and Chen, C.-F. (2018). *Chin. J. Org. Chem.* 38: 541–554. (c) Fang, L., Lin, W.-B., and Chen, C.-F. (2018). *Chin. Chem. Lett.* 29: 1223–1225.

19 Field, J.E., Muller, G., Riehl, J.P., and Venkataraman, D. (2003). *J. Am. Chem. Soc.* 125: 11808–11809.

20 Sawada, Y., Furumi, S., Takai, A. et al. (2012). *J. Am. Chem. Soc.* 134: 4080–4083.

21 Nakamura, K., Furumi, S., Takeuchi, M. et al. (2014). *J. Am. Chem. Soc.* 136: 5555–5558.

22 Fujise, K., Tsurumaki, E., Fukuhara, G. et al. (2020). *Chem. Asian J.* 15: 2456–2461.

23 Kubo, H., Shimizu, D., Hirose, T., and Matsuda, K. (2020). *Org. Lett.* 22: 9276–9281.

24 Sakai, H., Shinto, S., Kumar, J. et al. (2015). *J. Phys. Chem. C* 119: 13937–13947.

25 Sakai, H., Kubota, T., Yuasa, J. et al. (2016). *J. Phys. Chem. C* 120: 7860–7869.

26 Oyama, H., Akiyama, M., Nakano, K. et al. (2016). *Org. Lett.* 18: 3654–3657.

27 Yang, W., Longhi, G., Abbate, S. et al. (2017). *J. Am. Chem. Soc.* 139: 13102–13109.

28 Dhbaibi, K., Favereau, L., Hooper, M.S. et al. (2018). *Chem. Sci.* 9: 735–742.

29 Shen, C., Gan, F., Zhang, G. et al. (2020). *Mater. Chem. Front.* 4: 837–844.

30 Cruz, C.M., Castro-Fernández, S., Macôas, E. et al. (2018). *Angew. Chem., Int. Ed.* 57: 14782–14786.

31 Cruz, C.M., Márquez, I.R., Mariz, I.F.A. et al. (2018). *Chem. Sci.* (9): 3917–3924.
32 Tanaka, H., Kato, Y., Fujiki, M. et al. (2018). *J. Phys. Chem. A* 122: 7378–7384.
33 Shen, Y., Lu, H.-Y., and Chen, C.-F. (2014). *Angew. Chem. Int. Ed.* 53: 4648–4651.
34 Li, M., Lu, H.-Y., Zhang, C. et al. (2016). *Chem. Commun.* 52: 9921–9924.
35 Li, M., Zhang, C., Fang, L. et al. (2018). *ACS Appl. Mater. Interfaces* 10: 8225–8230.
36 He, D.-Q., Lu, H.-Y., Li, M., and Chen, C.-F. (2017). *Chem. Commun.* 53: 6093–6096.
37 Fang, L., Li, M., Lin, W.-B. et al. (2017). *J. Org. Chem.* 82: 7402–7409.
38 Fang, L., Li, M., Lin, W.-B., and Chen, C.-F. (2018). *Tetrahedron* 74: 7164–7172.
39 Lin, W.-B., Li, M., Fang, L. et al. (2017). *Chem. –Asian J.* (12): 86–94.
40 (a) Zhang, D.-W., Li, M., and Chen, C.-F. (2020). *Chem. Soc. Rev.* 49: 1331–1343. (b) Li, M., Wang, Y.-F., Zhang, D.-W. et al. (2020). *Sci. China Mater.* https://doi.org/10.1007/s40843-020-1496-7.
41 Yavari, K., Delaunay, W., De Rycke, N. et al. (2019). *Chem. –Eur. J.* 25: 5303–5310.
42 (a) Liu, B., Böckmann, M., Jiang, W. et al. (2020). *J. Am. Chem. Soc.* 142: 7092–7099. (b) Ma, Z., Winands, T., Liang, N. et al. (2019). *Sci. China Chem.* 62: 208–214.
43 Dhbaibi, K., Favereau, L., and Crassous, J. (2019). *Chem. Rev.* 119: 8846–8953.
44 Oyama, H., Nakano, K., Harada, T. et al. (2013). *Org. Lett.* 15: 2104–2107.
45 Murayama, K., Oike, Y., Furumi, S. et al. (2015). *Eur. J. Org. Chem.* 7: 1409–1414.
46 Katayama, T., Nakatsuka, S., Hirai, H. et al. (2016). *J. Am. Chem. Soc.* 138: 5210–5213.
47 Alnoman, R.B., Rihn, S., O'Connor, D.C. et al. (2016). *Chem. –Eur. J.* 22: 93–96.
48 Saikawa, M., Nakamura, T., Uchida, J. et al. (2016). *Chem. Commun.* 52: 10727–10730.
49 Clarke, R., Ho, K.L., Alsimaree, A.A. et al. (2017). *ChemPhotoChem* 1: 513–517.
50 Ray, C., Sanchez-Carnerero, E.M., Moreno, F. et al. (2016). *Chem. –Eur. J.* 22: 8805–8808.
51 Maeda, C., Nagahata, K., Shirakawa, T., and Ema, T. (2020). *Angew. Chem., Int. Ed.* 59: 7813–7817.
52 Shen, C., Srebro-Hooper, M., Jean, M. et al. (2017). *Chem. –Eur. J.* 23: 407–418.
53 Domínguez, Z., López-Rodríguez, R., Álvarez, E. et al. (2018). *Chem. –Eur. J.* 24: 12660–12668.
54 Goto, K., Yamaguchi, R., Hiroto, S. et al. (2012). *Angew. Chem., Int. Ed.* 51: 10333–10336.
55 Ushiyama, A., Hiroto, S., Yuasa, J. et al. (2017). *Org. Chem. Front.* 4: 664–667.
56 Sundar, M.S., Talele, H.R., Mande, H.M. et al. (2014). *Tetrahedron Lett.* 55: 1760–1764.
57 Bosson, J., Labrador, G.M., Pascal, S. et al. (2016). *Chem. –Eur. J.* 22: 18394–18403.
58 Hernández Delgado, I., Pascal, S., Wallabregue, A. et al. (2016). *Chem. Sci.* 7: 4685–4693.
59 Duwald, R., Bosson, J., Pascal, S. et al. (2020). *Chem. Sci.* 11: 1165–1169.

60 (a) Sang, Y., Han, J., Zhao, T. et al. (2019). *Adv. Mater.* 32: 1900110. (b) Wang, X., Duan, P., and Liu, M. (2014). *Chem. –Asian J.* 9: 770–778. (c) Sang, Y., Duan, P., and Liu, M. (2018). *Chem. Commun.* 54: 4025–4028.

61 Kumar, J., Nakashima, T., and Kawai, T. (2015). *J. Phys. Chem. Lett.* 6: 3445–3452.

62 Li, H., Li, B.S., and Tang, B.Z. (2019). *Chem. –Asian J.* 14: 674–688.

63 Kaseyama, T., Furumi, S., Zhang, X. et al. (2011). *Angew. Chem. Int. Ed.* 50: 3684–3687.

64 Yamano, R., Hara, J., Murayama, K. et al. (2017). *Org. Lett.* 19: 42–45.

65 Longhi, G., Castiglioni, E., Villani, C. et al. (2016). *J. Photochem. Photobiol. A* 331: 138–145.

66 Yamamoto, Y., Sakai, H., Yuasa, J. et al. (2016). *J. Phys. Chem. C* 120: 7421–7427.

67 Yamamoto, Y., Sakai, H., Yuasa, J. et al. (2016). *Chem. –Eur. J.* 22: 4263–4273.

68 Shen, C., Anger, E., Srebro, M. et al. (2014). *Chem. Sci.* 5: 1915–1927.

69 Brandt, J.R., Wang, X., Yang, Y. et al. (2016). *J. Am. Chem. Soc.* 138: 9743–9746.

70 Yan, Z.-P., Luo, X.-F., Liao, K. et al. (2020). *Front. Chem.* 8: 501–509.

71 Saleh, N., Srebro, M., Reynaldo, T. et al. (2015). *Chem. Commun.* 51: 3754–3757.

72 Gauthier, E.S., Abella, L., Hellou, N. et al. (2020). *Angew. Chem. Int. Ed.* 59: 8394–8400.

73 Isla, H., Srebro-Hooper, M., Jean, M. et al. (2016). *Chem. Commun.* 52: 5932–5935.

74 (a) Hellou, N., Srebro-Hooper, M., Favereau, L. et al. (2017). *Angew. Chem. Int. Ed.* 56: 8236–8239. (b) Macé, A., Hellou, N., Hammoud, J. et al. (2019). *Helv. Chim. Acta* 102: e1900044.

75 Biet, T., Cauchy, T., Sun, Q. et al. (2017). *Chem. Commun.* 53: 9210–9213.

76 Hiroaki, I., Hayato, S., Yoshinori, O. et al. (2018). *Chem. –Eur. J.* 24: 16889–16894.

77 Saleh, N., Moore, B. II, Srebro, M. et al. (2015). *Chem. –Eur. J.* 21: 1673–1681.

78 Sakai, H., Kubota, T., Yuasa, J. et al. (2016). *Org. Biomol. Chem.* 14: 6738–6743.

79 Pascal, S., Besnard, C., Zinna, F. et al. (2016). *Org. Biomol. Chem.* 14: 4590–4594.

80 Otani, T., Tsuyuki, A., Iwachi, T. et al. (2017). *Angew. Chem. Int. Ed.* 56: 3906–3910.

81 Lin, W.-B., He, D.-Q., Lu, H.-Y. et al. (2020). *Chem. Commun.* 56: 1863–1866.

14

Helicenes and Nonlinear Optical Properties: A Good Match?

Benoît Champagne

Theoretical Chemistry Laboratory, Unit of Theoretical and Structural Physical Chemistry, Namur Institute of Structured Matter (NISM), University of Namur, Namur, Belgium

14.1 Introduction

Nonlinear optics investigates the behavior of light in media in which the polarizations respond nonlinearly to electromagnetic fields [1]. The dominant response originates usually from the effects of the electric field of light on the electric polarization. Materials with large nonlinear optical (NLO) responses have attracted much interest because of foreseeable applications in optoelectronics, including frequency mixing (a specific case is frequency doubling, such as the conversion of the 1064 nm output from Neodymium-doped Yttrium Aluminum Garnet (Nd:YAG) lasers into green light at 532 nm) and electro-optical switching. Many applications are based on inorganic materials, like potassium titanyl phosphate (KTP) used in the green laser pointer [2]. Over the years, organic materials have gained increased interest, owing to their large and fast (electronic) responses combined with the flexibility of organic synthesis. This has stimulated a broad range of studies to understand the relationships between the structures and the NLO properties, combining chemical design, synthesis, experimental characterizations, and quantum chemical calculations [3]. Moreover, organic materials have extended the fields of applications of NLO from frequency mixing to bioimaging [4–6], to sensing [7], and to molecular logics [8–10]. At the molecular scale, nonlinear optics tackles the induced electric dipole moment ($\Delta\mu$) in terms of external electric fields oscillating at angular frequencies ω_n with $n = 1, 2, \ldots$:

$$\Delta\mu_\zeta(-\omega_\sigma) = \alpha_{\zeta\eta}(-\omega_\sigma;\omega_1)E_\eta(\omega_1) + \frac{1}{2}\,\beta_{\zeta\eta\chi}(-\omega_\sigma;\omega_1,\omega_2)E_\eta(\omega_1)E_\chi(\omega_2)$$
$$+ \frac{1}{6}\,\gamma_{\zeta\eta\chi\xi}(-\omega_\sigma;\omega_1,\omega_2,\omega_3)E_\eta(\omega_1)E_\chi(\omega_2)E_\xi(\omega_3) + \cdots$$

where summations over repeated indices (molecular axes coordinates, ζ, η, χ, and ξ) are assumed. This equation, adopting the T convention, defines the polarizability (α), the first (β), and the second (γ) hyperpolarizability tensor elements. Each term represents an induced dipole oscillating at the angular frequency $\omega_\sigma = \sum_n \omega_n$. At second order, which is the main target of this chapter, NLO phenomena encompass

Helicenes: Synthesis, Properties, and Applications, First Edition.
Edited by Jeanne Crassous, Irena G. Stará, and Ivo Starý.

sum-frequency generation (SFG) with $\omega_1 \neq \omega_2$ (a special case being $\omega_1 = 2\omega_2$), second harmonic generation (SHG) with $\omega_1 = \omega_2 = \omega$, dc-Pockels (dc-P) effect ($\omega_1 = \omega$ and $\omega_2 = 0$, or *vice versa*), and optical rectification (OR) where $\omega_1 = -\omega_2$ [11]. At third-order, third harmonic generation (THG) with $\omega_1 = \omega_2 = \omega_3 = \omega$, (static) electric field-induced SHG (EFISHG) where $\omega_1 = \omega_2$ and $\omega_3 = 0$, and degenerate four-wave mixing (DFWM) with $\omega_1 = -\omega_2 = \omega_3$ are typical phenomena.

Several techniques have been developed to measure the hyperpolarizabilities of molecules and ions in solutions. A classification could be based on the origin of the probed phenomenon, purely electric or combining electric and magnetic effects. We refer the reader to Sections 14.4 and 14.5 for the mixed responses of molecules and interfaces while the EFISHG, the hyper-Rayleigh Scattering (HRS), and the SFG (in the case where $\omega_1 = 2\omega_2$) techniques are briefly discussed here, focusing mostly on the quantities that can be determined. In EFISHG, an external static electric field is applied to the sample to create preferential molecular orientations and therefore to break the isotropicity of the medium. Using the EFISHG setup enables to probe the SHG response, which contains both a β and a γ contribution [12]:

$$\gamma_{\text{EFISHG}} = \gamma_{//}(-2\omega; \omega, \omega, 0) + \frac{\mu\beta_{//}(-2\omega; \omega, \omega)}{3kT}$$

Thus, to disentangle the second- and third-order responses, measurements must be performed at different temperatures while the norm of the dipole moment (μ) is also needed to access the β amplitude. The $\beta_{//}$ and $\gamma_{//}$ quantities are combinations of tensor elements where $\beta_{//}$ is related to the projection of the vectorial representation of the β tensor on the dipole moment (again assuming Einstein summations):

$$\beta_{//}(-2\omega; \omega, \omega) = \beta_{//} = \frac{3}{5}\frac{\mu_\zeta \beta_\zeta}{\mu}$$

with

$$\beta_\zeta = \beta_{\zeta\eta\eta} + 2\beta_{\eta\zeta\eta}$$

$\gamma_{//}$ corresponds to the isotropic invariant of the γ tensor:

$$\gamma_{//}(-2\omega; \omega, \omega, 0) = \gamma_{//} = \frac{1}{15}(2\gamma_{\zeta\zeta\eta\eta} + \gamma_{\zeta\eta\eta\zeta})$$

In HRS, the probed quantity is solely the SHG β tensor [13]. Typical HRS experiments collect the vertically polarized scattered light at an angle of 90° with respect to the incident light direction. The scattered light intensity is proportional to $\beta^2_{\text{HRS}}(-2\omega; \omega, \omega)$, which contains two terms, $\langle \beta^2_{\text{ZZZ}} \rangle$ and $\langle \beta^2_{\text{ZXX}} \rangle$, respectively, associated with vertically and horizontally polarized incident light.

$$\beta_{\text{HRS}}(-2\omega; \omega, \omega) = \beta_{\text{HRS}} = \sqrt{\langle \beta^2_{\text{ZZZ}} \rangle + \langle \beta^2_{\text{ZXX}} \rangle}$$

The relationships between these quantities defined in the laboratory frame (upper-case indices) and those in the molecular coordinates systems are:

$$\langle \beta^2_{\text{ZZZ}} \rangle = \frac{1}{105}[2\beta_{\zeta\eta\chi}\beta_{\zeta\eta\chi} + \beta_{\zeta\eta\eta}\beta_{\zeta\chi\chi} + 4(\beta_{\zeta\zeta\eta}\beta_{\eta\chi\chi} + \beta_{\zeta\zeta\eta}\beta_{\chi\chi\eta} + \beta_{\zeta\eta\chi}\beta_{\eta\zeta\chi})]$$

$$\langle \beta^2_{\text{ZXX}} \rangle = \frac{1}{105}[6\beta_{\zeta\eta\chi}\beta_{\zeta\eta\chi} + 3\beta_{\zeta\eta\eta}\beta_{\zeta\chi\chi} - 2(\beta_{\zeta\zeta\eta}\beta_{\eta\chi\chi} + \beta_{\zeta\zeta\eta}\beta_{\chi\chi\eta} + \beta_{\zeta\eta\chi}\beta_{\eta\zeta\chi})]$$

From these, a useful quantity can be evaluated, the Depolarization Ratio (DR)

$$\mathrm{DR} = \frac{\langle \beta^2_{ZZZ} \rangle}{\langle \beta^2_{ZXX} \rangle}$$

DR tells about the shape of the NLOphore. In the static limit, DR ranges from 1.5, which corresponds to an ideal octupolar response, to 9, which is related to the dipolar limit. When the β tensor is dominated by one diagonal element, which determines the charge transfer (CT) axis and is parallel to the dipole moment, DR = 5 and the system is said to be a one-dimensional (1-D) NLOphore. The antisymmetric isotropic component of $\beta(-3\omega; 2\omega, \omega)$,

$$\overline{\beta}_{\mathrm{SFG}}(-3\omega; 2\omega, \omega) = \overline{\beta} = \frac{1}{6}[\varepsilon_{\zeta\eta\chi}\beta_{\zeta\eta\chi}]$$

where $\varepsilon_{\zeta\eta\chi}$ is the unit skew-symmetric tensor and is a quantity that can be determined from coherent SFG measurements. In fact, for isotropic media, in the electric-dipole approximation, coherent second-order three-wave mixing ($\omega_1 \neq \omega_2$) is only symmetry-allowed for chiral molecules, provided the mixture is not racemic [14, 15]. This quantity changes sign with the enantiomer and it is usually small, smaller than originally thought. Moreover, it tends to zero in the limit of a zero-frequency excitation source but gets larger in (near) resonance conditions. Note that it would also vanish in the limit of degenerate frequencies.

As shown by the very first equation of this chapter, the T convention was used to define the first and second hyperpolarizabilities [16]. This implies that quantities reported in this chapter and taken from works where another convention was originally employed have been corrected. For β, the atomic units are defined such that 1 a.u. of $\beta = 3.6212 \times 10^{-42}\ \mathrm{m^4\ V^{-1}} = 3.2064 \times 10^{-53}\ \mathrm{C^3\ m^3\ J^{-2}} = 8.639 \times 10^{-33}$ esu.

The prototypical second-order NLO molecules are built from a pair of donor (D) and acceptor (A) groups attached to a π-conjugated segment [17], forming a one-dimensional (1D) D-π-A NLOphore (Scheme 14.1a). Many studies, encompassing simple models, have been devoted to these 1D push–pull π-conjugated systems, highlighting the interplay between the strengths of the D and A moieties

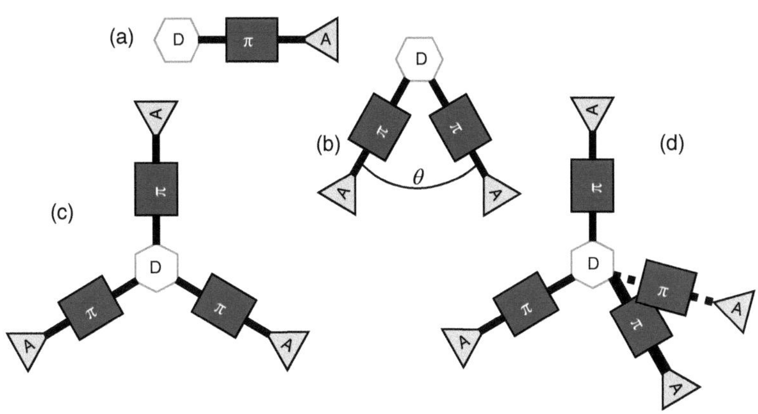

Scheme 14.1 Typical molecular architectures of second-order NLO compounds.

and the size and nature of the π-linker on the first hyperpolarizability [3, 18]. Large second-order NLO responses have also been evidenced for molecules adopting a Λ or V shape where two donor (acceptor) groups interact with the same acceptor (donor) group (Scheme 14.1b) [19]. In this case, besides the characteristics described here for the 1D systems, the first hyperpolarizability also depends on the angle (θ) between the two segments. Another category of molecules that have been extensively studied have an octupolar architecture (Scheme 14.1c), distinguishable from the two first categories because they have a vanishing dipole moment (if the symmetry is strictly satisfied) while β might be sizable [20]. Moreover, molecules with other remarkable shapes have also been proposed as efficient second-order NLO compounds, including tetrahedral ones (Scheme 14.1d) [21] and distorted octahedral compounds.

The study of the NLO properties of helicenes began more than two decades ago, owing to the seminal work by Verbiest and co-workers [22] where the chiral supramolecular organization of tetrasubstituted helicenebisquinones makes the second-order NLO response about 30 times larger than for the corresponding racemic material. Chirality was demonstrated to be at the origin of the dominant SHG responses while this helicene derivative does not fulfill the classical D-π-A requirement for achieving a high β response. However, this work has stimulated many experimental and quantum chemical investigations on the NLO responses of helicene derivatives, with the aim of a better understanding and, thereof, of their optimization. At a modest level, our group took part in these investigations, focusing mostly on the electric SHG response and these studies constitute the backbone of this chapter. In addition, representative results from the literature have been selected. The ensemble is presented and analyzed in the following sections.

14.2 Electric Second-Order NLO Responses

Despite the seminal works of Verbiest and co-workers [22], which concern both electric and mixed electric-magnetic responses, most investigations have tackled the electric component of the second-order NLO responses. Three ways to tune these responses have mainly been studied, the choice (i) of the nature and position of the substituents, (ii) of the nature of the helicene backbone, as well as (iii) of the size of the helicene. Although these effects are sometimes intertwined, they have been split into the following three paragraphs. The last two paragraphs deal with the less traditional three-wave mixing response as well as with other, though related, architectures.

14.2.1 Effect of the Nature and Position of the Substituents

Although this quantity cannot be measured, the first investigation reported the norm of the vector component of the β tensor, $|\beta|$, defined as

$$|\beta| = (\beta_\zeta \beta_\zeta)^{1/2}$$

These have been evaluated for different mono- and disubstituted tetrathia[7]helicene either at the coupled-perturbed Kohn-Sham (CPKS) level (B3LYP exchange-correlation functional) or using a summation-over-states (SOS) expression with spectroscopic quantities evaluated at the semiempirical complete neglect of differential overlap/spectroscopy (CNDO)/S level [23]. When the substituents (R_1, R_2) are (H, H), the $|\beta|$ response is small and depends much on the level of calculation (including which geometry was used). As soon as a donor or an acceptor substitutes one of the terminal positions (Scheme 14.2), $|\beta|$ increases and the effect is further enhanced when both are present, as a result of the well-known push–pull effect. So, the largest $|\beta|$ value is achieved for (R_1, R_2) = (NO_2, NH_2). Looking at the SOS/CNDO/S results, changing the donor/acceptor (D/A) pair has little impact on the static $|\beta|$, while CPKS/B3LYP appears more suitable to account for the differences in $|\beta|$ as a function of the D/A strengths. However, including the effects of frequency dispersion in the SOS/CNDO/S treatment better reveals the $|\beta|$ enhancement when (R_1, R_2) = (NO_2, NH_2). Such substitution topology therefore assumed that the dominant β component is parallel to the helical axis.

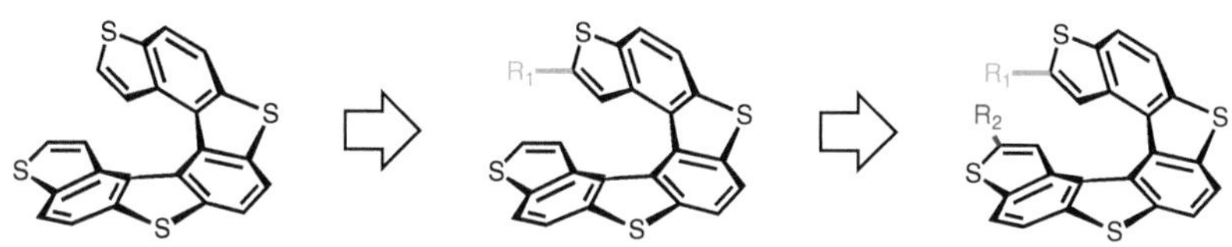

Scheme 14.2 Tetrathia[7]helicene scaffold and its most common mono- and disubstitutions.

A subsequent work has reported on the measurement of β_{HRS} of the non-substituted tetrathia[7]helicene in chloroform at $\lambda = 800$ nm [24]. A β_{HRS} value of $(960 \pm 120) \times 10^{-30}$ esu was obtained. Since the maximum absorption wavelength of this compound is 388 nm, this β_{HRS} value is exalted by resonance. Therefore, to provide an intrinsic response, the simple two-state approximation was employed to extrapolate β_{HRS} to the static limit, giving a value of $(44 \pm 6) \times 10^{-30}$ esu. This work demonstrated first that, even without D and/or A substituent, β_{HRS} of such a poorly polar molecule is nonzero, simply because it is non-centrosymmetric. Even though the comparison between the $|\beta|$ and β_{HRS} responses is not straightforward, it appears that the calculated values of Ref. [23] are at least one order of magnitude smaller.

Shortly later, the time-dependent Hartree–Fock (TDHF) method was enacted in combination with the Austin Model 1 (AM1) semiempirical Hamiltonian to investigate the effects of multiple substitutions by donor/acceptor groups on the tetrathia[7]helicene scaffold and to highlight the preferential substitution positions [25] (Scheme 14.3). This was achieved by considering the NH_2/NO_2 groups, the best D/A substituents of Ref. [23]. In the absence of substituent (case I), β is small, dominated by its radial component, and the DR is close to 1.5. D/A substitutions at the extremities (positions 2 and 13, case II) enhance the response (by a factor of 5 if considering β_{HRS} at 1064 nm), which is now dominated by its axial component and characterized by a DR value of 4. Then, a D/A pair was inserted in such a way that

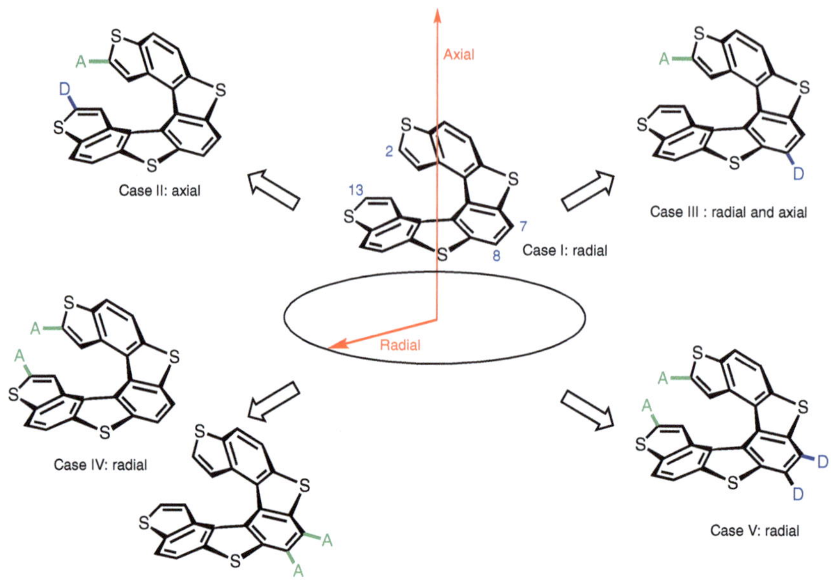

Scheme 14.3 Evolution of the nature (axial *versus* radial) of the β response of tetrathia[7]helicene as a function of the position of the multiple substitutions by donor and/or acceptor groups. Similar results would be obtained by permuting all the donor and acceptor moieties.

one occupies one of the extremities of the helicene while the other one is located on the central benzene ring (case III). Depending on the efficiency of the resonance between the D and A groups, β_{HRS} increases by 50–150% but in all the cases, the radial component is the largest and DR ranges from 4 to 5. The next configuration (case IV) corresponds to a pair of D or A groups both at the extremities or at the center of the scaffold. This restores a fully radial response, which, in most cases, is of similar amplitude to what was observed with the previous D/A substitutions. In this case, the largest responses are obtained (i) with a pair of NO_2 group (in comparison to a pair of NH_2 groups), demonstrating that the helicene scaffold plays the role of an electron donor and (ii) when the substituents are placed on the central benzene ring. Optimizing both aspects leads to compounds (case IV) with a β_{HRS} value that is about twice larger than in the case of the traditional D/A substituents in position 2/13 (case I) and a DR larger than 5.5, highlighting a clear dipolar character. Finally, the latter two strategies (cases III and IV) have been combined in order to achieve much larger β values (case V). In such a case, β_{HRS} increases by at least 300% with respect to case I and it is dominated by its radial component.

The reliability of using a semiempirical Hamiltonian has been assessed by comparing, for a selection of helicenes, the static TDHF (or rather coupled-perturbed Hartree–Fock (CPHF)) AM1 values with reference data obtained at the second-order Møller–Plesset pertubation Theory (MP2)/6-31G level, where the first hyperpolarizability tensor elements are obtained from finite field numerical differentiations of the field-dependent energy. So, for tetrathia[7]helicene, the

2,13-diamino-substituted, and the 2-amino-13-nitro-substituted derivatives, the β_{HRS} values are in the 0.18 : 0.74 : 1.00 ratio whereas the CPHF/AM1 values are in the 0.24 : 0.68 : 1.00 ratio.

Bossi and co-workers [26] have reported on a systematic experimental and quantum chemical investigation of the EFISHG response of substituted tetrathia[7]helicenes, carrying out the NO_2, CH=CHCN, and $COCF_3$ units as accepting groups, either at the terminal thiophene position or on the central benzene ring. The EFISHG measurements, performed at 1907 nm, were analyzed in the light of quantum chemical calculations, performed at the TDHF/6-311+G(d) level with a description of the solvent (chloroform) effects using the polarizable continuum model (PCM). Both the $\beta(-2\omega;\ \omega,\ \omega)$ and $\gamma(-2\omega;\ \omega,\ \omega,\ 0)$ tensor components have been evaluated at that level. Then, in order to provide more quantitative β values, additional calculations were performed to get the static MP2 responses. In this way, using the following expression – known as the multiplicative approximation – best estimates of the dynamic responses have been obtained:

$$\beta(-2\omega;\omega,\omega) = \beta_{\text{TDHF,IEFPCM}}(-2\omega;\omega,\omega)\frac{\beta_{\text{MP2,vacuo}}(0)}{\beta_{\text{HF,vacuo}}(0)}$$

Comparing the EFISHG responses (Scheme 14.4), at both experimental and theoretical levels, leads to the following conclusions: (i) CH=CH-NO_2 is a better acceptor group than CO-CF_3 (from comparing the monosubstituted compounds), (ii) CH=CH-CN is also a better acceptor group than CO-CF_3 (from comparing the disubstituted compounds), (iii) disubstitution enhances the response (from the species with one and two CO-CF_3 substituents), and (iv) substituting the rings at the extremities is more favorable than substituting the central ring. Calculations further enable to show that – without accounting for electron correlation on γ – the third-order contribution to EFISHG amounts to about 10%, with the exception of the parent tetrathia[7]helicene where this γ contribution is 25 times larger than the β contribution.

Si and Yang [27] have studied the effects of substitution on the β_{HRS} of a hexahelicene modified by the presence of a BN unit (Scheme 14.5). They calculated the static response using the time-dependent density functional theory (TDDFT) method in combination with the B3LYP XC functional, which has been selected because the UV/visible absorption and circular dichroism spectra are well-reproduced using TDDFT/B3LYP vertical excitation energies. When (R, R′) = (H, H), the β_{HRS} response is small, though larger than that of the reference urea molecule. Then, they found that β_{HRS} increases by a factor of 2 when adding a pair of amino donor groups. The factor goes to 3 when substituting by a pair of nitro groups while, finally, combining a donor and an acceptor results in an enhancement by a factor larger than 7. These results further highlight that the push–pull strategy is efficient for heterohelicenes to achieve large second-order NLO responses.

Unpublished results by Zhang and Champagne [28] are presented in Scheme 14.6. They highlight that the donor/acceptor role of rings fused to the tetrathia[7]helicene scaffold is an alternative to using substituents.

Compound	Measured	Calculated
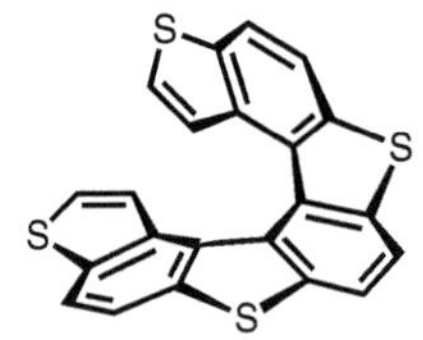	–0.19	0.03
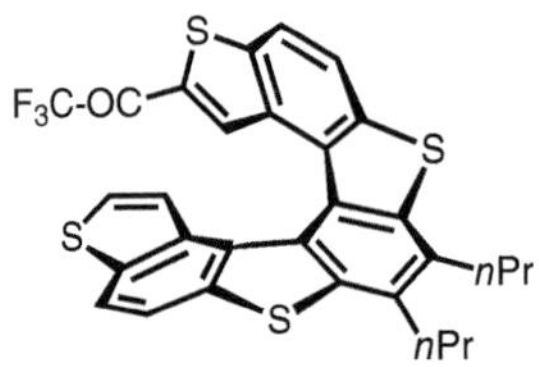	0.51	0.24
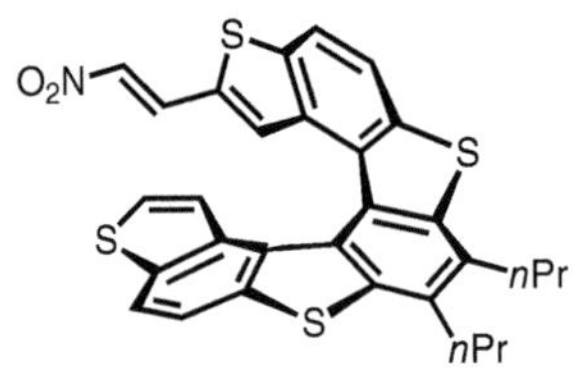	0.83	1.08
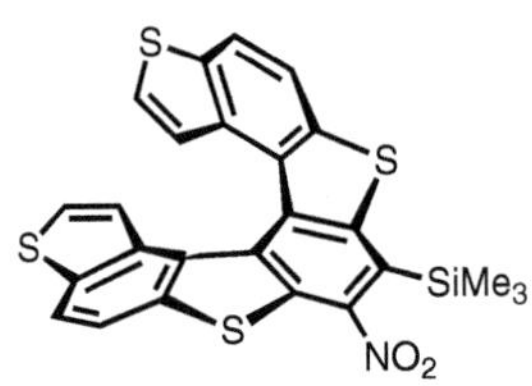	0.55	0.56
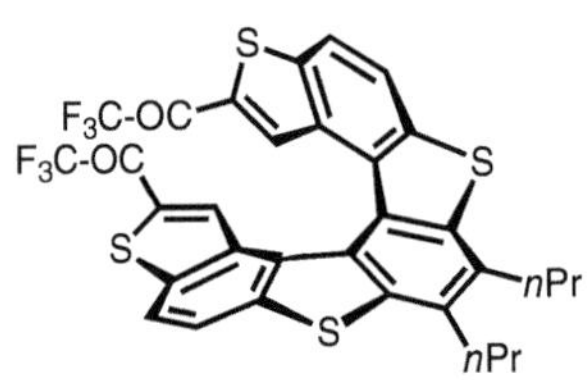	0.79	0.59
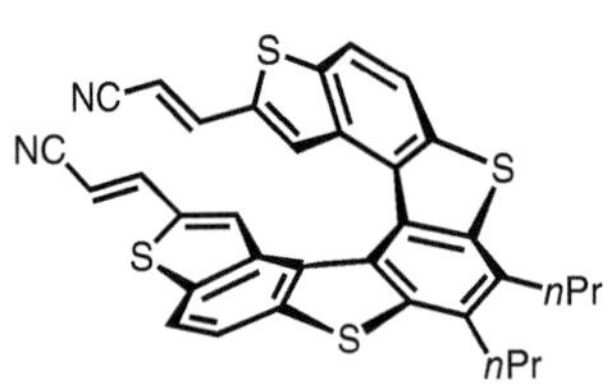	1.00	1.00

Scheme 14.4 Comparison between measured (chloroform, $\lambda = 1907$ nm) and calculated (full expression with MP2 corrections on β) EFISHG responses of tetrathia[7]helicene derivatives. The values are relative to those of the last compound.

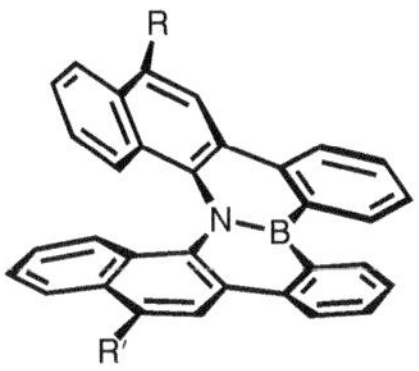

R, R′	β_{HRS}
H, H	2.4
Br, Br	2.4
NH_2, NH_2	5.8
NO_2, NO_2	7.8
NO_2, NH_2	17.5

Scheme 14.5 Substituted hexahelicene containing a BN unit and its static β_{HRS} responses (10^{-30} esu) as a function of the nature of the substituents [27]. Source: Adapted from [27].

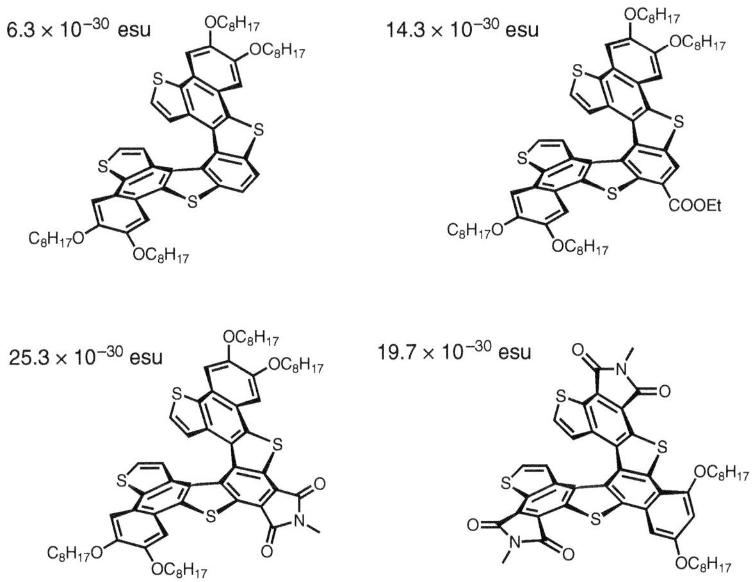

Scheme 14.6 β_{HRS} responses ($\lambda = 1907$ nm) as evaluated at the TDDFT/LC-BLYP/6-311+G* level for derivatives of tetrathia[7]helicenes bearing fused rings. Solvent effects (chloroform) were described with PCM.

14.2.2 Effect of the Nature of the Helicene Backbone, Including Oxidation Effects

The D/A effects described in the previous paragraph can be tuned by modifying the helical backbone. A straightforward modification consists in changing the system by oxidation or reduction. The oxidation effects have been addressed for a collection of homo- and hetero-helicenes in Ref. [29]. These helicenes present different backbones (benzene, thiophene, or both units) and different substitution patterns, though these are always symmetric so that the compounds are related to the categories I and IV of Scheme 14.3. This study was also carried out at the TDHF/AM1 level but, for open-shell systems, using an extended TDHF scheme [30], where the total responses are evaluated as the sum of the α and β spin responses. Besides the first hyperpolarizability, that study also calculated the frequency-dependent polarizability in order to deduce an upper bound to the first excitation energies. For all

helicenes, oxidation leads to an increase of β_{HRS} by one to two orders of magnitude, which is associated with a strong reduction of the first excitation energies. In most cases, this is accompanied by a modification of the DR, which tends toward the typical 27/11 value characterizing the model of the free electron in a helix [31]. In the case of the dithia- and tetrathia[7]helicenes with or without substituents, the oxidation modifies the character of the β response in the sense that for the neutral compound, it is purely radial, whereas for the oxidized one, the axial component is of the same order of magnitude, though smaller. In Ref. [29], the effects of the counteranion were also investigated. So, in the case of tetrathia[7]helicene, the presence of a chloride anion reduces the positive charge on the helicene and subsequently reduces its first hyperpolarizability, by a factor of 18 with respect to the oxidized species without counteranion. Still, the first hyperpolarizability remains 23 times larger than in the parent neutral helicene. This anion effect is explained by the reduction of the charge on the helicene (+0.3 e instead of +1.0 e), which goes in pair with a much smaller reduction of the first excitation energies. These results show that the control of the nature of the anion as well as of the cation–anion interactions can be used to tune the first hyperpolarizability, as also demonstrated for stilbazolium derivatives [32, 33].

The effects of the nature of the helicene backbone on the first hyperpolarizability have been further addressed by considering 7-ring helicenes built from symmetric combinations of thiophene and benzene rings [34]. Indeed, based on previous experimental investigations (e.g. the lowest-energy peak in UV/visible absorption spectra is redshifted when starting from pure thiophene or carbon–sulfur helicenes and increasing the number of benzene units [35]), the number of benzene/thiophene rings controls the delocalization along the helix because the thiophene units lead to cross-conjugated π systems and smaller electron delocalization. Figure 14.1 collects

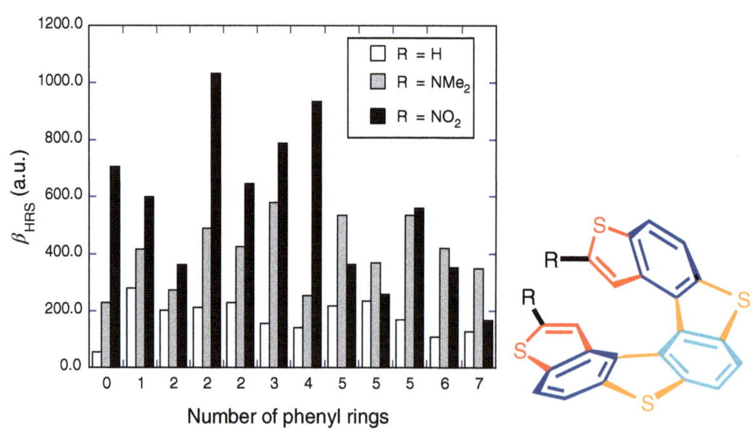

Figure 14.1 Evolution of the β_{HRS} response (wavelength of 1064 nm) as a function of the number of benzene rings in 7-ring helicenes. In addition to helicenes without any substituents, the results are also presented when a pair of NMe_2 or NO_2 substituents are placed at the helix extremities (R = H, NMe_2, or NO_2). The calculations were performed at the TDHF/6-311+G(d) level of approximation [34]. The general structure shows that each colored set of rings is either made of benzene or thiophene. Source: Adapted from [34].

the β_{HRS} values ($\lambda = 1064$ nm) obtained at the TDHF/6-311+G(d) level of approximation as a function of the number of benzene rings ($N = 0$–7, which means that, in parallel, the number of thiophene rings goes from 7 to 0). In the absence of substituents, the first hyperpolarizabilities are small. The smallest β_{HRS} value is found for the helicene containing only thiophene cycles while small values also characterize the helicenes with a high content of benzene rings. On the other hand, no specific pattern was observed in the case of $\beta_{//}$ because this quantity also depends on the orientation of the dipole moment. When the helicenes are substituted by a pair of strong donor groups (NMe_2), the hyperpolarizabilities increase and this increase is enhanced in the case of strong acceptor groups (NO_2). Looking at these variations upon substitutions, one can draw conclusions on the responsivity or the helicene backbone to polarization by substituents. Yet, these β_{HRS} variations are not straightforward to interpret. For the NO_2 substitutions, the presence of benzene rings at the extremities of the helix is usually detrimental to a large β_{HRS} value while alternating benzene/thiophene rings is a good strategy. On the other hand, when the substituents are NMe_2 groups, in addition to the detrimental role of the benzene rings, which are less polarizable, and the cross-π-conjugation, the auxiliary acceptor character of the thiophene groups can damp the first hyperpolarizability.

These structure–property relationships enable to propose tetrathia[7]helicene, (Scheme 14.7) as an efficiently polarizable scaffold and to conjecture that substituting the thiophene rings by furans, which are less aromatic and therefore more polarizable, should be beneficial to achieve large β responses. This was verified by TDHF calculations and a substantial increase (+60%) of β_{HRS} for the helicene substituted by two nitro groups in terminal positions was found.

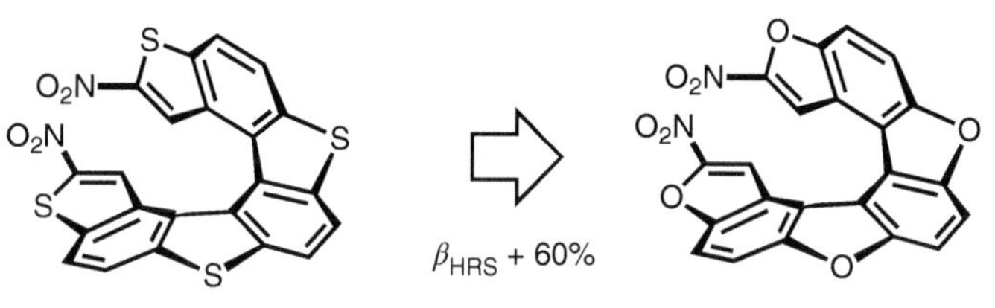

Scheme 14.7 Strategy to enhance the β_{HRS} response of tetrathia[7]helicene.

In the course of that study [34], two aspects of the calculations have been assessed, (i) the effects of the solvent as evaluated by using the PCM and (ii) the effect of electron correlation by employing the long-range corrected Becke–Lee–Yang–Parr (LC-BLYP) range-separated hybrid exchange-correlation functional within time-dependent DFT, in comparison to the TDHF method. Selected results in the case of the helicenes substituted by NO_2 groups are shown in Figure 14.2.

For the isolated (*in vacuo*) helicenes, the TDHF and TDDFT/LC-BLYP values are similar, with a systematic increase of 33% from Hartree–Fock (HF) to LC-BLYP. Then, accounting for solvent effects increases the β_{HRS} responses, especially at the LC-BLYP level. Indeed, the slope of the HF (DCM = dichloromethane) *versus* HF(isolated) linear regression equation amounts to 2.6, whereas, for LC-BLYP

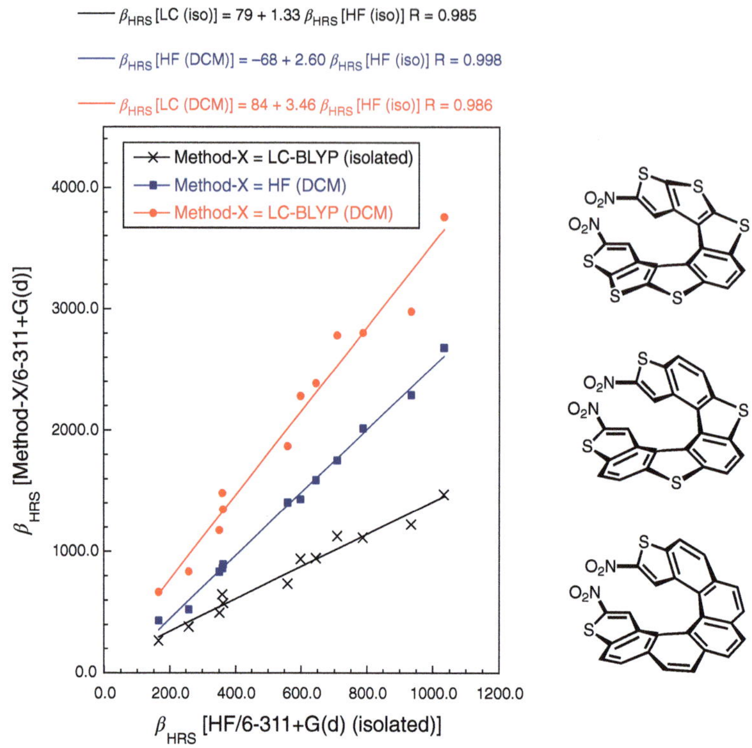

Figure 14.2 β_{HRS} responses of NO_2-substituted helicenes as determined at different levels of approximation (wavelength of 1064 nm) [34], together with selected structures containing 2, 4, and 6 thiophene rings. Source: Adapted from [34].

(DCM = dichloromethane) versus HF (isolated), it reaches 3.5. In all the cases, the correlation coefficient is close to one, demonstrating that a good insight into the NLO responses of substituted helicenes can already be obtained by performing calculations at the TDHF level without accounting for solvent effects. Similar relationships have been found for the unsubstituted and NMe_2-substituted analogs [34]. Yet, from a quantitative point of view, solvent effects need to be included while electron correlation should be treated at an adequate level.

14.2.3 Effect of the Size of the Helicene

Botek et al. [36] have employed the TDHF/AM1 scheme to investigate the evolution of the first hyperpolarizabilities of [n]-helicenes as a function of their size. $\beta_{//}$ is positive (meaning that the μ and β vectors point in the same direction) and presents quasiperiodic oscillations with n. The maximum values correspond to $n = 8$ and 14, whereas the minimum values to $n = 11$ and 17. Yet, these maximum values are small, of the order of a few 10^2 a.u. (a few 10^{-30} esu). This behavior has been rationalized by decomposing the μ and β vectors into radial and axial contributions, showing that $\beta_{//}$ is dominated by the radial component and that this component behaves similarly with n. The same investigation also highlighted that the effect of D/A substituents at the extremities of the [n]-helicenes could be favorable or detrimental to the $\beta_{//}$ value, depending on the helicene size.

14.2.4 Three-Wave Mixing

To our knowledge, measurements of $\overline{\beta}$ have not yet been performed for helicenes. On the other hand, a few quantum chemical investigations have been reported. The first calculation was an application illustrating the newly developed TDHF scheme, which allows the evaluation of the SFG β responses by resorting to the 2n + 1 rule [37]. Results on the frequency dispersion of $\overline{\beta}$ highlighted that the SFG responses of [4]helicene and [5]helicene have different signs but remains of the order of 10^{-2} a.u. for energies of incident photons 0.5 eV off resonance. These amplitudes are similar to those of the prototypical SFG molecule, (*R*)-(+)-propylene oxide despite the presence of π-delocalization in the helicenes. These calculations were later extended by investigating the evolution of $\overline{\beta}$ as a function of the [n]helicene [36]. So, using the TDHF/AM1 scheme, $\overline{\beta}$ increases monotonically with the size of the system and reaches, at $\hbar\omega = 0.5$ eV, a value of −0.04 a.u. for $n = 13$.

14.2.5 Other Helicenes and Related Helical Systems

Stimulated by the synthesis of some of their analogs [38], the β_{HRS} responses of double helicenes have been calculated by Si et al. [39] and compared to the responses of corresponding single helicenes. These calculations, performed at the TDDFT/B3LYP6-31+G(d), focused on different substitution patterns. They highlighted that double helicenes substituted by donor (NH_2) or acceptor (NO_2) groups at their extremities present β_{HRS} responses much smaller than 10^{-30} esu. Then, when combining donor and acceptor substituents on the same double helicene scaffold, the β_{HRS} values are enhanced by up to two orders of magnitude (the largest value amounts to 30×10^{-30} esu). Note that the parent single helicenes substituted by either two donor or two acceptor groups present intermediate β_{HRS} values between those of the symmetrically and asymmetrically substituted double helicenes. Therefore, provided the regioselectivity of the substitutions is controlled, the double helicenes are original structures to reach large first hyperpolarizabilities. Yet, it would also be interesting to compare these results obtained on the chiral double helicenes with those of the achiral meso form.

Coe, Teplý, and co-workers have introduced helquat dyes as efficient second-order NLO systems [40], in particular when combined with donor amino groups. Helquats are hybrids of the diquat (1,1′-ethylene-2,2′-bipyridyldiylium) and helicene structural motifs, thus playing the role of electron acceptor. *Per se*, these are not helicenes since the helix contains nonaromatic rings but the combination of strong donor and acceptor groups can lead to large first hyperpolarizabilities while the helical structure ensures chiroptical properties. Then, as a function of the type of amino substituent, β_{HRS} responses as large as $500–1300 \times 10^{-30}$ esu have been measured at $\lambda = 800$ nm.

14.3 Third-Order NLO Responses

The static $\gamma_{//}$ responses of helicenes bearing at the extremities of their helical scaffolds radicaloid moieties have been calculated using the CPKS/LC-BLYP/6-31G(d) method [41]. These open-shell systems present an intermediate/large diradical character (0.5–0.9) and display intramolecular pancake bonds. The $\gamma_{//}$ responses attain 10^5 a.u. when the pair of terminal radicaloid rings are different, cyclopentadienyl and tropylium, enabling an asymmetric charge distribution. The corresponding closed-shell helicenes present $\gamma_{//}$ values that are 2–7 times smaller than those of the diradicaloids. Such enhancement of $\gamma_{//}$ has been associated with a substantial decrease of the first excitation energy.

14.4 Beyond the Electric Dipole Approximation

The two-photon circular dichroism (TPCD) of simple helicenes ([6]helicene, dithia[7]helicene and its tetramethoxybisquinone derivative, as well as tetrathia[7]helicene) has been calculated at the TDDFT/B3LYP/aug-cc-pVDZ level, using an origin-invariant quadratic response approach [42]. TPCD, which arises from the difference in two-photon absorption (TPA) of left- and right-circularly polarized light, is a chiral sensitive effect, for which helicenes were considered as good candidates because they present large optical rotation and (one-photon) circular dichroism. These calculations demonstrated that the TPCD signatures are large, suitable for a first experimental observation of TPCD. Among these, the largest TPCD response was found for the tetramethoxybisquinone derivative of dithia-[7]helicene and was associated with the combined effect of electron delocalization and charge-transfer character of the excited states, similarly to what is usually highlighted for the pure electric first hyperpolarizabilities. In addition, among the different contributions to the overall response, the calculations revealed that the TPCD intensities are mostly governed by the electric and magnetic dipole contributions, rather than the electric quadrupole terms. That work also pointed out the interest to employ range-separated hybrids instead of global hybrids, like B3LYP. Although the first experimental observation of TPCD was carried out on the enantiomers of 1,1′-bi(2-naphthol) and supported by TDDFT calculations [43], experimental TPCD spectra have later been reported for helicene derivatives [44]. That study has highlighted not only the role of intramolecular CTs (ICT) on the amplitude of the TPCD but also TPA responses. So, the extension of the electronic delocalization beyond the helicene core – by adding π-conjugated linkers with a terminal cyano group – enhances the TPCD response and, to a larger extent, the TPA response (Scheme 14.8). These structure–property relationships were qualitatively reproduced by TDDFT/CAM-B3LYP calculations, although solvent effects were not taken into account. These TDDFT calculations further help rationalizing the effect of intramolecular CTs within the helicene scaffold *versus* those between the helicene and its π-conjugated substituents. Indeed, they show that electric (dipole and quadrupole) transition moments are mostly affected by the

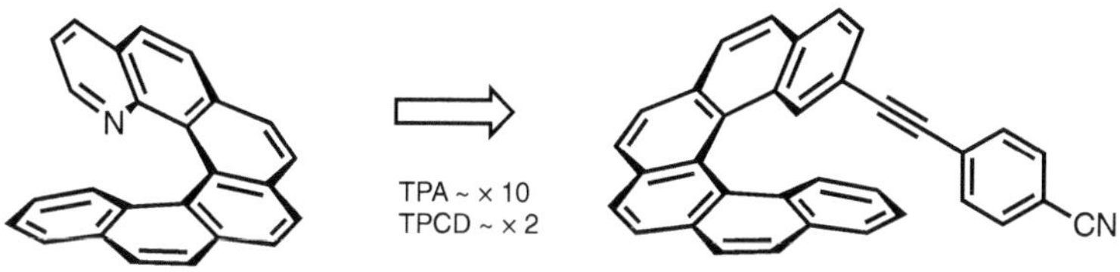

Scheme 14.8 ICT effect on the TPA and TPCD responses of helicene derivatives [44]. Source: Adapted from [44].

extension of the π-conjugation beyond the helicene core whereas the extension of the helicene scaffold principally affects the magnetic dipole transition moments.

Later on, within an investigation of chiral molecules, Rizzo and co-workers [45] studied the magnetic field-induced second harmonic generation (MFISHG) of [6]helicene. This third-order phenomenon corresponds to the change of sign of the SHG response between enantiomers when an external magnetic field is applied but, since the intensities are related to their squares, its recovering requires combining the induced polarization with an SHG wave of similar intensity. Their TDDFT/B3LYP calculations highlighted that the simple apolar [6]helicene presents MFISHG responses of the same order of magnitude as push–pull π-conjugated twisted biphenyl derivatives, suggesting that the substitution of this helicene by well-chosen donor/acceptor groups would result in an enhanced response. In addition, based on numerical values obtained from a representative set of molecules, this work also concluded that, currently, this effect is at the limits of measurability.

14.5 Helicene Aggregates and Self-Assembled Monolayers

As pointed out in the Introduction of this Chapter, the use of helicenes as efficient second-order NLO materials began with an investigation by Verbiest and co-workers [22] on Langmuir–Blodgett films of a chiral helicene (Scheme 14.9). Indeed, their NLO responses are much higher than that of films of the corresponding racemic mixture of the enantiomers. This helicene (if only one enantiomer is present) forms columnar stacks, which assemble into long fibers and can be deposited as Langmuir–Blodgett films. These films present large $\chi^{(2)}$ responses (the macroscopic equivalent

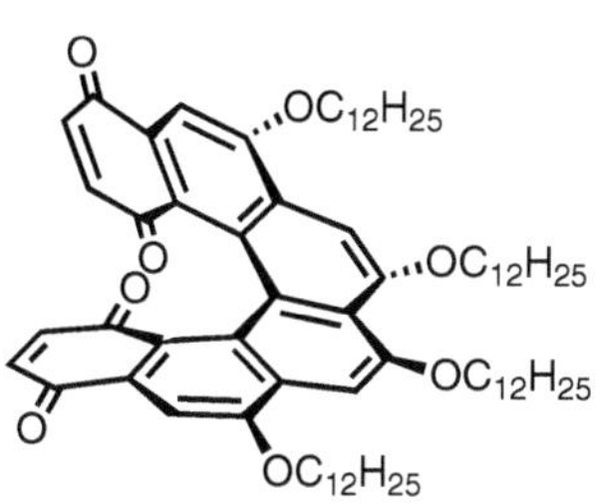

Scheme 14.9 Helicenebisquinone derivative of Ref. [22]. Source: Adapted from [22].

of β), about 30 times larger than for the films obtained from the racemic mixture of the same helicene, suggesting that the large response of the nonracemic samples is related to chirality-specific $\chi^{(2)}$ components. Analysis of the polarization-dependent SHG responses allows confirming this hypothesis and attributes the large responses to the $\chi^{(2)}_{XYZ}$-type components, which change sign between enantiomers.

Subsequently, Verbiest and co-workers [46] studied the electric field modulation of the second-order NLO responses of discotic nematic liquid crystals built from a nonracemic helicene similar to the one of Scheme 14.9. In absence of an orienting electric field, these helical columns organize in such a disordered way that only the $\chi^{(2)}_{XYZ}$ components can differ from zero. On the other hand, when an electric field is applied perpendicularly (along the Z axis) to the sample plane, the constitutive helical columns reorganize, orienting themselves parallel to the electric field (their dipole moment is parallel to their long axes). This lowers the symmetry so that additional $\chi^{(2)}$ components are different from zero, $\chi^{(2)}_{XYZ}$, $\chi^{(2)}_{XXZ}$, $\chi^{(2)}_{ZXX}$, and $\chi^{(2)}_{ZZZ}$. Moreover, the field-oriented material exhibits circular-difference effects of opposite sign when flipping the polarity of the orienting electric field (i.e. it is not necessary to use the molecule with the opposite handedness to change the sign of the response).

Stimulated by these studies [22, 46], Botek et al. [47] simulated the $\chi^{(2)}$ response of oriented films of helicenes. A two-step method was employed where (i) in a first step, the β tensors are calculated and then, (ii) the $\chi^{(2)}$ tensor components are evaluated as simple tensor sums of the molecular responses while also taking into account the symmetry of the molecular organization in the film. In particular, in this study, not only both the pure electric quadratic responses, $\chi^{(2),\,eee}$, but also their mixed electric-magnetic counterparts, $\chi^{(2)eem}$ and $\chi^{(2)mee}$, have been determined. This work provided preliminary insight into the importance of the mixed contribution to the $\chi^{(2)}$ response as a function of the nature of the helicene. In particular, in absence of D/A substituents, the pure electric response is small and the mixed electric-magnetic response can be of the same order of magnitude. On the other hand, D/A substituents enhance much more the electric response than the mixed one so that the mixed/electric ratio gets much smaller. Yet, several aspects of the theoretical approach deserve to be addressed, including (i) the hypothesis that the electric quadrupolar contribution is negligible, (ii) the gauge-invariance (typically by using gauge including atomic orbitals), (iii) electron correlation effects because Ref. [47] was carried out at the TDHF (or random phase approximation) level, and, probably the most important, (iv) going beyond the tensor sum of the molecular responses to evaluate the macroscopic responses.

14.6 Conclusions and Outlook

Owing to the combination of π-conjugation and lack of centrosymmetry, helicenes have demonstrated a natural interest in achieving large second-order NLO responses, either at the molecular or macroscopic (so-far, functionalized surfaces) levels. Yet, as shown by the examples given in the previous sections, optimizing this combination is not straightforward because the helical scaffold is often associated with a reduction of π-conjugation within (part of) the helical structure. So, when comparing the (electric) first hyperpolarizabilities of donor/acceptor-substituted helicenes with those of reference molecules like Disperse Red 1 (its static β_{HRS} value amounts to 66×10^{-30} esu or 76×10^{2} a.u.; whereas at $\lambda = 1064$ nm, it is one order of magnitude larger [48]), the balance is in favor of the latter, demonstrating that improvements are still possible. Reviewing the experimental and quantum chemical results from literature has unraveled design rules to optimize the first hyperpolarizabilities, mostly in what concerns the nature and position of the D/A acceptor substituents. On the other hand, less could be concluded concerning the nature of the helicene scaffold and the charge effects. In parallel, the specific molecular architecture of helicenes shows they are well-suited to achieve large mixed electric-magnetic nonlinear responses while, again, several molecular modification directions (substituents, hetero-/homo-backbone, charges) have still to be disclosed. Aside from the numerous, still incomplete, quantum chemical investigations, fewer experimental works have shed light on the potential of helicenes for achieving large NLO responses, leaving space for additional systematic investigations of the relationships between the second- and third-order NLO responses and the structure of the helicenes.

Moreover, as evidenced by the type of quantum chemical methods that have been employed, ranging from semiempirical schemes, to the Hartree–Fock method, and to DFT with a narrow range of exchange-correlation functionals, there remains a need for performing calculations at higher levels, either with wavefunction methods with explicit electron correlation or with DFT with more recent and more accurate exchange-correlation functionals. In addition, so far, the interpretations of the electric but also mixed electric-magnetic responses have relied on simple schemes, based on the topology of the frontier orbitals or on atomic charge distributions while improved tools might be used or extended to the specific nature of helicenes. Among these, the excitation-induced electronic reorganizations and the transition densities could provide additional insights to select the best helicene scaffolds and their substituents. Some of these tools have already been employed to investigate the electronic CD responses of helicenes [49] as well as the electric first hyperpolarizabilities of helical collagen [50].

Acknowledgments

This chapter results from collaborations on the NLO responses of helicenes with J.M. André, I. Asselberghs, A. Bossi, E. Botek, K. Clays, B. Jansík, S. Labidi, E. Licandro, S. Maiorana, O. Quinet, A. Persoons, C. Rigamonti, S. Righetto, A. Rizzo, M. Spassova, G.R. Stephenson, M. Turki, Th. Verbiest, Y. Zhang, and H. Ågren to whom the author is much grateful for insightful discussions. The calculations were performed on the computers of the « Consortium des Équipements de Calcul Intensif (CÉCI) » (http://www.ceci-hpc.be), including those of the « UNamur Technological Platform of High-Performance Computing (PTCI) » (http://www.ptci.unamur.be), for which we gratefully acknowledge the financial support from the FNRS-FRFC, the Walloon Region, and the University of Namur (Conventions No. 2.5020.11, GEQ U.G006.15, U.G018.19, 1610468, and RW/GEQ2016).

References

1 Wagnière, G.H. (1993). *Linear and Nonlinear Optical Properties of Molecules*. Basel: Verlag Helvetica Chimica Acta.

2 Bierlein, J.D. and Vanherzeele, H. (1989). Potassium titanyl phosphate: properties and new applications. *J. Opt. Soc. Am. B* 6: 622–633.

3 Kanis, D.R., Ratner, M.A., and Marks, T.J. (1994). Design and construction of molecular assemblies with large second-order optical nonlinearities. Quantum Chemical Aspects. *Chem. Rev.* 94: 195–242.

4 Campagnola, P.J., Wei, M.D., and Loew, L.M. (1999). High-resolution nonlinear optical imaging of live cells by second harmonic generation. *Biophys. J.* 77: 3341–3349.

5 Moreaux, L., Sandre, O., Blanchard-Desce, M., and Mertz, J. (2000). Membrane imaging by simultaneous second-harmonic generation and two-photon microscopy. *Opt. Lett.* 25: 320–322.

6 Reeve, J.E., Anderson, H.L., and Clays, K. (2010). Dyes for biological second harmonic generation imaging. *Phys. Chem. Chem. Phys.* 12: 13484–13498.

7 Champagne, B., Plaquet, A., Pozzo, J.L. et al. (2012). Nonlinear optical molecular switches as selective cation sensors. *J. Am. Chem. Soc.* 134: 8101–8103.

8 Coe, B.J. (1999). Molecular materials possessing switchable quadratic nonlinear optical properties. *Chem. Eur. J.* 5: 2464–2471.

9 Delaire, J.A. and Nakatani, K. (2000). Linear and nonlinear optical properties of photochromic molecules and materials. *Chem. Rev.* 100: 1817–1845.

10 Castet, F., Rodriguez, V., Pozzo, J.-L. et al. (2013). Design and characterization of molecular nonlinear optical switches. *Acc. Chem. Res.* 46: 2656–2665.

11 Verbiest, T., Clays, K., and Rodriguez, V. (2009). *Second-Order Nonlinear Optical Characterization Techniques: An Introduction*. New York: CRC Press.

12 Ledoux, I. and Zyss, J. (1982). Influence of the molecular environment in solution measurements of the second-order optical susceptibility for urea and derivatives. *Chem. Phys.* 73: 203–213.

13 Hendrickx, E., Clays, K., and Persoons, A. (1998). Hyper-Rayleigh scattering in isotropic solution. *Acc. Chem. Res.* 31: 675–683.

14 Giordmaine, J.A. (1965). Nonlinear optical properties of liquids. *Phys. Rev.* 138: A1599–A1606.

15 Fischer, P., Wiersma, D.S., Righini, R. et al. (2000). Three-wave mixing in chiral liquids. *Phys. Rev. Lett.* 85: 4253–4256.

16 Reis, H. (2006). Problems in the comparison of theoretical and experimental hyperpolarizabilities revisited. *J. Chem. Phys.* 125: 014506.

17 Marder, S.R., Cheng, L.-T., Tiemann, B.G. et al. (1994). Large first hyperpolarizabilities in push-pull polyenes by tuning of the bond length alternation and aromaticity. *Science* 263: 511–514.

18 Lu, D., Chen, G., Perry, J.W., and Goddard, W.A. (1994). Valence-bond charge-transfer model for nonlinear optical properties of charge-transfer organic molecules. *J. Am. Chem. Soc.* 116: 10679–10685.
19 Wortmann, R., Krämer, P., Glania, C. et al. (1993). Deviations from Kleinman symmetry of the second-order polarizability tensor in molecules with low-lying perpendicular electronic bands. *Chem. Phys.* 173: 99–108.
20 Ledoux, I., Zyss, J., Siegel, J.S. et al. (1990). Second-harmonic generation from non-dipolar non-centrosymmetric aromatic charge-transfer molecules. *Chem. Phys. Lett.* 172: 440–444.
21 Cho, M., An, S.-Y., and Lee, H. (2002). Nonlinear optical properties of tetrahedral donor-acceptor octupolar molecules: effective five-state model approach. *J. Chem. Phys.* 116: 9165–9173.
22 Verbiest, T., Van Elshocht, S., Kauranen, M. et al. (1998). Strong enhancement of nonlinear optical properties through supramolecular chirality. *Science* 282: 913–915.
23 Daul, C.A., Ciofini, I., and Weber, V. (2003). Investigation of NLO proprieties of substituted (M)-tetrathia-[7]-helicenes by semiempirical and DFT methods. *Int. J. Quantum Chem.* 91: 297–302.
24 Clays, K., Wostyn, K., Persoons, A. et al. (2003). Experimental study of the second-order non-linear optical properties of tetrathia-[7]-helicene. *Chem. Phys. Lett.* 372: 438–442.
25 Champagne, B., André, J.-M., Botek, E. et al. (2004). Theoretical design of substituted tetrathia-[7]-helicene with large second-order nonlinear optical responses. *ChemPhysChem* 5: 1438–1442.
26 Bossi, A., Licandro, E., Maiorana, S. et al. (2008). Theoretical and experimental investigation of electric field induced second harmonic generation in tetrathia[7]helicenes. *J. Phys. Chem. C* 112: 7900–7907.
27 Si, Y. and Yang, G. (2013). Photophysical properties of azaboradibenzo[6]helicene derivatives. *J. Mater. Chem. C.* 1: 2354–2361.
28 Champagne, B. and Zhang, Y. unpublished.
29 Botek, E., Spassova, M., Champagne, B. et al. (2005). Hyper-Rayleigh scattering of neutral and charged helicenes. *Chem. Phys. Lett.* 412: 274–279.
30 Botek, E. and Champagne, B. (2004). Semiempirical time-dependent unrestricted Hartree–Fock approach for the determination of molecular linear and nonlinear optical properties. *Chem. Phys. Lett.* 387: 130–135.
31 Maki, J.J. and Persoons, A. (1996). One-electron second-order optical activity of a helix. *J. Chem. Phys.* 104: 9340–9348.
32 Tessore, F., Cariati, E., Cariati, F. et al. (2010). The role of ion pairs in the second-order NLO response of 4-X-1-methylpiridinium salts. *ChemPhysChem* 11: 495–507.
33 Ramos, T.N., Castet, F., and Champagne, B. (2021). Second harmonic generation responses revealing the formation of dimers of ion-pairs. *J. Phys. Chem. B* 125: 3386–3397.
34 Champagne, B. and Labidi, S.N. (2016). Second-order nonlinear optical responses of heptahelicene and heptathiahelicene derivatives. *Chem. Phys. Lett.* 644: 195–200.

35 Rajca, A., Pink, M., Xiao, S. et al. (2009). Functionalized thiophene-based [7]helicene: chirooptical properties versus electron delocalization. *J. Org. Chem.* 74: 7504–7513.

36 Botek, E., Champagne, B., Turki, M., and André, J.-M. (2004). Theoretical study of the second-order nonlinear optical properties of [N]helicenes and [N]phenylenes. *J. Chem. Phys.* 120: 2042–2048.

37 Quinet, O. and Champagne, B. (2001). Sum-frequency generation first hyperpolarizability from time-dependent Hartree-Fock method. *Int. J. Quantum Chem.* 85: 463–468.

38 Liu, X., Yu, P., Xu, L. et al. (2013). Synthesis for the mesomer and racemate of thiophene-based double helicene under irradiation. *J. Org. Chem.* 78: 6316–6321.

39 Si, Y., Zhao, X., Su, Z., and Yang, G. (2016). Theoretical study on thiophene-based double helicenes with intrinsic large second-order nonlinear optical response. *RSC Adv.* 6: 84705–84711.

40 Coe, B.J., Rusanova, D., Joshi, V.D. et al. (2016). Helquat dyes: helicene-like push-pull systems with large second-order nonlinear optical responses. *J. Org. Chem.* 81: 1912–1920.

41 Takamuku, S. and Nakano, M. (2019). Theoretical study on second hyperpolarizabilities of intramolecular Pancake-Bonded diradicaloids with helical scaffold. *ACS Omega* 4: 2741–2749.

42 Jansík, B., Rizzo, A., Ågren, H., and Champagne, B. (2008). Strong two-photon circular dichroism in helicenes: a theoretical investigation. *J. Chem. Theory Comput.* 4: 457–467.

43 Toro, C., De Boni, L., Lin, N. et al. (2010). Two-photon absorption circular dichroism: a new twist in nonlinear spectroscopy. *Chem. Eur. J.* 16: 3504–3509.

44 Diaz, C., Vesga, Y., Echevarria, L. et al. (2015). Two-photon absorption and two-photon circular dichroism of hexahelicene derivatives: a study of the effect of the nature of intramolecular charge transfer. *RSC Adv.* 5: 17429–17437.

45 Rizzo, A., Rikken, G.L.J.A., and Mathevet, R. (2016). Ab initio study of the enantio-selective magnetic-field-induced second harmonic generation in chiral molecules. *Phys. Chem. Chem. Phys.* 18: 1846–1858.

46 Verbiest, T., Sioncke, S., Persoons, A. et al. (2002). Electric-field-modulated circular-difference effects in second-harmonic generation from a chiral liquid crystal. *Angew. Chem. Int. Ed.* 41: 3882–3884.

47 Botek, E., André, J.-M., and Champagne, B. (2005). Mixed electric-magnetic second-order nonlinear response of helicenes. *J. Chem. Phys.* 122: 234713.

48 Campo, J., Desmet, F., Wenseleers, W., and Goovaerts, E. (2009). Highly sensitive setup for tunable wavelength hyper-Rayleigh scattering with parallel detection and calibration data for various solvents. *Opt. Express* 17: 4587–4604.

49 Spassova, M., Asselberghs, I., Verbiest, T. et al. (2007). Theoretical investigation on bridged triarylamine helicenes: UV/visible and circular dichroism spectra. *Chem. Phys. Lett.* 439: 213–218.

50 de Wergifosse, M., de Ruyck, J., and Champagne, B. (2014). How the second-order nonlinear optical response of collagen triple helix appears: a theoretical investigation. *J. Phys. Chem. C* 118: 8595–8602.

15

Helicenes for Optoelectronic Applications and Devices

Seán T. J. Ryan and Matthew J. Fuchter

15.1 Introduction

According to IUPAC, helicenes are the set of ortho-fused polycyclic aromatic or heteroaromatic compounds in which all rings (minimum five) are angularly arranged so as to give helically shaped molecules, which are thus chiral [1]. Owing to their helical shape, helicenes are chiral molecules and exist as a pair of enantiomers designated as ***P*** (+, plus) and ***M*** (−, minus) (see Figure 15.1). Similar to other conjugated aromatic structures, helicenes have the potential to act as small-molecule organic semiconductors (OSCs) for the development of electronic devices. Compared to silicon, OSCs can be used to realize low-cost, lightweight devices, with simple fabrication requirements and with tunable molecular and material properties, including HOMO/LUMO energy levels and charge carrier mobility [2]. One advantage of helicenes over more conventional small-molecule OSCs, such as pentacene, is that the helical geometry can impart unique molecular and material properties, such as the ability to generate circularly polarized luminescence (CPL) in organic light-emitting diodes (OLEDs) (see Section 15.2.3) and to tune charge carrier mobility by variation of chiral composition (see Section 15.3.1). Furthermore, the intrinsic chirality of helical aromatics in principle affords a new degree of freedom by which to exert control over electron spin (see Section 15.4). Thus, while much less developed than mainstream OSC materials, it is possible that helical OSCs may provide unique opportunities in optoelectronic applications in the future.

It is only recently that scientists have taken significant steps to realize the potential of chiral molecules in the field of optoelectronics [3]. This chapter aims to highlight selected examples of exciting new technological advances stemming from the utilization of helicenes. We shall present examples where helicenes are used in both their racemic and enantiopure forms in order to demonstrate the advantages offered by either chiral compositions. To focus our discussion, we shall not explore tuning of electronic circular dichroism (CD) and CPL in solution through helicene structure–property relationships, as these have already been explored in a number of extensive reviews [4–9]. Our discussion shall concentrate on device applications,

Helicenes: Synthesis, Properties, and Applications, First Edition.
Edited by Jeanne Crassous, Irena G. Stará, and Ivo Starý.

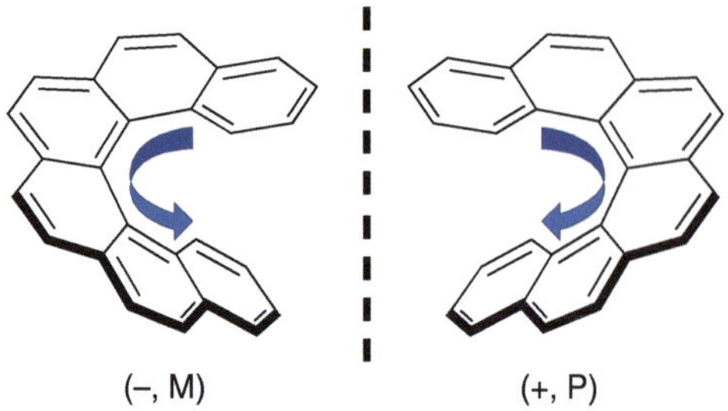

Figure 15.1 Illustration of (−, **M**) and (+, **P**) enantiomers of carbo[6] helicene.

centered predominantly around recent advances in the fields of OLEDs, organic field effect transistors (OFETs), and photovoltaics (PVs), especially where the chirality of the helicene imparts a novel function. Finally, we discuss selected cases in which the potential for novel function in helicene-based optoelectronic devices has been proposed, though not yet fully realized.

15.2 OLED Applications

Since the first demonstration of a practical OLED by Tang and VanSlyke [10] in 1987, OLEDs have revolutionized the lighting and display industry and are now a key component of smartphone displays. OLEDs typically comprise an electroluminescent (EL) layer sandwiched between a transparent anode and reflective cathode. This architecture allows for the possibility to include additional charge injection, transport, and blocking layers to improve efficiency. The simple processing requirements of OSCs mean that OLEDs can be mass produced on a wide range of substrates (see Figure 15.2 and Information Box 15.1). In this section we discuss the application of helicenes to OLEDs for a range of functions in multilayered architectures and highlight how the unique properties of helicenes impart OLED function not accessible with non-chiral OSCs.

15.2.1 Helicene Racemates as Emissive OLED Materials

The first example of a helicene in an OLED application in the academic literature appeared in 2010 by Sooksimuang and coworkers [11], with earlier examples of [5]helicenes in the patent literature [12]. 3,12-Dimethoxy-7,8-dicyano-[5]helicene

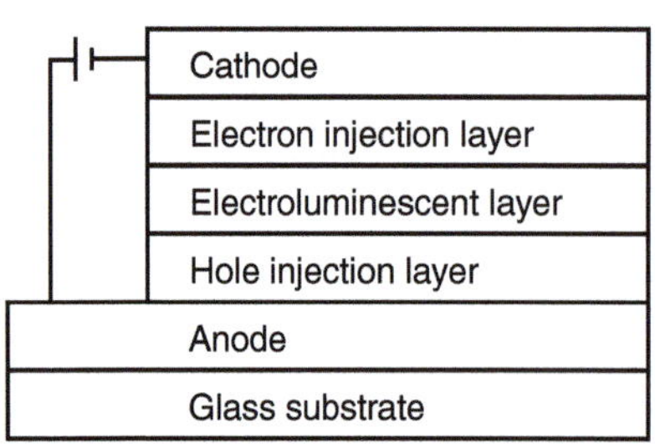

Figure 15.2 Schematic illustration of an OLED device.

Information Box 15.1

Organic Light-Emitting Diodes: Fundamental Operation Principles

Upon application of a suitably high driving voltage in forward bias, electrons are injected from the cathode and holes from the anode and move through the multilayered device. If the charges recombine (forming an exciton) in the EL layer, the electrical energy is converted into light. The electronically generated excitons relax to their ground states via radiative (light-emitting) or non-radiative (non-light-emitting) pathways. The possibility to simply tune the EL layer to emit light at specific or multiple wavelengths, together with the ability to optimize device performance (see below), renders OLEDs vital components of many modern technologies.

On account of heat generation and dissipation during OLED operation, organic materials within OLEDs must possess certain thermal stability requirements to ensure high performance at elevated temperatures and a long lifetime of the device. Relevant thermal properties include melting temperature, T_m, glass transition temperature, T_g, and decomposition temperature, T_d. For reference, *N*,*N*′-di(1-naphthyl)-*N*,*N*′-diphenyl-(1,1′-biphenyl)-4,4′-diamine (NPB, see Figure 15.4) is a typical, well-known organic material for OLEDs possessing $T_m = 279\,°C$, $T_g = 95\,°C$ and $T_d = 345\,°C$.

The four principle parameters describing OLED device performance are maximum brightness, turn-on voltage, current efficiency, and power efficiency described as follows:

- Maximum brightness ($cd\,m^{-2}$) describes the peak luminance of the device, where cd (= candela), the SI unit of luminous intensity, describes the measure of power emitted by a light source per unit solid angle weighted with respect to the sensitivity of the human eye.
- Turn-on voltage is defined as the applied voltage, which imparts a luminance of $1\,cd\,m^{-2}$.
- Current efficiency (or luminous efficiency; SI unit, $cd\,A^{\ 1}$) is defined as the ratio of the luminance to the current density (SI unit, $A\,m^{-2}$).
- Power efficiency (SI unit, $lm\,W^{-1}$) is defined as the luminous flux (SI unit, lm = cd sr; sr = steradian, SI unit of solid angle) per unit of power, where the luminous flux is calculated for OLED emission considered as Lambertian (or plane) emission.

In order to afford the reader some reference points with respect to the magnitudes of the device parameters quoted in the text, here we detail some real-world device performance requirements, taking a standard smartphone as an example.

(Continued)

Information Box 15.1 (Continued)

For ideal viewing conditions, an OLED screen must produce a brightness of ≈200 cd m^{-2}, requiring a power consumption of ≈600 mW. For a standard ≈ 5.1″ smartphone screen with a surface area of ≈0.008 m^2, this translates to a current efficiency of ≈8 lm W^{-1}. However, on account of losses endured as a part of light extraction (i.e. the light produced by the bare OLED exiting the device), as much as 70% of light is absorbed or otherwise dissipated by the use of anti-glare filters and total internal reflection, as well as unfavorable wave guide modes. Therefore, in order to be described as "useful" in this context, an OLED must in fact be capable of a brightness of ≈1500 cd m^{-2} with an applied voltage of less than 5 V (ideally ≈3 V) with a current efficiency of ≈50 lm W^{-1}.

As different spin statistics limit the maximum theoretical efficiencies of fluorescent and phosphorescent OLEDs, the following device parameters offer a rough guideline as to what constitutes a "good" laboratory OLED device:

Fluorescent OLED: turn-on voltage <3 V, current efficiency ≈10 cd A^{-1}, power efficiency ≈10 lm W^{-1}, maximum brightness ≈2000 cd m^{-2}.

Phosphorescent OLED: turn-on voltage <5 V, current efficiency ≈30 cd A^{-1}, power efficiency ≈23 lm W^{-1}, maximum brightness ≈5000 cd m^{-2}.

Further information regarding the device performance of well-known, conventional, and commercial OLED materials may be found at Ref. [85].

(**1**, see Figure 15.3) was used as the EL layer, displaying thin film photoluminescence (PL) (λ_{max} = 498 nm) with a bandgap of 2.6 eV. Helicene **1** demonstrated thermal properties compatible with device operation: melting temperature, T_m = 307 °C, glass transition temperature, T_g = 130 °C, and decomposition temperature, T_d = 330 °C. The performance of the resulting device (device **A**; see Figure 15.3) is shown in Table 15.1, with additional device components listed in Figure 15.4.

Tao and coworkers [13] subsequently manufactured an OLED using diaza[7]helicene **2** (see Figure 15.3) dispersed within a host (4,4′-bis(*N*-carbazolyl)-1,1′-biphenyl, CBP; see Figure 15.4). In guest–host systems, an emissive species is doped into a host matrix, which minimizes their aggregation, mitigating π–π stacking-induced exciton quenching and generally resulting in devices with improved performance. The diaza[7]helicene-CBP blend films are thermally stable (T_d = 372 °C and T_g = 203 °C) with a thin film optical bandgap of 2.79 eV. Device **B** demonstrated blue light emission (λ_{max} = 446 nm) with performance shown in Table 15.1. This relatively poor device performance was attributed to the high energy barrier (0.93 eV) between the 2,9-dimethyl-4,7-diphenyl-1,10-phenanthroline (BCP) electron transport layer and the EL layer, which restricts electron transport.

In 2015, Liu and coworkers [14] investigated the electroluminescence properties of monoaza[6]helicene **3** (see Figure 15.3), with thermal stability properties, T_d = 297 °C and T_m = 131 °C, and a bandgap of 2.92 eV. The authors compared the performance of **3** in both neat (device **C**) and in guest–host (device **D**) EL layer device architectures. The doped emitter architecture was found to exhibit a

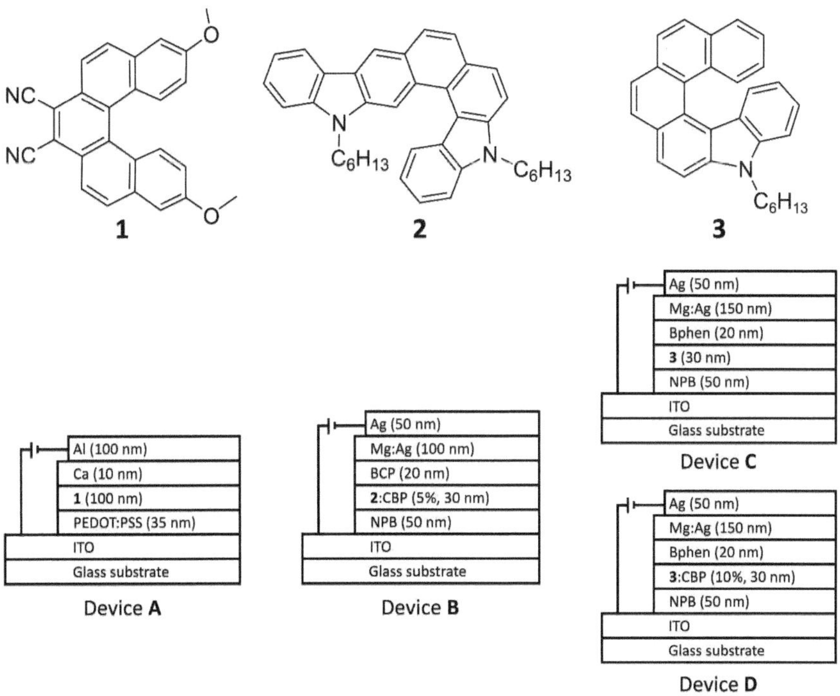

Figure 15.3 Molecular structures of helicenes **1–3** (top) with schematic depictions of the devices **A–D** to which they have been applied as electroluminescent materials (bottom).

slightly improved device performance relative to the neat emitter architecture (see Table 15.1), which was attributed to rapid energy transfer from the CBP host to **3,** as well as suppression of the aggregation-induced quenching of neat films (evidenced by a low quantum yield).

Moorthy and coworkers [15] synthesized a series of eight carbo[5]helicenes **4–6** and monoaza[5]helicenes **7–11** functionalized with carbazoles, diphenylamines, and triphenylamines for application as EL materials in OLEDs (see Figure 15.5). The helicenes all exhibited excellent thermal stabilities (T_d = 383–514 °C, T_m = 217–440 °C, and T_g = 115–214 °C). An inspection of the device data (see Table 15.1) reveals the influence of chemical structure and device architecture/composition on device performance. Carbo[5]helicene devices were found to exhibit lower turn-on voltages (3–4 V) relative to the aza[5]helicene devices (3.5–6.5 V), while guest–host devices demonstrated better efficiencies than their neat analogs. The best overall performance was achieved using helicene **11** in device **F**.

15.2.2 Helicene Racemates as OLED Device Components

Moorthy and coworkers further developed their study of helicenes **4**, **7–11**, demonstrating their utility as components in multilayered devices (see Figure 15.5). The HOMO energy levels of these materials (5.04–5.4 eV) are suitable to enable efficient

Table 15.1 OLED device performance data from devices A–H for helicene racemates **1–11** acting as emissive species.

Helicene	Device/EL layer	Turn-on voltage (V)	Current efficiency (cd A^{-1})	Power efficiency (lm W^{-1})	Maximum brightness (cd m^{-2})
1	**A**/neat	3.7	0.64	0.29	1587
2	**B**/guest–host	5.3	0.22	0.09	2365
3	**C**/neat	5.4	0.42	0.11	1209
	D/guest–host	5.1	0.68	0.17	3245
4	**E**/neat	4.0	0.41	0.32	1520
5	**E**/neat	3.0	0.39	0.31	2300
	G/guest–host	4.0	1.41	1.11	1150
6	**E**/neat	3.0	1.33	1.40	3220
	G/guest–host	3.5	3.72	3.34	1810
7	**F**/neat	4.5	1.53	0.91	1190
8	**E**/neat	4.0	0.87	0.42	2050
	F/neat	3.5	1.43	1.00	1440
	G/guest–host	3.5	2.65	2.08	3040
9	**E**/neat	3.5	1.06	0.91	2190
	H/guest–host	6.0	1.63	0.85	935
11	**F**/neat	3.5	4.39	3.06	2940
	H/guest–host	6.5	1.01	0.48	752

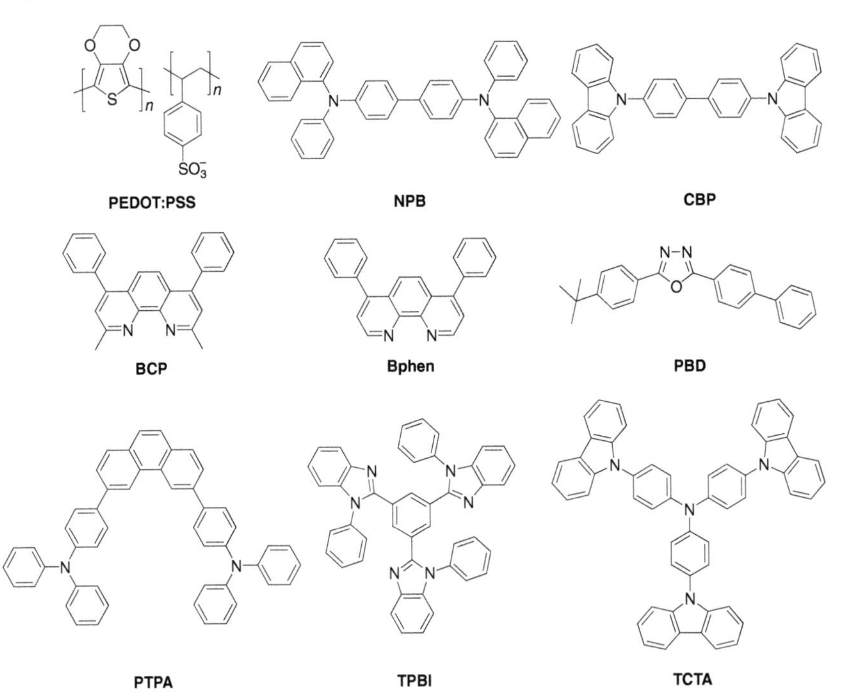

Figure 15.4 Molecular structures of device component materials used for helicene-based OLEDs.

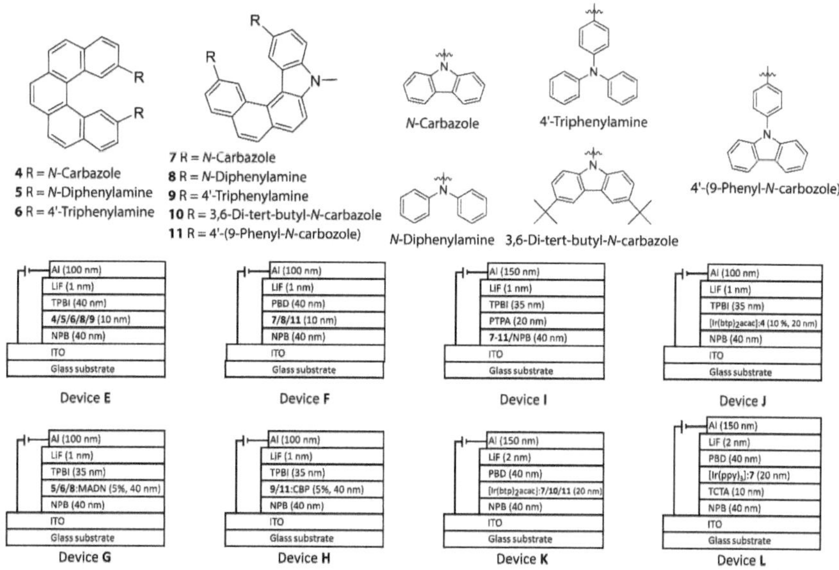

Figure 15.5 Molecular structures of helicenes **4–11** (top left) and amine functional groups (top right) with schematic depictions of the devices **E–L** to which they have been applied as electroluminescent, HTLs, and matrix host materials (bottom).

hole injection, rendering them attractive candidates for hole transport layers (HTLs) [15]. Helicenes **7–11** as HTLs and 3,6-bis(triphenylamino)phenanthrene (PTPA, see Figure 15.4) as the EL layer were incorporated into device **I**. The performance of helicene HTLs was compared to conventional *N,N′*-di(1-naphthyl)-*N,N′*-diphenyl-(1,1′-biphenyl)-4,4′-diamine (NPB; see Figure 15.4), with helicene-based devices achieving higher maximum brightness and greater current and power efficiencies (see Table 15.2).

Such helicenes can also serve as host materials for phosphorescent metal complexes in phosphorescent OLEDs (PHOLEDs) (see Section 15.2.3.2) as they possess triplet energy levels higher than those of common heavy metal complexes, which facilitates efficient energy transfer. Red emissive devices using iridium complex [Ir(btp)$_2$acac] were fabricated using **4** (device **J**; see Figure 15.5) and **7**, **10**, and **11** (device **K**), while a green emissive device using dopant [Ir(ppy)$_3$] and host **7** was incorporated into device **L**. Of the red emissive devices, **11** demonstrated the best performance as shown in Table 15.2, while device **L**, although possessing a higher turn-on voltage, exhibited otherwise superior performance.

While the helicene racemates reported may not yet compete with conventional state-of-the-art organic materials for devices, these studies demonstrate the versatility of helicenes in high performance OLEDs, from the emitting species to HTLs and even host matrices. Further investigations into structure bulk–property relationships, processing procedures, and device architectures are expected to achieve improved performance in the future.

Table 15.2 OLED device performance data from devices I–L for helicene racemates **4**, **7–11** acting as HTLs and host matrices.

Helicene	Device/helicene function	Turn-on voltage (V)	Current efficiency (cd A^{-1})	Power efficiency (lm W^{-1})	Maximum brightness (cd m^{-2})
4	J/host matrix	4.5	5.53	3.47	2640
7	I/HTL	3.5	1.3	0.73	1740
	K/host matrix	4.0	3.89	3.04	1860
	L/host matrix	4.5	29.0	18.2	4200
8	I/HTL	3.5	1.48	1.33	1990
9	I/HTL	3.5	1.25	0.91	911
10	I/HTL	3.5	1.76	1.58	1930
	K/host matrix	4.5	2.48	1.30	976
11	I/HTL	4.0	1.69	1.18	2020
	K/host matrix	3.5	4.66	2.93	1560
NPB	I/HTL	3.0	1.12	1.00	1590

15.2.3 Helicenes as Enantiopure Materials for OLEDs

The chiroptical responses of enantiopure helicenes are well-known [6, 16]. Such effects originate from differences in their interaction with left-handed (LH) and right-handed (RH) circularly polarized (CP) light. CP light can be described as two plane polarized waves of equal amplitude, at right angles to one another, but with a quadrature phase relationship. The resultant summed electric field vector is of constant magnitude and rotates at a fixed rate, tracing a helix as the wave travels through space (see Figure 15.6). CP light is also described as chiral light, with two "enantiomeric" LH and RH forms. The circular polarization of light can be quantified by the dissymmetry factor (or g factor), defined as follows:

$$g = \frac{I_L - I_R}{{}^1\!/_2(I_L + I_R)}$$

where I_L and I_R represent the intensity of LH and RH CP light, respectively. As can be seen from the preceding equation, fully LH CP light has g factor of +2; fully RH CP light has a g factor of −2; and nonpolarized light has a g factor of 0. Typically, a subscript is used to describe which dissymmetric optical process is being described, g_y, where y = *abs*, *PL* or *EL* for *absorbed light*, *photoluminescence*, or *electroluminescence*.

CP light is of great technological interest for many cutting-edge technologies, including but not limited to three-dimensional (3D) displays [17], quantum computing [18], optical spintronics [19], biological imaging [20], and sensing [21]. Among these is the realization of CP-OLEDs, which would increase the efficiency of state-of-the-art display technologies. Currently, OLEDs use anti-glare filters (optical components comprising a quarter wave plate and linear polarizer) to improve contrast (see Figure 15.6 and Information Box 15.2) [22]. As conventional

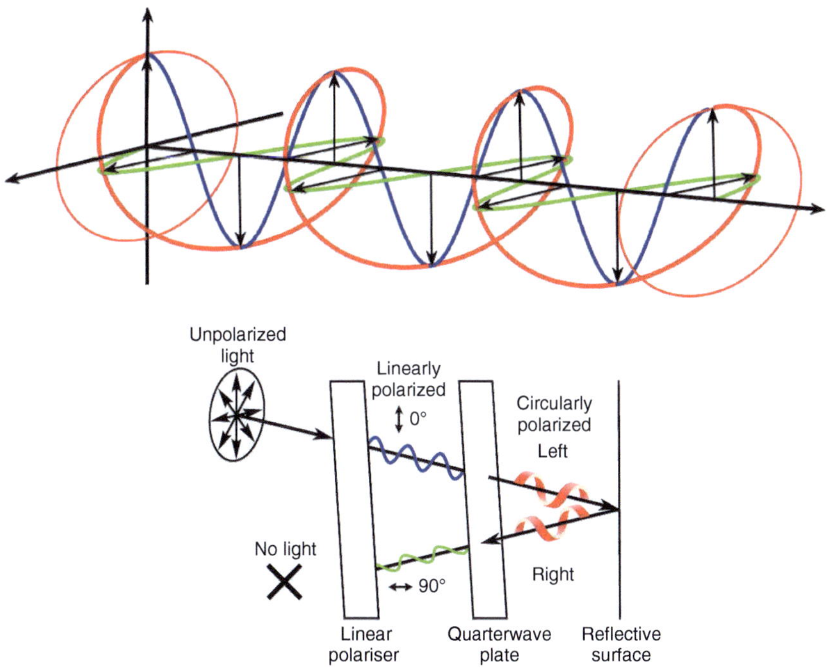

Figure 15.6 Schematic illustration of the vector composition of CP light (top) and the basic operation principles of an anti-glare filter (bottom).

Information Box 15.2

Anti-glare Filters

Glare is generated when ambient, external light enters an OLED device and is then reflected off the terminal cathode or other reflective surfaces. Anti-glare filters are comprised of a linear polarizer and quarter wave plate appended to the front of an OLED display. Ambient light is first linearly polarized by the polarizer before being converted into circularly polarized light by the quarter wave plate. Upon reflection off the cathode, the handedness of the CP light is inverted. Upon passing back through the quarter wave plate, the light is converted back into linearly polarized light but with a plane of polarization now orientated at 90° to the linear polarizer, which is then able to absorb the light and prevent it from exiting the device as glare.

Source: Based on [22].

(non-CP) OLEDs emit unpolarized light, such anti-glare filters absorb 50% of the emitted light, resulting in reduced brightness and shorter operational lifetimes. If the OLED could be engineered to emit CP light, it would pass straight through the anti-glare filter, increasing the display efficiency. As an example, a CP-OLED achieving a dissymmetry factor $|g_{EL}| = 0.5$ would increase brightness by 25%.

The dual semiconducting and chiroptical properties of enantiopure helicenes makes them attractive candidates for the fabrication of CP-OLEDs. In this section we explore the use of helicenes as chiral materials for the design of both fluorescent and phosphorescent CP-OLEDs.

15.2.3.1 Helicenes as Small-Molecule Dopants for CP-OLEDs

Before the incorporation of helicenes, CP-OLEDs made use of light-emitting polymers (LEPs) bearing chiral side chains as the EL layer [23–27]. The first example of a CP-OLED containing a helicene was reported in 2013 by Fuchter, Campbell, and coworkers [28] who combined aza[6]helicene **12** with the well-known, device optimized LEP, poly[9,9-dioctylfluorene-co-benzothiadiazole] (F8BT) (see Figure 15.7). Different helicene:polymer blend ratios were investigated to evaluate the impact of the helicene additive on morphological and spectroscopic properties of the EL layer. The authors found that the helicene induced highly dissymmetric absorption in the polymer absorption band ($\lambda = 475$ nm), which was sensitive to the helicene loading ratio (7%: $|g_{abs}| = 0.03$; 53%: $|g_{abs}| = 0.2$), similar in magnitude to thin films comprising polymers bearing chiral side chains [25]. The PL was also found to be strongly CP (7%: $|g_{PL}| = 0.2$; 53%: $|g_{PL}| = 0.5$). The dissymmetry of EL was evaluated in single layer OLEDs (device **M**), which achieved a power efficiency and maximum brightness of 1.1 lm W^{-1} and 3000 cd m^{-2}. The CP-EL spectra of the device were also very similar to the CP-PL spectra with a $|g_{EL}| = 0.2$.

This system was revisited by Fuchter, Campbell, and coworkers [29] in 2019, who found that the handedness of CP-EL and CP-PL was sensitive to active layer

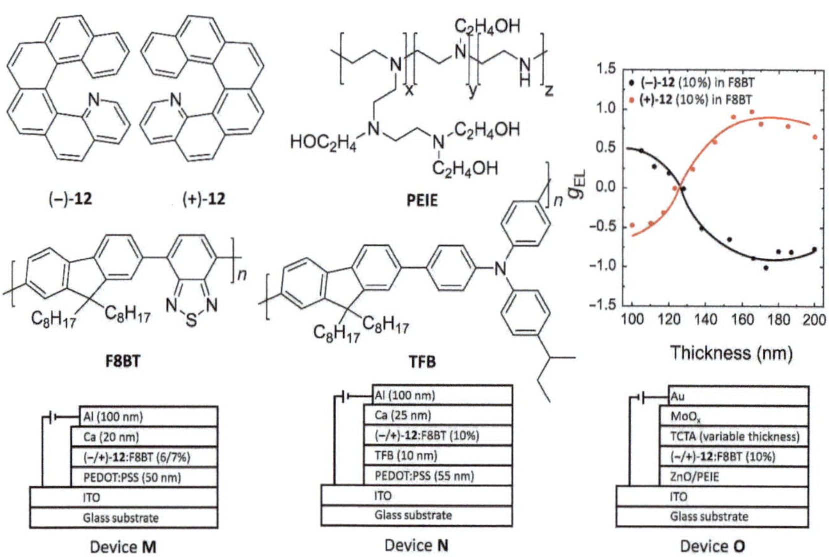

Figure 15.7 Molecular structures of helicenes (*M*, −)-**12** and (*P*, +)-**12** and polymers, F8BT, TFB, and PEIE (top left). Schematic depictions of the devices **M**, **N**, and **O** to which helicene **12** has been applied as EL layer chiral additives (bottom). Graphic illustration of g_{EL} as a function of EL layer thickness in device **N** (top right).

thickness. In the as-cast state, the polymer absorption band in the blend system (F8BT:**(+)-12** (10%)) was found to be CD silent, but strong chiroptical effects were observed in the annealed systems ($|g_{abs}| = 0.85$). Increasing the film thickness from 100 to 200 nm resulted in a sign inversion of g_{PL} and g_{EL}, with **(−)-12** and **(+)-12** giving equal and opposite responses. Devices were fabricated with varying emissive layer thicknesses (device **N**). The extracted g_{EL} behavior exhibited the same trend as that observed for g_{PL}, whereby for the **(−)-12** (10%) in F8BT device, g_{EL} varies from +0.5 (100 nm) to −1.0 (200 nm) with the inversion point occurring just above 120 nm. Mirror responses were observed for **(+)-12**, and the highest g_{EL} was observed to be −1.05 for **(−)-12** in the thick device regime (140 nm). The influence of emissive layer thickness on device performance was then examined for F8BT: **(-)-12** (10%). The best CP-OLED performance was achieved with “thick” EL layers (160 nm) (see Table 15.3). The inversion of handedness as a function of thickness was proposed to be due to the interplay of two distinct mechanisms in the different thickness regimes: localized CP-EL originating from the molecular chirality of individual polymer chromophores, induced by the helicene, and the nonlocal impact of light propagation through a chiral medium. It should also be noted that non-helicene chiral additives can also be used to induce effects in F8BT [30].

In 2020, Fuchter, Campbell, and coworkers employed an inverted device architecture with a TCTA blocking layer for F8BT:**(+/−)-12** (10%) CP-OLEDs (device **O**), achieving remarkably improved performance [31]. The inverted architecture also displayed a maximum $|g_{EL}| = 0.57$ with opposite CP handedness compared with the conventional noninverted device **N**. The improved performance was found to be highly dependent both on the inverted architecture and on the TCTA layer with significant improvement upon increasing the TCTA thickness from 5 to 40 nm (see Table 15.3). In conventional F8BT: **(+/−)-12** (10%) CP-OLEDs, **(+/−)-12** aggregates at the interface between the TFB hole injection and EL layers, resulting in hole trapping on the helicene on account of its higher HOMO (−5.4 eV) compared with F8BT (−5.8 eV). In the inverted architecture, holes are injected from the opposite direction but still become trapped on **(+/−)-12** as they traverse the EL

Table 15.3 CP-OLED device performance data from devices N and O for helicene **(+/−)-12** acting as a chiral dopant for F8BT.

Helicene	Device	Turn-on voltage (V)	Current efficiency (cd A^{-1})	Power efficiency (lm W^{-1})	Maximum brightness (cd m^{-2})	Electroluminescence dissymmetry factor ($\|g_{EL}\|$)
(+/−)-12	**M**	5.9	—	1.1	3 000	0.2
(+/−)-12	**N (EL layer 160 nm)**	5.2	4.0	1.3	2 054	0.5–1.0
(+/−)-12	**O (TCTA 5 nm)**	7.4	0.02	0.006	192	0.12
(+/−)-12	**O (TCTA 40 nm)**	2.3	16.4	16.6	28 500	0.57

layer toward the ZnO/polyethylenimine ethoxylated (PEIE) electron transport layer. Addition of TCTA shifts the recombination zone away from the ZnO/PEIE interface and aggregated **(+/−)-12** so that excitons are predominantly formed on the F8BT chiral phase, enhancing radiative recombination and increasing $|g_{EL}|$, resulting in the first CP-OLED to exhibit simultaneous high device performance and high electroluminescence dissymmetry approaching the performance requirements of real-world CP-OLED technologies.

Fuchter, Campbell, and coworkers subsequently undertook a mechanistic study to elucidate the origin of the intense chiroptical properties in helicene additive/LEP CP-OLEDs [32]. Using a combination of techniques, including atomic force microscopy (AFM), resonant soft X-ray scattering, and Mueller matrix spectroscopic ellipsometry, the authors were able to demonstrate that the strong polymer chiroptical properties were mainly attributable to so-called natural optical activity [33], arising from magnetoelectric coupling rather than previously assumed mesoscopic structural medium chirality [34]. It was further suggested that this intense magnetoelectric coupling originates from the formation of in-plane helical periodic modulation of polymer fibrils in a double twist cylinder-type blue phase [35, 36], the first such report in a conjugated polymer.

15.2.3.2 Helicene Complexes for Circularly Polarized Phosphorescent Light-Emitting Diodes (CP-PHOLEDs)

The recombination of electrons and holes in fluorescent materials, such as the LEPs described in Section 15.2.3.1, limits their internal quantum efficiency (IQE) to 25% [37]. In contrast, phosphorescent materials can achieve IQEs approaching 100%, owing to their ability to emit light from triplet states (so-called phosphorescence). This is most easily realized by the use of heavy metal complexes (Eu, Ir, Pt, Zn) as the luminescent entity, which have high levels of spin–orbit coupling [38]. Such materials have been demonstrated as PHOLEDs with external quantum efficiencies exceeding 30% [39, 40]. Early work to introduce chirality into such heavy metal complexes for the realization of OLEDs that emit CP phosphorescence (CP-PHOLEDs) made use of chiral ligands with axial or point chirality that are attached to the metal center [41–43], configurational chirality at the metal center [41, 44], or a combination of the two (see Figure 15.8). These early attempts, however, suffered from poor device performance and low $|g_{EL}|$.

In 2010, Autschbach, Crassous, Réau, and coworkers [45], drawing on the well-known cyclometalation behavior of 2-phenylpyridine ligands to Pt metal centers and the phosphorescent nature of their resultant complexes, prepared three chiral metallahelicenes using a series of 4-(2-pyridyl)-benzo[g]phenanthrenes **13a–13c** (see Figure 15.9). Helicenes **13a–13c** were conformationally unstable at room temperature, and as such could not be enantiomerically resolved. Conformational stability could be achieved upon cyclometalation to form Pt metallahelicenes **14a–14c**. The metallahelicenes displayed emission behavior considerably different to their parent helicenes. While **13a–13c** displayed blue fluorescence in dichloromethane solution at room temperature and long-lived green phosphorescence at 77 K, **14a–14c** displayed only red phosphorescence at

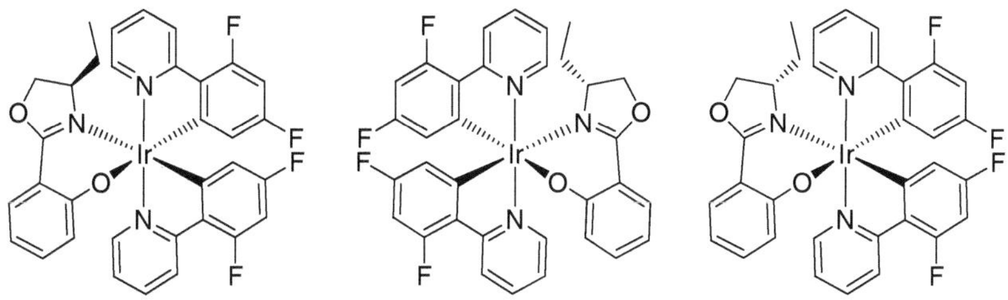

Δ-Ir(dfppy)$_2$-R-(edp) Λ-Ir(dfppy)$_2$-R-(edp) Δ-Ir(dfppy)$_2$-S-(edp)

Figure 15.8 Molecular structures of Ir complexes demonstrating the simultaneous introduction of stereogenic point (*R*, *S*) chirality and octahedral (Δ, Λ) chirality [44].

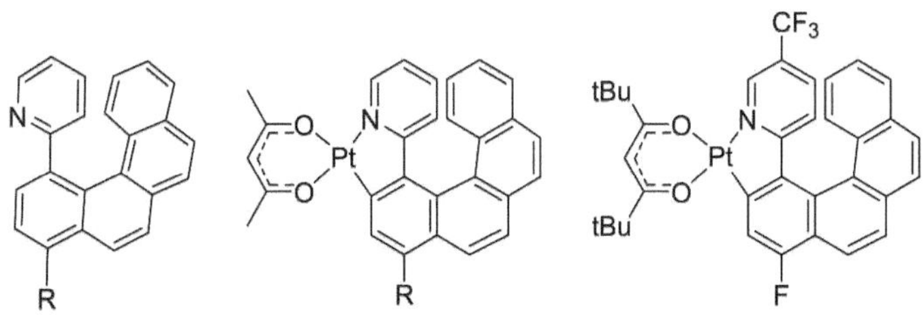

13a R = OMe
13b R = F
13c R = H

14a R = OMe
14b R = F
14c R = H

15

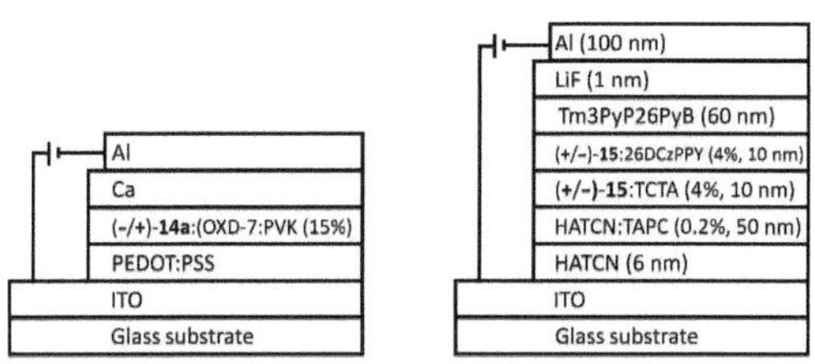

Device **P** Device **Q**

Figure 15.9 Molecular structures of (metalla)helicenes **13–15** (top) with schematic depictions of the devices **P** and **Q** to which **14** and **15** have been applied as electroluminescent materials (bottom).

room temperature with a maximum luminescence quantum yield of 10% and mirror image CPL spectra with $|g_{lum}| \approx 10^{-2}$ [46].

Building on these complexes, in 2016, Fuchter and coworkers [22] were the first to demonstrate the use of a chiral metallahelicene in a working CP-PHOLED. CP-PHOLEDs were fabricated using complex **14a** (device **P**). Devices fabricated with enantiopure and racemic (*rac*) **14a** displayed similar performance parameters (see Table 15.4). The higher solubility of the enantiopure complexes and their resultant lower tendency to phase separate in the EL layer host matrix (see Figure 15.10) yielded enhanced device-optimized performance compared to their racemic counterpart (see Table 15.4) achieving g_{EL} of +0.22 and −0.38. It is worth noting that the $|g_{EL}|$ measured here in the solid state is an order of magnitude

Table 15.4 PHOLED device performance data from devices O and P for metallahelicenes **14** and **15** acting as phosphorescence emitters.

Helicene	Device	Turn-on voltage (V)	Current efficiency (cd A^{-1})	Power efficiency (lm W^{-1})	Maximum brightness (cd m^{-2})	Electroluminescence dissymmetry factor ($\lvert g_{EL} \rvert$)
(*rac*)-14a	**P**	4.4	0.33	0.13	122	0
(−)-14a	**P** (optimized)	3.6	0.49	0.23	374	0.22–0.38
(−)-15	**Q**	3.4	22.52	18.62	11590	1.4×10^{-3}
(+)-15	**Q**	3.2	19.52	15.39	9732	1.6×10^{-3}

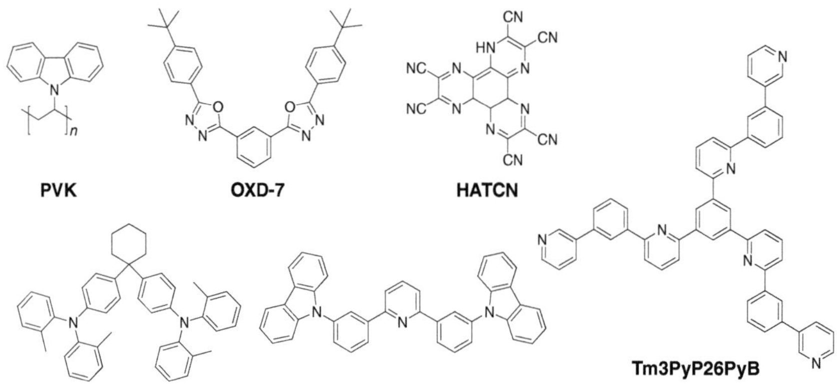

Figure 15.10 Molecular structures of device component materials used for CP-PHOLEDs.

greater than $\lvert g_{lum} \rvert$ measured in solution for the same complex by Autschbach, Réau, Crassous, and coworkers [46], which suggests that dissymmetry factors measured in solution are not always a reliable indicator of solid-state chiroptical properties.

In 2019, Zheng and coworkers [47] modified **14a** in an attempt to improve its conformational stability for vacuum sublimation. Fluorine and trifluoromethyl groups were appended to the helical ligand, and the bidentate acetylacetonate ligand was replaced with bis(trimethylated) derivative to give complexes **(+)-15** and **(−)-15**. The best device performances were found using a double EL layer architecture (device **Q**), which was found to exhibit improved performance over device **P** (see Table 15.4). While the device **Q** displayed significantly improved performance, $\lvert g_{EL} \rvert$ values were found to be on the order of 10^{-3}, similar to the $\lvert g_{lum} \rvert$ values found in dichloromethane solution but substantially lower than the studies by Fuchter and coworkers [22].

The disconnection between the $\lvert g_{EL} \rvert$ values for Pt-helicene-derived CP-PHOLEDs reported by Fuchter and coworkers [22] and Zheng and coworkers [47] is curious and has not yet been reconciled. It should be noted that the former involved solution processing using a polymeric host, whereas the latter employed vacuum sublimation. In the case of fluorescent materials, particularly LEPs (Section 15.3.1), solution processing and post deposition treatment play key roles in the macromolecular structure of the active layer and the strength of $|g_{EL}|$. Whether the same is true for the phosphorescent materials and guest–host systems remains to be seen. How such differences influence the measured outcome, especially $|g_{EL}|$ from CP-OLEDs, is worthy of further study.

15.3 Enantiomers vs. Racemates: Bulk Charge Transport of Helicenes and Application to Optoelectronic Devices

While different chiral compositions (racemic, scalemic, enantiopure) of chiral molecules have mostly identical molecular properties in solution (e.g. NMR resonances, redox potentials, and optical extinction coefficients of nonpolarized light) [48], very different materials properties may emerge in the solid state. Different chiral compositions may lead to distinctly different solid-state packing structures, which govern the proximity and orientation of neighboring molecules. This packing can have a critical influence on device-relevant properties, such as charge injection and transport.

In this section we discuss examples where helicene chiral composition has been exploited to modulate charge carrier mobilities. We further highlight examples where helicene chiral composition is used to tune device function related to light absorption for CP light detection and OPVs.

15.3.1 Helicene Bulk Charge Transport and OFETs

OFETs are three-terminal devices, which constitute the basic building blocks of integrated circuits. They consist of an OSC connecting a drain and a source electrode, in either top- or bottom-contact modes; the movement of charges between which is moderated by an applied gate electrode voltage (see Figure 15.11 and Information Box 15.3).

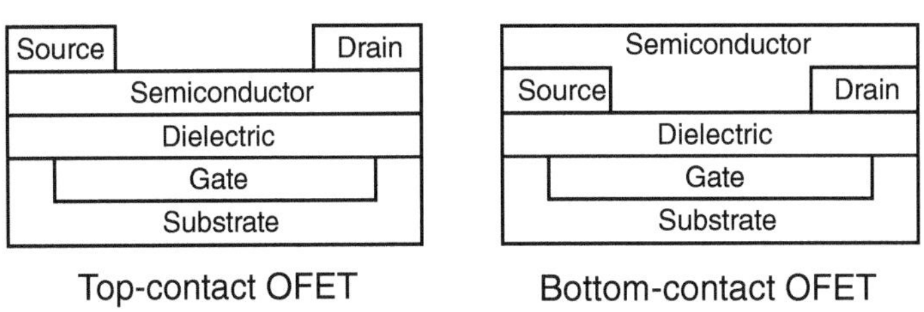

Figure 15.11 Schematic depictions of bottom-gate OFETs in top-contact (left) and bottom-contact (right) modes.

Information Box 15.3

Organic Field Effect Transistors: Fundamental Operation Principles

Organic field effect transistors (OFETs) regulate the current flow between the source and the drain electrodes as a function of applied gate–source voltage (V_{gs}) between the gate and source electrodes. When no V_{gs} is applied, the device is said to be in its "off" state. When a V_{gs} is applied, the device is in its "on" state. The current ratio between these two states (on–off ratio, $I_{on} : I_{off}$) is the defining parameter governing the switch behavior of the device along with the electron and hole charge career mobilities (μ_e and μ_h), where higher values of $I_{on} : I_{off}$, μ_e and μ_h translate to better device performance.

Further information regarding the device performance of well-known, conventional, and commercial OFET materials may be found at Ref. [86, 87].

The earliest demonstration of a helicene applied as an OFET material came in 2009, when Facchetti, Licandro, Muccini, and coworkers [49] designed and synthesized two tetrathia[7]helicenes, **16a** and **16b** (see Figure 15.12). Top-contact OFET devices were fabricated from racemic material via vapor deposition on silica and hexamethyldisilazane-treated silica substrates using Au deposited source and drain electrodes. Helicene **16a** showed no thin film transistor behavior, while helicene **16b** revealed poor hole mobility, $\mu_h = 1.7 \times 10^{-7}$ cm^2 V^{-1} s^{-1}, and an on–off ratio, $I_{on} : I_{off} = 10^2$, which were attributed to non-optimized morphology. In 2013, Li, Nuckolls, and coworkers [50] reported **17**, bearing six [4]helicenes and two [5]helicenes at its exterior. OFETs of **17** were prepared, which exhibited a much greater hole mobility of $\mu_h = 2 \times 10^{-3}$ cm^2 V^{-1} s^{-1}. In both examples, only racemic mixtures were investigated.

The first investigation into differential bulk charge transport exhibited by racemic and enantiopure helicenes occurred in 2012, when Hatakeyama, Nakamura,

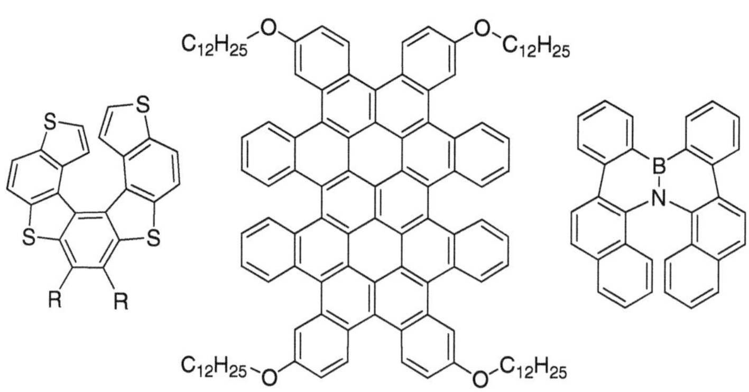

16a R = H
16b R = C_3H_7

17

18

Figure 15.12 Molecular structures of helicenes **16–18**.

and coworkers [51] designed and synthesized **18**. Thin films of **(*rac*)-18**, and **(+)-18** were fabricated using vacuum deposition. **(*rac*)-18** exhibited a hole mobility of $\mu_h = 4.6 \times 10^{-4}$ cm^2 V^{-1} s^{-1}, but exhibited no measurable electron mobility. On the other hand, **(+)-18** exhibited an enhanced hole mobility, $\mu_h = 7.9 \times 10^{-4}$ cm^2 V^{-1} s^{-1}, as well as an electron mobility of even greater magnitude, $\mu_e = 4.5 \times 10^{-3}$ cm^2 V^{-1} s^{-1}. The electronic couplings between nearest neighbors were calculated from the crystal structures, which revealed that the HOMO coupling in **(*rac*)-18** was six times greater than the LUMO coupling, while in **(+)-18** the maximum HOMO coupling was only two fifths that of the LUMO coupling. These results are in good agreement with the mobilities observed in the thin films, suggesting that the molecular orientations and intermolecular orbital coupling of the amorphous films are similar to those observed in the crystals.

In 2017, Campbell, Fuchter, and coworkers [52] further investigated the emergent properties of helicene semiconductors as a function of molecular chirality. Tapping mode AFM revealed films of **(+)-12** and **(*rac*)-12** to differ in morphology and evolve over time. Freshly prepared bottom gate–bottom-contact OFETs incorporating **(*rac*)-12** exhibited hole mobility of $\mu_h = 2.0 \times 10^{-3}$ cm^2 V^{-1} s^{-1}, 80 times greater than that displayed by **(+)-12**, $\mu_h = 2.4 \times 10^{-5}$ cm^2 V^{-1} s^{-1}. Consistent with temporal effects observed via AFM, both devices demonstrated enhanced device mobilities after seven days under ambient conditions with **(+)-12** showing a 30-fold increase ($\mu_h = 7.8 \times 10^{-4}$ cm^2 V^{-1} s^{-1}) and **(*rac*)-12** a 1.4-fold increase ($\mu_h = 2.8 \times 10^{-3}$ cm^2 V^{-1} s^{-1}). To rationalize the stark difference in device performance between **(+)-12** and **(*rac*)-12**, the authors conducted a comparative study using grazing incidence wide angle X-ray scattering, X-ray diffraction, and single crystal X-ray diffraction data in combination with crystal structure prediction methods. These studies assigned likely packing motifs to thin films of **(+)-12** and **(*rac*)-12** and qualitatively supported the device data. The highest hole mobility calculated was for **(*rac*)-12** that occurs in a plane of alternating chiral handedness, which may only occur in a racemic sample composition.

The orders of magnitude difference in charge mobilities between racemic and enantiopure materials demonstrate that the impact chiral composition can have on the carrier mobilities for solid-state electronic devices. Further structure–property relationship studies that can rationalize the impact of chiral composition on molecular packing and resultant bulk properties could provide a set of design rules for high performance materials [53, 54].

15.3.2 Helicenes for CP Photodetectors

As discussed in Section 15.2.3, CP light is central to many technologies, such as CP tomography [55, 56], spin-encoded optical communication [19], and quantum-based optical computing [57]. For such technologies, the development of devices that can detect CP light is equally as important as those that can emit CP light. Using chiral OSCs to prepare intrinsic CP detectors could lead to a range of opportunities, including the creation of arrays of micrometer-scale CP light-sensitive devices, as well as integration with complementary metal-oxide-semiconductor electronics.

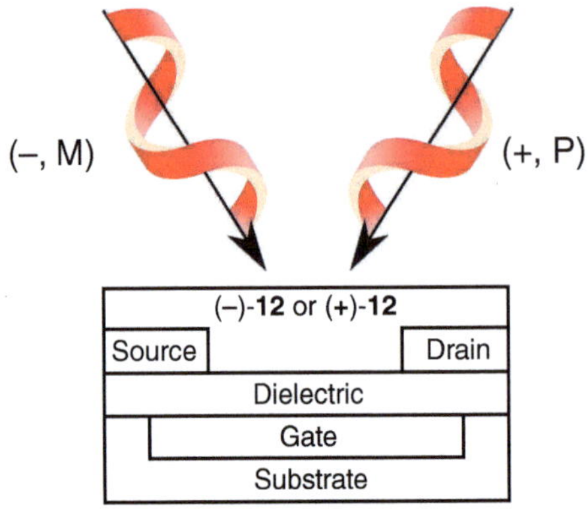

Figure 15.13 Schematic depiction of bottom-gate-bottom-contact organic phototransistor using helicene **12** for the detection of LH and RH CP light.

In 2013, Fuchter, Campbell, and coworkers [58] fabricated CP phototransistors based on helicene **12**. **(+)-12** and **(−)-12** were incorporated into bottom-contact devices as the semiconducting element (see Figure 15.13). The enantiomers displayed hole mobilities of $\mu_h = 1 \times 10^{-4}$ cm^2 V^{-1} s^{-1} and an on–off ratio, $I_{on} : I_{off} = 1 \times 10^3$. The authors then investigated the device responses upon exposure to CP light (365 nm, 10 mW cm^{-2}). When exposed to RH CP light, the off current of the **(+)-12** device increased one order of magnitude, 1×10^{-10} A to 1×10^{-9} A, while exposure to LH CP light resulted in no measurable change. The **(−)-12** device exhibited an equal and opposite response, while **(*rac*)-12** displayed no selectivity in I_{off} when exposed to LH or RH CP light. The observed CP photoresponse was reversible when the illumination source was removed (rise time of 2.6 ms, decay time 6.2 ms), showing the possibility of neat helicenes for dynamic CP light detection.

15.3.3 Helicenes for Organic Photovoltaics

Photovoltaics (PVs) provide the means to convert solar energy into electrical energy (see Information Box 15.4). While silicon-based PVs typically achieve high power conversion efficiencies (PCEs), they do not efficiently absorb across the entire solar spectrum and require extreme fabrication conditions. Organic PVs (OPVs), however, can be tuned to absorb specific wavelengths using chemical synthesis and as they can be engineered for operation under specific light conditions (e.g. infrared [IR] irradiation), offer the potential for cheap, large-area mass production on flexible substrates. The tuning of light-absorbing material properties using chiral composition has the potential to impact the function and performance of OPVs [59, 60]. Beyond OPVs, perovskite solar cells (PSCs) are a recent PV technology. Based on the ABX_3 structure, where A and B are cations, B typically representing a heavy metal such as lead and X representing a halogen anion, perovskites have enjoyed unprecedented success in attaining high PCEs (> 20 %) in a very short time compared with other PV technologies. The design of novel hole extraction layers (HELs) remains a principle strategy to improve PSC performance. The set of criteria for successful small-molecule HELs include (i) high thermal/photochemical stability, (ii) a HOMO higher in energy than the perovskite valence band, (iii) high solubility for homogeneous film formation, and (iv) a high hole mobility.

Information Box 15.4

Organic Photovoltaics: Fundamental Operation Principles

Organic solar cells operate on the basic principle of photoinduced charge transfer and separation across the interface of an electron donor (e.g. electron-rich OSC) and an electron acceptor (e.g. electron-poor OSC) sandwiched between a cathode and clear anode. Separated charges are then extracted at the cathode and anode into an external circuit through which they dissipate their energy, generating power, before returning to the solar cell. The principle parameters by which solar cells are characterized are the open circuit voltage (V_{OC}), short circuit current density (J_{SC}), fill factor (FF), and power conversion efficiency (PCE). V_{OC} is defined as the maximum achievable voltage upon solar irradiation, and is found when no external load is connected between the terminals. J_{SC} is defined as the current density, which flows through the solar cell upon irradiation when the voltage across the cell equals zero (i.e. when the solar cell is short circuited). The product of ($V_{OC} \times J_{SC}$) corresponds to the theoretical maximum power of a solar cell under ideal conditions. The ratio of the actual maximum obtainable power (P_{MP}) to this product is known as the fill factor:

$$\mathrm{FF} = \frac{P_{\mathrm{MP}}}{V_{\mathrm{OC}} \times J_{\mathrm{SC}}}$$

Finally, the PCE is defined as the fraction incident solar power (P_{in}) converted into electricity, such that higher values of V_{OC}, J_{SC}, and FF correspond to higher PCEs:

$$\mathrm{PCE} = \frac{V_{\mathrm{OC}} \times J_{\mathrm{SC}} \times \mathrm{FF}}{P_{\mathrm{in}}}$$

The bulk heterojunction architecture, consisting of the donor and acceptor materials blended together in the same layer to create distinct donor and acceptor domains, is critical for efficient OPV operation. Optimization to create donor and acceptor domains with dimensions smaller than the exciton diffusion length (i.e. the length traversed by a photoexciton within its excited state lifetime) is necessary to allow for efficient conversion of photons into free electrons and holes at the donor/acceptor interfaces. The donor material must also have good charge collection pathways to the anode and similarly for the acceptor material to the cathode to ensure efficient collection of photogenerated charges.

Further information regarding the device operation and performance of well-known, conventional, and commercial OPV materials may be found at Ref. [88].

In 2013 Li, Nuckolls and coworkers [50] demonstrated the utility of **17** in OPV applications. **17**/$PC_{70}BM$ blends were incorporated into device **R** (see Figure 15.14), achieving performance shown in Table 15.5. It should be noted that although the device demonstrated a high V_{OC}, the PCE is low due to the small fill factor (FF).

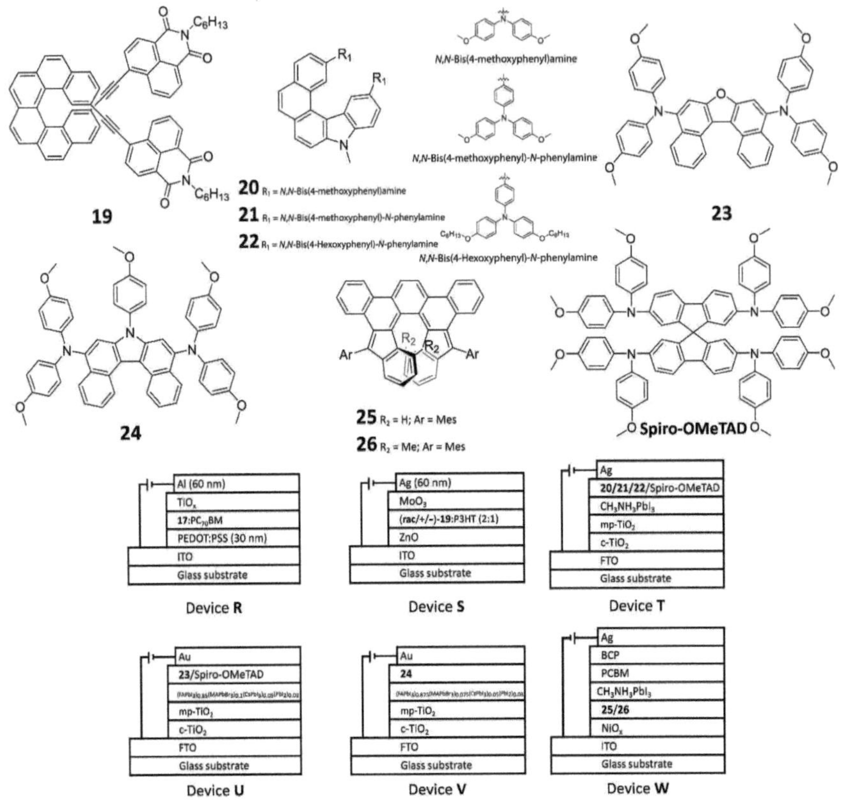

Figure 15.14 Molecular structures of helicenes **19–26** and Spiro-OMeTAD (top) and schematic depictions of the photovoltaic devices **R–W** to which they have been applied as electron acceptors and HELs (bottom).

In 2017, Favereau, Blanchard, Cabanetos, Crassous, and coworkers [61] demonstrated the potential of helicenes to act as electron acceptor components in OPVs (see Figure 15.14). Racemic and enantiopure **19**, consisting of two naphthalene diimides appended to a carbo[6]helicene scaffold, were blended with the polymer, poly(3-hexylthiophene-2,5-diyl) (P3HT), which were then incorporated into device **S**. The performance data are compared in Table 15.5, which show all-round improved performance in the case of the enantiopure material. The authors attributed the lower J_{SC} in the racemic device to the reduced electron mobility ($\mu_e = 1.8 \times 10^{-4}$ cm^2 V^{-1} s^{-1}) compared with that found in the enantiopure blend devices ($\mu_e = 4 \times 10^{-4}$ cm^2 V^{-1} s^{-1}), which contributes to overall reduced performance. This was in turn attributed to the specific interactions of nearest neighbor chiral molecules in the different blends.

The ability of helicenes to act as HELs for PSCs was demonstrated in 2018 by Lin, Tao, Sun, and coworkers [62]. Aza[5]helicene derivatives **20**, **21**, and **22** (see Figure 15.14) were synthesized and exhibited HOMO energy levels of −4.82, −4.95, and −4.94 eV, respectively, compatible for operation with the well-known perovskite, $CH_3NH_3PbI_3$. Substantially variable material properties were observed,

Table 15.5 OPV device performance data from devices Q and R for helicenes **17–26**.

Helicene	Device	Open-circuit voltage (V)	Short-circuit current ($mA\ cm^{-2}$)	Fill factor	Power conversion efficiency (%)
(*rac*)-17	**R**	0.98	7.9	0.37	2.86
(*rac*)-19	**S**	0.41	3.48	0.29	0.41
(−)-19	**S**	0.62	6.91	0.49	2.09
(+)-19	**S**	0.62	6.85	0.47	2.00
20a	**T**	1.01	22.73	0.71	16.20
20b	**T**	0.97	21.71	0.68	14.46
20c	**T**	0.93	9.73	0.28	2.50
Spiro-OMeTAD	**T**	1.00	21.83	0.76	16.57
23	**U**	1.08	24.48	0.79	21.03
Spiro-OMeTAD	**U**	1.09	24.53	0.76	20.44
24	**V**	1.12	24.31	0.77	21.0
NiO_x/25	**W**	1.03	21.85	0.80	18.1
NiO_x/26	**W**	1.03	22.48	0.79	18.2
NiO_x/no helicene	**W**	1.04	19.39	0.77	15.6

with hole mobilities of $\mu_h = 2.55 \times 10^{-5}$, 3.46×10^{-5}, and $2.93 \times 10^{-7}\ cm^2\ V^{-1}\ s^{-1}$ and glass transition temperatures of $T_g = 116$, 133, and 49 °C, respectively. Helicenes **20–22** were incorporated into device **T**, and their performances were compared to the benchmark HEL material, Spiro-OMeTAD (see Table 15.5). Although Spiro-OMeTAD exhibited slightly higher PCEs, helicenes **20** and **21** demonstrated far improved device stability, retaining approximately 80% of their initial device efficiency upon exposure to harsh conditions (80% humidity at 35 °C for 72 hours) compared with approximately 42% for Spiro-OMeTAD. This improved stability imparted by **20** and **21** was attributed to higher thin film hydrophobicity, which was confirmed by water contact angle measurements, providing greater protection for the perovskite layer from moisture and oxygen. The inferior device performance of **22** was attributed to poor film morphology arising from its low T_g.

In 2019, De Angelis, Wang, and coworkers applied oxa[5]helicene **23** in HELs for PSCs [63], exhibiting an HOMO energy level of −5.29 eV and hole mobility of $\mu_h = 3.3 \times 10^{-5}\ cm^2\ V^{-1}\ s^{-1}$, five times greater than Spiro-OMeTAD. **23** was incorporated into device **U** with perovskite $(FABbI_3)_{0.85}(MAPbBr_3)_{0.10}(CsPbI_3)_{0.05}(PbI_2)_{0.03}$ (see Table 15.5). Not only did **23** exhibit improved device performance compared to Spiro-OMeTAD with an excellent PCE but also higher device stability under harsh conditions (illumination under 1 sun at 60 °C), which was attributed to superior photo- and thermal stability of **23** and superior film morphology. Wang and coworkers simultaneously demonstrated another aza[5]helicene, **24**, as an efficient HEL [64] for perovskite $(FABbI_3)_{0.85}(MAPbBr_3)_{0.10}(CsPbI_3)_{0.05}(PbI_2)_{0.03}$ (see Table 15.5), with an HOMO energy level of −5.21 eV and hole mobility $4.1 \times 10^{-5}\ cm^2\ V^{-1}\ s^{-1}$.

Device **V** exhibited an excellent PCE of 21%, higher than the 20.3% obtained for Spiro-OMeTAD, which was attributed to greater hole mobility and attenuated interfacial hole recombination.

Wu, Chueh, and coworkers applied singlet biradical helicenes **25** and **26** as surface modifiers to the HTL, NiO_x, for PSCs [65]. The helicene-modified NiO_x layers exhibited higher hydrophobicity promoting larger perovskite crystal domains, greater homogeneous film formation, and more efficient perovskite photoluminescence quenching relative to unmodified NiO_x. The improved device performance in device **W** (see Table 15.5) was attributed to their stable partial biradical character, which provides interfacial defect passivation enhancing charge extraction.

Helicenes clearly demonstrate excellent potential for PV device function. Although OPV devices achieve performances well below the current state of the art, the ability of helicenes to be used in energy harvesting applications is still shown. In particular, the study by Favereau, Blanchard, Cabanetos, Crassous, and coworkers [61] demonstrates that chiral composition may be used to tune device performance, where, in this case, enantiopure materials exhibit superior energy conversion efficiencies compared to racemic. The recent advances in helicene application for HELs in perovskite technologies are further cause for optimism regarding the future for helicenes in PV applications. Their 3D molecular structures and thin film packing clearly offer unique means to tune hole mobilities and influence perovskite stability, morphology, and interfacial charge transfer efficiencies.

15.4 Helicenes and the Chiral-Induced Spin Selectivity Effect

The relationship between chirality and magnetism is well established through effects such as magneto-chiral dichroism [66] and electrical magneto-chiral anisotropy [67]. Interestingly, the equations governing these effects may also be used to describe the relationship between chirality and electron spin. This has implications for the field of organic spintronics, in which the electron spin degree of freedom ($\mu_S = +½, -½$), in addition to that of the electron charge, is exploited for encoding information [68]. As the energy required to affect electron spin flipping is low, higher power densities may be achieved in spin-based devices, with concurrent lower power consumptions.

OSCs are typically composed of light elements with low spin–orbit coupling, which results in long spin relaxation times [69]. Consequently, spin information is thought to be retained in OSCs for longer than their inorganic counterparts, which renders them attractive candidates for passive elements in spintronic applications. However, this low spin–orbit coupling makes them unsuitable for spin-polarized current injection and organic spintronic devices typically require ferromagnetic electrodes [68].

Chiral molecules have recently been proposed to offer an exciting new strategy to realize spin-polarized electron currents by means of the chiral-induced spin selectivity (CISS) effect [70]. Several reviews [71–74] have been written explicating the CISS effect along with a number of theoretical models [75–79], which attempt to explain

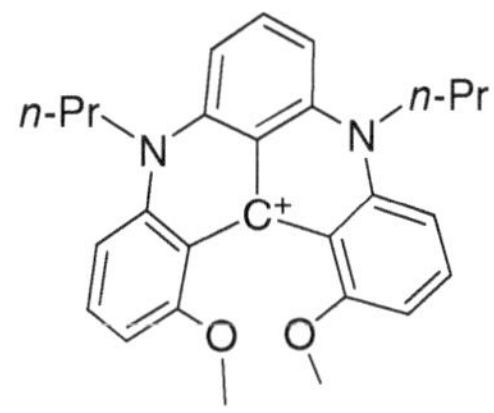

27

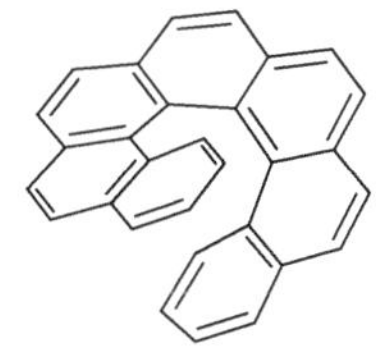

28

Figure 15.15 Molecular structures of helicenes **27** and **28.**

its fundamental origins. However, as the true origins of the CISS effect are still contested, here we simply state its result, which is to couple an electron's spin to its direction of motion. CISS, therefore, enables chiral molecules to realize spin-polarized currents by spin filtering out those electrons with spins coupled opposite to the direction of current flow, which may itself be controlled by voltage bias.

Though first demonstrated in polymeric biological (and insulating) materials, such as DNA [70, 80] and peptides [81], in 2016, Naaman, Lacour, and coworkers [82] demonstrated the ability of small-molecule helicenes to act as spin filters (see Figure 15.15). Magnetic conductive probe AFM was used to measure the conduction through spin cast films of enantiopure helicene **27** deposited on atomically flat highly orientated pyrolytic graphite substrates. Spin selectivity studies were then carried out using a magnetic (iron-coated silicon) tip for the injection of spin-polarized electrons. Upon application of externally applied parallel and antiparallel magnetic fields, spin polarization at a 1.0 V bias was found to be $+(49 \pm 3)\%$ and $-(45 \pm 3)\%$ for **(+)-27** and **(−)-27**, respectively (similar to the 60% achieved for DNA [70]).

In 2018, Ernst, Zacharias, and coworkers [83] further investigated the electron spin filtering effect of helicenes, employing the pure hydrocarbon, carbo[7]helicene **28** (see Figure 15.15). Monolayers of **28** were deposited on copper surfaces. The spin-polarized transmission of photoelectrons, P_z, generated by ultraviolet laser pulses through the monolayers was then monitored using a Mott polarimeter. Upon excitation with nonpolarized and LH and RH CP light, bare copper surfaces emitted electrons with essentially zero spin polarization. However, copper surfaces coated in **(+)-28** and **(−)-28** gave spin polarizations $P_z = +(5.5 \pm 0.6)\%$ and $P_z = -(6.7 \pm 0.6)\%$, respectively. Further experiments with LH and RH CP light and with metal substrates (silver and gold) with stronger degrees of intrinsic spin–orbit coupling confirmed that the photoelectron spin polarization is dictated by the chirality of the helicene and that substrate-mediated spin–orbit coupling affords no significant contribution. Furthermore, theoretical and experimental studies on the hybridization of molecular and metallic substrate states confirmed that the photoelectrons are not emitted from occupied **28** states, providing further confirmation that the spin polarization is a result of transmission through the helicene monolayers.

The proposal that helicenes may act as spin filters generates a wide range of spintronic device applications, potentially eliminating the need for complicated ferromagnetic electrodes and magnetic elements. The bifunctional ability of helicenes to act as charge transport OSCs as well as spin filters could also have further implications in the field of spin-OLEDs. In such devices spin-polarized charge injection could increase the probability of singlet exciton formation, enhancing the theoretical maximum IQE from 25 to 50%, in the case of perfect spin polarization [84].

15.5 Conclusion

To date, the exploration of helicenes for organic electronic devices has been one driven mostly by curiosity. However, as summarized in this chapter, helicenes offer both excellent performance and novel functionality including the emission and detection of CP light, the tuning of charge carrier mobilities and the spin polarization of electron currents. Not only does chirality and chiral composition offer additional degrees of freedom when designing new OSCs from a molecular standpoint, importantly, it allows for alteration of material properties too. Multidisciplinary collaborative efforts between chemists, physicists, and computational and materials scientists are essential to further understand and exploit the emergent properties of helicenes. Particularly exciting is their dual capability to act as small-molecule OSC and to control photon and electron spin at room temperature. Indeed, it is in this domain that helicenes may provide their greatest impact, enabling scientists to simultaneously manipulate the electronic and spintronic properties of OSCs for the realization of next-generation optoelectronic devices.

Abbreviations

26DCzppy	2,6-bis[3-(9*H*-carbazol-9-yl)phenyl]pyridine
AFM	atomic force microscopy
BCP	2,9-dimethyl-4,7-diphenyl-1,10-phenanthroline
BPhen	4,7-diphenyl-1,10-phenanthroline
CBP	4,4′-bis(*N*-carbazolyl)-1,1′-biphenyl
CD	circular dichroism
CISS	chiral induced spin selectivity effect
CP	circularly polarized
CP-OLED	circularly polarized organic light-emitting diode
CP-PhOLED	circularly polarized phosphorescence organic light emitting diode
EL	electroluminescence
eV	electron volt
F8BT	poly[9,9-dioctylfluorene-co-benzothiadiazole]
FF	fill factor
g Factor	dissymmetry factor
HATCH	dipyrazino[2,3-*f*:2′,3′-*h*]quinoxaline-2,3,6,7,10,11-hexacarbonitrile
HEL	hole extraction layer
HOMO	highest occupied molecular orbital
HPLC	high performance liquid chromatography

HTL	hole transport layer
HTM	hole transport material
I_{off}	off current
I_{on}	on current
IQE	internal quantum efficiency
ITO	indium tin oxide
J_{SC}	short circuit current density
LEP	light-emitting polymer
LiF	lithium fluoride
LUMO	lowest occupied molecular orbital
MoO_3	molybdenum trioxide
NPB	*N*,*N*′-di(1-naphthyl)-*N*,*N*′-diphenyl-(1,1′-biphenyl)-4,4′-diamine
OFET	organic field effect transistor
OLED	organic light-emitting diode
OSC	organic semiconductor
OXD-7:PVK	1,3-bis[2-(4-tert-butylphenyl)-1,3,4-oxadiazo-5-yl]benzene:poly (9-vinylcarbazole)
P3HT	poly(3-hexylthiophene-2,5-diyl)
PBD	2-(4-tert-butylphenyl)-5-(4-biphenylyl)-1,3,4-oxadiazole
$PC_{70}BM$	[6,6]-phenyl C_{70} butyric acid methyl ester
PCE	power conversion efficiency
PEDOT:PSS	poly(3,4-ethylenedioxythiophene):polystyrene sulfonate
PEIE	polyethylenimine ethoxylate
PhOLED	phosphorescence light-emitting diode
PL	photoluminescence
PSC	perovskite solar cell
PTPA	3,6-bis(triphenylamino)phenanthrene
PV	photovoltaic
TAPC	(di-[4-(*N*,*N*-ditolyl-amino)-phenyl] cyclohexane
TCTA	4,4′,4″-tris(carbazol-9-yl)triphenylamine
T_d	degradation temperature
TFB	poly(9,9-dioctylfluorene-alt-*N*-(4-sec-butylphenyl) diphenylamine)
T_g	glass transition temperature
T_m	melting temperature
Tm3PyP26PyB	1,3,5-tris(6-(3-(pyridin-3-yl)phenyl)pyri-din-2-yl)benzene
TPBI	2,2′,2″-(1,3,5-benzinetriyl)tris(1-phenyl-1*H*-benzimidazole)
V_{gs}	gate-source voltage
V_{OC}	open circuit voltage
wt%	weight percent
ZnO	zinc oxide
λ_{max}	peak maximum wavelength
μ_e	electron field effect mobility
μ_h	hole field effect mobility

References

1 Shen, Y. and Chen, C.F. (2012). Helicenes: synthesis and applications. *Chem. Rev.* 112 (3): 1463–1535.

2 Köhler, A. and Bässler, H. (2015). *Electronic Processes in Organic Semiconductors*. Weinheim, Germany: Wiley-VCH Verlag GmbH & Co. KGaA.

3 Brandt, J.R., Salerno, F., and Fuchter, M.J. (2017). The added value of small-molecule chirality in technological applications. *Nat. Rev. Chem.* 1 (6): 0045.

4 Chen, N. and Yan, B. (2018). Recent theoretical and experimental progress in circularly polarized luminescence of small organic molecules. *Molecules* 23 (12): 3376.

5 Kumar, J., Nakashima, T., and Kawai, T. (2015). Circularly polarized luminescence in chiral molecules and supramolecular assemblies. *J. Phys. Chem. Lett.* 6 (17): 3445–3452.

6 Sánchez-Carnerero, E.M., Agarrabeitia, A.R., Moreno, F. et al. (2015). Circularly polarized luminescence from simple organic molecules. *Chem. Eur. J.* 21 (39): 13488–13500.

7 Zhao, W.L., Li, M., Lu, H.Y., and Chen, C.F. (2019). Advances in helicene derivatives with circularly polarized luminescence. *Chem. Commun.* 55 (92): 13793–13803.

8 Isla, H. and Crassous, J. (2016). Helicene-based chiroptical switches. *Comptes Rendus Chim.* 19 (1–2): 39–49.

9 Dhbaibi, K., Favereau, L., and Crassous, J. (2019). Enantioenriched helicenes and helicenoids containing main-group elements (B, Si, N, P). *Chem. Rev.* 119 (14): 8846–8953.

10 Tang, C.W. and Vanslyke, S.A. (1987). Organic electroluminescent diodes. *Appl. Phys. Lett.* 51 (12): 913–915.

11 Sahasithiwat, S., Mophuang, T., Menbangpung, L. et al. (2010). 3,12-Dimethoxy-7,8-dicyano-[5]helicene as a novel emissive material for organic light-emitting diode. *Synth. Met.* 160: 1148–1152.

12 Chen, J.P. (2003). Helicene and dibenzofluorene materials for use in organic light emitting devices. US 2003.0143422A1, issued 2003.

13 Shi, L., Liu, Z., Dong, G. et al. (2012). Synthesis, structure, properties, and application of a carbazole-based diaza[7]helicene in a deep-blue-emitting OLED. *Chem. Eur. J.* 18 (26): 8092–8099.

14 Hua, W., Liu, Z., Duan, L. et al. (2015). Deep-blue electroluminescence from nondoped and doped organic light-emitting diodes (OLEDs) based on a new monoaza[6]helicene. *RSC Adv.* 5 (1): 75–84.

15 Jhulki, S., Mishra, A.K., Chow, T.J., and Moorthy, J.N. (2016). Helicenes as all-in-one organic materials for application in OLEDs: synthesis and diverse applications of carbo- and aza[5]helical diamines. *Chem. Eur. J.* 22 (27): 9375–9386.

16 Tanaka, H., Inoue, Y., and Mori, T. (2018). Circularly polarized luminescence and circular dichroisms in small organic molecules: correlation between excitation and emission dissymmetry factors. *ChemPhotoChem* 2 (5): 386–402.

17 Kim, D.-Y. (2006). Potential application of spintronic light-emitting diode to binocular vision for three-dimensional display technology. *J. Korean Phys. Soc.* 49 (9): 505.

18 Wagenknecht, C., Li, C.M., Reingruber, A. et al. (2010). Experimental demonstration of a heralded entanglement source. *Nat. Photonics* 4 (8): 549–552.

19 Farshchi, R., Ramsteiner, M., Herfort, J. et al. (2011). Optical communication of spin information between light emitting diodes. *Appl. Phys. Lett.* 98 (16): 162508.

20 Heffern, M.C., Matosziuk, L.M., and Meade, T.J. (2014). Lanthanide probes for bioresponsive imaging. *Chem. Rev.* 114 (8): 4496–4539.

21 Zhang, X., Yin, J., and Yoon, J. (2014). Recent advances in development of chiral fluorescent and colorimetric sensors. *Chem. Rev.* 114 (9): 4918–4959.

22 Brandt, J.R., Wang, X., Yang, Y. et al. (2016). Circularly polarized phosphorescent electroluminescence with a high dissymmetry factor from PHOLEDs based on a platinahelicene. *J. Am. Chem. Soc.* 138 (31): 9743–9746.

23 Peeters, E., Christiaans, M.P.T., Janssen, R.A.J. et al. (1997). Circularly polarized electroluminescence from a polymer light-emitting diode. *J. Am. Chem. Soc.* 119 (41): 9909–9910.

24 Geng, Y., Trajkovska, A., Culligan, S.W. et al. (2003). Origin of strong chiroptical activities in films of nonafluorenes with a varying extent of pendant chirality. *J. Am. Chem. Soc.* 125 (46): 14032–14038.

25 Oda, M., Nothofer, H.-G., Lieser, G. et al. (2000). Circularly polarized electroluminescence from liquid-crystalline chiral polyfluorenes. *Adv. Mater.* 12 (5): 362–365.

26 Wilson, J.N., Steffen, W., McKenzie, T.G. et al. (2002). Chiroptical properties of poly(p-phenyleneethynylene) copolymers in thin films: large g-values. *J. Am. Chem. Soc.* 124 (24): 6830–6831.

27 Meskers, S.C.J., Peeters, E., Langeveld-Voss, B.M.W., and Janssen, R.A.J. (2000). Circular polarization of the fluorescence from films of poly(p-phenylene vinylene) and polythiophene with chiral side chains. *Adv. Mater.* 12 (8): 589–594.

28 Yang, Y., da Costa, R.C., Smilgies, D.-M. et al. (2013). Induction of circularly polarized electroluminescence from an achiral light-emitting polymer via a chiral small-molecule dopant. *Adv. Mater.* 25 (18): 2624–2628.

29 Wan, L., Wade, J., Salerno, F. et al. (2019). Inverting the handedness of circularly polarized luminescence from light-emitting polymers using film thickness. *ACS Nano* 13 (7): 8099–8105.

30 Lee, D.-M., Song, J.-W., Lee, Y.-J. et al. (2017). Control of circularly polarized electroluminescence in induced twist structure of conjugate polymer. *Adv. Mater.* 29 (2): 1700907.

31 Wan, L., Wade, J., Shi, X. et al. (2020). Highly efficient inverted circularly polarized organic light-emitting diodes. *ACS Appl. Mater. Interfaces* 12 (35): 39471–39478.

32 Wade, J., Hilfiker, J.N., Brandt, J.R. et al. (2020). Natural optical activity as the origin of the large chiroptical properties in π-conjugated polymer thin films. *Nat. Commun.* 11: 6137.

33 Arteaga, O. (2016). Natural optical activity vs circular Bragg reflection studied by Mueller matrix ellipsometry. *Thin Solid Films* 617: 14–19.

34 Lakhwani, G. and Meskers, S.C.J. (2011). Circular selective reflection of light proving cholesteric ordering in thin layers of chiral fluorene polymers. *J. Phys. Chem. Lett.* 2 (13): 1497–1501.

35 Wright, D.C. and Mermin, N.D. (1989). Crystalline liquids: the blue phases. *Rev. Mod. Phys.* 61 (2): 385–432.

36 Hamley, I.W. (2010). Liquid crystal phase formation by biopolymers. *Soft Matter* 6 (9): 1863–1871.

37 Forrest, S.R. and O'Brien, D.F. (1999). Excitonic singlet-triplet ratio in a semi-conducting organic thin film. *Phys. Rev. B* 60 (20): 14422–14428.

38 Baldo, M.A., O'Brien, D.F., You, Y. et al. (1998). Highly efficient phosphorescent emission from organic electroluminescent devices. *Nature* 395 (6698): 151–154.

39 Zaen, R., Park, K., Lee, K.H. et al. (2019). Blue phosphorescent Ir(III) complexes achieved with over 30% external quantum efficiency. *Adv. Opt. Mater.* 7 (24): 1901387.

40 Lee, C.W. and Lee, J.Y. (2013). Above 30% external quantum efficiency in blue phosphorescent organic light-emitting diodes using pyrido[2,3-b]indole derivatives as host materials. *Adv. Mater.* 25 (38): 5450–5454.

41 Yan, Z.P., Liao, K., Han, H.B. et al. (2019). Chiral iridium(III) complexes with four-membered Ir-S-P-S chelating rings for high-performance circularly polarized OLEDs. *Chem. Commun.* 55 (57): 8215–8218.

42 Han, J., Guo, S., Wang, J. et al. (2017). Circularly polarized phosphorescent electroluminescence from chiral cationic iridium(III) isocyanide complexes. *Adv. Opt. Mater.* 5 (22): 1700359.

43 Zinna, F., Giovanella, U., and Di Bari, L. (2015). Highly circularly polarized electroluminescence from a chiral europium complex. *Adv. Mater.* 27 (10): 1791–1795.

44 Li, T.Y., Jing, Y.M., Liu, X. et al. (2015). Circularly polarised phosphorescent photoluminescence and electroluminescence of iridium complexes. *Sci. Rep.* 5 (1): 14912.

45 Norel, L., Rudolph, M., Vanthuyne, N. et al. (2010). Metallahelicenes: easily accessible helicene derivatives with large and tunable chiroptical properties. *Angew. Chem. Int. Ed.* 49 (1): 99–102.

46 Shen, C., Anger, E., Srebro, M. et al. (2014). Straightforward access to mono- and bis-cycloplatinated helicenes displaying circularly polarized phosphorescence by using crystallization resolution methods. *Chem. Sci.* 5 (5): 1915–1927.

47 Yan, Z., Luo, X., Liu, W. et al. (2019). Configurationally stable platinahelicene enantiomers for efficient circularly polarized phosphorescent organic light-emitting diodes. *Chem. Eur. J.* 25 (22): 5672–5676.

48 Jaques, A.C.J., Collet, A., and Wilen, S.H. (1981). *Enantiomers, Racemates, and Resolutions*. New York: Wiley.

49 Kim, C., Marks, T.J., Facchetti, A. et al. (2009). Synthesis, characterization, and transistor response of tetrathia-[7]-helicene precursors and derivatives. *Org. Electron.* 10: 1511–1520.

50 Xiao, S., Kang, S.J., Wu, Y. et al. (2013). Supersized contorted aromatics. *Chem. Sci.* 4 (5): 2018–2023.

51 Hatakeyama, T., Hashimoto, S., Oba, T., and Nakamura, M. (2012). Azabo-radibenzo[6]helicene: carrier inversion induced by helical homochirality. *J. Am. Chem. Soc.* 134 (48): 19600–19603.

52 Yang, Y., Rice, B., Shi, X. et al. (2017). Emergent properties of an organic semiconductor driven by its molecular chirality. *ACS Nano* 11 (8): 8329–8338.

53 Salerno, F., Rice, B., Schmidt, J.A. et al. (2019). The influence of nitrogen position on charge carrier mobility in enantiopure aza[6]helicene crystals. *Phys. Chem. Chem. Phys.* 21 (9): 5059–5067.

54 Rice, B., Leblanc, L.M., Otero-De-La-Roza, A. et al. (2018). A computational exploration of the crystal energy and charge-carrier mobility landscapes of the chiral [6]helicene molecule. *Nanoscale* 10 (4): 1865–1876.

55 Yu, C.J., Lin, C.E., Yu, L.P., and Chou, C. (2009). Paired circularly polarized heterodyne ellipsometer. *Appl. Opt.* 48 (4): 758–764.

56 Jan, C.-M., Lee, Y.-H., Wu, K.-C., and Lee, C.-K. (2011). Integrating fault tolerance algorithm and circularly polarized ellipsometer for point-of-care applications. *Opt. Express* 19 (6): 5431–5441.

57 Sherson, J.F., Krauter, H., Olsson, R.K. et al. (2006). Quantum teleportation between light and matter. *Nature* 443 (7111): 557–560.

58 Yang, Y., Correa Da Costa, R., Fuchter, M.J., and Campbell, A.J. (2013). Circularly polarized light detection by a chiral organic semiconductor transistor. *Nat. Photonics* 7: 634–638.

59 Zerdan, R.B., Shewmon, N.T., Zhu, Y. et al. (2014). The influence of solubilizing chain stereochemistry on small molecule photovoltaics. *Adv. Funct. Mater.* 24 (38): 5993–6004.

60 Sugawara, K., Nakamura, N., Yamane, Y. et al. (2016). Influence of chirality on the cyclohexene-fused C60 fullerene derivatives as an accepter partner in a photovoltaic cell. *Green Energy Environ.* 1 (2): 149–155.

61 Josse, P., Favereau, L., Shen, C. et al. (2017). Enantiopure versus racemic naphthalimide end-capped helicenic non-fullerene electron acceptors: impact on organic photovoltaics performance. *Chem. Eur. J.* 23 (26): 6277–6281.

62 Lin, Y.S., Abate, S.Y., Lai, K.W. et al. (2018). New helicene-type hole-transporting molecules for high-performance and durable perovskite solar cells. *ACS Appl. Mater. Interfaces* 10 (48): 41439–41449.

63 Xu, N., Li, Y., Ricciarelli, D. et al. (2019). An oxa[5]helicene-based racemic semiconducting glassy film for photothermally stable perovskite solar cells. *iScience* 15: 234–242.

64 Wang, J., Shi, H., Xu, N. et al. (2020). Aza[5]helicene rivals N-annulated perylene as π-linker of D–π–D typed hole-transporters for perovskite solar cells. *Adv. Funct. Mater.* 30 (30): 2002114.

65 Lee, C.C., Chen, C.I., Te Fang, C. et al. (2019). Improving performance of perovskite solar cells using [7]helicenes with stable partial biradical characters as the hole-extraction layers. *Adv. Funct. Mater.* 29 (13): 1808625.

66 Train, C., Gruselle, M., and Verdaguer, M. (2011). The fruitful introduction of chirality and control of absolute configurations in molecular magnets. *Chem. Soc. Rev.* 40 (6): 3297–3312.

67 Pop, F., Auban-Senzier, P., Canadell, E. et al. (2014). Electrical magnetochiral anisotropy in a bulk chiral molecular conductor. *Nat. Commun.* 5: 3757.

68 Sanvito, S. (2011). Molecular spintronics. *Chem. Soc. Rev.* 40 (6): 3336–3355.

69 Wang, S.J., Venkateshvaran, D., Mahani, M.R. et al. (2019). Long spin diffusion lengths in doped conjugated polymers due to enhanced exchange coupling. *Nat. Electron.* 2 (3): 98–107.

70 Göhler, B., Hamelbeck, V., Markus, T.Z. et al. (2011). Spin selectivity in electron transmission through self-assembled monolayers of double-stranded DNA. *Science* 331: 894–897.

71 Naaman, R. and Waldeck, D.H. (2012). Chiral-induced spin selectivity effect. *J. Phys. Chem. Lett.* 3 (16): 2178–2187.

72 Naaman, R. and Waldeck, D.H. (2015). Spintronics and chirality: spin selectivity in electron transport through chiral molecules. *Annu. Rev. Phys. Chem.* 66 (1): 263–281.

73 Naaman, R., Paltiel, Y., and Waldeck, D.H. (2019). Chiral molecules and the electron spin. *Nat. Rev. Chem.* 3 (4): 250–260.

74 Naaman, R., Paltiel, Y., and Waldeck, D.H. (2020). Chiral molecules and the spin selectivity effect. *J. Phys. Chem. Lett.* 11: 3660–3666.

75 Eremko, A.A. and Loktev, V.M. (2013). Spin sensitive electron transmission through helical potentials. *Phys. Rev. B* 88: 165409.

76 Dalum, S. and Hedegård, P. (2019). Theory of chiral induced spin selectivity. *Nano Lett.* 19 (8): 5253–5259.

77 Pan, T.-R., Guo, A.-M., and Sun, Q.-F. (2016). Spin-polarized electron transport through helicene molecular junctions. *Phys. Rev. B* 94: 235448.

78 Matityahu, S., Utsumi, Y., Aharony, A. et al. (2016). Spin-dependent transport through a chiral molecule in the presence of spin-orbit interaction and nonunitary effects. *Phys. Rev. B* 93: 75407.

79 Varela, S., Mujica, V., and Medina, E. (2016). Effective spin-orbit couplings in an analytical tight-binding model of DNA: spin filtering and chiral spin transport. *Phys. Rev. B* 93: 155436.

80 Zwang, T.J., Hürlimann, S., Hill, M.G., and Barton, J.K. (2016). Helix-dependent spin filtering through the DNA duplex. *J. Am. Chem. Soc.* 138 (48): 15551–15554.

81 Aragonès, A.C., Medina, E., Ferrer-Huerta, M. et al. (2017). Measuring the spin-polarization power of a single chiral molecule. *Small* 13 (2): 1602519.

82 Kiran, V., Mathew, S.P., Cohen, S.R. et al. (2016). Helicenes-a new class of organic spin filter. *Adv. Mater.* 28 (10): 1957–1962.

83 Kettner, M., Maslyuk, V.V., Nürenberg, D. et al. (2018). Chirality-dependent electron spin filtering by molecular monolayers of helicenes. *J. Phys. Chem. Lett.* 9 (8): 2025–2030.

84 Alek Dediu, V., Hueso, L.E., Bergenti, I., and Taliani, C. (2009). Spin routes in organic semiconductors. *Nat. Mater.* 8: 707–716.

85 Koden, M. (2016). *OLED Display and Lighting*. Chichester, UK: Wiley-IEEE Press.

86 Wang, C., Jiang, L., and Hu, W. (2013). *Organic/Polymeric Field-Effect Transistors, in Organic Optoelectronics*, 95–170. Weinheim, Germany: Wiley-VCH Verlag GmbH & Co. KGaA.

87 Janasz, Ł., Ulański, J., and Pisula, W. (2019). Organic field-effect transistors based on nanostructured blends. In: *Solution-processable Components for Organic Electronic Devices*, 365–412. Weinheim, Germany: Wiley-VCH Verlag GmbH & Co. KGaA.

88 Brabec, C., Scherf, U., and Dyakonov, V. (2014). *Organic Photovoltaics*. Weinheim, Germany: Wiley-VCH Verlag GmbH & Co. KGaA.

16

Miscellaneous Applications of Helicenes: Surface Science and Asymmetric Catalysis

Jeanne Crassous

Institut des Sciences Chimiques de Rennes, CNRS, ISCR – UMR 6226, Université de Rennes 1 Rennes, France

16.1 Introduction

Being prototypic functionnalized helical polyaromatic hydrocarbons, helicenes and heteroatom-containing derivatives may exhibit enantiodiscrimination that appears appealing to study in different domains. For example, the homochiral/heterochiral assembly of helicenes on metallic surfaces or the degree of enantiodiscriminination that can be reached when using helicenes as chiral catalytic species are of great interest, not only from a fundamental viewpoint but also for applicative reasons. In the first part of this chapter, few selected works addressing the most important features of self-assembly of helicenes onto metallic surfaces, especially on copper, gold, and silver, are presented. In the second part, the catalytic activity of a diversity of helicenic derivatives (bearing C, P, S, N, and O atoms) used in selected examples of metal-based catalysis or in organocatalysis are presented. Through this chapter, the potentiality of helicenes in these two different fields of research is thus demonstrated.

16.2 Self-Assembly of Helicenes onto Surfaces

Helicenes have recently revealed interest as semiconductors for applications in organic electronic devices, such as transistors [1], CP-OLEDs [2, 3], or as spin filters [4, 5] (see Chapter 15). For this reason, the way helicenes can self-assemble onto surfaces is attracting much attention. Aside from applicative domains, there are interesting fundamental features to study regarding the 2D self-assembly of helical molecules. In their pure crystalline forms, carbo[n]helicenes ($n = 4 - 9$) are known to crystallize as conglomerates [6–8]. How this behaviour translates onto different types of surfaces is thus of great interest. The technique of choice for such studies is scanning tunneling microscopy (STM) coupled with density functional theory (DFT) calculations. Herein, we describe a few selected examples of self-assembly of racemic or enantiopure helicenic samples. This part is not exhaustive at all and only aims at highlighting a few aspects in this domain.

Helicenes: Synthesis, Properties, and Applications, First Edition.
Edited by Jeanne Crassous, Irena G. Stará, and Ivo Starý.

16.2.1 Self-Assembly of Heptahelicene

The assembly of heptahelicene (**[7]H**, Figure 16.1) on different surfaces, especially copper, silver, and gold, has been extensively studied by the group of Karl Ernst [9]. Fully carbonated helicenes are expected to interact via van der Waals forces. On most metal surfaces, **[7]H** adsorbs such that a maximum of phenyl rings become aligned parallel to the surface. For each metal surface lattice and symmetry, the closest part of the helicene will be responsible for the stereochemical molecule–surface recognition while the farthest part will give the contrast in STM. At coverages close to the saturated monolayer ($\theta = 1$), it was shown that the self-assembly of enantiopure (*M*)-**[7]H** leads to long-range order (at temperature as low as 50 K to avoid mobility of molecules), with very similar chiral motifs observed on Cu(111), Ag(111), and Au(111) surfaces, with all adsorbate lattices displaying an oblique tilt relative to the substrate lattice. For instance, in the case of Cu(111), a typical three-dot figure was observed in STM at monolayer saturation coverage, representing a homochiral triplet cluster in which the three molecules in the unit cell are rotated by 120° with respect to each other, causing the distal parts to be close to each other (Figure 16.2a) [12]. Identical motifs on different surfaces suggest that the aggregation is dominated by lateral intermolecular interaction. At lower coverage ($\theta = 0.9$), different chiral cluster motifs were observed on Cu(111) with the formation of a "6 + 3-structure" (Figure 16.2b) [12]. The unit cell includes a six-molecule pinwheel appearance that actually reflects the handedness of the single molecule. The distal parts of the six molecules appear either in a clockwise or counterclockwise sense (Figure 16.2b inserts), thus highlighting the transmission of single-molecule handedness into self-assembled molecular layers through distinct azimuthal orientations.

In the case of racemic **[7]H** on Cu(111) and at monolayer saturation coverage, rotational domains (due to the 3-fold symmetric substrate) and mirror domains were observed [11]. Note that only one type of mirror domain was found on the same terrace (Figure 16.2c) with mirror domain boundaries at step edges of the metal surface. Higher-resolution STM images revealed zigzag rows with an oblique tilt with respect to the $[1\bar{1}0]$ direction of the Cu(111) surface (Figure 16.2d). Through theoretical modeling, the zigzag rows were identified as alternating *M*/*P* enantiomers [13]. The formation of racemic mirror domains comes from the fact that there are two equal but enantiomorphous alignments of *M*/*P* enantiomeric in a pair (Figure 16.2e, f).

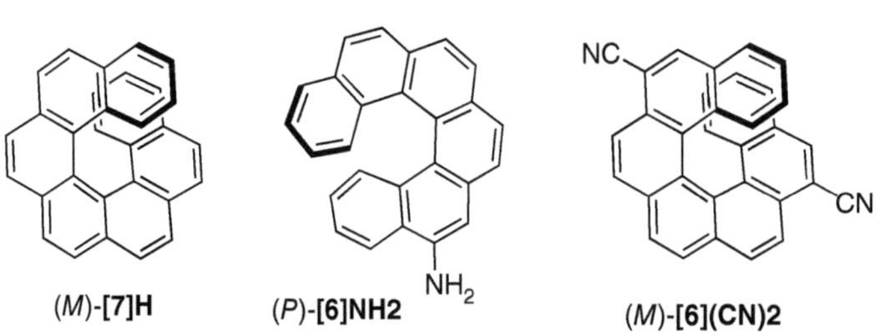

Figure 16.1 Selected helicenes whose self-assembly onto surfaces was studied by STM.

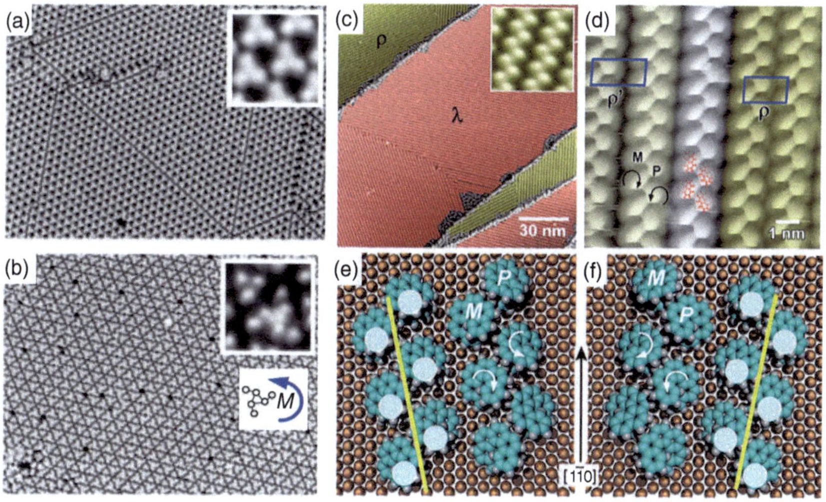

Figure 16.2 STM images acquired at 50 K of long-range ordered monolayers of (*M*)-**[7]H** on Cu(111). (a) The 3-structure of the complete monolayer for a saturated monolayer ($\theta = 1$). The insets show the molecular cluster units of the structures. (b) The 6+3-structure at 95% ($\theta = 0.95$) of the saturated monolayer coverage. (c) Self-assembly of (*rac*)-**[7]H** at monolayer saturation phase showing mirror domains with zigzag rows as a structural motif ($150 \times 150\ nm^2$; inset, $5 \times 5\ nm^2$). (d) Higher-magnification STM image ($10 \times 10\ nm^2$) that suggests alternation of both enantiomers in a single zigzag row. Double row distances (ρ and ρ') vary at intermediate coverages. (e, f) Structure models for both mirror domains. White filled circles mark the distal parts of the molecules that dominate the STM contrast. Circular arrows mimic the increasing brightness in the STM image (d). The [11$\bar{}$0] direction of the substrate surface is indicated as an arrow. Source: Reproduced with permission from references [10, 11]. Copyright 2008 and 2014 American Chemical Society.

On Ag(111) and Au(111) zigzag rows of (*rac*)-**[7]H** were also observed [10], but with some pronounced differences such as no enantiomorphism due to oblique adsorbate lattice vector tilt and a relatively low degree of long-range order.

Interestingly, cooperative response to small chiral bias was observed, leading to large magnification effects in supramolecular systems. Similar effects to the so-called "sergeant-and-soldiers" and "majority-rule" effects, are thus also present in 2D systems [13, 14]. Furthermore, the formation of helicenes assembled on surfaces enabled to observe chirality-dependent electron spin filtering through the low-energy photoemission of polarized electron spins using a Mott polarimeter [5].

16.2.2 Self-Assembly of 5-Amino-Hexahelicene

The self-assembly of heterohelicenes and carbohelicenes functionalized with polar groups, such as hexathia[11]helicene [15], heptahelicene-2-carboxylic acid [16, 17], 6-13-dicyano[7]helicene, or 5-aminohelicene (*vide infra*) on metallic surfaces was studied. For instance, De Feyter *and coll.* examined the self-assembly of racemic 5-aminohelicene (*rac*)-**[6]NH2** at liquid/solid interfaces [18]. A solution of (*rac*)-**[6]NH2** in 1,2,4-trichlorobenzene (TCB) onto Au(111) surface led to the formation of a "three-dot" p3-(*P*3) pattern (see trimeric assembly on Figure 16.3a1), with partial spontaneous resolution of (*M*) and (*P*) domains on the surface.

The assembly of (*M*)-**[6]NH2** on Cu(100) and Au(111) under ultra-high vacuum (UHV) was also studied by Kellog, Lingenfelder *and coll.* The authors examined the interplay between van der Waals and H bonding NH_2 groups and their influence on the chiral footprint of the enantiopure adsorbates on a solid surface. The main difference between the two surfaces was found in the origin of the molecule–surface interaction. While the C6 rings-surface interaction dominates in the case of Cu(100), the amino–surface interaction is crucial on Au(111). In both cases, the amino group does not induce polar interactions *via* hydrogen bonding but rather maximizes van der Waals interactions and drives the self-assembly [20]. The self-assembly of enantiopure (*M*)-**[6]NH2** on Au(111) under UHV conditions (a in Figure 16.3a2) [19] revealed two rotational domains formed by rows of dimers oriented along the <1-11> Au crystallographic directions. According to DFT calculations, the NH_2 groups govern the interaction with the surface (i.e. "N down"), whereas the C6 rings pack closely to maximize van der Waals interactions between neighboring molecules. (*rac*)-**[6]NH2** showed the emergence of two enantiomorphous domains which are rotated by 6° with respect to the <1-11> crystallographic directions and have a double-row structure (b and c in Figure 16.3a2). DFT calculations suggested that the rows can be described as a sequence of dimers formed by large protrusions

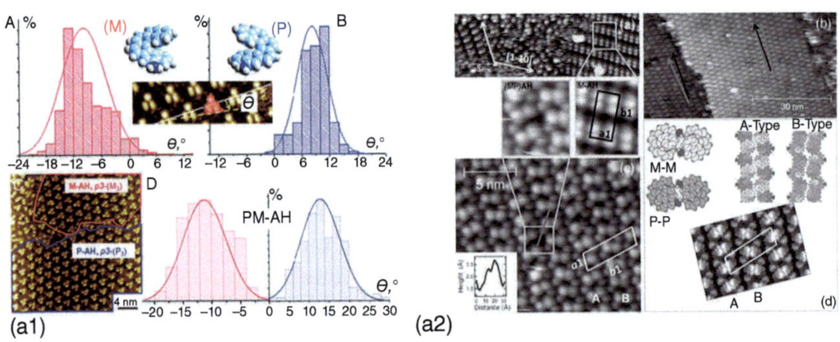

Figure 16.3 (a1) Self-assembly of 5-aminohelicene (*rac*)-**[6]NH2** formed at Au(111)/TCB interface. (A), (B), and (D) Distribution of tilt angles (Θ) between the sides of helicene trimers and the unit cell vectors in p3-(**[6]NH2**$_3$) patterns of (*M*)-, (*P*)-, and (*rac*)-**[6]NH2**, respectively. (C) STM image of homochiral conglomerates (p3-(M_3) and p3-(P_3), left part and right part, respectively) formed from (*rac*)-**[6]NH2**. Source: [18], Balandina, T, et al. / reproduced with permission from Royal Society of Chemistry. (a2) (a) STM image of (*M*)-**[6]NH2** on Au(111) at UHV. (b, c) STM images of (*rac*)-**[6]NH2** on Au(111). (b) Two enantiomorphous domains separated by a monoatomic step of the surface. (c) High-resolution image of the molecular structure developed by the racemate. The white parallelogram indicates the unit cell. The square highlights the presence of a defect (zoomed in the inset). The gray line indicates the line profile shown in the inset. (d) DFT-based molecular model of the racemic structure in (c) that highlights the presence of *M*–*M* and *P*–*P* dimers in the A- and B-type molecular-row models. Different colors indicate different chirality: Light gray for (*M*)-**[6]NH2** and dark gray for (*P*)-**[6]NH2**. Constant-current STM image simulation of the fully relaxed structure. Source: [19], Van der Meijden et al. / reproduced with permission from Wiley.

(N-up) and connected by small protrusions (N-down). The structural model consists of rows formed by the mentioned N-up dimers, linked by N-down molecules of the opposite chirality (d in Figure 16.3a2). The dimers are M–M in the A-type row, whereas the dimers are P–P in the B-type row. This assembly of double rows under UHV thus appeared remarkably different from the triangular (p3) structures formed at the Au(111)/1,2,4-TBC interface, where the solvent played an important role.

16.2.3 Self-Assembly of 6,13-Di-Cyano-[7]Helicene

In 2011, Diederich *and coll.* prepared and studied the self-assembly of racemic and enantiopure 6,13-di-cyano-[7]Helicene (**[7](CN)2**), Figure 16.1) onto Cu(111) surface [21]. STM and DFT studies revealed a spontaneous chiral resolution process of racemic helicene on the surface (Figure 16.4a). **[7](CN)2** formed fully segregated domains of pure enantiomers (2D conglomerate) on Cu(111) through optimized intermolecular CN···HC(Ar) hydrogen bonding and CN···CN dipolar interactions. Jung *and coll.* reported the self-assembly of (*P*)- and (*M*)-(**[7](CN)2** on Cu(111) for which STM images revealed the formation of mirror-imaged H-bonded chains (Figure 16.4b) [22]. Locally, the symmetry of the H-bonded dimers is mirrored when the helicene of opposite chirality sense is used. Upon annealing for 1 hour at 300 K, coordination with Cu adatoms occurred, thus creating Cu-coordinated chains. In this case, the direction of helicene chains is independent of the chirality (*P*) or (*M*) sense of the molecular building blocks. The tolerance to symmetry was thus observed to increase considerably and no spontaneous resolution was observed for such 1D arrangements formed by Cu coordination, contrary to more common situations found on STM of helicene on surfaces.

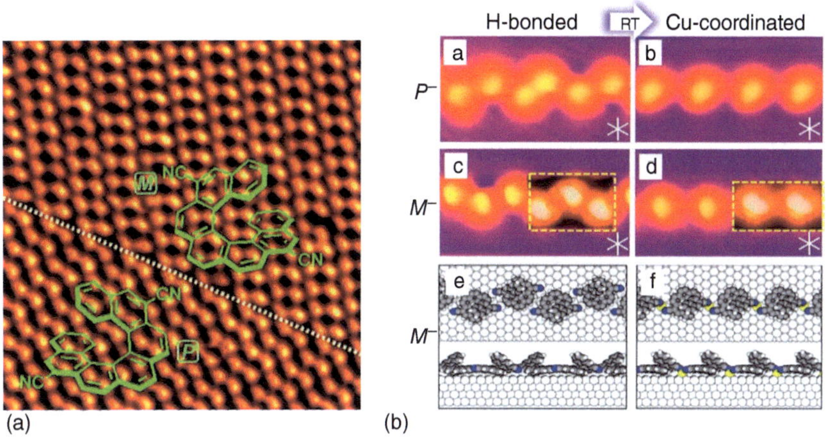

Figure 16.4 (a) Spontaneous resolution of racemic bis-cyano-heptahelicene (**[7](CN)2**) observed by STM on a Cu(111) surface. Source: [21], Stöhr, M, et al. / reproduced with permission from Wiley. (b) Formation of 1D-chain and transformation to Cu-coordinated chain observed by STM on the same Cu(111) surface from enantiopure (*P*)- and (*M*)-(**[7](CN)2**). Source: [22], Shchyrba, A, et al. / reproduced with permission from American Chemical Society.

16.3 Applications of Helicenes in Enantioselective Catalysis

Helicene derivatives have been explored in asymmetric catalysis and have show satisfying results in some specific reactions. A variety of helicenic molecules have been employed as organocatalysts, organometallic catalysts, or as chiral inducers. Helicenes containing phosphine, phosphite, phosphinite, *N*-heterocyclic carbene (NHC), phosphole, pyridine, pyridine-oxide, bipyridine, aminopyridine, thiophene unit, or unsubstituted carbohelicenes showed efficiency in stereoselective catalysis, where the rigidity, the bulkiness, the dissymmetric environment, and match-mismatch effects, appeared as important features for the stereocontrol of the reaction. The development of novel helical systems thus gives new opportunities to develop efficient catalysts that can enrich the existing catalytic strategies for asymmetric reactions which are of great interest for a synthetic chemist at the laboratory and the industry levels. For selected reviews on helicenes in asymmetric catalysis, see references [23–26].

16.3.1 Asymmetric Transition Metal Catalysis

16.3.1.1 P-containing Helicenes in Enantioselective Rh(I)-promoted Hydrogenation

A pioneering example of asymmetric catalysis using a helicene-based complex is the asymmetric hydrogenation of di-methyl itaconic acid ester **1** to enantioenriched (*S*)-**2** (reaction (1), Table 16.1), developed in 1997 by Reetz *and coll.*, using rhodium(I) complex of 2,15-bis(diphenylphosphino)-carbo[6]helicene ligand (*M*)-**L1** as the chiral catalyst (see structures of **L1**-**L8** in Figure 16.5). A 39% enantiomeric excess (*ee*) was obtained (Table 16.1, entry 1) [27]. The intramolecular P-P distance (6.48 Å) obtained from the X-ray crystallographic structure being too large for **L1** to act as a ditopic chelating ligand, it rather behaved as a monodentate one. In 2011, the enantiopure Rh(I) complex of bis(diphenylphosphino)-thiahelicene ligand (*M*)-**L2** was tested by Licandro *and coll.* in reaction (1) (Table 16.1, entry 2), and in reaction (2) (Table 16.1, entries 4,5). The hydrogenated (*S*)-**2** and (*S*)-**4** enantiomers were obtained from **1** and **3**, respectively with moderate *ee*s (31–40%). The chelated complex [(*M*)-**L2**-Rh(COD)$_2$][BF$_4$] was found to be stable and presumed to be the active species [28]. In 2003, Yamaguchi *and coll.* used phosphites such as (*M*,*M*,*S*,1*R*,2*S*,5*R*)-**L3**, displaying helical, axial, and central chirality, as effective ligands in reaction (1) and *ee*s up to 96% were obtained (Table 16.1, entry 3) [29]. The stereochemistry of the helicene moiety played an important role in the asymmetric induction, and matched/mismatched phenomena were observed between helical and axial chirality.

16.3.1.2 P-containing Helicenes in Enantioselective Pd(II)-Catalyzed Allylic Substitution

In 2000, the (*P*) enantiomer of the helical diphosphane **L1** was employed in the Pd-catalyzed allylic substitution of acetate by dimethylmalonate in

Table 16.1 Enantioselective hydrogenation of 2-methylene-succinate **1** and methyl-2-acetamidoacrylate **3**.

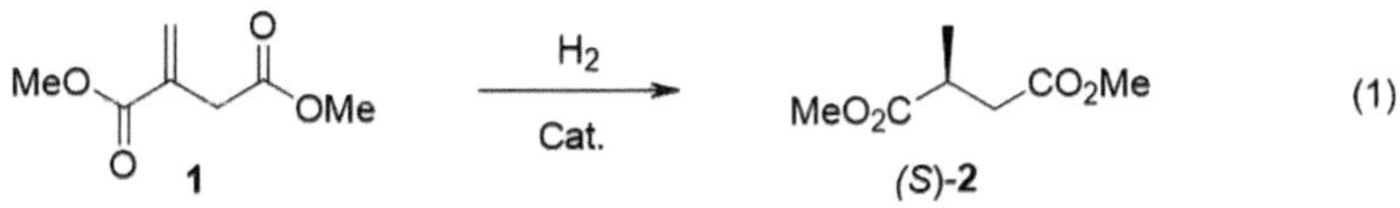

Entry	Catalyst	Conditions H_2 / % catalyst	*ee* [%]	Yield [%]	Ref.
1	[(*M*)-**L1**-Rh(COD)$_2$][BF$_4$]	1 atm, 0.1 mol% cat., rt	39	54	[27]
2	[(*M*)-**L2**-Rh(COD)$_2$][BF$_4$]	5 bars, 0.2 mol% cat., rt	31	27	[28]
3	[(*M*,*M*)-**L3**-Rh(COD)$_2$][BF$_4$]	90 bars, 1 mol% cat., −78 to 20 °C	96	Quant.	[29]
	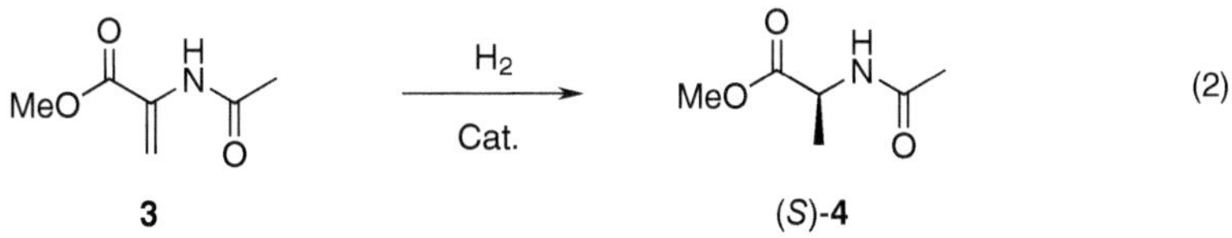				
4	[(*M*)-**L2**-Rh(COD)$_2$][BF$_4$]	5 bars, 0.2 mol% cat., rt, 5 h	32	67 (conv.)	[28]
5	[(*M*)-**L2**-Rh(COD)$_2$][BF$_4$]	5 bars, 0.2 mol% cat., 24 h, higher conc.	40	Quant.	[28]

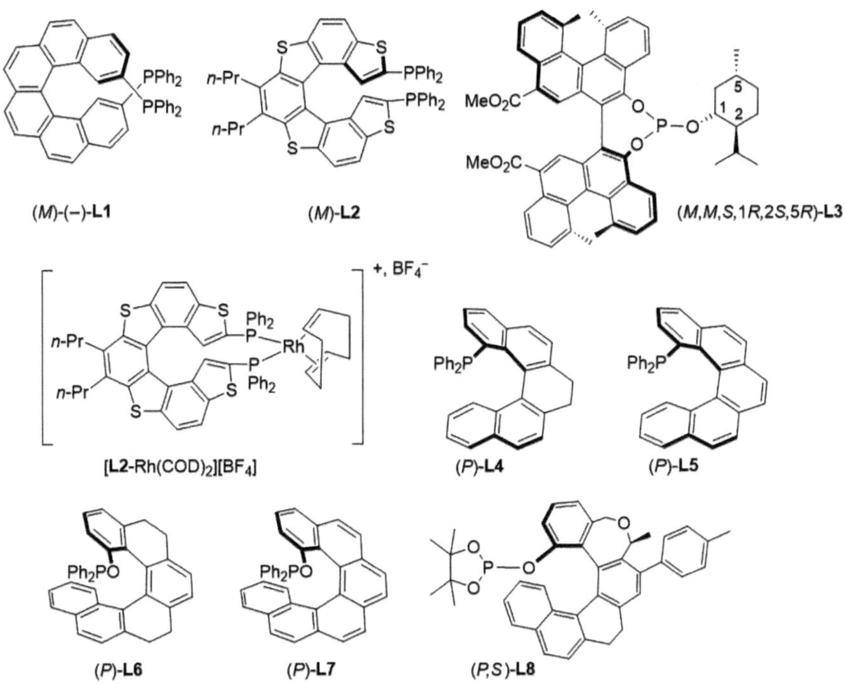

Figure 16.5 Phosphorated helicenic ligands **L1**-**L8** used in the asymmetric enantioselective transition metal-catalyzed hydrogenation, allylic substitution, and Suzuki–Miyaura coupling reactions.

1,3-diphenylpropenylacetate **5** (reaction (3), Table 16.2, entry 1). Under these conditions, kinetic resolution occurred and the left starting material **5** displayed *ee*s up to 99% while the allylic malonate product **7** showed an *ee* of 82.7% [30]. Note that the Pd allyl complex was structurally undefined and most probably not chelated (*vide supra*).

The two 1-(diphenylphosphino)-pentahelicenic ligands **L4** and **L5** (see Figure 16.5) proved very effective in asymmetric catalysis [31]. Reaction (3) was first studied (Table 16.2, entries 2,3) as a model reaction. Partially hydrogenated ligand (*M*)-**L4** was highly effective in this reaction, affording (*S*)-**7** in 99% yield with 94% *ee*, while fully aromatic ligand (*M*)-**L5** afforded (*S*)-**7** in 99% yield with only 71% *ee*. Similarly, (*M*)-**L4** proved highly effective in the asymmetric allylation

Table 16.2 Pd(II)-catalyzed enantioselective allylic substitution reactions.

(*rac*)-**5** **6** (*R*)-**7** (*R*)-**5** (3)

Entry	Ligand	Conditions H_2 / % catalyst	*ee* [%] (conf.)	Yield [%]	Ref.
1	(*P*)-**L1**	BSA, KOAc, 0.5 mol% cat. ligand-metal (2 : 1), 2 h, rt	82.7 ((*R*)-**7**) and 99 ((*R*)-**5**)	98	[30]
2	(*M*)-**L4**	Cs_2CO_3, 0.5 mol% cat., rt	94 ((*S*)-**7**)	99	[31]
3	(*M*)-**L5**	Cs_2CO_3, 0.5 mol% cat., rt	71 ((*S*)-**7**)	99	[31]
4	(*P*)-**L6**	BSA, LiOAc, 10 mol% ligand, 2.5 mol% [Pd], rt	90 ((*R*)-**7**)	96	[32]
5	(*P*)-**L7**	BSA, LiOAc, 10 mol% ligand, 2.5 mol% [Pd], rt	84 ((*R*)-**7**)	97	[32]
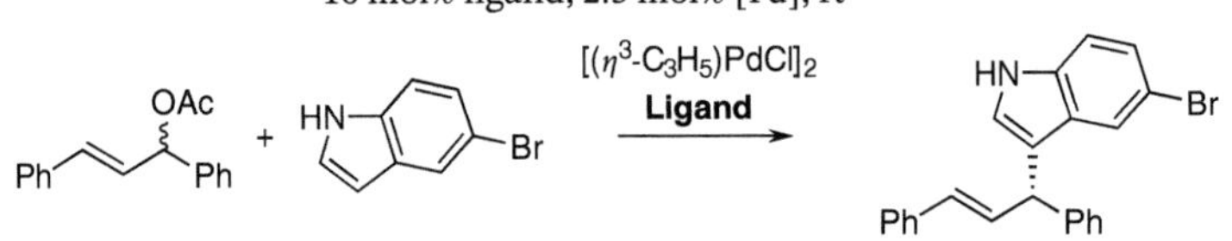 (*rac*)-**5** **8** (*S*)-**9** (4)					
6	(*M*)-**L4**	Cs_2CO_3, 1 mol% cat. CH_2Cl_2, rt	99 (*S*)	99	[31]
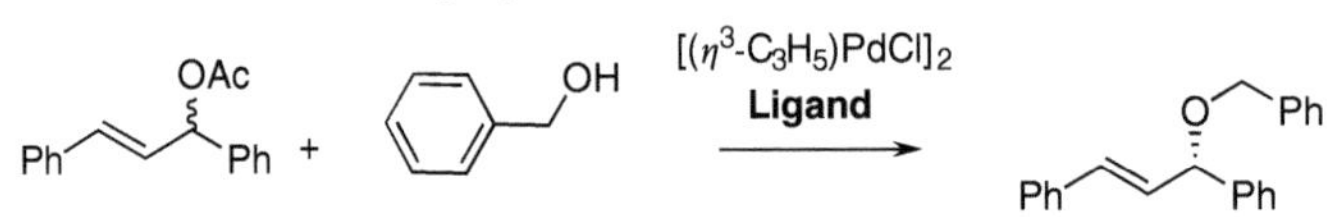 (*rac*)-**5** **10** (*S*)-**11** (5)					
7	(*M*)-**L4**	Cs_2CO_3, 1 mol% cat. CH_2Cl_2, rt	96 (*S*)	95	[31]

of indoles with 1,3-diphenylallyl acetate (up to 99% *ee*, reaction (4), Table 16.2, entry 6), and in the etherification of alcohols (up to 96% *ee*, reaction (5), Table 16.2, entry 7). The helical partially hydrogenated and fully aromatic phosphinites, namely 1-diphenylphosphinoxy-5,6,9,10-tetrahydro[6]helicene (*P*)-**L6** and 1-diphenylphosphinoxy-carbo[6]helicene (*P*)-**L7** (Figure 16.5), respectively, were also used as enantiopure chiral ligands in reaction (3) and gave *ee*s as high as 97% (Table 16.2, entries 4,5) [32].

16.3.1.3 Helicenic Phosphite in Enantioselective Ir(I)-Catalyzed Allylic Amination

Starý *and coll.* used chiral ligands such as (*P*,*S*)-**L8** (Figure 16.5) bearing a pendant phosphite moiety in the Ir-catalyzed allylic amination reaction depicted on Scheme 16.1. Ees as high as 90% were obtained for (*S*)-**14** obtained from cinnamyl methyl carbonate **12** and benzylamine **13**, together with very few amounts of linear product **15** [33]. Note that the same ligand gave moderate *ee*s up to 32% in asymmetric Rh-catalyzed hydroformylation reactions [33].

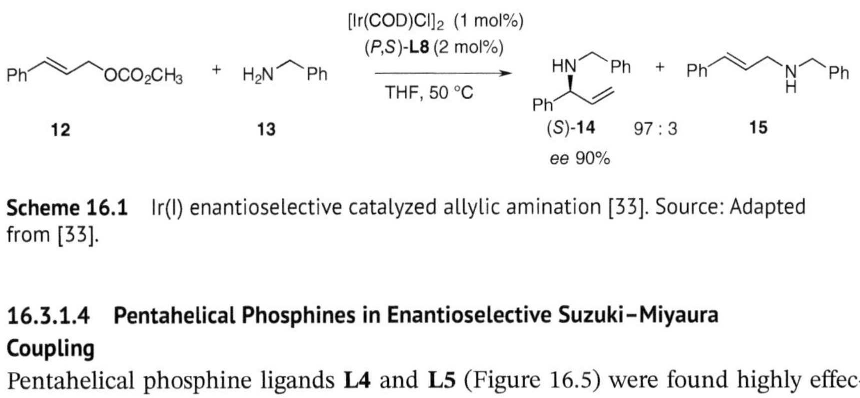

Scheme 16.1 Ir(I) enantioselective catalyzed allylic amination [33]. Source: Adapted from [33].

16.3.1.4 Pentahelical Phosphines in Enantioselective Suzuki–Miyaura Coupling

Pentahelical phosphine ligands **L4** and **L5** (Figure 16.5) were found highly effective in the stereocontrol of axial chirality in Suzuki–Miyaura coupling reactions, as illustrated in entries 1 and 2 of Table 16.3, in the coupling between diisopropyl (1-bromonaphthalen-2-yl) phosphonate **16** and *o*-tolylboronic acid **17** with 95% *ee* for product (*R*)-**18** (reaction 6) [31].

Table 16.3 Pd(II)-catalyzed enantioselective Suzuki–Miyaura coupling.

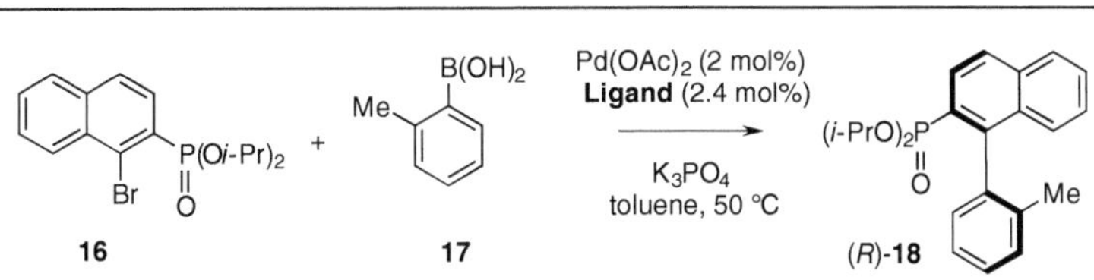

(6)

Entry	Ligand	*ee* [%] (conf.)	Yield [%]	Ref.
1	(*P*)-**L4**	81	87	[31]
2	(*P*)-**L5**	95	97	[31]

16.3.1.5 Helical NHCs in Enantioselective Ni(0)-Catalyzed [2+2+2] Cycloisomerization

Enantiopure helical NHC ligands generated from imidazolium salts (*M*,*R*,*R*)-(−)-**L9** and (*M*,*R*,*R*),(*M*,*R*,*R*)-(−)-**L10a-g** in Figure 16.6 were tested in the enantioselective Ni(0)-catalyzed [2+2+2] cycloisomerization of **19** to dibenzo[6]helicene (*P*)-(+)-**20** (reaction (7), Table 16.4) [34]. High yields and good *ee*s were obtained for the mono-oxa[5]helicenic unsymmetrical imidazolium salt **L19** (Table 16.4, entry 1) while symmetrical 1,3-bisoxa[5]helicenyl imidazolium salts gave much higher *ee*s. The presence of bulky aryl group at the opposite terminus of the oxa[5]helicene backbone with respect to the position of the imidazolium unit led to an increase of the *ee* of (*P*)-(+)-**20**: from 41% *ee* (Table 16.4, entry 2) to 66% (Table 16.4, entry 6). The highest level of chirality transfer from a helical NHC ligand to a helical product was achieved in [2+2+2] cycloisomerization of the aromatic triyne **21** to dibenzo[7]helicene (*P*)-(+)-**22** (reaction (8), Table 16.5), with lower yield but with *ee*s up to 86% when using NHC precursor **L10c**.

16.3.1.6 Helical NHCs in Enantioselective Ru-Catalyzed Metathesis

Using the same approach, an imidazolium salt bearing a pendant carbo[6] unit was also prepared (see Figure 16.6), and its corresponding Ru-NHC complex **L11-[Ru]** obtained from the first-generation Hoveyda–Grubbs catalyst isolated and identified [35]. Complex **L11-[Ru]** was evaluated in asymmetric ring-closing metathesis (RCM) and ring-opening metathesis-cross metathesis (ROM/CM) reactions, which proceeded with reasonably good enantioselectivities (Scheme 16.2).

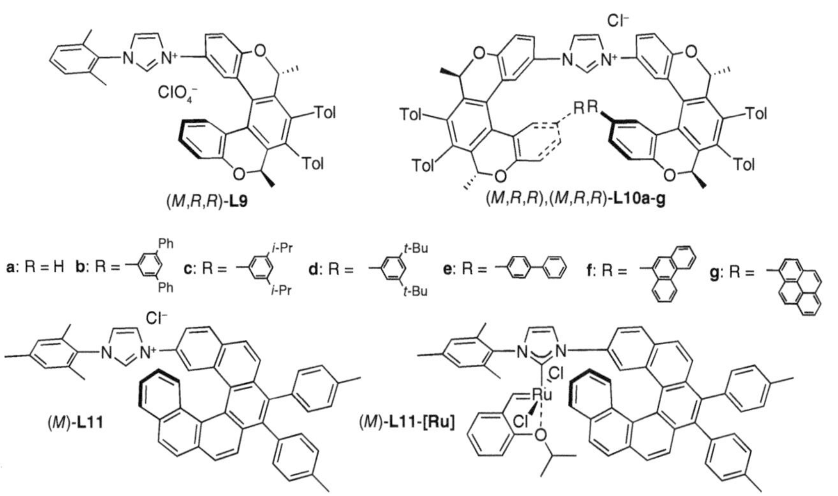

Figure 16.6 Imidazolium precursors of NHC ligands used in Ni(0)-catalyzed enantioselective [2+2+2] cycloisomerization reactions (**L9-L10a-g**) and in Ru(II)-promoted Hoveyda–Grubbs enantioselective metathesis (**L11** and corresponding NHC-complex **L11-[Ru]**) [34, 35]. Source: Adapted from [34, 35].

Table 16.4 Enantioselective [2+2+2] cycloisomerization of triyne **19** to dibenzo[6]helicene **20** in the presence of enantiopure NHC ligands generated from (*M*,*R*,*R*)-(−)-**L9** and (*M*,*R*,*R*),(*M*,*R*,*R*)-(−)-**L10a**-**g** [34].

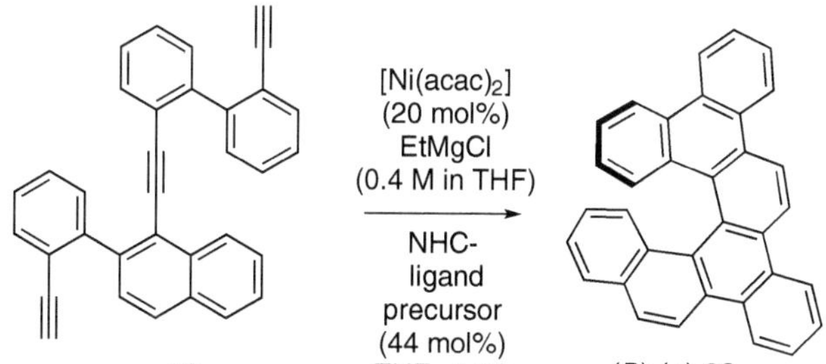

19 (*P*)-(+)-**20** (7)

Entry	NHC precursor	*ee* [%][a)]	Conv [%][b)]
1	(*M*,*R*,*R*)-**L9**	17	>90
2	(*M*,*R*,*R*),(*M*,*R*,*R*)-**L10a**	41	>90
3	(*M*,*R*,*R*),(*M*,*R*,*R*)-**L10h**	59	>90
4	(*M*,*R*,*R*),(*M*,*R*,*R*)-**L10b**	61	90
5	(*M*,*R*,*R*),(*M*,*R*,*R*)-**L10c**	64	90
6	(*M*,*R*,*R*),(*M*,*R*,*R*)-**L10d**	66	81
7	(*M*,*R*,*R*),(*M*,*R*,*R*)-**L10e**	44	89
8	(*M*,*R*,*R*),(*M*,*R*,*R*)-**L10f**	56	73
9	(*M*,*R*,*R*),(*M*,*R*,*R*)-**L10g**	47	90

a) Determined by HPLC on a Chiralpak IA column.
b) Estimated by HPLC.
Source: Adapted from [34].

Table 16.5 Enantioselective [2+2+2] cycloisomerization of triyne **21** to dibenzo[7]helicene **22** in the presence of enantiopure NHC ligands generated from (*M*,*R*,*R*),(*M*,*R*,*R*)-(−)-**L10a,c** [34].

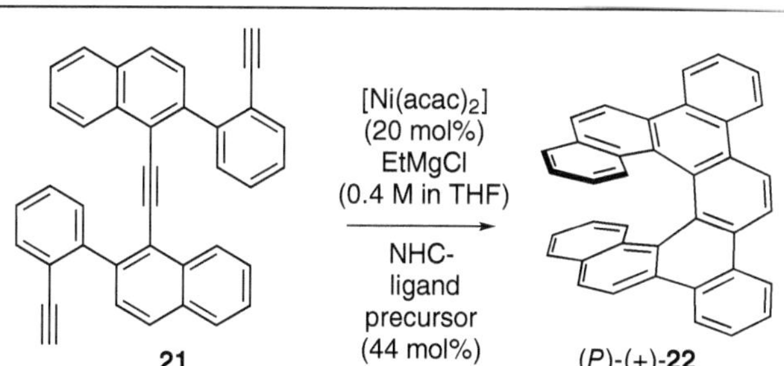

21 (*P*)-(+)-**22** (8)

Entry	NHC precursor	*ee* [%][a)]	Conv [%][b)]
1	(*M*,*R*,*R*),(*M*,*R*,*R*)-**L10a**	72 (+)	64
2	(*M*,*R*,*R*),(*M*,*R*,*R*)-**L10c**	86 (+)	86

a) Determined by HPLC on a Chiralpak IA column.
b) Estimated by HPLC.
Source: Adapted from [34].

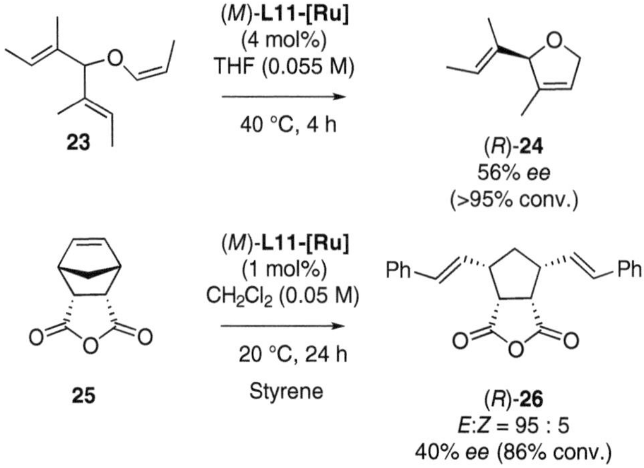

Scheme 16.2 Enantioselective RCM (a) and ROM-CM (b) catalyzed by (*M*)-**L11-[Ru]** [35]. Source: Adapted from [35].

16.3.1.7 Phosphahelicenes in Enantioselective Au(I) Cycloisomerization of Enynes

Phosphahelicenes are efficient chiral platforms for asymmetric gold-catalyzed cycloisomerizations processes. Several gold complexes have been prepared by Marinetti, Voituriez, and coworkers. The fine-tuning of the phosphahelicene ligands furnished efficient catalytic systems such as (S_P,*P*)-*endo*-**L12a** and **L13** (Figure 16.7) displaying high activity and giving high *ee*s [36, 37]. In these complexes, the P atom is embedded within the helical structure and the *endo* stems for the gold atom pointing toward the helical groove. This topology is well-fitted for efficient enantioselectivity despite the linear Au^I environment. Complexes **L12a** and **L13** were used as precatalysts in a benchmark reaction, i.e. the cycloisomerization of the *N*-tethered 1,6-enynes **27** into aza-bicyclo-[4.1.0]heptenes **28** (reaction (9) in Table 16.6). Good catalytic activity at room temperature was observed after activation with $AgBF_4$. Changing the activating agent from $AgBF_4$ to $AgNTf_2$ did not change the enantioselectivity level (75% *ee*), while other silver salts such as AgOTf or $AgSbF_6$ decreased the *ee* to 45 and 63%, respectively. The influence of the helical structure and the stereochemistry of both the P atom and of the helical core (while the configuration of Men* remained unchanged), and the influence of the P pointing inward (*endo*) or outward (*exo*) the helical groove, were examined in detail on the asymmetric efficiency of **27** to **28** (compare complexes on Figure 16.7 and results in Table 16.6) [36]. Cycloisomerization of other classes of enynes was also investigated. Dienynes **29a,b** that display conjugated enyne moieties were considered. Depending on the nature of the R substituent, the gold-catalyzed cycloisomerization afforded either the aza-bicyclo[4.1.0]heptene **30** (for R=H) or the tricyclic derivative **31** (for R=Ph), which resulted from a vinylcyclopropane-cyclopentene rearrangement of the intermediate aza-bicyclo[4.1.0]heptene. In the cycloisomerization of **29b** (R=Ph), the nature of the silver salt was shown to have a remarkable effect on the enantioselectivity level, going from a moderate 65% *ee* for $AgBF_4$ to an excellent

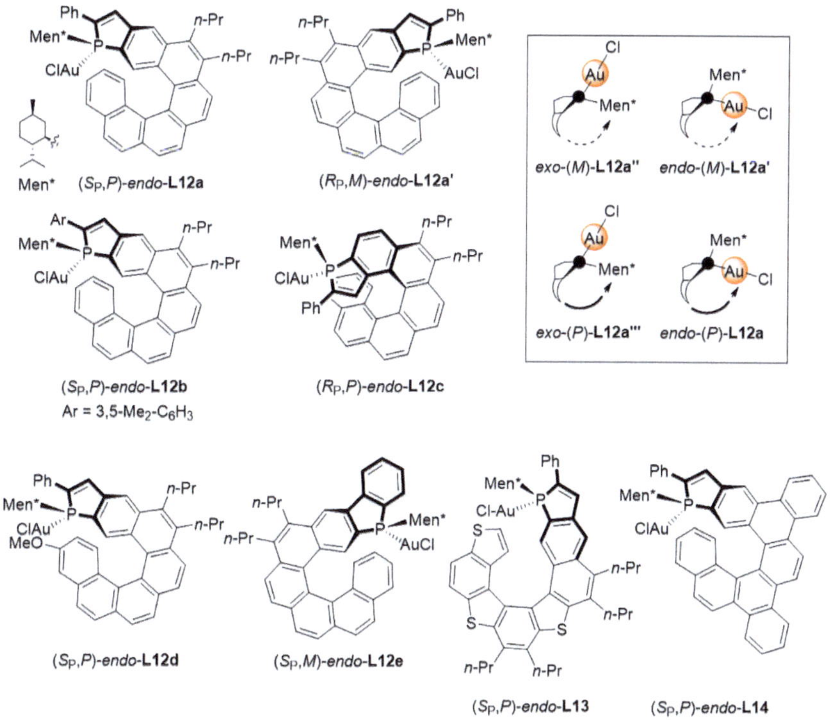

Figure 16.7 Typical gold complexes with different structures and stereochemistries studied in the asymmetric intramolecular cyclization of **27** to **28** [36–38]. Source: Adapted from [36–38].

96% *ee* for $AgNTf_2$. The thiaphosphahelicene–Au^I catalyst (S_P,*P*)-*endo*-**L13** gave higher *ee*s in such cycloisomerization reactions. As shown in Scheme 16.3, similar complexes ((S_P,*P*)-*endo*-**L14** bearing a dibenzo fused ligand depicted in Figure 16.7) were tested as catalysts in the intramolecular [2+2] cyclization of 1,6-allenene **32** into bicyclo[3.2.0]-heptane derivative **33**, with 88% *ee*, when catalyzed by *endo*-(*P*)-**L14** stereoisomer [38].

16.3.1.8 Helicenes as Chiral Inducers in the Autocatalytic Soai Reaction

In 2017, Soai *and coll.* reported the use of enantiomerically pure (*P*)- or (*M*)-**L15a,b** [39] and **L18** [40] depicted in Figure 16.8 as chiral inducers in the famous autocatalytic "Soai reaction" corresponding to the addition of diisopropyl-zinc to pyrimidine aldehyde **34** yielding secondary alcohols, either (*R*)- or (*S*)-**36** (reaction (10), Table 16.7). In this process, with only a few percentage of enantiomeric excess of helicenic chiral inducer used, very high *ee*s of pyrimidine-alkanol are obtained thanks to chiral amplification through asymmetric autocatalysis. The same group observed reversal of the sense of enantioselectivity between 1- and 2-aza[6]helicenes, i.e. (*P*)-**L16** and (*P*)-**L17** chiral inducers (entries 4,5 in Table 16.7) [41]. In the case of sulfur- or nitrogen-containing helicene inducers, the asymmetric reaction is proposed to proceed via coordination of the heteroatom to i-Pr_2Zn

Table 16.6 Stereochemical influence of phosphahelicene-gold complexes in enantioselective enyne cycloisomerizations.[a)] [36–38]

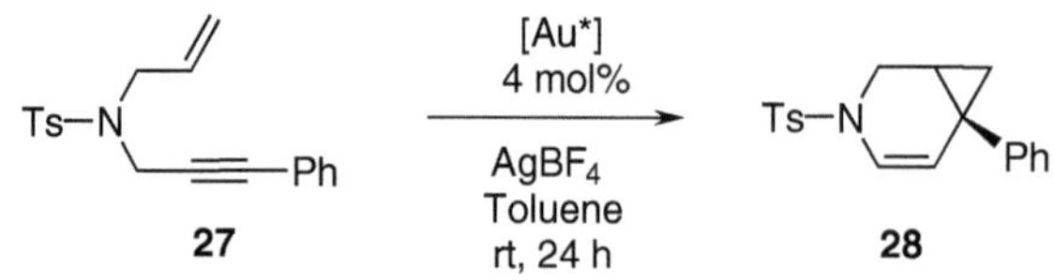

(9)

Entry	Catalyst	*ee* [%][b)]	Conv. [%]	Config.
1	(S_P,*P*)-*endo*-**L12a**	81	>95	(1*R*,6*S*)
2	(S_P,*P*)-*endo*-**L12b**	84	>95	(1*R*,6*S*)
3	(R_P,*P*)-*endo*-**L12c**	35	43	(1*S*,6*R*)
4	(R_P,*M*)-*endo*-**L12a'**	42	90	(1*S*,6*R*)
5	(S_P,*M*)-*exo*-**L12a"**	n.d.	<10	—
6	(R_P,*P*)-*exo*-**L12a'''**	n.d.	<5	—
7	(S_P,*P*)-*endo*-**L12d**	82	>95	(1*R*,6*S*)
8	(S_P,*M*)-*endo*-**L12e**	7	70	(1*S*,6*R*)
9	(S_P,*P*)-*endo*-**L13**	74	n.d.	(1*R*,6*S*)

a) Ts: toluene-4-sulfonyl.
b) n.d.: not determined.
Source: Adapted from [36–38].

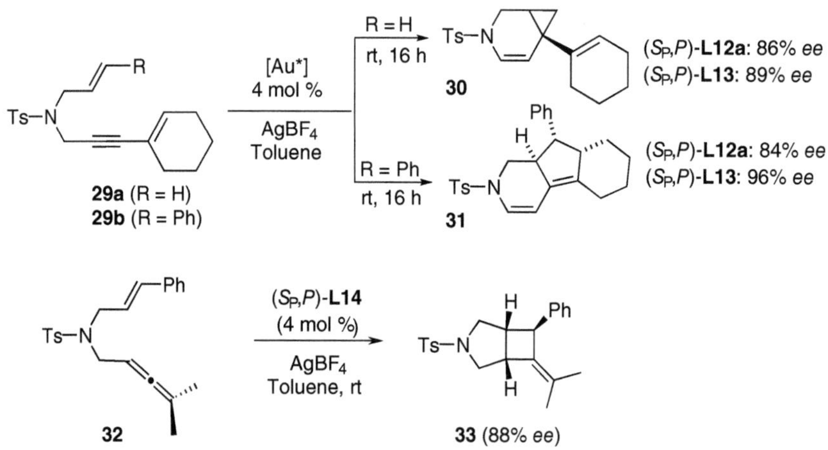

Scheme 16.3 Enantioselective catalysis with gold(I) complexes **L12a**, **L13**, and **L14** [38]. Source: Adapted from [38].

to form a chiral active zinc alkoxide species. In the case of carbo[6]helicene, a charge-transfer complex between the helicene and the pyrimidine unit is assumed to be formed, with some preferential diastereomeric association. When *i*-Pr_2Zn **35** is added to aldehyde **34**, a nonracemic zinc alkoxide of alkanol **36** is generated.

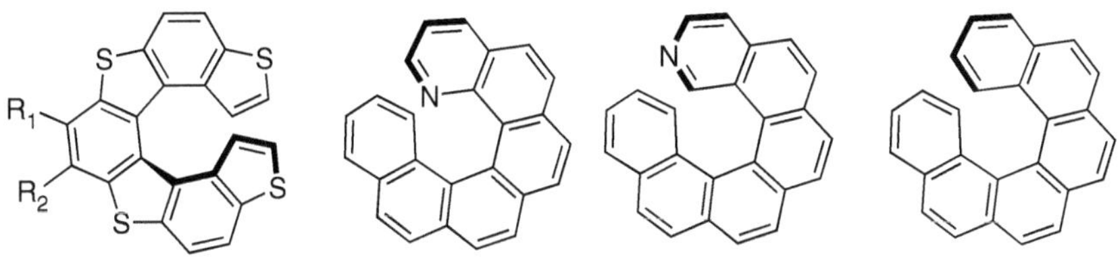

Figure 16.8 Helicenic chiral inducers used in the Soai autocatalytic reaction.

Table 16.7 Asymmetric autocatalytic Soai reaction using diverse helicenic catalytic stereo-inducers [39–41].

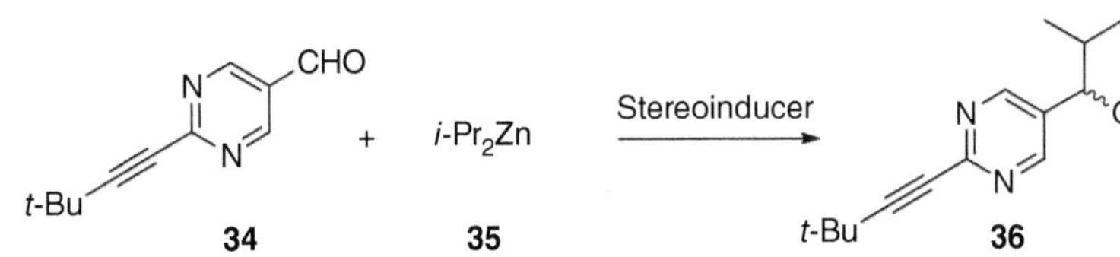

(10)

Entry	Chiral inducer	*ee* [%]	Conditions **L/34/35**	*ee* [%] (conf. **36**)	Yield [%]	Ref.
1	(*P*)-**L15a**	2	0.02 : 0.42 : 0.86	83 (*S*)	92	[39]
2	(*M*)-**L15a**	3	0.02 : 0.42 : 0.86	85 (*S*)	89	[39]
3	(*P*)-**L15b**	2	0.02 : 0.42 : 0.86	83 (*S*)	88	[39]
4	(*P*)-**L16**	33	0.4 : 1 : 6.7	99 (*S*)	91	[41]
5	(*P*)-**L17**	>99	0.4 : 1 : 6.7	93 (*R*)	88	[41]
6	(*P*)-**L18**	>99.5	6.2 mol% of **L18**	95 (*S*)	95	[40]
7	(*P*)-**L18**	~0.13	11 mol% of **L18**	56 (*S*)	88	[40]

Source: Adapted from [39–41].

16.3.2 Asymmetric Organocatalysis

16.3.2.1 Helicene N-oxide Catalyzed Enantioselective Desymmetrization of *Meso* Epoxides

Enantiopure helicene-N-oxides **L19**-**L20** were prepared (Figure 16.9) by Takenaka *and coll.* and tested as catalysts in the desymmetrization of *meso* epoxides with chlorosilanes [42, 43]. *Cis*-stilbene epoxide **37** reacted with $SiCl_4$ and gave the corresponding (*R*,*R*)-chlorohydrin **38** with high *ee* values (reaction (11), Table 16.8). The ring-opening gave higher enantioselectivity for substrates bearing aromatic substituents rather than alkyl groups (entries 1 and 2 versus 3 and 4). Catalyst **L19** provided better *ee* values than **L20** for acyclic epoxides (entries 1–3), but the opposite was true for the cyclic epoxide (entry 4). The scope of the reaction was additionally probed with catalyst **L21** and the enantioselectivity was shown to be sensitive to electronic effects (entries 5–7). Finally, the ring-opening proceeded

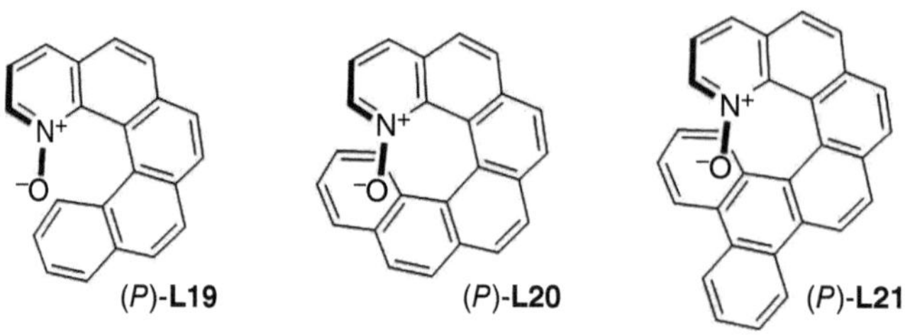

Figure 16.9 Aza[5] and [6]helicene oxides **L19-21** ((*P*) enantiomers shown) used as organocatalysts in the asymmetric ring-opening of *cis*-stilbene epoxide **37** with $SiCl_4$ [42, 43]. Source: Adapted from [42, 43].

Table 16.8 Desymmetrization of *meso* epoxides **37** by (*P*)-helical pyridine *N*-oxides (*P*)-**L19-L21** [42, 43].

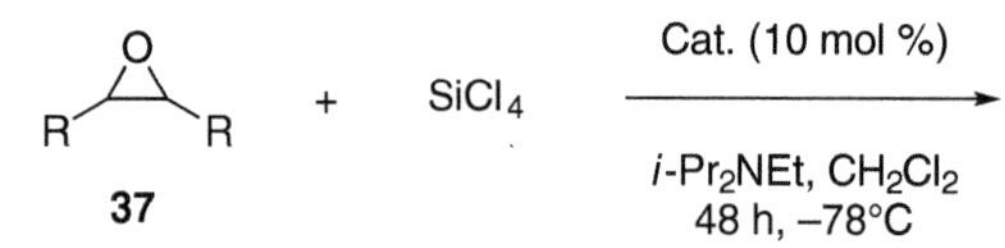

(11)

Entry	R	(*P*)-**L19**	(*P*)-**L20**	(*P*)-**L21**
1	Ph	77 %, 93 % *ee*	80 %, 92 % *ee*	77 %, 94 % *ee*
2	2-Naphthyl	79 %, 81 % *ee*	77 %, 73 % *ee*	76 %, 92 % *ee*
3	$BnOCH_2$	71 %, 49 % *ee*	68 %, 42 % *ee*	72 %, 65 % *ee*
4	cyclooctene	70 %, 0 % *ee*	68 %, 22 % *ee*	74 %, 33 % *ee*
5	$4\text{-}ClC_6H_4$			84 %, 94 % *ee*
6	$4\text{-}CF_3C_6H_4$			83 %, 92 % *ee*
7	$4\text{-}CH_3C_6H_4$			83 %, 87 % *ee*
8	$CH_2O(CH_2)_3Ph$			63 %, 72 % *ee*
9	$\text{-}CH_2OCH_2\text{-}$			64 %, 33 % *ee*

Source: Adapted from [42, 43].

with a moderate *ee* value for an acyclic alkyl-substituted epoxide (entry 8), but with modest *ee*s for a cyclic substrate (entry 9).

16.3.2.2 Helical Amino-Pyridium Catalyzed Asymmetric Addition to Nitroalkenes

The helical chiral 2-aminopyridinium ions (*P*) or (*M*)-[**L22a-f**.H^+]$TFPB^-$ and [**L16**.H^+]$TFPB^-$ (Figures 16.8 and 16.10) were reported in 2010 as a new class of hydrogen-bond donor catalysts [44]. The approach of merging a 2-aminopyridinium core into the helical framework enabled to position an inherently chiral element at the H-bonding site. These helically chiral organic catalysts were tested on the addition of 4,7-dihydroindole **39** to nitroalkene **40** affording *β*-nitro-indol-2-yl product **41** after subsequent oxidation (equation (12), Table 16.9). Single H-bond donor (*P*)-[**L16**.H^+]$TFPB^-$ was found inefficient (entry 1), indicating that the

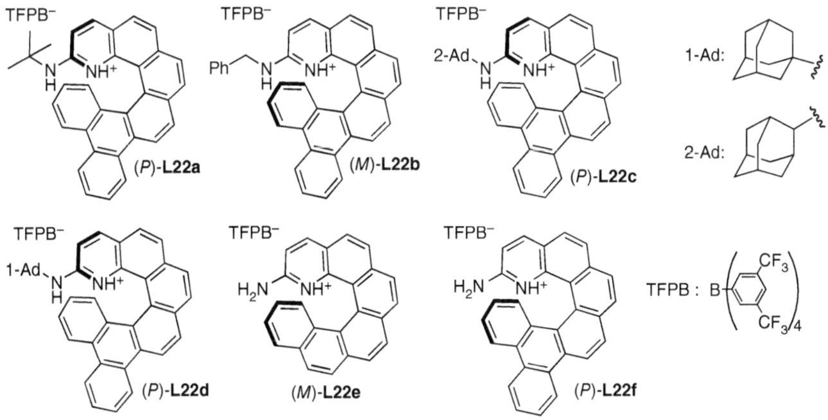

Figure 16.10 Helical pyridinium salts from **L22a-f** used as chiral hydrogen-bond donor catalysts [44]. Source: Adapted from [44].

Table 16.9 Evaluation of helical chiral 2-aminopyridinium catalysts (*P*)-**L22a-f** and (*P*)-[**L16**.H$^+$]TFPB$^-$ in the asymmetric addition of 4,7-dihydroindole **39** to nitroalkene **40** yielding (*R*)-**41** [44].

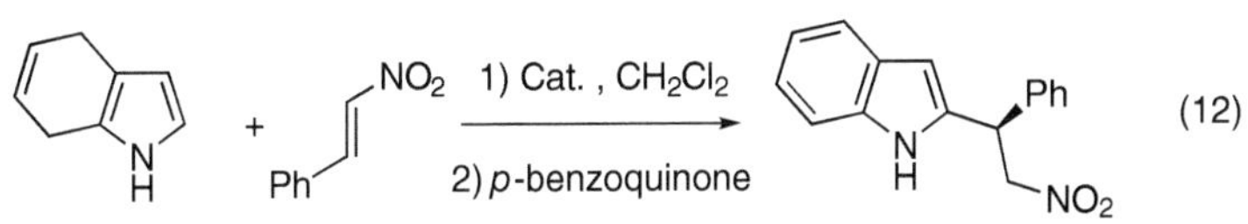

Entry	Catalyst	Mol [%]	Temp [°C]	Time [hr]	Yield [%]	*er*[a)]
1	(*P*)-[**L16**.H$^+$]TFPB$^-$	10	−40	20	65	53 : 47
2	(*P*)-**L22a**	2	−40	20	79	92 : 8
3	(*P*)-**L22b**	2	−40	20	72	69 : 31
4	(*P*)-**L22c**	2	−40	20	73	83 : 17
5	(*P*)-**L22d**	2	−40	20	88	93 : 7
6	(*P*)-**L22d**	0.5	−40	48	80	92 : 8
7	(*P*)-**L22e**	10	−40	20	80	64 : 36
8	(*P*)-**L22f**	10	−40	20	85	69 : 31
9	(*P*)-**L22f**	2	−40	20	85	69 : 31

a) Enantiomeric ratio in favor of the (*R*) enantiomer.
Source: Adapted from [44].

2-amino group is required for asymmetric induction. Benzo-fused derivative (*P*)-**L22f** appeared more selective than (*P*)-**L22e** (entries 7 and 8), even with 2 mol% catalyst loading (entry 9). *N*-alkylated catalysts **L22a-d** were also evaluated (entries 2–6) and the enantioselectivity was gradually improved as the alkyl substitution increased. A high level of enantioselectivity was maintained even when the catalyst loading was reduced to 0.5 mol% (entry 6).

16.3.2.3 Helical Bipyridine *N*-monoxide-Catalyzed Enantioselective Propargylation of Aldehydes

Takenaka *and coll.* used helical chiral 2,2′-bipyridine *N*-monoxides **L23** in the enantioselective catalytic propargylation of aldehydes **42a,b** with allenyltrichlorosilane **43** to yield enantioenriched **44a,b** (Scheme 16.4) [45]. The authors first examined the reactivity of allenyltrichlorosilane in Lewis base-catalyzed addition to benzaldehyde using as catalysts pyridyl-N-oxides (*P*)-L**19**-**L21** (see Figure 16.9) and chiral 2,2′-bipyridine N-monoxides (*P*)-**L23** and found that **L23** was the best catalyst. Then they applied the optimized conditions to a range of substituted benzaldehydes; for instance, the 2-substituted benzaldehydes gave the best yield and *ee*s (see **44b** in Scheme 16.4). The authors proposed a mechanism involving coordination of the helical-bipy-N-oxide to allenyltrichlorosilane and accounting for the stereochemistry of the process.

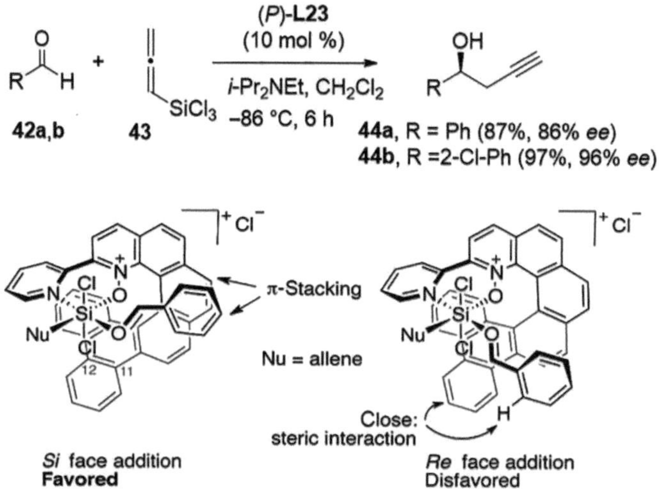

Scheme 16.4 Enantioselective catalytic propargylation of aldehydes **42a,b** with allenyltrichlorosilane **43** to **44a,b** using chiral 2,2′-bipyridine *N*-monoxides **L23**. Proposed model for explaining the preferred approach of catalyst (*P*)-**L23**. Source: Reproduced from ref. [45]. Copyright 2011, American Chemical Society.

16.3.2.4 Azahelicene-Catalyzed Enantioselective Acyl Transfer Reaction

Stará and Starý reported the utilization of enantiopure 2-aza[6]helicene (*M*)-**L17** as organocatalyst in the kinetic resolution of racemic secondary alcohol 1-phenylethanol (*rac*)-**45**, through an asymmetric acyl transfer reaction to ester (*S*)-**46** (Scheme 16.5) [46]. Moderate reactivity and selectivity factor ($s = 9, 10$) were observed.

16.3.2.5 Phosphahelicene-Catalyzed Enantioselective [3+2] Annulation

Phosphahelicenes were also revealed efficient chiral organocatalysts in the enantioselective [3+2] annulation of arylidenemalononitriles with allenoates (equation (13), Table 16.10 and Figure 16.11) [47]. The reaction proved to be successful with a large series of γ-substituted allenoates and γ-substituted buta-2,3-dienenitriles with

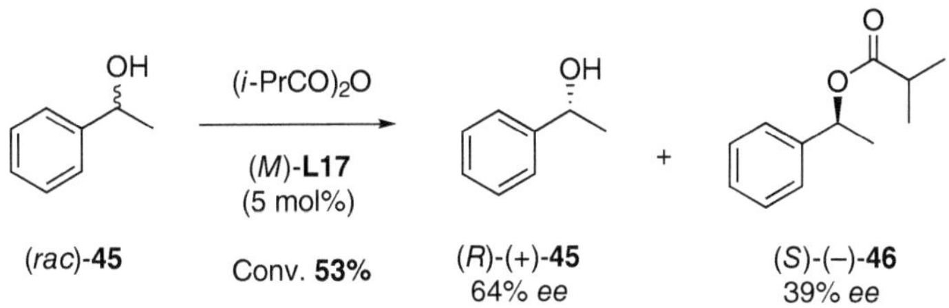

Scheme 16.5 Kinetic resolution of secondary alcohol (*rac*)-**45** catalyzed by aza[6]helicene (*M*)-**L17**. Conditions: (*i*-PrCO)$_2$O (1.0 equiv.), (*M*)-**L17** (5 mol %), *N*-ethyldiisopropylamine (0.75 equiv.), chloroform, rt, 110 hours, **45** : **46** = 47 : 53 (GC) [46]. Source: [46] / Institute of Organic Chemistry and Biochemistry.

Table 16.10 Screening of the HelPhos catalysts in an organocatalytic [3+2] cyclization reaction yielding enantioenriched **49** [47, 48].

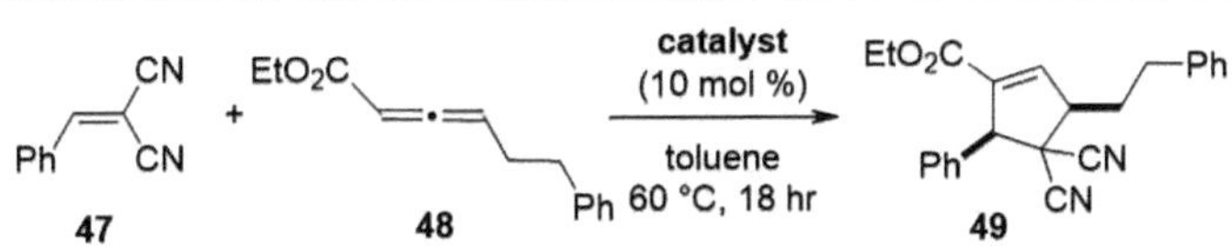

(13)

Entry	Catalyst	*dr* (%)	Yield [%]	*ee* [%] (optical rotation sign)
1	(*P*)-Men*-HelPhos (*P*)-**L24**	>95 : 5	30	89 (+)
2	(*P*)-Ipc*-HelPhos (*P*)-**L25**	>95 : 5	37	95 (+)
3[a),b)]	(*P*)-Ipc*-HelPhos (*P*)-**L25**	>95 : 5	91	96 (+)
4	(*P*)-Ipc*-HelPhos (*M*)-**L25**	85 : 15	83	68 (−)

a) Reaction temperature 80 °C.
b) As an additional experiment, the reaction in entry 3 has been carried out at a 5 mol% catalyst loading: total conversion was attained after 48 hours at 80 °C, yielding **49** in 96% *ee*.
Source: Adapted from [47, 48].

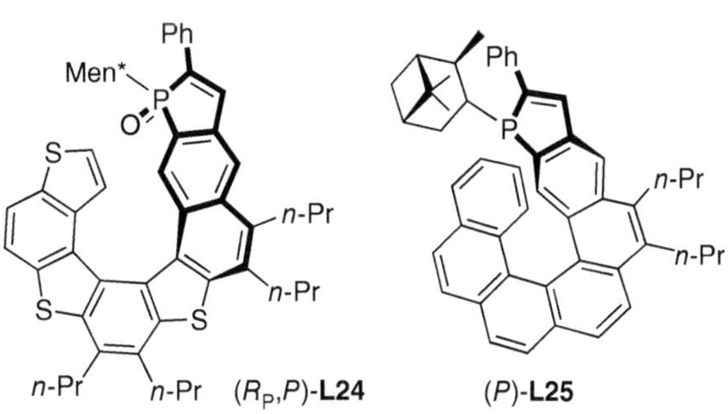

Figure 16.11 Examples of phosphahelicenes utilized in asymmetric organocatalysis.

very high diastereoselectivities (>95 : 5 *dr*) and enantioselectivities (20 examples, *ee*s 85–97%). Benzophosphahelicene analogues such as the reduced form **L25** proved also efficient in such of enantioselective organocatalytic reaction [48]. In Chapter 6 are reported other examples of reactions catalyzed by thiahelicenes.

16.4 Conclusion

It was shown that different types of racemic or enantiopure helicenes, such as unsubstituted carbo[7]helicene, and functionalized molecules such as amino-carbo[6]helicene and dicyano-carbo[7]helicene, can self-assemble onto metallic surfaces and give access to well-defined organized patterns, sometimes with spontaneous resolution and through amplification processes, as observed by STM. The self-assembly is controlled by several parameters, including the structure of the surface, the deposition method, the number of fused phenyl rings, the attached functional group, the enantiopurity of the helicene. DFT calculations help analyze how the molecules organize relative to the surface and between each other. This 2D organization gives access to chiral layers on surfaces displaying some appealing properties such as electron spin polarization and exhibiting promising applications for spintronics and optoelectronics. For this purpose, novel helicenic systems such as open-shell helical radicals are currently of interest [49]. Other important applicative areas such as on-surface reactivity of helicenes can be targeted [50, 51]. Single-molecule properties such as charge transport and piezoelectric effect can also be accessed [52]. Helicene derivatives also display interesting catalytic activity as chiral ligands for asymmetric metal-catalyzed reaction, chiral organocatalysts, or chiral inducers. However, these helically chiral systems currently do not compete with other classes of chiral catalysts such as axially chiral ones. This is probably explained by the strong rigidity of the helical backbone, and the difficult access to the catalytically active site. Interestingly, it was observed that the combination of different stereochemical elements (helical combined with central and axial chirality) can give access to highly efficient enantioselective processes. Multihelicenic catalytic systems and less rigid partially hydrogenated ones have also shown higher efficiency and are certainly the strategic goals to pursue for further developments. Finally, it would be advantageous to use self-assembled layers of helicenes on surfaces for asymmetric heterogeneous catalytic activity.

List of Abbreviations

DFT	density functional theory
dr	diastereomeric ratio
ee	enantiomeric excess
er	enantiomeric ratio
Ment	menthyl
NHC	*N*-heterocyclic carbene
RCM	ring-closing metathesis and
ROM/CM	ring-opening metathesis-cross metathesis
rt	room temperature
STM	scanning tunnel microscopy
TCB	1,2,4-trichlorobenzene
Tol	*p*-tolyl
Ts	toluene-4-sulfonyl

UHV ultra-high vacuum

References

1 Yang, Y., da Costa, R.C., Fuchter, M.J., and Campbell, A.J. (2013). *Nat. Photonics* 7: 634–638.

2 Yang, Y., da Costa, R.C., Smilgies, D.-M. et al. (2013). *Adv. Mater.* 25: 2624–2628.

3 Dhbaibi, K., Abella, L., Meunier-Della-Gatta, S. et al. (2021). *Chem. Sci.* 12: 5522–5533.

4 Kiran, V., Mathew, S.P., Cohen, S.R. et al. (2016). *Adv. Mater.* 28: 1957–1962.

5 Kettner, M., Maslyuk, V.V., Nürenberg, D. et al. (2018). *J. Phys. Chem. Lett.* 9: 2025–2030.

6 Brock, C.P., Schweizer, W.B., and Dunitz, J.D. (1991). *J. Am. Chem. Soc.* 113: 9811–9820.

7 Martin, R.H. and Marchant, M.J. (1974). *Tetrahedron* 30: 343–345.

8 Jacques, J., Collet, A., and Wilen, S.H. (1994). *Enantiomers, Racemates, and Resolutions*. Malabar, FL: Krieger Pub. Co.

9 Ernst, K.-H. (2016). *Acc. Chem. Res.* 49: 1182–1190.

10 Seibel, J., Parschau, M., and Ernst, K.-H. (2014). *J. Phys. Chem. C* 118: 29135–29141.

11 Parschau, M., Fasel, R., and Ernst, K.-H. (2008). *Cryst. Growth Des.* 8: 1890–1896.

12 Fasel, R., Parschau, M., and Ernst, K.-H. (2003). *Angew. Chem. Int. Ed.* 42: 5178–5181.

13 Fasel, R., Parschau, M., and Ernst, K.-H. (2006). *Nature* 439: 449–452.

14 Ernst, K.-H. (2010). *Origins Life Evol. Biospheres* 40: 41–50.

15 Taniguchi, M., Nakagawa, H., Yamagishi, A., and Yamada, K. (2003). *J. Mol. Catal. A Chem.* 199: 65–71.

16 Rahe, P., Nimmrich, M., Greuling, A. et al. (2010). *J. Phys. Chem. C* 114: 1547–1552.

17 Hauke, C.M., Rahe, P., Nimmrich, M. et al. (2012). *J. Phys. Chem. C* 116: 4637–4641.

18 Balandina, T., van der Meijden, M.W., Ivasenko, O. et al. (2013). *Chem. Comm.* 49: 2207–2209.

19 van der Meijden, M.W., Gelens, E., Quirós, N.M. et al. (2016). *Chem. - Eur. J.* 22: 1484–1492.

20 Ascolani, H., van der Meijden, M.W., Cristina, L.J. et al. (2014). *Chem. Comm.* 50: 13907–13909.

21 Stöhr, M., Boz, S., Schär, M. et al. (2011). *Angew. Chem. Int. Ed.* 50: 9982–9986.

22 Shchyrba, A., Nguyen, M.-T., Wäckerlin, C. et al. (2013). *J. Am. Chem. Soc.* 135: 15270–15273.

23 Narcis, M.J. and Takenaka, N. (2014). *Eur. J. Org. Chem.* 2014: 21–34.

24 Aillard, P., Voituriez, A., and Marinetti, A. (2014). *Dalton Trans.* 43: 15263–15278.

25 Hasan, M. and Borovkov, V. (2018). *Symmetry* 10: 10.

26 Demmer, C.S., Voituriez, A., and Marinetti, A. (2017). *C. R. Chimie* 20: 860–879.

27 Reetz, M.T., Beuttenmüller, E.W., and Goddard, R. (1997). *Tetrahedron Lett.* 38: 3211–3214.

28 Monteforte, M., Cauteruccio, S., Maiorana, S. et al. (2011). *Eur. J. Org. Chem.* 5649–5658.

29 Nakano, D. and Yamaguchi, M. (2003). *Tetrahedron Lett.* 44: 4969–4971.

30 Reetz, M.T. and Sostmann, S. (2000). *J. Organomet. Chem.* 603: 105–109.

31 Yamamoto, K., Shimizu, T., Igawa, K. et al. (2016). *Sci. Rep.* 6: 36211.

32 Tsujihara, T., Inada-Nozaki, N., Takehara, T. et al. (2016). *Eur. J. Org. Chem.* 2016: 4948–4952.

33 Krausová, Z., Sehnal, P., Bondzic, B.P. et al. (2011). *Eur. J. Org. Chem.* 2011: 3849–3385.

34 Sanchez, I.G., Samal, M., Nejedly, J. et al. (2017). *Chem. Comm.* 53: 4370–4373.

35 Karras, M., Dąbrowski, M., Pohl, R. et al. (2018). *Chem. – Eur. J.* 24: 10994–10998.

36 Yavari, K., Aillard, P., Zhang, Y. et al. (2014). *Angew. Chem. Int. Ed.* 53: 861–865.

37 Aillard, P., Voituriez, A., Dova, D. et al. (2014). *Chem. – Eur. J.* 20: 12373–12376.

38 Aillard, P., Retailleau, P., Voituriez, A., and Marinetti, A. (2015). *Chem. – Eur. J.* 21: 11989–11993.

39 Kawasaki, T., Suzuki, K., Licandro, E. et al. (2006). *Tetrahedron: Asymmetry* 17: 2050–2053.

40 Sato, I., Ryutaro Yamashima, R., Kadowaki, K. et al. (2001). *Angew. Chem. Int. Ed.* 40: 1096–1098.

41 Matsumoto, A., Yonemitsu, K., Ozaki, H. et al. (2017). *Org. Biomol. Chem.* 15: 1321–1324.

42 Takenaka, N., Sarangthem, R.S., and Captain, B. (2008). *Angew. Chem. Int. Ed.* 47: 9708–9710.

43 Chen, J. and Takenaka, N. (2009). *Chem. – Eur. J.* 15: 7268–7276.

44 Takenaka, N., Chen, J., Captain, B. et al. (2010). *J. Am. Chem. Soc.* 132: 4536–4537.

45 Chen, J., Captain, B., and Takenaka, N. (2011). *Org. Lett.* 13: 1654–1657.

46 Samal, M., Misek, J., Stará, I.G., and Starý, I. (2009). *Collect. Czech. Chem. Commun.* 74: 1151–1159.

47 Gicquel, M., Zhang, Y., Aillard, P. et al. (2015). *Angew. Chem. Int. Ed.* 54: 5470–5547.

48 Aillard, P., Gicquel, M., Yavari, K. et al. (2018). *Eur. J. Org. Chem.* 2018: 5853–5860.

49 Giaconi, N., Sorrentino, A.L., Poggini, L. et al. (2021). *Angew. Chem. Int. Ed.* 60: 15276–15280.

50 Mairena, A., Wäckerlin, C., Wienke, M. et al. (2018). *J. Am. Chem. Soc.* 140: 15186–15189.

51 Stetsovych, O., Švec, M., Vacek, J. et al. (2017). *Nature Chem.* 9: 213–218.

52 Stetsovych, O., Mutombo, P., Švec, M. et al. (2018). *J. Am. Chem. Soc.* 140: 940–946.

Index

a

absorption coefficients 147, 148, 397
acene-bridged PDI dimers 348
acene-linked PDIs 348
acetylated diaza [4]helicene 137
achiral or racemic helical nanographenes 292–297
achiral polyfluorene-based polymers 338
acridinium 130, 142
acridinium chromophore 144, 145
acridinium system 128
1,8-bis(1-adamantyl)naphthalenol (*P*) 242
1,8-bis(1-adamantyl) structure 241
Adiabatic Hessian or the Vertical Hessian model 386
alkyl diphosphine oxide 226
alkyne trimerization 329
aminoacridinium intermediate 145
amino-carbo[6]helicene 524
amino diaza [4]helicene 159
5-aminohelicene 507, 508
5-amino-hexahelicene 507–509
aminomethylhelicene 246
amino-pyridinium 520, 521
annulative π-extension (APEX) 268, 285
anti-glare filters 476, 480, 481
arene-linked triynes 104
aromatic ketone 121, 122
aromatic ortho-fused polycyclic compounds 424
aryl lithium 345
aryl-substituted alkyne 266
asymmetrically fused double [7][5]helicene 379
asymmetric catalysis 237, 246–247, 329, 434, 505–524
asymmetric hydrogenation 224
asymmetric induction 247
asymmetric organocatalysis
 azahelicene catalyzed enantioselective acyl transfer reaction 522
 helical amino-pyridium catalyzed asymmetric addition to nitroalkenes 520–521
 helical chiral 2,2′-bipyridine *N*-monoxides 522
 helicene *N*-oxide catalyzed enantioselective desymmetrization of *meso* epoxides 519
 phosphahelicene catalyzed enantioselective [3+2] annulation 522, 523
asymmetric photosynthesis 1
asymmetric synthesis of double helicenes 109
asymmetric synthesis of nonracemic helicenes
 catalytic enantioselective synthesis 82–84
 stoichiometric diastereoselective synthesis 76–82
asymmetric transition metal catalysis 510–519
Au(I)-catalyzed quadruple intramolecular hydroarylation 115
Au(111)/1,2,4-trichlorobenzene (TBC) 509
autocatalytic Soai reaction 517–519
axial chirality 66, 205, 329, 436, 447, 510, 513, 524

Helicenes: Synthesis, Properties, and Applications, First Edition.
Edited by Jeanne Crassous, Irena G. Stará, and Ivo Starý.

axially chiral biaryls 110, 115, 336
axially chiral bis(aryllithium)-Li_2 121
axially fused carbo[6]helicenes 362
aza-bicyclo-[4.1.0]heptenes 171, 516
azabora 378
azabora[5]helicene dyes 387
azaborahelicenes 28, 343
azaborahexahelicenes 343
azadioxatriangulenium (ADOTA) 130
aza[4]helicenes 338, 339, 439, 440
4-aza[6]helicene 16, 169
aza[6]helicene-phosphole 176, 400, 402
aza[6]helicenes 6, 16–18, 115, 116, 168–171, 176, 177, 335, 338, 345, 400, 439, 482, 517, 522, 523
aza[9]helicene 122
azahelicene 15–23, 63–68, 115–117, 128, 145, 147, 176, 177, 188, 335–341, 343, 361, 384, 400–402, 404, 424, 437–440, 522
  catalyzed enantioselective acyl transfer reaction 522
azahelicene-phosphole Pd(II) and Cu(I) complexes
  multitopic helicenic ligands 402–404
  π-conjugated helicenic P,N bidentate ligands 400–402
azahelicene-phospholes 176, 177, 400, 401, 404
5-azahexahelicene 385
azaoxa 130, 131
azaoxa [5]helicene 142
azaoxa[6]helicene 67, 79, 439
azaoxahelicenes 66, 72, 378
azathia [4]helicene 131
azoniahelicenes 15, 19, 20, 68–71, 345

b
3,3′-bis(benzo[1,2-*b* +4,3-*b*′]dithiophene) derivatives 200, 204
3,3′-bis(benzo[1,2-*b* +4,3-*b*′] dithiophene)dihaldehyde 205
benzo[1,2-*b* +4,3-*b*′]dithiophene units 205
1,2-bis-(2-benzo[1,2-*b* +4,3-*b*′] dithiophenyl) ethene 202
benzocarbo[*n*]helicenes 11, 12
1,2-bis(benzodithienyl)ethenes 200, 201
1,2-bis(benzodithienyl)ethenes 200, 201
benzodithiophene 202–205
benzo-fused diazatriangulenium dyes (BDATA) 133
benzo[*ghi*]perylenes 7, 9, 15, 17, 20, 31, 32
benzo[5]helicenes 11, 12, 15, 55, 305, 429, 432
benzo[*c*]phenanthren-2-ols 120
1,4-benzoquinone 53, 341
benzoyl peroxide (BPO) 154
benzylamine 513
biaryl bisphosphine 105, 106, 110
bicyclic helicenes 241–242, 257
bidentate helicenes 171–177
bihelicenol 246, 247
bihelicenol phosphite ligand 247
5,5′-bihelicenol (*P*,*P*,*R*/*S*) 239
(*R*)-BINAP complex-catalyzed intramolecular [2 + 2 + 2] cycloadditions 108
binaphthyl-linked terminal tetrayne 111
biphenyl-linked terminal tetrayne 110
2,2′-bipyridine 66, 171, 275, 301, 522
bipyridine-helicene-based multifunctional pH-switchable platform 408–410
bipyridine-helicene-rhenium complexes 410–411
bipyridine-helicene-ruthenium 173
bisoxazine-based helicene derivatives 439
2,20-bipyridine 444
2,2′-bithiophene (α-T_2) 212
blue CPL [7]helicene-like compound 426
borahelicenes 343
borohelicenes 435–438
boronated arylquinoline skeleton 437
boron-azahelicenes 437, 438
boron-fused double [5]helicene 435
borylated arylisoquinoline skeleton-based helicenes 387
bowl-shaped corannulene core 382
bowl-shaped corannulene derivatives 270
2,7-Bpin anthracene 348
2,7-Bpin naphthalene 348
brominated benzo[a]xanthone 135
brominated dioxa [5]helicene 135
2-bromo-3,3′-bibenzo[1,2-*b* +4,3-*b*′]dithiophenes 205
2-bromobenzoic acid 138

bromo-functionalized cationic [5]helicenes 134
bromohelicenes 32
2-bromohexahelicene 374, 375, 385
Buchwald–Hartwig cross-coupling reactions 138
$[Bu_4N][PF_6]$ 156

c

C^N chelate ligands (Pt, Ir, Os) 182–185
carbocation-based heterohelicenes 345
carbo[5]helicene 10, 170, 290, 292, 294–301, 304, 305, 313, 477
carbo[6]helicene 10, 11, 115, 116, 118, 119, 170, 175, 179–181, 184, 224, 290, 295, 299, 301, 331, 333, 358, 362, 474, 492, 510, 513, 518, 524
carbo[7]helicene 110, 111, 191, 290, 298, 299, 301–304, 495, 524
carbo[9]helicene 111, 302
carbo[*n*]helicenes 329, 376, 505
carbohelicenes 62, 108, 268, 424, 427
carbon 290–292
  achiral or racemic 292
  Chiral or enantioenriched 297
carbon helical nanographenes
  achiral or racemic 292–297
  chiral and enantioenriched 297–305
carboxaldehyde 136, 137
carboxylic derivative 152
cata-condensed systems 374
catalytic enantioselective synthesis 82–84, 103, 104, 120
cationic π-systems 127
cationic [4]helicene 128, 131, 145, 154, 157
cationic [5]helicene 133–135, 137
cationic [*n*]helicene 127, 161
cationic Au(I)/(*R*)-BINAP complex 115
cationic Au(I)/(*R*)-BINAP complex-catalyzed enantioselective sequential intramolecular hydroarylation 115
cationic Au(I)/(*R*)-BINAP complex-catalyzed quadruple hydroarylation 116
cationic Au(I)/(*R*)-Difluorphos catalyst 116
cationic Au(I)/(*R*)-MeO-DTB-BIPHEP complex-catalyzed intramolecular hydroarylation 116
cationic diaza [4]helicenes 135, 136, 138, 151, 155, 158, 161
cationic diaza [6]helicenes 135
cationic dibromo diaza [6]helicene 156
cationic dioxa[6]helicene 439
cationic heterohelicenes 343
bis cationic imidazolium salt 160
α-cationic phosphonites 118–120
cationic Rh(I)/(*R*)-H_8-BINAP 108, 110
cationic Rh(I)/(*R*)-Segphos 110
cationic Rh(I)/(*S*)-Difluorophos complex 111
cationic Rh(I)/(*S*)-H_8-BINAP complex 112
cationic Rh(I)/(*S*)-Segphos catalyst 112
cationic Rh(I)/(*S*)-xyl-Segphos catalyst 110
cationic rhodium (I) complexes (±) 224
cationic triarylcarbenium helicenes
  absorption and fluorescence properties 144–148
  cellular imaging 151–153
  central nucleophilic additions 140–141
  chiroptical properties 148–150
  early-stage equatorial functionalization of [4]helicene 141–142
  electrochemiluminescence (ECL) 154–155
  generalities 135
  helical pitch and configurational stability 142–143
  interaction with DNA 151–153
  north functionalization 135–137
  organometallic chemistry and photocatalysis 159–161
  south functionalization 137–139
  spintronics 161
  stable radicals 157–158
  synthesis and resolution of 128–135
  thin film voltammetry 156–157
  tropical functionalization 139–140
cationic triarylmethyl Ar_3C^+ dyes 127
cationic triarylmethyl helicene series 145
CH_2Cl_2 / CH_3CN 146

chalcogen-containing helicene derivatives 27
charge transfer (CT) complexation 244–246
chiral amine [(–)-sparteine] 121
chiral and enantioenriched helical nanographenes 297–305
chiral BODIPY derivatives 137, 436
Chiralcel® OD column 210
chiral formamides 121, 211
chiral induced spin selectivity effect 494–496
chirality-induced spin selectivity (CISS) 161
chiral lanthanide complexes 424
chiral ligand [AuCl(SMe_2)/(*S*)-BINAP] 336
chiral metal diamine dithiolene complex 444
chiral molecules 199, 244, 329, 384, 397, 455, 467, 473, 487, 492, 494, 495
chiral nanographene embedding hextuple helicene units 277
chiral nanographenes 86, 264, 275, 277
chiral neomenthylindene Co(I) complex 105
Chiralpack® IA column 210
chiral peropyrene enantiomers 427
chiral phosphoric acid catalyst 121, 122
chiral polymers 424
chiral S-shaped double helicene-like molecule 109
chiral tetrathiahelicene-based alkyl phosphine-borane complexes 219–220
chiral vanadium complex 120
chiroptical switching 149, 167, 168, 173, 177, 341, 342, 345, 406
Chlamydomonas reinhardtii 152
4-chloro isomer 142
2,13-bis(chloromethyl)[7] thiaheterohelicene 208
2-chloro-phenyl boronic acid 270
cinnamyl methyl carbonate 513
circular dichroism (CD) 14, 149, 167, 215, 219, 289, 301, 333, 341, 374, 395, 398, 459, 466, 473
circularly polarized electro-luminescence (CPEL) 432, 448
circularly polarized light (CPLight) 3, 148, 220, 289, 332, 361, 397, 398, 423, 424, 466, 481
circularly polarized luminescence (CPL)
  carbohelicenes 424–434
  defined 423
  heterohelicenes 434–442
  spectra 374
circularly polarized organic light-emitting diode (CP-OLED) 432, 442, 482
circularly polarized phosphorescent light-emitting diodes (CP-PHOLEDs) 484–487
classic electron acceptors 345
cobalt-mediated/catalyzed [2 + 2 + 2] cycloisomerisation 66
π-conjugated diketopyrrolopyrrole helicene derivatives 427
π-conjugated helicenic P,N bidentate ligands 400–402
π-conjugated long molecular structure 374
cooling-water circulation 3
coordination chemistry
  bidentate helicenes 171–177
  monodentate helicenes 168–171
  polydentate helicenes 177–179
corannulene 12, 27, 263, 270, 294, 301, 315, 353, 356, 382
Cotton effects 150, 302, 303, 305, 312, 346, 348, 350, 352, 353, 355, 365, 380
coumarin-fused helicene 116, 118
coupled-perturbed Kohn-Sham (CPKS) level 457
$CpCo(CO)_2/PPh_3$ 56
C_2-symmetric fused and annulated double [6]helicenes 379
cyclic amide monomer (*P*) 241
cyclic anhydride tetramer (*P*) 240
[2 + 2 + 2] cycloaddition strategy 103
cyclobis[4]helicene 274
cycloborylated hexa-, octa-, and deca-helicenes 343
cyclodextrins (CDx) 242, 243

cycloiridiated complexes with helicenic N-heterocyclic carbene 411–413
cyclometallated helicenic complexes
C^N chelate ligands (Pt, Ir, Os) 182–188
NHCs (Ir, Re) and related complexes 184–188
sandwich-type helicenic complexes 188–192
cyclopentadienyl (Cp) anions 188
[2+2+2] cyclotrimerization 266

d

D-band for PAH 375
degenerate four-wave mixing (DFWM) 454
dehydrative p-extension (DPEX) 285
density functional theory (DFT) 150, 216, 333, 374, 398, 459, 505
deoxynucleosides 250
Depolarization Ratio (DR) 455
(*M*)-diacetate 210
(*P*)-diacetate 210
diacetylenic carbo[6]helicene 331
dialdehyde 201, 204, 205, 223, 240, 335, 346
dialkyl acetylene dicarboxylates 115
N,*N*′-dialkylated PDI 348
1,2-dialkylquinoxalinefused [7]helicene 426
dialkynyl ketones 110–111
dialkynylphosphine oxides 110, 112
diamino benzene derivative 346
diaminohclicene 244
1,8-diaminonaphthalene-based tetrayne 116
2,7-diaminonaphthalene-based tetrayne 116
diaryltetrayne 435
diaza derivative 131, 135, 155
diaza [4]helicene 127, 136, 137, 141, 142, 145, 148, 149, 151, 155, 157–159, 161, 162, 439, 440
diaza [5]helicene 17, 133, 135, 143
diaza [6]helicene 16, 132, 136–138, 143, 147, 156, 345, 439
diaza[7]helicene 18, 308, 346, 476
8,11-diaza[7]heptahelicene 20
diaza[10]helicene 116
diazaoxatriangulenium (DAOTA) 128
diazapentacenes 148
diazatetracene 148
dibenzo(C,P)chrysene 264
dibenzo(hetero)[5]helicenes 54
dibenzo[6]helicene 57, 58, 82, 104, 105, 514, 515
dibenzo[7]helicene 82, 84, 105, 108, 514, 515
dibenzochrysene 268, 274
dibenzocoronene diimide core 353
dibenzopyrido[n]helicenes 64
(*Z*)-diboronic acid esters 203
B,*B*-dibromo-azaborole system 343
2,11-dibromo-[4]helicene 274
7,8-dibromo[5]helicene 275, 277
9,10-dibromo-[7]helicene 277
dibromomaleimide 203
6,13-dicyano[7]helicene 507, 509
Diels–Alder reaction of adamantylbenzyne 241
differential molar extinction coefficient 377
(*S*)-Difluorphos 108
dihalogenated biphenyl substrate 268
dihydro-aza[6]helicene 121, 122
dihydro-aza[7]helicene 121
dihydro-dioxa[6]helicene 122
dihydrohelicenes 1, 28, 30
4a,4b-dihydrophenanthrene (DHP) 2
9,10-diketo[7]helicene 32
diketopyrrolopyrrole 427
diisopropyl (1-bromonaphthalen-2-yl) phosphonate 513
di*iso*propylethylamine (DIPEA) 139
dimethoxychromenoacridinium (DMCA) 130
dimethoxychromenoxanthenium (DMCX) 130
dimethoxy [4]helicene DMQA 142
2,6-dimethoxyphenyllithium 131, 133
dimethoxyquinacridinium (DMQA) 127
dimethylacetylene dicarboxylate 114
(*P*)-2,13-bis(dimethylaminomethyl)-tetrathia[7]helicene 227

1,12-dimethylbenzo[*c*]phenanthrene-5,8-dicarboxylaic acid 238
1,12-dimethylbenzo[*c*]phenanthrenes 239
dimethyl-substituted diyne 119
dinaphtho[2,1-*i* +1′,2′-*l*]hexahelicene 379
N,N′-di(1-naphthyl)-*N,N*′-diphenyl-(1,1′-biphenyl)-4,4′-diamine 479
dinitro derivative 137, 239
dioxa [4]helicene 130, 142, 143, 145
dioxa [5]helicene 133, 135
dioxa [6]helicene 122, 131, 139, 140, 147, 148, 150
diphenanthro[3,4-*c* +3′,4′-*l*]chrysene 379
2,15-bis(diphenylphosphino)-carbo[6]helicene ligand (*M*) 510
1-(diphenylphosphino)-pentahelicenic ligands 512
bis(diphenylphosphino)-thiahelicene ligand (*P*) 510
1-diphenylphosphinoxy-carbo[6]helicene (*P*) 513
1-diphenylphosphinoxy-5,6,9,10-tetrahydro[6]helicene (*P*) 513
1,3-diphenylpropenylacetate 512
7,8-dipropyltetrathia[7]helicene 121, 222
dispersed silica (*P*)-nanoparticles 252, 256
displays 127, 149, 187, 228, 302, 346, 403, 423
dissymmetry factor 171, 289, 299, 301, 305, 332, 335, 338, 350, 357, 389, 396, 397, 434, 480, 481, 483, 486
1,12-disubstituted carbo[4]helicenes 119
3,6-disubstituted phenanthrene linker 350
7,8-disubstituted 7-THs 205
dithia-and tetrathia[7]helicenes 462
dithia[6]helicene 23, 313–315, 444
dithia[7]helicene 17, 223, 466
1,2-dithienylbenzenes 24
dithienylethene (DTE) 180
diversity-oriented helicene synthesis 104
divinyl benzene 341
1,3-diynes 104, 110
diynes 60, 71–74, 78, 83, 84, 104, 110, 112, 115, 118, 119, 122, 124, 295
bis(diynones) 116
donor–acceptor helicenes 331
double azahelicene 115–117, 335, 336, 384, 424
double BODIPY (bis(boron-dipyrromethene)) complex 436
double carbo[6]helicenes 116, 298, 299
double cross-coupling Suzuki–Miyaura reactions 268
double Fischer indolization 122
double[8]helicenes 350, 352
double Wittig olefination 335
Duschinsky and Herzberg–Teller effects 386
Dy(III) ion 171
dynamic light scattering (DLS) 251

e

(*M*)-enantiomers 161
(*P*)-enantiomer 386
eight-shaped azahelicene dimer 438
8-shaped helicene dimer 425
electric field-induced second harmonic (EFISH) technique 221
electric second-order NLO responses 456–465
electrochemiluminescence (ECL) 154–155
electrodeless discharge lamps (EDLs) 5, 6
electroluminescent dissymmetric factor 432
electron-accepting azahelicenes 338
electron-accepting borahelicenes 343
electron-accepting π-extended frameworks embedded with helicenes
  extended helicenes with naphthalimide units 346–348
  perylene diimide (PDI) 348–363
  terrylene-based helicenes 363–365
electron-accepting heterohelicenes
  azahelicenes 335–341
  borahelicenes 343
  cationic heterohelicenes 343–345
  incorporation of oxygen atoms 341–343

electron donating carbo[6]helicene core 331
electron hole 405, 407
electronic circular dichroism (ECD) 14, 149, 150, 167, 219, 289, 333, 374, 395, 396, 401, 402, 473
electron-poor dialkynylketones 110
electron-poor ester-linked tetraynes 110
electron-poor internal tetrayne 111
electron-rich terminal tetrayne 111
electron-rich tetraynes 110
electrophilic aromatic substitutions involving aromatic halides 273–274
electrospray ionization mass spectrometry (ESIMS) 191
elegant bilayer helical nanographene 263
elusive *o*-quinone analogue 341
enantiomeric helical aromatic imides 429
enantiomerically pure substituted (*M*)-tetrathia-[7]-helicenes 221
enantiomerically stable PDI-based [5]helicenes 352
enantiopure aza[6]helicene 338
enantiopure BINPHAT anion 129, 132
enantiopure 9,10-dibromo-[7]helicene 277
enantiopure 2,13-dimethyltetrathia[7] helicene 205
enantiopure helicene 76, 79, 152, 161, 177, 179, 184, 188, 263, 480, 482, 488, 495, 519, 524
enantiopure mono-cycloplatinated-[8] helicene 442
enantiopure or racemic 4-aza[6]helicene 168
enantiopure 7-TH tertiary diphosphine oxides 225
enantioselective Ni(0)-catalyzed [2+2+2] cycloisomerization 514
enzymatic resolution 210
enzyme inhibitor 250
expanded helicenes 53, 87, 88, 286, 287, 384
extended helicenes with naphthalimide units 346–348
π-extended nanohelicene 429
π-extended PDI-heterohelicenes 361
extended transition state–natural orbitals for chemical valence (ETS-NOCV) 412
external chamber photoreactor 3
bis(2-ethylhexyl)sebacate 156

f

Fc-based terminal alkyne 208
ferroelectricity 248–249
fill factor (FF) 491, 493
fluorescence quantum yield 299, 303, 307, 313, 363, 378, 384, 423, 424, 448
2-formyltetrathia[7]helicene (±) 228
fourfold alkyne cyclization reaction 427
Friedel–Crafts acylation 296, 329
functionalized helicenes 62, 76, 207, 221, 309, 334, 440
functionalized dioxa [5]helicenes 135
fused double-chrysene derivative 268
fused double [7]helicene 191, 379
fused triple [5]helicenes 380, 381

g

G-band for polycyclic aromatic hydrocarbons (PAH) 375
gold(I) complexes 118, 119, 180, 181, 224, 225, 518
gold(I) of phosphahelicene 171
graphene 86, 264, 279, 283–287, 290, 305, 382

h

α-halogenated benzoic acids 137
halloysite nanotubes (HNTs) 228
Hartree–Fock (HF) perturbation theory 398
(*R*)-H_8-BINAP catalyst 110
Heck–Mizoroki reactions 265
Heck–Mirozoki/Suzuki–Miyaura cross-coupling reactions 266
helical amino-pyridium catalyzed asymmetric addition to nitroalkenes 520–521
helical chiral *N,N,O,O*-boron chelated dipyrromethenes 435
helical chiral 2,2′-bipyridine *N*-monoxides 522

helical diphosphane 510
helically chiral aza[6]helicenes 115
helically chiral fully benzenoid carbo[6]helicenes 115
helically chiral helicene 378
helically twisted graphenes 382
helical nanographenes
  carbon 290–292
  heteroatom-containing 305
  structural and photophysical properties 316–321
helical NHCs in enantioselective Ru-catalyzed metathesis 514–516
π-helical non-fullerene acceptors 333
helical π-conjugated systems 331
[4]helicene 127–131, 136–138, 141–149, 151, 152, 154, 155, 157–159, 161, 162, 180, 187, 264, 265, 268, 274, 275, 289–291, 346, 361, 447, 465, 488
[5]helicene 6, 7, 9, 10, 13–17, 19, 20, 22, 31, 32, 60, 61, 67, 68, 79, 82, 114, 115, 128, 133–135, 137, 138, 142, 143, 146, 150, 187, 188, 191, 264, 268, 275, 277, 279, 289, 300, 301, 306, 313, 346, 352–354, 364, 376, 380, 381, 384, 425, 426, 429, 433, 435, 446, 465, 474, 488
[5]helicene-containing TDI 364
[6]helicene
  precursor 350
  skeleton 427
[7]helicene-based PDI dimers 350
[9]helicene-like molecules 108, 441
[11]helicene-like molecule 109
helicene
  advanced helicene architectures 85–88
  advantage of 473
  asymmetric catalysis 246–247
  asymmetric synthesis of nonracemic 76–84
  bicyclic helicenes 241–242
  carbo-and heterohelicenes
    azahelicenes 63–68
    azoniahelicenes 68–71
    carbohelicenes 62–63
    oxahelicenes 71–72
    sila-, phospha-and thiahelicenes 73–75
  chiral-induced spin selectivity effect 494–496
  coordination chemistry
    bidentate helicenes 171–177
    monodentate helicenes 168–171
    polydentate helicenes 177–179
  ECD spectra 385–389
  electric second-order NLO responses 456–465
  electronic properties
    ferroelectricity 248–249
    self-assemblymonolayer (SAM) method 247–248
  enantiomers *vs.* racemates
    CP photodetectors 489–490
    helicene bulk charge transport and OFETs 487–489
    organic photovoltaics 490–494
  functionalized 76
  grafted nanoparticles
    aggregation of nanoparticles 251
    gold nanoparticles 251
    metal-catalyzed chemical equilibrium 252
    molecular recognition and separation 252–256
    silica nanoparticles 251–252
    time event 256–257
  helical length dependence of chiroptical properties of 376–378
  helicene aggregates and self-assembled monolayers 467–468
  interactions with biological macromolecules
    enzyme inhibitor 250
    nucleic acid binding 249–250
  noncovalent bond interactions and chiral recognition
    charge transfer (CT) complexation 244–246
    Langmuir–Blodgett (LB) technique 244
    oligosaccharides in water 242–243
    self-assembly gel formation 246
  OLED 474
  optical activity of 395

organometallic chemistry
cyclometallated helicenic complexes 182
metal-ethynyl-helicene complexes 179–181
metal-vinyl-helicene complexes 179
photochemical transformations 31–33
of π-electron systems
intermolecular alkyne [2 + 2 + 2] cycloisomerisation 59–61
intramolecular [2 + 2 + 2] cycloisomerisation 54–59
racemic and enantiopure forms 473
with remote electron-withdrawing functionalities 330–335
small numbers of benzene rings 237–238
symmetry impact 378–384
synthesis of 53
tetracyclic helicenes 238–241
third-order NLO responses 466
helicene backbones 9, 23, 27, 35, 57, 89, 330, 348, 456, 461–464, 514
helicene-based azacarbene 444
helicene-based nanostructures 378
helicene-based NGs 290
helicene bis-*p*-quinones 341
helicenebisquinone 341, 342, 456, 467
helicenediamide 248
helicenediamine 249
helicenedimethanol 245
helicenedi(methylamine)s 249
helicene-ethynyl-iron complexes 180
helicene nanoribbons 53, 362, 363
helicene N-oxide catalyzed enantioselective desymmetrization of *meso* epoxides 519–520
helicene optical activity 396–400
helicene quinone 341
helicene synthesis
metal-catalyzed hydroarylation 115–120
metal-catalyzed oxidative annulation 120
Ni-and Co-catalyzed [2 + 2 + 2] cycloaddition 104–105
organocatalyzed annulation 121–123
Pd-catalyzed [2 + 2 + 2] cycloaddition 114–115
Rh-and Ir-catalyzed [2 + 2 + 2] cycloaddition 105–114
helicenethiols 251
helicenic electron acceptors 329, 330, 333, 365
helicenic phosphite in enantioselective Ir(I)-catalyzed allylic amination 513
bis-helicenic terpyridine ligand 443
helicenoids 28, 31, 362
helicenyl arynes 60, 115
HelPhos catalysts 523
heptahelicene 20, 506–507, 509
(hetero)arylboronic acids 205
heteroatom-containing helical nanographenes
heteroatoms of group 15 (N, P) 306–312
heteroatoms of group 16 (O, S) 313–322
heteroditopic ligands 176
hetero-double-helices 254
hetero[6]helicenes 330
hetero[*n*]helicenes 34, 378
heterohelicenes
azahelicenes 438–440
borohelicenes 435–438
metal helicene derivatives 442–445
oxalohelicenes 440–441
regulation of CPL 446
silahelicenes 435
thiohelicenes 441–442
hexabenzo[7]helicene 104, 297, 382, 383
hexabenzotriphenylene 60, 299, 313, 429
hexahelicene 151, 191, 220, 302, 373–375, 379, 385, 389, 459, 461, 507–509
hexa-*peri*-hexabenzocoronenes (HBC) 285, 346
hexa-*peri*-hexabenzo[7]helicene 297, 382, 393
hexaphenylbenzene core 346
hexapyrrolohexaazacoronene (HPHAC) 308, 309
hexathia[7]helicene 26
hexathia[11]helicene 24, 507

hexathiahelicene 211, 213
hexayne 68, 69, 82, 84, 88, 109
highest occupied and lowest unoccupied molecular orbitals (HOMO-LUMO) 212, 213, 294, 295, 297, 338, 389, 409, 410, 473
highly oriented pyrolytic graphite (HOPG) 286
histamine *N*-methyl transferase (HNMT) 250
homophthalimide 346
Hoveyda–Grubbs catalyst 514
Hückel approximation 374
Hünig's base 139, 140, 345
hydrazines 121, 122, 129
hydrogenated (*S*)-2|b and (*S*)-4|b enantiomers 510
hydrohelicenes 447
5-hydroxyhelicene 239
hydroxyl-containing dioxa 135
2,13-(hydroxymethyl)tetrathia[7]helicene 210
hyperpolarizability tensor elements 453, 458
hyper-Rayleigh scattering (HRS) 222, 454

i

immersion well photoreactor 3, 5
imine tetramer (*P*) 240
indoline-derived enamines 140
inherently dissymmetric chromophore 374
in situ acylation/S_NAr sequence 137
in situ drop-coating method 355
intermediate neglect of differential overlap for spectroscopic modification (INDO/S) method 377
intermolecular alkyne [2 + 2 + 2] cycloisomerisation 59
intraligand charge transfer (ILCT) 403
intramolecular charge transfer (ICT) 333, 448
2-iodo-benzodithiophene 203
ionization potential (IE) 217, 218, 399
ion mobility separation (IMS) 191
iron-ethynyl-helicenes 406–407
Ir(I)/(*S,S*)-Et-FerroTANE complex 112

k

Katz's conditions 3, 16
Knoevenagel condensations 20, 71, 346
Kohn–Sham (KS) molecular orbitals (MOs) 358
Kramers–Kronig relation 374, 389

l

ladder copolymers 179
Langmuir–Blodgett (LB) films 177, 244, 341, 467
laterally-extended expanded helicenes 287
lead photocyclizations 3
leuco adducts 140, 141
ligand-to-ligand CT (LLCT) 403
ligand-to-metal CT (LMCT) 405
light-emitting polymers (LEPs) 482
luminescence dissymmetric factor 423, 432
LUMO coupling 489

m

magnetic flux density 377
majority-rule effects 507
Mallory photocyclization 329
Mallory reaction 2, 8, 201, 335
metal-catalyzed homocouplings with aromatic halides 274–279
metal-catalyzed hydroarylation 115–120
metal-catalyzed oxidative annulation 120
metal-ethynyl-helicene complexes 179–181
metal helicene derivatives 442–445, 448
metal-to-ligand CT (MLCT) 403
metal-vinyl-helicene complexes 179
methoxy-diaza [5] helicene 143
2-methoxy-1-naphthaldehyde 131
(2-methoxynaphthalen-1-yl)lithium 131
2-methoxyphenyllithium 133
methyl-2-acetamidoacrylate 511
2-methylene-succinate 511
2-methylhexahelicene 385
Mizoroki–Heck couplings 266
molar extinction coefficient 147, 377
molecular spring 208, 223
molecularly defined PAHs 268
Møller-Plesset perturbation theory 216

molten pyridine hydrochloride 133
(*M*)-monoacetate 210
mono-and bis-cycloplatinated helicenes 407–408
mono-and bispentacenes 139
mono-and bistetracenes 139
mono-and disubstituted tetrathia[7]helicene 457
mono and disubstituted 7-TH nitriles 207
monoaza[5]helicenes 477
monoazahelicenes 336
mono-/bisaddition products 139
mono carboxaldehyde 136
monodentate helicenes 168–171
mono[8]helicene 350
mono-oxa[5]helicenic unsymmetrical imidazolium salt 514
monosubstituted 7-THs 205
multihelicenic platforms 263–279
multi-hydrogen-functionalized helicene derivatives 440
multiple carbo[*n*]helicenes 13–14, 33
multipole helicenes 53, 60
multitopic 2,6-bis(aza[6] helicenephosphole)-pyridine 177
multitopic helicenic ligands 402–404

n

nanographenes (NGs) 86, 263–265, 275, 277, 279, 283–322, 350, 353–355, 388, 429
naphthalene diimide (NDI) 308, 345, 346, 492
naphthalene monoimide (NMI) units 309
naphthalimide monoanhydride 346
naphthalimide units 333
near-infrared (NIR) fluorophores 345
NHCs (Ir, Re) and related complexes 184–188
[*n*]helicenes 1–3, 7, 9, 127, 161, 167, 287, 290, 291, 300, 347, 348, 376–378, 424, 465
N-heterocyclic carbenes (NHCs) 159, 184, 187, 188, 411–413, 510
Ni(0)/chiral NHC catalyst 105
Ni(0)/(*M,R,R*)-(*M,R,R*)-NHC complex 105
Ni(0)/(*S*)-MOP complex-catalyzed intramolecular [2 + 2 + 2] cycloaddition 104
Ni(0)/PPh_3 complex 112
$Ni(CO)_2(PPh_3)_2$ 57
Ni-and Co-catalyzed [2 + 2 + 2] cycloaddition 104–105
nickel-catalyzed alkyne [2 + 2 + 2] cycloisomerization of triyne 341
nitroalkenes 122, 520–521
nitrogen-containing helicenes 19, 517
noncovalent bond interactions of helicenes 242–246, 251, 257
non-fullerene electron acceptor materials 333
nonlinear optical responses 177, 453
nonlinear optical materials 220, 341, 395, 424
non photochemical oxidative cyclization 200, 203
nonracemic compounds 54
nonracemic helicene 3, 54, 76–85, 88, 341, 468
nonracemic helicene bis-*p*-quinone 341
nonracemic tetrahydro[6] and [7]helicenes 105
nucleic acid binding 249–250

o

octa-and decahelicenes 343
octabenzo[5]helicenes 305
octyl ester [4]helicene 152
oligothiophene-based [7]helicene 121
one-dimensional (1D) D-π-A NLOphore 455
optical activity (OA) 167, 192, 220, 374, 378, 379, 386, 395–400, 402, 403, 484
optically active 2-[[(2,4,5,7-tetranitro-9-fluorenylidene)amino]oxy] propionic acid 376
optically active triple helicenes 115
optically pure disulfide [6]helicene derivative (*P*) 441
optically pure [7]helicene 376
optically stable enantiomeric tetrahydro-benzo[5]helicene derivatives 432

optically stable enantiomeric tetrahydro[5]helicene derivatives 431
optical rotations (ORs) 54, 167, 263, 288, 333, 373, 395, 466, 523
optical rotatory dispersion (ORD) 223, 333, 374
orange-red fluorescence 429
organic field effect transistors (OFETs) 290, 343, 346, 474, 487, 488
organic light emitting diodes (OLEDs) 474
  electroluminescent (EL) layer sandwiched 474
  fundamental operation principles 475–476
  helicene racemates
    device components 477–480
    emissive 476
  helicenes as enantiopure materials
    CP-PHOLEDs 484–487
    small-molecule dopants, CP-OLEDs 482–484
  lighting and display industry 474
  technology 395
organic photovoltaics (OPVs) 290, 333, 474, 490–494
organic semiconductors (OSCs) 211, 248, 292, 345, 361, 473
organocatalysis 53, 214, 225–226, 519, 523
organocatalyzed annulation 121–124
organometallic alkynyl-gold(I) complexes 180
organometallic catalysis 209, 224–225
organometallic chemistry and photocatalysis 159–161
ortho-annulated aromatic rings 329
ortho-dibrominated arenes 268
ortho-dibromoarenes 274
ortho-fused benzene rings 53, 329, 424, 429
oxahelicenes 27, 68, 71–72, 79–81, 84, 88
oxa[5]helicenol 122
oxalohelicenes 440–441
oxidative cyclization of bis(benzodithienyl)ethenes 200
oxidative photocyclization 335
  benzocarbo[*n*]helicenes 11–12
  multiple carbo[*n*]helicenes 13–14
  substituted carbo[*n*]helicenes 9–11
  unsubstituted carbo[*n*]helicenes 6–9
oxidative photocyclodehydrogenation method 1, 263
oxidative Scholl cyclizations 266

p

palladium-catalyzed Sonogashira and Suzuki coupling reactions 137
palladium-mediated cross-coupling reactions 266–272
palladium-catalyzed Heck coupling reactions 273
P-containing helicenes
  in enantioselective Pd(II)-catalyzed allylic substitution 510–513
  in enantioselective Rh(I)-promoted hydrogenation 510
P-containing helicenic ligands 170
Pd(0)/(*R*)-BINAP catalyst 114
Pd(0)/(*R*)-QUINAP complex-catalyzed enantioselective intermolecular [2 + 2 + 2] cycloaddition 115
Pd-catalysed coupling method 203
Pd-catalyzed [2 + 2 + 2] cycloaddition 114
PDI-azahelicene 361
PDI-embedded [8]helicene 350
pentacenes 138, 292, 473
pentahelical phosphines in enantioselective Suzuki–Miyaura coupling 513
penta-[6]helicene 270, 353
pentahelicene (PH) units 429
pentathia[9]helicene 23
(+) and (−)-pentathia[9]heterohelicene 386
persistent diaza [4]helicene radicals 157
perylene diimide (PDI) 263, 311, 345, 348
perylene imide 433
perylene units 263
3-phenanthrenol-linked triynes 108, 441
phenylated dioxa [5]helicene 135
N-phenylbenzohydroxamic acid 22
2-phenylnaphthalene-linked triyne 108
phenyl-naphthalene-linked triyne 105

phospha[7]-and [9]helicene-like molecules 110
phospha[7]helicenes 73, 79, 112, 113
phosphahelicenes 75
  catalyzed enantioselective [3+2] annulation 522–524
  in enantioselective Au(I) cycloisomerization of enynes 516–517
phosphine-borane adducts 209, 224
phosphine oxides 209, 217, 218, 224, 226, 306
phosphine selenides 209, 214, 224
phosphole-containing helicene 28, 32
phosphole helicene derivatives 432
phosphonium bromide 335
phosphorescence 171, 180, 182, 184, 187, 188, 396, 399, 407–413, 442–444, 484, 486
phosphorescent OLEDs (PHOLEDs) 476, 479
photochemically induced quaternization 20
photochemical oxidative cyclization 200–204
photochemical preparation, helicene-like molecules 28–31
photochemical preparation, carbohelicenes
  eliminative photocyclizations of carbo[*n*]helicenes 14–15
  oxidative photocyclizations
    benzocarbo[*n*]helicenes 11–12
    multiple carbo[*n*]helicenes 13–14
    substituted carbo[*n*]helicenes 9–11
    unsubstituted carbo[*n*]helicenes 6–9
photochemical preparation, heterohelicenes
  azahelicenes 15–23
  phosphole-containing helicene 27
  thiahelicenes 23–27
photochemical transformations, helicenes 31–33
photocyclodehydrogenation 1, 2, 19, 23, 31, 53, 54, 62, 68, 85, 86, 89, 263, 346
cis/*trans* photoisomerization of stilbenes 2
photoreactors 3, 5, 6, 34
photoswitches 223
photovoltaics (PVs) 263, 290, 311, 333, 474, 490–494
phthalocyanines 177, 178
plain dioxa [5]helicene 133
platinahelicenes
  bipyridine-helicene-based multifunctional pH-switchable platform 408–410
  mono-and bis-cycloplatinated helicenes 407–408
platinum-catalyzed double-cyclization reaction 426
(*P*)-monoacetate 210
polarizability 221, 398, 453, 461
polarizable continuum model (PCM) 217, 399, 459, 463
polyaromatic hydrocarbons (PAHs) 329, 505
polyaza 378
polycyclic aromatic hydrocarbons (PAH) 263, 268, 285, 375
polycyclic derivatives 241, 242
polydentate helicenes 177–179
poly[9,9-dioctylfluorene-co-benzothiadiazole] (F8BT) 482
polyethylenimine ethoxylate (PEIE) 484
polyfluorene-based polymer 338, 340
poly(lactic-co-glycolic acid) (PLGA) 228
polyurethane 156
positively charged quaternary sp^2 nitrogen atoms 343
potassium titanyl phosphate (KTP) 453
power conversion efficiency (PCE) 292, 333, 490, 491, 493
(*P*,*P*,*P*)-**16** (hexabenzotriphenylene) 429
pristine carbo[*n*]helicenes 378
propyl and ethylene-2-pyridine side chains 159
N-propyl-4-aza [4]helicene 151
protonated azahelicenes 343
prototypical [6]helicene 330
prototypical helicenes 329
Pseudomonas cepacia-mediated transesterification 210
Pt-catalyzed cycloisomerization 268
Pummerer fragmentation 129

pyrene 6, 12, 87, 242, 244, 245, 248, 263, 299, 388, 389
pyreno[a]pyrene-based helicene 389
pyreno[a]pyrene helicene system 388
Pyrex-filtered light 20
pyridine containing heterohelicene 336
pyridine-fused helicenes 116
pyridohelicenes 54, 64–67, 168, 335, 336, 338, 343
bis(3-(pyridylethynyl)phenyl) ligand 175
pyrrole (carbazole) 15, 19

q
quinacridinium DMQA 145
quinone-like substructures 341
quinoxaline-tetrathia[9]helicene 442

r
racemic and optically active [7]thiaheterohelicenes 208
racemic/enantiopure helicenes 263, 524
racemic or enantiopure 5-aza[6]helicene 4 ligands 170
racemization behavior 379
Raman optical activity (ROA) 167, 220, 374, 395, 398
range-separated functionals (RSFs) 399
Re(I) and Ir(III) helicenic phosphors
  bipyridine-helicene-rhenium complexes 410–411
  cycloiridiated complexes with helicenic N-heterocyclic carbene 411–413
redshifted cyanine-like dioxa [6]helicenes 150
redshifted dye 150
relatively simple pristine (unsubstituted and hydrocarbon) helicenes 379
Rh(I)/cyclobutadiene complex 108
Rh-and Ir-catalyzed [2 + 2 + 2] cycloaddition 105–114
rhenium-based enantiomers 443
rhodacycle 108
rhodium-catalyzed [2 + 2 + 2] and [2 + 1 + 2 + 1] cycloadditions 106
rhodium-catalyzed double intramolecular [2 + 2 + 2] cycloadditions 109
rhodium-catalyzed hydrogenation of itaconate 247
right-right/left-left rule 242, 245, 257
rollover cycloplatinated derivative 446
Ru-and Fe-[6]helicene redox-triggered chiroptical switches
  iron-ethynyl-helicenes 406–407
  ruthenium-vinyl-helicenes 404–405
rylene diimides 345, 346, 363

s
sandwich-type helicenic complexes 188–192
scanning tunneling microscopy (STM) 222, 505
Scholl reaction 24, 53, 85, 86, 285, 294, 295, 297–299, 308, 309, 311–314, 322
Se adduct 160
self-assembly monolayer (SAM) method 247
self-assembly onto surfaces 505–509
sergeant-and-soldiers 507
S-shaped double azahelicenes 115, 116, 336
short circuit current density (J_{SC}) 491
sila[7]helicene 73, 84, 112, 113, 435
silahelicenes 60, 73, 74, 435
silicon-linked bis(propargylic alcohol) 112, 435
single crystal X-ray diffraction (SCXRD) analysis 115, 350, 489
single molecule magnet (SMM) 171
six stereoisomeric double[8]helicenes 350
sodium ionophore 156
sodium tetrakis 3.5-bis(trifluoromethyl) phenyl borate 156
Sonogashira coupling 62, 65, 205, 208, 268, 294, 302, 331
(–)-sparteine-mediated annelation 121
(–)-sparteine-mediated asymmetric silylation 121
spin filters 161, 329, 495, 496, 505
SPINOL [1,1′-spirobiindane-7,7′-diol]-derived chiral phosphoric acid (*S*) 121
spintronics 89, 161, 480, 494–496, 524
squalene-[4]helicene construct 152
S-shaped C_2 symmetric fused double [6]helicene 379
stable radicals 157–159, 214

stairwise kinetics 256
stereodefined *Z*-diboronic esters 202
stoichiometric diastereoselective synthesis 76–82
Stone–Wales rearrangement 286
substituted carbo[*n*]helicenes 9–11
meta-substituted distyrylnaphthalene derivatives 9
sulfoxides 129, 140, 182
sulfur bridged cationic [4]helicene 131
sum-over-states (SOS) 374, 389, 401
superacenes 285
superacid/Lewis acid-mediated S_EAr 274
superhelicenes 294, 302, 303, 305, 314
supertwistacene 303
Suzuki coupling/photocyclization sequence 350
Suzuki–Miyaura cross-coupling reactions 135, 266, 270, 348
switches 167, 192, 223, 329, 406
sydnone 336, 337
symmetrical 1,3-bisoxa[5]helicenyl imidazolium salts 514
symmetry-allowed photoreaction 2

t

targeted sila[7]helicene 112
p-terphenyl-2,2′′,6,6′′-tetrayne derivative 427
terpyridine (Terpy) ligands 177
terrylene-based helicenes 363–365
terrylene diimide (TDI) 364
terrylene diimide-based double[5]helicene 364
terrylene diimide-based helicenes 363
2,2′ +5′,2′′-terthiophene (α-T_3) ones 212
tetraaza[7]helicene 447
tetrabrominated corannulene substrate 270
tetrabrominated TDI 364
tetrabromo cyclooctatetrathiophene 315–322
tetracenes 138, 440
tetracyclic helicenes 238–241
tetrahydro[6]helicene 82, 83, 104, 106, 513
tetrahydro[7]helicene 104, 106
1-tetralone 264
tetramethoxybisquinone derivative 466
tetraphenyl ethylene (TPE) 427
tetrasubstituted helicenebisquinones 456
tetrathia[7]helicene 23, 25, 191, 199, 203, 205, 207, 209, 210, 222, 224, 227, 457–463, 466, 488
tetrathia[9]helicene skeleton 442
tetrathiahelicene (7-TH) scaffold
- 3,3′-bis(benzo[1,2-*b*+4,3-*b*′] dithiophene) derivatives 204–205
- computational studies 216–220
- diphosphane (±) 224
- electrochemical properties and applications 211–215
- ferrocene derivatives 213
- functionalization 206–209
- interaction with biomolecules 226–229
- mono and diphosphane (±) 224
- nonlinear optic (NLO) materials 220–222
- organocatalysis 225–226
- organometallic catalysis 224–225
- organometallic complexes 207, 208
- oxidative cyclization of bis(benzodithienyl)ethenes 200
- phosphorus derivatives 213, 216–219
- resolution of enantiomers 209

tetrayne 59, 71–75, 83, 104, 110–112, 115, 116, 124, 427
thiahelicenes 16, 23–27, 73–75, 170, 191, 203, 206, 211, 221–223, 225, 229, 313, 386, 510, 523
(*M*)-thiahelicenophane 227
thiahelicenophanes 227
(thio)xanthone 137
thickness-dependent photoluminescence and electroluminescence spectra 341
thin film voltammetry 156–157
thiohelicenes 441–442
thiophene-containing propeller-shaped nanographene 315
thioxanthenium 131
third-order NLO responses 466, 469
time-dependent density functional theory (TD-DFT) 150, 216, 333, 358, 398, 399, 408, 413, 459

time-dependent Hartree–Fock (TDHF) method 222, 398, 457
o-tolaldehyde 238
(*R*)-tol-BINAP catalyst 110
o-tolylboronic acid 513
TPA upconverted CPL (TP-UCPL) 289
trans-alkene 202
trans effect 176
transition-metal-catalyzed aromatic ring construction 103
transition-metal-catalyzed [2 + 2 + 2] cycloaddition of alkynes 103, 104, 115, 124
triangulenium dyes triazatriangulenium (TATA) 128
triarylmethylium π systems 144
1,2,3-triazolium-based gold(I) complex 119
1,2,4-triazolium-based chiral gold(I) complex 119
tribenzo[*b,n,p,qr*]perylene substructures 383
tribenzoperylene-extended double [6]helicene 383
1,2,4-trichlorobenzene (TCB) 507
trifluoromethanesulfonate-substituted naphthalene 114
trifluoromethylbenzene 254, 256
1,3-bis(trifluoromethyl)benzene 252
1,3,3-trimethyl-2-methyleneindoline 139
trimethylsilyl and trifluoromethanesulfo nate-substituted [5]helicenes 115
o-trimethylsilyl-triflate benzene 364
trinaphthylenes 274, 275
trioxatriangulenium (TOTA) system 131
3,6-bis(triphenylamino)phenanthrene 479
triphenylenes 191, 274, 300, 301
triple [5]helicene 13, 14, 60, 300, 380, 381, 384
tripropylamine (TPA) 154
trithia[5]helicene 23
trithia[7]helicene 23
triynes 54, 55, 58, 59, 62–73, 75, 76, 78–84, 87, 88, 103–106, 108, 112, 124, 299, 341, 441, 514, 515
Truly chiral 215
two-photon absorption (TPA) 289, 429, 466
two-photon circular dichroism (2PACD) 167, 289, 466
two-photon circular dichroism (TPCD) 289, 466
two-photon emission (TPE) 429
two-photon upconverted emission (TPE) 289

u

undecabenzo[7]helicene unit 304
unsubstituted carbo[*n*]helicenes 6–9
unsymmetrical diaza [4]helicene 159
unsymmetrically substituted diaza [6]helicenes 132

v

vacant-defects 286
van der Waals interactions 177, 257, 508
vibrational circular dichroism (VCD) 167, 301, 333, 374, 395
vibrational Raman optical activity (VROA) spectroscopy 220
vibrational transitions (VCD) 398
vibronically resolved ECD 374
vibronic (chir)optical spectra 412
vibronic coupling 398
vibronic contribution 399
vibronic effects 375
Vilsmeier–Haack reactions 136

w

water-soluble helicene dyes 155
Weinreb amide 206
Wilkinson's catalyst $Rh(PPh_3)_3Cl$ 57
Wittig reaction 3, 23, 24, 201, 202

x

X-and S-type triple [5]helicenes 380
xanthone 131, 133, 135, 137
X-shaped double [6]helicene 383, 384
X-type triple [5]helicene 380
xyl-Segphos 84, 106, 107, 110, 112

z

Zwitterionic dye 149